The Genetics and Biology of
DROSOPHILA

The Genetics and Biology of
DROSOPHILA
VOLUME 3c

Edited by

M. ASHBURNER
*Department of Genetics, University of Cambridge,
Cambridge, England*

H. L. CARSON
*Department of Genetics, University of Hawaii,
Honolulu, Hawaii, U.S.A.*

and

J. N. THOMPSON, Jr.
*Department of Zoology, University of Oklahoma,
Norman, Oklahoma, U.S.A.*

1983

ACADEMIC PRESS
A subsidiary of Harcourt Brace Jovanovich, Publishers

LONDON   NEW YORK
PARIS   SAN DIEGO   SAN FRANCISCO
SÃO PAULO   SYDNEY   TOKYO   TORONTO

ACADEMIC PRESS INC. (LONDON) LTD.
24/28 Oval Road
London NW1

United States Edition published by
ACADEMIC PRESS INC.
111 Fifth Avenue
New York, New York 10003

Copyright © 1983 by
ACADEMIC PRESS INC. (LONDON) LTD.

All Rights Reserved
No part of this book may be reproduced in any form by photostat, microfilm,
or any other means, without written permission from the publishers.

British Library Cataloguing in Publication Data

The genetics and biology of drosophila
Vol. 3c
1. Drosophila—Genetics   2. Insects—Genetics
I. Ashburner, M.   II. Carson, H. L.
III. Thompson, J. N. jr.
595.77′4   QH470.D7   75—19614
ISBN 0 12 064947–0

Printed in Great Britain
at the Alden Press, Oxford

Contributors

BARKER, J. S. F., *Department of Animal Science, University of New England, Armidale, N.S.W. 2351, Australia.*

CROW, J. F., *Department of Genetics, University of Wisconsin, Madison, Wisconsin 53706, U.S.A.*

FITZ-EARLE, M., *Department of Biology, Capilano College, North Vancouver, B.C., Canada.*

HOLM, D. G., *Department of Zoology, The University of British Columbia, Vancouver, B.C., Canada.*

KIDWELL, M. G., *Division of Biology and Medicine, Brown University, Providence, Rhode Island 02912, U.S.A.*

MATHER, K., *Department of Genetics, University of Birmingham, P.O. Box 363, Birmingham B15 2TT, England.*

RINGO, J. M., *Department of Zoology, University of Maine, Orono, Maine 04469, U.S.A.*

SIMMONS, M. J., *Department of Genetics and Cell Biology, University of Minnesota, St. Paul, Minnesota 55108, U.S.A.*

SLATKO, B. E., *Department of Biology, Williams College, Williamstown, Massachusetts 01267, U.S.A.*

SPIETH, H. T., *Department of Zoology, University of California, Davis, California 95616, U.S.A.*

TEMPLETON, A. R., *Department of Biology, Washington University, St. Louis, Missouri 63130, U.S.A.*

THOMPSON, J. N. jr., *Department of Zoology, University of Oklahoma, Norman, Oklahoma 73019, U.S.A.*

WOODRUFF, R. C., *Department of Biological Sciences, Bowling Green State University, Bowling Green, Ohio 43403, U.S.A.*

Editorial Note on Taxonomic Usage

The use of specific names of Drosophilidae in these volumes has been made uniform with the catalogue by M. R. Wheeler in Chapter 1. Authority names are cited only in Chapter 1. In addition, although we respect the opinion of those authors who believe it is incorrect to do so, it has also been our policy to italicize the names of species groups in all chapters. Any exceptions to these policies are explained in the appropriate chapters. In particular, certain taxonomic revisions were in press at the time the catalogue was finalized for publication (May 1, 1981) or have appeared since it was published. Thus, new species not named in Wheeler's catalogue are identified by footnotes in those chapters where they are mentioned. Furthermore, some taxonomic problems, such as the status of *D. arizonensis* and *D. mojavensis* (see Chapter 1), are beyond the scope of this series of reviews. In order to avoid confusion, some reviewers have retained the original names in their discussion of the relevant literature, but have noted the existence of these unresolved problems in footnotes.

Preface

Thirty years have passed since the publication of that landmark volume, "Evolution in the Genus *Drosophila*" by Patterson and Stone. Partly because of the galvanic effect of this book, the literature on the population biology, ecology, evolutionary biology and taxonomy of *Drosophila* has grown enormously in the intervening years. The use of *Drosophila* has permitted exploration of the genetic components in an unprecedented manner in these subjects. This is the key to the great advancement these years have witnessed. As Patterson and Stone so clearly saw, techniques for studying genetic variability within and between populations provided a new and clarifying dimension for assessing the mode of origin of both species and adaptations.

Every biologist and, indeed, virtually every intelligent layman, wants intellectual fuel for the understanding of the origin of man and life. These great central biological questions have stimulated the observations and experiments described in this volume.

The revolution in evolutionary genetics which has made this volume possible has been a quiet one. The molecular genetics of protein synthesis has been generally acclaimed as the greatest triumph of modern biology. Yet, in the long run, the discoveries summarized here appear to be leading to a synthetic view of life which has an even more extraordinary philosophical depth.

The role of this volume, however, is not to make that ultimate synthesis. Rather we have asked the authors to display for the inquiring reader some of the most exciting data which have been emerging from the new *Drosophila* work. The reader will find authoritative accounts, necessarily in considerable technical detail, of the evolutionary and population genetics of these flies. These articles will be of particular interest to those evolutionary biologists who deal with plants and animals which are less amenable to genetic analysis than *Drosophila*. Among the great variety of species discussed will be found useful analogues for diverse sorts of biogeographical and genecological research. The reader will find, perhaps to his surprise, that the sciences dealing with the systematics and ecology of the Drosophilidae have come of age in the last thirty years.

The first section of this volume deals with the world fauna. For the first time anywhere, we offer here a catalogue which covers the Drosophilidae of the world. It summarizes the extraordinary wealth of species belonging to numerous genera in the two subfamilies. As the rest of the volume indicates, here lies an enormous scientific resource, the extent of which has been only vaguely realized. We then proceed to a series of geographically-oriented overviews of the various faunal realms. Such treatments are united for the first time under a single cover.

Following this is found up-to-date accounts of the genetic biology of the more intensively-worked groups of species. The research on some of these groups (*D. virilis*, for example) has not been reviewed since Patterson and Stone's work. Although the molecular and chromosomal data have been brought together, it still remains difficult for the reader to traverse the boundaries of the species groups. Nevertheless, we have not asked the authors to codify their approaches into a uniform scheme of presentation. This is a task not to be approached by editorial fiat. Indeed, we consider it one of the functions of this volume to reveal such major problems. We hope that this collection of reviews will inspire someone to undertake the reduction of these extraordinary biological data.

For the series of chapters that come after the species groups, we have solicited a series of essays summarizing the advances, most not reviewed heretofore, dealing with the ecology of *Drosophila*. Even a cursory examination of these chapters reveals the adaptive exuberance that these flies display. We find that the members of this worldwide family have evolved into a great variety of niches; convergent evolution is abundantly demonstrated.

Because of its key importance, we present an extensive section on genetic variability in populations. Here we see the depths to which analysis of chromosomal and polygenic variability have been developed in recent years. In 1950, molecular genetics of *Drosophila* barely existed. Now we are able to provide a perceptive assessment of the data coming out of the "decade of electrophoresis" (1966–1976). The articles in this section attempt to draw together, at least in a preliminary way, the data from the different species groups.

The other sections of the volume feature articles on the application of *Drosophila* technologies to the understanding of fundamental evolutionary problems, such as mutation, selection, competition and population structure. Here the perceptive reader will glimpse the real frontiers, especially in those areas where the theoretical has led way beyond the experimental. As one reads these essays, *Drosophila* emerges as the paramount fast-breeding diploid bisexual organism for the experimentalist.

In some of the final essays we are even able to introduce the ultimate: extraordinary developments in nucleic acid chemistry now make it possible to deal directly with DNA itself. Locked in each species is a code representing the historical accomplishment of selective and stochastic forces operating in the past. As biochemical techniques render this code more and more legible, perhaps we can dimly discern here the exciting directions in which future work will lead this subject.

Norman, Oklahoma MICHAEL ASHBURNER,
January 1981 HAMPTON L. CARSON
    and
JAMES N. THOMPSON, JR.

Acknowledgement

The editors would have been unable to complete these volumes without the unfailing support of their families and friends. We also owe a great debt to our many colleagues who have not only made many valuable suggestions but have also given us considerable practical help, especially by the reviewing of manuscripts.

At Academic Press, Roger Farrand, Anthony Watkinson and Jenny Mugridge have, by their patience and advice, been an invaluable help.

M. Ashburner
H. L. Carson
J. N. Thompson, jr.

Contents

Contributors	v
Preface	vii
Contents of volumes 3a, b, d, and e	xvii

19. The Mutation Load in *Drosophila*
JAMES F. CROW AND MICHAEL J. SIMMONS

I. Mutation Load Theory	2
II. Measuring the Mutation Rate	8
A. Lethal Mutations	8
B. Minor Viability Mutations	9
C. Induced Mutations	14
D. Hybrid Dysgenesis	16
E. Nature of Minor Viability and Fitness Mutations	18
F. Conclusions About Mutation Rates	18
III. Dominance	19
A. Induced Mutations	20
B. Spontaneous Mutations	23
C. Mutations from Natural Populations	25
D. Partial Dominance of Mutations Affecting Fitness	27
E. Conclusions About Dominance	29
IV. Epistasis	29
References	31

20. Factors Affecting Mutation Rates in Natural Populations
R. C. WOODRUFF, BARTON E. SLATKO AND JAMES N. THOMPSON, JR.

I. Introduction	37
II. Measurement of Mutation Rates	39
A. Methods to Determine Lethal and Visible Mutation Rates	47
B. Methods to Determine the Frequency of Chromosome Breakage	50
C. Precautions in Measuring the Rates of Mutation and Chromosome Breakage in Natural Population Lines	52
III. Factors Affecting Mutation Potential	54
A. Extrinsic Factors	54
B. Intrinsic Factors	60
IV. Genetic Mutator Factors	63
A. Characteristics of Some Well-Known Mutator Systems from Natural Populations	64
B. Characteristics of Other Mutator Systems from Natural Populations	101
V. Effect of Mobile DNA Elements on Spontaneous Mutation Rates	102
VI. Conclusions	104
Acknowledgements	107
References	107

21. Intraspecific Hybrid Sterility
MARGARET G. KIDWELL

- I. Introduction 125
- II. *Drosophila pseudoobscura* 127
- III. The *Drosophila willistoni* Group 128
  - A. *D. willistoni* 128
  - B. *D. equinoxialis* 129
  - C. *D. tropicalis* 129
  - D. *D. paulistorum* 129
- IV. *Drosophila melanogaster* 130
  - A. *SF* Sterility 131
  - B. *GD* Sterility 134
  - C. Comparison of *SF* and *GD* Sterility Systems 140
- V. Other Species 146
- VI. Discussion 147
- Acknowledgements 148
- References 149

22. Response to Selection
KENNETH MATHER

- I. Introduction 155
- II. General Considerations 157
  - A. Selection and Characters 157
  - B. Types of Selection 158
  - C. Kinds of Variation 162
  - D. Hidden Variability 163
  - E. Inbreeding and Outbreeding 168
  - F. Recombination 170
  - G. Effects of Mutation 171
- III. Directional Selection 175
  - A. Early Generations 175
  - B. Selection Response in Later Generations 182
  - C. Chaeta Number and Fertility 186
  - D. Correlated Responses and Recombination 188
- IV. Stabilizing Selection 191
  - A. Additive Genetic Variation 191
  - B. Canalization 196
- V. Disruptive Selection 199
  - A. Types of Disruptive Selection 199
  - B. D^- Selection: Polymorphism 201
  - C. D^+ Selection: Divergence 203
  - D. Isolation 206
  - E. Conclusion 209
- VI. Genetic Architecture of Characters 210
  - A. Balance and Natural Selection 210
  - B. Types of Character 212

	Acknowledgements	215
	References	215

23. Mating Behavior and Sexual Isolation in *Drosophila*
HERMAN T. SPIETH AND JOHN M. RINGO

- I. Historical Development 224
- II. Basic Nature of *Drosophila* Mating Behavior 225
- III. Stimuli Involved in Mating Behavior 229
  - A. Auditory Stimuli 230
  - B. Olfactory Stimuli 233
  - C. Tactile Stimuli 234
  - D. Visual Stimuli 236
- IV. Ontogeny 237
  - A. Male Displays and Mating Propensity 238
  - B. Female Behavior 239
  - C. Fate Mapping 242
  - D. Experience and Mating Preferences 243
- V. Measurement of Sexual Activity 244
  - A. Some Commonly Used Experimental Designs . . . 245
  - B. Measurements of Nonrandom Mating 247
  - C. Models of Mating Behavior 251
- VI. Adaptiveness of Mating Behavior 251
  - A. Courtship and Life History 251
  - B. Sexual Selection 254
- VII. Phylogeny 257
  - A. Microevolutionary and Macroevolutionary Changes . . 257
  - B. Value of Behavior in Determining Phylogenetic Relationships . 260
- VIII. Sexual Isolation 261
  - A. Degree of Sexual Isolation at Various Stages of Population Differentiation 261
  - B. Role of Males and Females 265
  - C. Environmental Factors Influencing Sexual Isolation . . 266
  - D. Evolutionary Origin of Sexual Isolation 268
- IX. Summary 270
- References 270

24. Interspecific Competition
J. S. F. BARKER

- I. Introduction 285
  - A. The Concept of Competition 286
  - B. The Nature of Competition 286
- II. Competitive Exclusion or Coexistence 287
  - A. Principle of Competitive Exclusion 287
  - B. Possible Outcome of Competition 289
  - C. Exclusion and Coexistence in Laboratory Populations of *Drosophila* 291
- III. Competition in Natural Populations of *Drosophila* . . . 296
  - A. Detection and Importance of Competition . . . 296
  - B. Indirect Evidence of Competition 298

    C. Direct Evidence of Competition 300
    D. Microhabitat Divergence and Competition 303
  IV. *Drosophila* as a Laboratory Population Model 305
    A. Ecology of Competition 305
    B. Competitive Ability as a Measure of Population Fitness . . 320
    C. Changes in Competitive Ability in Two-Species Populations . 325
    D. Competition Involving More Than Two Species . . 328
  V. Implications in Evolution, Ecology and Biological Control . . 330
  Acknowledgements 333
  References 333

25. Natural and Experimental Parthenogenesis
ALAN R. TEMPLETON

  I. Introduction 343
  II. The Incidence of Parthenogenesis in the genus *Drosophila* . . 345
  III. Mechanisms of Parthenogenesis and the Genetic Consequences . . 348
    A. General Considerations 348
    B. Central Fusion 349
    C. Terminal Fusion 352
    D. Gamete (Pronuclear) Duplication 353
    E. Mixtures of Mechanisms 354
    F. Non-Disjunction and Production of Males 356
    G. Sexual Reproduction and Parthenogenesis 360
  IV. The Genetic Basis of Parthenogenesis 360
    A. Evidence from Selection Experiments 360
    B. Evidence from Crosses Between Unisexual and Bisexual Strains . 363
    C. Evidence from Contrasts of Isogenic Lines 363
    D. Evidence from Screening Experiments 363
    E. Components of the Genetic Basis of Parthenogenesis . . 364
  V. Behavioral Genetics 369
    A. Sexual Behavior and Isolation 369
    B. Other Behavioral Traits 372
  VI. Population Genetics 373
    A. Population Genetic Theory for Automictic Thelytoky . . 373
    B. Clonal Selection in Experimental Populations . . . 376
    C. The Unit of Selection in Experimental Populations . . 381
    D. Total Homozygosity and Isozyme Variability . . . 385
  VII. Speciation 386
    A. The Origin of Parthenogenetic Species and the Evolution of Sex . 386
    B. Parthenogenesis as an Experimental Model of Speciation in Sexual Populations 393
  VIII. Prospects for Parthenogenetic *Drosophila* 395
  References 395

26. *Drosophila melanogaster* Models for the Control of Insect Pests
MALCOLM FITZ-EARLE AND DAVID G. HOLM

  I. Introduction 399
  II. Theoretical Models for Population Suppression and Manipulation . 402

III.	Population Suppression and Replacement by Translocation Lines	403
IV.	Population Replacement by Compound Autosome Strains	405
	A. Meiotic Behaviour	405
	B. Laboratory and Field Studies	408
V.	Population Replacement by Compound; Free-Arm Stocks	410
VI.	Controlling or Eradication Factors	413
VII.	Other Possible Genetic Mechanisms for Insect Control	414
VIII.	Genetic Control Methods Applied to Pest Insects Other Than *Drosophila melanogaster*	416
	A. Chromosome Rearrangements and Meiotic Drive	416
	B. Desirable Factors	418
	Acknowledgements	419
	References	419

Author Index . . . . *l*i
Subject Index . . . . *l*xv
Index of Genetic Variations . . . . *l*xxix

Contents of Volume 3a

Chapter 1: The Drosophilidae: A Taxonomic Overview
  Marshall R. Wheeler
Chapter 2: Geographical Survey of Drosophilidae: Nearctic Species
  Marshall R. Wheeler
Chapter 3: Drosophilidae of the Neotropical Region
  F. C. Val, C. R. Vilela and M. D. Marques
Chapter 4: Drosophilidae of the Palearctic Region
  Gerhard Bächli and M. Teresa Rocha Pité
Chapter 5: Composition and Biogeography of the Afrotropical
    Drosophilid Fauna
  Léonidas Tsacas, Daniel Lachaise and Jean R. David
Chapter 6: Oriental Species, Including New Guinea
  Toyohi Okada
Chapter 7: Species of Australia and New Zealand
  I. R. Bock and P. A. Parsons
Chapter 8: Drosophilidae of Pacific Oceania
  D. Elmo Hardy and K. Y. Kaneshiro
Chapter 9: Domesticated and Widespread Species
  P. A. Parsons and S. M. Stanley
Chapter 10: Entomophagous and Other Bizarre Drosophilidae
  Michael Ashburner
Author Index
Subject Index
Index of Drosophilid Taxa

Contents of Volume 3b

Chapter 11: Evolution and Speciation in the *Drosophila obscura* Group
  S. Lakovaara and A. Saura
Chapter 12: Evolution of the *repleta* Group
  M. Wasserman
Chapter 13: The *robusta* and *melanica* Groups
  M. Levitan
Chapter 14: The *Drosophila willistoni* Species Group
  L. Ehrman and J. R. Powell
Chapter 15: The *virilis* Species Group
  L. H. Throckmorton
Chapter 16: Genetics and Evolution of Hawaiian *Drosophila*
  H. L. Carson and J. S. Yoon
Chapter 17: Yeasts and *Drosophila*
  M. Begon
Chapter 18: The Breeding Sites of Temperate Woodland *Drosophila*
  B. Shorrocks
Author Index
Subject Index
Index of Drosophilid Taxa

Contents of Volume 3d

Chapter 27: Methods of Collecting *Drosophila*
  H. L. Carson and W. B. Heed
Chapter 28: Population Structure of *Drosophila*: **Genetics and Ecology**
  C. E. Taylor and J. R. Powell
Chapter 29: Selection and Measures of Fitness
  P. W. Hedrick and E. Murray
Chapter 30: Ecophysiology: Abiotic Factors
  J. David, R. Allemand, J. van Herrewege and Y. Cohet
Chapter 31: Seasonality and Diapause in *Drosophilids*
  J. Lumme and S. Lakovaara
Chapter 32: Breeding Sites of Tropical African Drosophilids
  D. Lachaise and L. Tsacas
Chapter 33: Ecology of Flower-breeding *Drosophila*
  D. Brncic
Author Index
Subject Index
Index of Drosophilid Taxa

Contents of Volume 3e

Parasitic wasps of *Drosophila*.
Drosophila melanogaster species group.
Chromosomal polymorphism.
Chromosomal Evolution: An overview.
Polygenic Variation in Natural and Experimental Populations
Protein polymorphisms in natural populations.
Protein evolution.
Ecology of desert *Drosophila*.
Cumulative indices for vols 1, 2 and 3.

19. The Mutation Load in Drosophila*

JAMES F. CROW
Department of Genetics
University of Wisconsin, Madison, Wisconsin, U.S.A.

and

MICHAEL J. SIMMONS
Department of Genetics and Cell Biology
University of Minnesota, St. Paul, Minnesota, U.S.A.

I. Mutation Load Theory	2
II. Measuring the Mutation Rate	8
A. Lethal Mutations	8
B. Minor Viability Mutations	9
C. Induced Mutations	14
D. Hybrid Dysgenesis	16
E. Nature of Minor Viability and Fitness Mutations	18
F. Conclusions About Mutation Rates	18
III. Dominance	19
A. Induced Mutations	20
B. Spontaneous Mutations	23
C. Mutations from Natural Populations	25
D. Partial Dominance of Mutations Affecting Fitness	27
E. Conclusions About Dominance	29
IV. Epistasis	29
References	31

Mutation is a *sine qua non* for evolution. Without the input of new mutations evolutionary change would eventually come to a standstill. Nevertheless, this may require an enormous time, for the Mendelian mechanism has remarkable variance-conserving efficiency. One would expect, from evolutionary considerations, that in a well adapted organism most mutant genes would be deleterious, or at best neutral, and this is abundantly observed experimentally. Paradoxically, recurrent mutation, the process on which improved fitness ultimately depends, has an immediate effect of reducing

* Paper number 2422 from the Laboratory of Genetics, University of Wisconsin. Supported in part by the Research Committee of the Graduate School, the National Institutes of Health (GM-22038 and RO1-ES01960), and the National Science Foundation (PCM-7903266).

fitness. It is the sieve of natural selection that screens the mutants, retaining only the minority whose average effect is to improve the fitness of their bearers.

This review deals with attempts to estimate the impact of mutation on the average fitness of the population. Only theoretical and experimental studies that bear on this specific question will be reviewed. Since the study of total fitness is very difficult, and sometimes an unattainable goal, we shall frequently make use of data on fitness components, such as viability and fertility. We shall not consider the very large number of studies of variability in natural populations, as revealed by analysis of concealed visible, lethal and sterilizing genes, nor with studies of protein variability and inversion polymorphisms. These have been the subject of other reviews (Lewontin, 1974; Dobzhansky, 1970).

I. Mutation Load Theory

J. B. S. Haldane (1937) was the first to find a relationship between the mutation rate and the average fitness of the population. He noted that:

> It is at once clear that in equilibrium such abnormal genes are wiped out by natural selection at exactly the same rate as they are produced by mutation. It does not matter whether the gene is lethal or almost harmless. In the first case, every individual carrying it, or if it is recessive, every individual homozygous for it, is wiped out. In the second the viability or fertility of such individuals may only be reduced by one-thousandth. In either case, however, the loss of fitness to the species depends entirely on the mutation rate and not at all on the effect of the gene on the fitness of the individual carrying it, provided this is large enough to keep the gene rare.

Haldane went on to note that the proportion by which the mean fitness is reduced by mutation is between one and two times the total mutation rate per gamete.

This is easily demonstrated. Consider the following single-locus, two-allele model where the parameter F allows for the possibility of nonrandom mating.

Genotype	AA	Aa	aa	
Zygotic frequency	$p^2(1-F)+pF$	$2pq(1-F)$	$q^2(1-F)+qF$	(1)
Fitness	w	wx	wy	

Mutation from A to a $= u$/generation; $y \leqslant x \leqslant 1$.

In this model, x and y are parameters which measure the fitness of the mutant homozygote and the heterozygote relative to the AA homozygote.

Then, letting p' be the frequency of the A allele next generation,

$$p' = \frac{w[p^2(1-F)+pF+pq(1-F)x](1-u)}{\bar{w}}, \qquad (2)$$

where

$$\bar{w} = [p^2(1-F)+pF]w + 2pq(1-F)wx + [q^2(1-F)+qF]wy. \qquad (3)$$

This latter quantity is the average fitness of the population. At equilibrium, $p'=p$, permitting $\bar{w}$ to be expressed as a function of u, F, w, x and y, which are regarded as constants.

If the mutant is recessive ($x=1$) then

$$\bar{w} = w(1-u), \qquad (4)$$

regardless of the values of F and y. If individual loci are independent in their effects on fitness and come to equilibrium frequencies independently, then the average fitness of the population is

$$\bar{W} = W\prod_i (1-u_i), \qquad (5)$$

where the product is taken over all relevant loci and W is the fitness of an equilibrium population with mutation rate zero. If the mutation rate, u, is small relative to the reciprocal of the number of relevant loci, then (5) can be satisfactorily approximated by

$$\bar{W} \approx We^{-U} \approx W(1-U), \qquad (6)$$

where $U=\Sigma u_i$ is the total gametic mutation rate for loci producing recessive deleterious mutants.

If the mutant gene is partially dominant, such that the heterozygote is the geometric mean of the two homozygotes ($y=x^2$) and $F=0$, then $\bar{w}=[w(p+qx)^2]$ and

$$\bar{w} = w(1-u)^2 \approx w(1-2u). \qquad (7)$$

For multiple loci, under the same assumptions as before,

$$\bar{W} = W\prod_i (1-u_i)^2 \approx W(e^{-2U}) \approx W(1-2U). \qquad (8)$$

Thus we see that, as Haldane said, the reduction in fitness caused by recurrent mutation is determined by the mutation rate and dominance, not the fitness of the individual mutant phenotypes. The proportion by which the fitness is reduced is approximately equal to the total gametic mutation rate for all relevant loci for recessive mutants and twice that if the heterozygote is the geometric mean of the two homozygotes. The fitness of the mutant homozygote makes almost no difference provided that there is some selection against the heterozygotes and the population does not depart greatly from Hardy–Weinberg ratios. To summarize, the proportion by which the fitness is reduced is k times the gametic mutation rate, where k is usually between 1 and 2, as Haldane has said.

Returning to the single locus model, and writing $x = 1 - hs$ and $y = 1 - s$,

$$\bar{w} = w[1 - u - \hat{q}hs(1-F)(1-u)], \qquad (9)$$

where $\hat{q}$ is the equilibrium frequency of the mutant allele (Crow, 1970, p. 137). We see that if $h = 0$ (complete recessivity) or $F = 1$ (complete homozygosity) then (9) reduces to (4). If h^2s is of higher order than the mutation rate (partial dominance) and F is close to zero (nearly random mating), then $\hat{q} \approx u/hs$ and (9) approximates (7).

Multiple mutant alleles do not alter the situation in any substantial way, if the individual mutants are rare enough that there is not an appreciable number of heterozygotes between two mutants; they need not be so rare if such heterozygotes are intermediate between the two mutant homozygotes, as is true for many phenotypically observable mutants, such as *Drosophila* eye colors. Neither does it matter much if the normal "allele" is really a population of alleles maintained by balancing selection, mutation balance among nearly neutral alleles, or some other variance-preserving mechanism. The mutation rate, u, is then the weighted average mutation rate of the normal alleles.

Under some circumstances k may be larger than 2, but such situations must surely be very rare in nature. This happens when selection is very weak and the heterozygote is inferior to both homozygotes (Kimura, 1961). Much more important is the possibility that k may be considerably less than one, as can happen if there is strong synergistic epistasis, i.e. such that the fitness of multiple-mutant genotypes is lower than if the mutants were independent in their effects on fitness. We shall discuss this later.

The word "load" entered the lexicon of genetics through H. J. Muller's (1950) discussion of "Our load of mutations." He noted that, in a population at genetic equilibrium and of constant size, each mutation is balanced by one gene elimination or "genetic death." This one-to-one relation holds unless two or more mutations can be disposed of in a single genetic death as would be the case if there is a completely recessive mutant or synergistic epistasis. This is simply another way of stating the Haldane principle. In Muller's usage a "load" or a "genetic death" was something to be avoided as far as possible. We use the words in a more general way and with no pejorative intent. However, we should note that the application of the load concept to natural populations has been controversial and that there is disagreement about the validity of certain models (see Wallace, 1970, for a discussion).

The mutation load may be defined in three ways, as:

1. The fraction by which the mean fitness is reduced in comparison to a reference genotype (or group of genotypes). Thus the load, L, is

$$L = \frac{W - \bar{W}}{W} \qquad (10)$$

where W is the fitness of the reference genotype and $\bar{W}$ the average fitness.

2. The fraction by which the population mean is reduced from a population that is otherwise identical, but has mutation rate zero. Clearly this leads to the same formulation as (10), where W is now the average fitness of the mutation-free population.

3. The amount of reproductive excess (that is, excess fitness) of the mutant-free genotype(s) over the average fitness, expressed as a fraction of the average fitness, required to maintain the entire population growth rate at the value $\bar{W}$. Thus, in order for there to be a mean fitness $\bar{W}$, W must be $\bar{W}/(1-L)$, so that $L = (W - \bar{W})/W$, as in (10).

Despite different interpretations, all three definitions lead to the same algebraic formulation. These concepts are dealt with more fully for traits other than fitness and for loads other than mutational by Crow (1970).

In the original formulation of the genetic load in terms of "lethal equivalents," Morton et al. (1956) defined a lethal equivalent as a "group of mutant genes of such number that, if dispersed in different individuals, they would cause on the average one death." This concept has been used in much of the analysis of genetic loads in natural *Drosophila* populations.

If the number of survivors, relative to the reference genotype, is S then the number of lethal equivalents is estimated as $-\ln S$. This is based on the assumption of independent effects such that if there are a number of drastic effects, $d_1, d_2, \ldots$ and a series of mild effects, $m_1, m_2, \ldots$ the probability of survival of an individual is

$$P = K\Pi(1-d_i)\Pi(1-m_j) = K\, e^{-(\Sigma d_i + \Sigma m_j)} = KS$$

where K is the environmental effect, which cancels out when the experimental and reference genotypes are competing in the same culture. We shall call the load defined in this way the *dispersed load*, to indicate that a Poisson correction has been made and to distinguish it from the load as calculated from (10). Thus:

$$\text{dispersed load} = \ln S. \qquad (11)$$

Note that with independent gene action, it is the dispersed load that is equal to the mutation rate, although if the value is small there is little distinction between the load as measured by (10) and the dispersed load. The dispersed load concept clearly makes sense with lethal chromosomes from natural populations where we often wish to make a distinction between the mean number of lethals per chromosome and the proportion of chromosomes with at least one lethal. For the separation of the dispersed load into drastic and mild components in homozygous and random combinations, see Greenberg and Crow (1960), Temin (1966), and Temin et al. (1969).

For deleterious genes that are independent in action, whether lethal or mild, the dispersed load corresponds to the gametic mutation rate, modified by a factor of two if the mutants are not completely recessive. When there is epistasis, defined as any departure from independent action, the analysis is more complicated.

Now to consider the effect of epistasis, we again use the relationships (1), assuming random mating ($F=0$) and letting dominance be arbitrary ($x=1-hs, y=1-s$). Then, following the pattern of equation (2), but fixing attention on the mutant allele rather than the normal, we write

$$q' = \frac{w[q - hspq - sq^2 + u(1-q)(1-hsq)]}{\bar{w}}, \qquad (12)$$

$$\bar{w} = w[1 - 2hspq - sq^2]. \qquad (12a)$$

For $h^2 s \gg u$, this leads to $\hat{q} \approx u/hs$.

In equation (12a) we note that $2hspq$ is the load component associated with heterozygotes and that sq^2 is the component associated with aa homozygotes. Letting n be the number of mutant alleles eliminated per genetic death and l be the load component, $2hspq + 2sq^2 = \Sigma n_i l_i = \bar{n}L$, where $L = \Sigma l_i$ or the total of the load components and $\bar{n} = \Sigma n_i l_i / \Sigma l_i$, the (weighted) average number of mutants eliminated per genetic death. Thus, we can rewrite (12) as

$$2q' = \frac{2q - \bar{n}L + 2u(1-q)(1-hsq)}{1-L}. \qquad (13)$$

At equilibrium, $q' = q = \hat{q}$, and replacing $hs\hat{q}$ by u, we obtain the approximation

$$L = \frac{2u(1-u)(1-\hat{q})}{\bar{n} - 2\hat{q}}. \qquad (14)$$

Note that when $\bar{n}=1$ (mutants eliminated through partial dominance) $L \approx 2u$ and when $\bar{n}=2$ and $h=0$ (recessive mutants), $L=u$, as expected from equations (7) and (4).

Equation (13) suggests an obvious extension to multiple loci. Let $\bar{x} = 2\Sigma q$ be the mean number of mutants per individual, summing over all mutants at all relevant loci. Then, by analogy with (13)

$$\bar{x}' = \frac{\bar{x} - \bar{n}L + 2\Sigma u_i(1-q_i)(1-q_i h_i s_i)}{1-L}. \tag{15}$$

If we neglect $q_i h_i s_i$, which is of the order of the mutation rate, and solve for the equilibrium by setting $\bar{x}' = \bar{x}$, we obtain

$$L \approx \frac{2U(1-\bar{q})}{\bar{n}-\bar{x}} \tag{16}$$

where $U (=\Sigma u)$ is the total mutation rate per gamete and $\bar{q} = \Sigma uq/\Sigma u$ is the average of the mutant allele frequencies, each being weighted by the associated mutation rate.

This equation was first given by King (1966). Since the mutant allele frequencies are ordinarily very small, we have the useful approximation

$$L \approx \frac{2U}{\bar{n}-\bar{x}}, \tag{17}$$

or in other words, the mutation load is equal to the total zygotic mutation rate divided by the difference between the number of mutant genes in the selectively eliminated group and in the whole population. This shows that, as was said above, if selection works so that there is a large difference in the number of mutants in the eliminated and noneliminated genotypes, the mutation load can be substantially reduced. The most effective selection is truncation selection, whereby all the individuals below a certain value in "fitness potential" leave no descendants while those above this threshold leave an equal number (Milkman, 1978; Wills, 1978; Kimura and Crow, 1978).

It is interesting to note that reduction of the load by multiple simultaneous gene eliminations is not possible in an asexual population. The only way that several mutants can be eliminated in one genetic death is if they arose in the same clone. Assuming (reasonably) that the mutations arise independently, the mutation load in a diploid asexual species is essentially

$$L = 1 - e^{-2U} \tag{18}$$

or the dispersed load is $2U$, regardless of the way the mutants act (Kimura and Maruyama, 1966; Crow, 1970, p. 147). This is one more argument for the advantage of Mendelian recombination.

From this theoretical discussion it is clear that, in order to assess the impact of mutation on the fitness of the population, it is necessary to consider three factors: (a) the total mutation rate; (b) dominance, since even slight partial dominance of harmful mutants doubles the load in comparison to complete recessivity; and (c) epistasis, since strong synergistic interaction—particularly something approaching truncation selection—can greatly reduce the load. We shall now consider these three factors in turn.

II. Measuring the Mutation Rate

A. LETHAL MUTATIONS

Recessive lethal mutations can easily be detected by mating schemes that utilize recombination-suppressing inversions to manipulate individual chromosomes as units. Data for studies in 1963 and earlier were reviewed by Crow and Temin (1964). The average lethal mutation rate for the X chromosome was 0·0025 for chromosomes taken from nature with only the male rate measured and 0·0026 for laboratory stocks where mutation rates in both sexes were measured. Recent data from more sophisticated experiments (Abrahamson *et al.*, 1980) suggest a slightly lower rate for females, 0·0015.

Experiments for autosomal rates are more difficult because of the inability to distinguish between a new mutant and one that pre-existed in the stock. Most autosomal experiments have used mating systems where mutant genes were allowed to accumulate for several generations in heterozygous males and then were later made homozygous. Earlier work with chromosome 2 was summarized by Crow and Temin (1964); later studies were done by Wallace (1968), Mukai *et al.* (1972), and Ohnishi (1977a). The lethal mutation rates vary between 0·004 and 0·006 per chromosome generation.

Since the second chromosome has about twice as many polytene chromosome bands as the X chromosome, the values are in rough agreement. Altogether, the (haploid) genomic rate in males is about 0·01, with no evidence for females being greatly different.

From the evidence offered by intensively studied chromosome regions, where there is an approximate one-to-one correspondence between the number of complementing lethal groups and the number of salivary chromosome bands (Judd *et al.*, 1972; Lim and Snyder, 1974; Liu and Lim, 1975), the total number of loci at which lethal mutations can occur is between 5,000 and 6,000. Thus, the lethal mutation rate is about 2×10^{-6} per locus generation.

Although we refer to such mutants as recessive, and shall continue to do so since they are ordinarily studied in the homozygous state, we should emphasize that there is abundant evidence, to be discussed later, that there is appreciable partial dominance so that the heterozygote is not as fit as the homozygous normal.

There is also a class of dominant lethals, thought to be due mostly to chromosome breakage and to nondisjunction and chromosome loss. A maximum estimate is given by the number of eggs that fail to hatch. The overall contribution of this class of events to the mutation load is to reduce the average fitness by about one percent.

The rate per locus for visible mutations has been estimated as $6-7 \times 10^{-6}$ (Schalet, 1960) and for isozymes as 4×10^{-6} (Mukai, 1970). Mukai and Cockerham (1977) present a higher isozyme mutation rate, $1 \cdot 2 \times 10^{-5}$. These are perhaps overestimates because of selection of loci at which previous mutants have occurred or which are polymorphic in natural populations. Since we have little idea of the number of such loci, it is not possible to assess their contribution to the mutation load.

B. Minor Viability Mutations

Attempts to measure the total genomic rate for viability mutations are mainly the work of Mukai and his associates (Mukai, 1964; Mukai et al., 1972) and Ohnishi (1977b). The principle is simple: a chromosome that has been tested when homozygous and found to have normal viability is kept

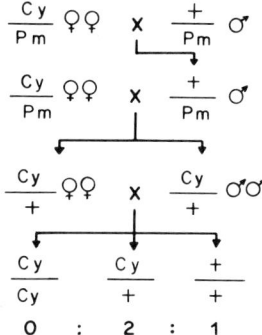

FIG. 1. Mating scheme for accumulating mutations on the *D. melanogaster* second chromosome, and for making the chromosome homozygous.

sequestered by heterozygosis for many generations during which mutations are allowed to accumulate. A typical mating scheme is shown in Fig. 1. Each generation a single $+/Pm$ male is mated to a few Cy/Pm females. The chromosome to be tested is the $+$ chromosome and the Cy/Pm females come, each generation, from a standard laboratory stock.

Because of the absence of male crossing over, the $+$ chromosome remains intact through the many generations of backcrossing. The flies are cultured under optimum nutritional conditions in uncrowded cultures to minimize any selection for heterozygous viability effects. Also, since only a single male is used, there is no selection for fertility. This ideal is not completely realized, however, for occasionally the male is sterile and a brother (or half-brother) from the same generation is substituted.

At intervals of several generations the chromosome is made homozygous by the mating shown at the bottom of Fig. 1. This scheme takes advantage of the recombination-suppressing properties of inversions in the Cy chromosome. Viability is measured by the relative frequency of $+/+$ to $Cy/+$ flies, counted throughout the period of emergence.

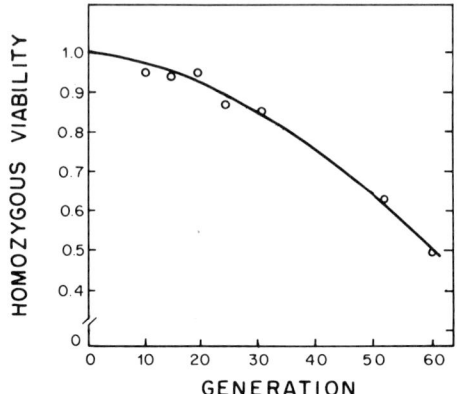

FIG. 2. Average viability of flies homozygous for second chromosomes that have accumulated mildly detrimental mutations for various lengths of time (from Mukai, 1969b).

Figure 2 shows the results of one experiment in which many chromosomes derived from the same progenitor were allowed to accumulate mutations independently. The data presented in the figure are averages for all the chromosomes tested, except that lethals have been excluded. There are two features of interest. One is that the decline in viability is essentially linear during the first 30 generations or so, although it becomes markedly curvilinear in later generations; there is some synergistic epistasis of the homozygous mutants. The second feature is that the experiment did indeed succeed in removing selection, at least partially, during the accumulation period. After 40 generations the homozygous viability of the chromosomes is considerably less than that found in natural populations at equilibrium.

The decline in viability alone does not provide a measure of the mutation rate, since what is measured is the total decline in viability from an unknown number of mutants with unknown individual effects. However, one can distinguish between a large number of mutations of very small effect and a smaller number with somewhat larger effects by the variation among chromosomes tested.

The statistics of interest are the mean and genetic variance of viability. As deleterious mutations accumulate the mean viability should decline at a rate proportional to the mutation rate per chromosome (Σu) and the average viability-depressing effect of a single mutation ($\bar{s}$). The decline, then, should equal $\bar{s}\Sigma u$. The variance of viability should increase each generation because the mutants occur independently. The approximate increase should be $\overline{s^2}\Sigma u$, where $\overline{s^2}$ is the mean square mutational effect on viability. The use of these linear assumptions is justified by the linear decrease in viability over the first 30 or 40 generations, which were all the data used for the calculations.

We now develop the statistics in more detail. Consider a single locus on the chromosome of interest. The symbols are as follows:

Genotype	Cy/Cy	$Cy/+$	$+/+$
Phenotype	Lethal	Curly (Cy)	Normal ($+$)
Theoretical ratio	1	2	1
Initial viability	0	A	B
Viability after mutation at the ith locus	0	$A(1-hs_i)$	$B(1-s_i)$

After t generations of mutation accumulation, the ratio of $+$ to Cy phenotypes in the culture will be

$$\frac{+_t}{Cy_t} = \frac{B(1-t\Sigma u_i s_i)}{2A(1-t\Sigma u_i h_i s_i)},$$

in comparison with the initial ratio (when $t=0$)

$$\frac{+_o}{Cy_o} = \frac{B}{2A}.$$

The summation is over all loci producing viability-affecting mutants.

The decrease in the ratio of $+$ to Cy phenotypes per generation will then be

$$\Delta M = \frac{1}{t}[1 - +_t/Cy_t \div +_o/Cy_o]$$

$$= \frac{(1-\bar{h})\Sigma u_i s_i}{1-t\bar{h}\Sigma u_i s_i}. \tag{19}$$

In this analysis, we have assumed that h is independent of s, an assumption that seems reasonable because of the small range of values of s.

The value of h cannot be larger than 1/2 without implying that the mutants are more nearly dominant than recessive. The value of $t\Sigma us$ when t is 30 generations or less is less than 0·1. Hence the denominator in equation

(19) is 0·95 or larger, and in view of other larger uncertainties of this analysis will be regarded as one. Hence equation (19) becomes approximately

$$\Delta M \approx (1-\bar{h})\Sigma u_i s_i. \tag{20}$$

Actually it is not necessary to assume that h is independent of s.

$$\Sigma u_i s_i (1-h_i) = (1-\bar{h})\Sigma u_i s_i$$

where

$$\bar{h} = \frac{\Sigma u_i s_i h_i}{\Sigma u_i s_i}.$$

Thus $\bar{h}$ is a weighted average where the h's are weighted by $u_i s_i$, as is appropriate since those mutants that occur most frequently and have the greatest effect should receive the most weight.

The genetic variance among replicates of the same chromosome during one generation of mutation accumulation will be binomially distributed at each locus (or for each class of mutants within a locus) and hence will be $u_i(1-u_i)s_i^2$, or to a sufficiently good approximation, $u_i s_i^2$. Making the same approximate correction for heterozygous effects in the Cy flies, the variance at a locus will be $(1-h)^2 u_i s_i^2$, and summing over all relevant loci, we have for the rate of accumulation of variance per generation

$$\Delta V \simeq E[(1-h_i)^2]\Sigma u_i s_i^2. \tag{21}$$

The two quantities, ΔM and ΔV, are experimentally observed. ΔM is the slope of the change in mean ratio of $+$ to Cy per generation, with the sign changed. ΔV is the rate of change per generation in the genetic variance of this ratio. In the experiments of Mukai and Ohnishi each mating was duplicated and the variance between such replicates was used to remove environmental and error variance.

We now note that

$$\Sigma u_i s_i = \bar{s}\Sigma u, \quad \bar{s} = \Sigma u_i s_i / \Sigma u_i \tag{22}$$

and

$$\Sigma u_i s_i^2 = \overline{s^2}\Sigma u, \quad \overline{s^2} = \Sigma u_i s_i^2 / \Sigma u_i$$
$$= (\bar{s}^2 + V_s)\Sigma u$$
$$= \bar{s}^2(1+C^2)\Sigma u, \tag{23}$$

where V_s is the variance in the individual values of s_i and C is the coefficient of variation of s_i.

We then use (20), (21), (22) and (23) to obtain two estimating equations. First, assuming the h_i's are all equal, so $h=\bar{h}$ is constant, we have

$$\frac{\Delta M^2}{\Delta V} = \frac{\Sigma u}{1+C^2} < \Sigma u \qquad (24)$$

$$\frac{\Delta V}{\Delta M} = \bar{s}(1-\bar{h})(1+C^2) > \bar{s}(1-\bar{h}). \qquad (25)$$

If h is not a constant, $E(1-h)^2$ in equation (21) is always equal to or larger than $(1-\bar{h})^2$; hence the inequalities in equations (24) and (25) are preserved (see Mukai, 1980, for another analysis).

In these formulae C is the troublesome parameter. Since the s_i values are not constant, C^2 must be greater than zero and therefore equation (24) provides only a minimum estimate of the mutation rate. If mutants are extremely variable in effect, then the true value of Σu can be far from the estimated value. For example, if the distribution of viabilities is exponential, then the standard deviation is equal to the mean, so $C=1$ and the mutation rate is twice as high as if C were 0.

One way of minimizing the problem is to base calculations on data from which chromosomes with extreme mutations have been excluded. This has the effect of reducing both ΔM and ΔV, but also of giving more reliable estimates of Σu and $\bar{s}(1-\bar{h})$, providing, of course, that one is interested in mutations with only slight effects on viability; for these, V_s is more nearly zero than it would otherwise be.

In their analyses, Mukai and Ohnishi disregarded chromosomes that reduced homozygous viability by more than 40–50%. These extremely debilitating chromosomes were thought to carry lethal or semilethal mutations rather than many milder mutants with a large cumulative effect. When the mean viability was plotted as a function of time, a linear decline was evident. The slope, ΔM, estimates $(1-\bar{h})\bar{s}\Sigma u$. Mukai (1964) and Mukai et al. (1972) obtained nearly identical values for this, 0·0038 and 0·0040, respectively. The former was estimated with 25 generations of mutation accumulation, while the latter was based on 40. In another 40-generation experiment, Ohnishi (1977b) observed a smaller decline in viability per generation, only 0·0017, which could perhaps be due to differences in culture conditions. He also recorded a change in the genetic variance of about 5×10^{-5} per generation, half as great as that seen in the other experiments. In all three cases the estimates for ΔV are less certain than those for ΔM. Genetic variance for viability is difficult to measure; moreover, in the experiments discussed here it did not increase uniformly, but fluctuated considerably. Thus our confidence in ΔV is less than in ΔM.

Putting the data together, we find the minimum mutation rates per second chromosome and maximum selective effects, according to experiment, to be

Σu	$\bar{s}(1-\bar{h})$	
>0·14	<0·027	Mukai (1964)
>0·17	<0·023	Mukai et al. (1972)
>0·06	<0·030	Ohnishi (1977b).

Remember, of course, that an effort has been made to exclude drastic viability mutations; these, if included, would increase the numbers somewhat. Lethals, for example, occur at a rate of about 0·006 per second chromosome per generation. Mukai (1964) and Mukai et al. (1972) verified this in their experiments, while Ohnishi (1977a) found a slightly lower value (0·004). The point is that the contribution of lethals to the total mutation rate is small compared to mutations with milder effects on viability.

Mutations with intermediate effects also occur, but these seem to be infrequent. Thus, the calculated mutation rate for genes affecting viability is not materially changed if lethal and semilethal mutations are counted in.

C. Induced Mutations

Induced mutation rates can be studied by the same techniques. There is a large body of literature on lethal mutations produced by radiation and chemicals (for a review, see Auerbach and Kilbey, 1971). Ohnishi (1977a) showed a linear relation between lethal mutation rate and EMS concentration in the food over a wide range of concentrations.

Radiation (Käfer, 1952; Friedman, 1964) produces a higher ratio of lethals to mildly detrimental mutants than is obtained for spontaneous mutations, as might be expected if a substantial part of the radiation effect is chromosome breakage. EMS-induced mutations are intermediate between spontaneous and X-ray induced in this regard (Ohnishi, 1977b). A concentration of 2.5×10^{-3} M fed for 24 hr induces about 0·3 recessive lethals, 0·1 delayed recessive lethals, and a minimum of 0·5 minor viability mutants per second chromosome.

Among the viability mutants, Ohnishi (1977b) found that the estimate of $\bar{s}(1-\bar{h})(1+C^2)$ from equation (25) is 0·09 for EMS-induced mutants in contrast to about 0·03 for spontaneous mutants. Unless this difference is attributable to greater variability in EMS-induced mutants, which seems the less likely alternative, $\bar{s}$ is greater for EMS-induced than for spontaneous mutants. (The range of permissible values of $\bar{h}$ is too small to permit it to be the major cause of the difference.) Thus, we conclude that EMS produces not only a larger fraction of drastic mutants than occur spontaneously, but also that the average effect within the mild category is greater.

Ohnishi (personal communication) has also studied the spectrum of effects of EMS-induced mutations on traits other than viability. The accumulation method just described, using the mean decrement and the variance increment per generation, gives a *minimum* estimate of the mutation rate. If the mutants do not all affect the trait in the same direction ΔM is correspondingly reduced, and the mutation rate is still further underestimated.

For viability, the isodirectionality assumption is reasonable. For other traits this needs to be investigated. Figure 3 shows the distribution of

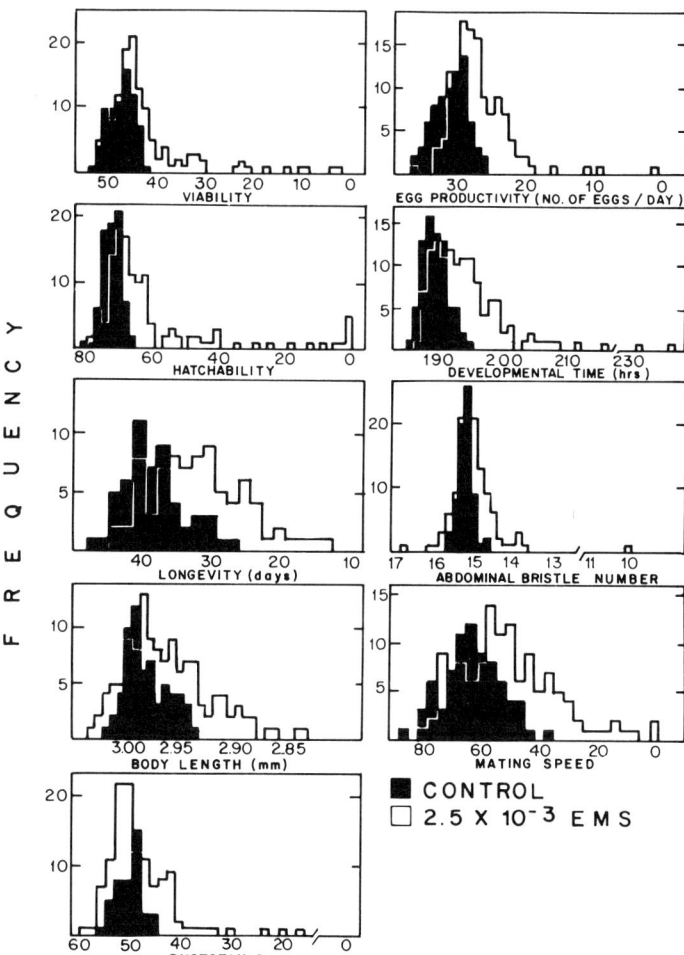

FIG. 3. Distributions of various traits among flies homozygous for second chromosomes treated with EMS (open blocks) or for untreated chromosomes (dark blocks) (from Ohnishi, unpublished observations).

phenotypes obtained when second chromosomes mutagenized with EMS are made homozygous. From the distribution it appears that mutants affecting viability, egg hatchability, egg productivity, developmental time, mating speed, and longevity are essentially isodirectional and in this regard satisfy the requirements of the estimation procedure. Furthermore the mutants are predominantly in the direction of reducing fitness, as expected.

On the other hand, mutants affecting body length, abdominal bristle number, and phototaxis were in both directions, as would be expected for a trait in which an intermediate value is optimum.

Ohnishi's extensive studies provide a minimum estimate of the mutation rate, subject of course to uncertainties about the measurements, particularly the variances. This is applicable only to those traits for which the mutants are essentially isodirectional. Otherwise, the inequality of equation (24) is greater and $\Delta M^2/\Delta V$ is even more of an underestimate of $\Sigma u/(1+C^2)$. As the mutants become nearly symmetrical, as in the case of bristle number, the method is useless and studies emphasizing the mutational component of the variance are more appropriate (Clayton and Robertson, 1955; Paxman, 1957). There is the distinct possibility that mutants with effects on traits that are components of fitness will be isodirectional if the effects are large but more nearly symmetrical when the size of the effect approaches zero, as emphasized by Fisher (1958). This further complicates the problem and reinforces the conclusion that such studies underestimate the rate of occurrence of mutants with extremely small effects.

The most striking of Ohnishi's results, from the standpoint of this review, is that several of the characters gave minimum mutation rate estimates higher than those for viability, and sometimes (e.g., egg production) considerably higher. This argues that the total mutation rate for all genes affecting fitness may indeed be very high. It is, however, not likely to be the sum of all the measurements because of pleiotropy. Indeed, Ohnishi's multivariate analysis shows a high correlation among chromosomes affecting traits connected with fitness; presumably the same mutant affects several traits. A number of studies (Knight and Robertson, 1957; Mitchell, 1977; Mitchell and Simmons, 1977; Simmons *et al.*, 1978, 1980; Sved, 1971; Sved and Ayala, 1970; Prout, 1971a,b; Bundgaard and Christiansen, 1972; Brittnacher, 1981) have shown that chromosomes with a small effect on viability have a larger effect on fitness through their effects on fertility and mating success.

D. Hybrid Dysgenesis

In addition to the statistical uncertainties discussed above, there are biological uncertainties connected with the methodology of estimating the mutation rate for genes affecting viability and other traits. These arise from a condition known as hybrid dysgenesis, in which the offspring of crosses between wild-caught *Drosophila* and laboratory strains exhibit several aberrant genetic phenomena (Kidwell *et al.*, 1977; Sved, 1979; Bregliano *et al.*, 1980; Kidwell, 1983). These include recombination in males, transmission ratio distortion, high mutability, and chromosome nondisjunction. In many cases the hybrids are so profoundly affected as to be sterile.

The abnormal traits are seen primarily in the offspring of crosses where the male parent carries chromosomes derived from wild-caught flies, and less so in the reciprocal (Engels, 1979). The significance of the condition is that the experiments of both Mukai and Ohnishi involved repeated crosses of the sort that might have produced dysgenic hybrids. The second chromosomes used for accumulating mutations in these experiments were derived from progenitors extracted from natural populations; these were then propagated through the male line in each generation by crossing to females with second chromosomes of laboratory origin. If dysgenesis was a factor then there are two possible consequences which bear on the estimation of the mutation rate.

One is that the rate itself was elevated. However, this does not seem to have been the case, for in each experiment the occurrence of lethal mutations on the second chromosome was monitored and found to be in agreement with rates customarily cited for the second chromosome, and for the X chromosome when the size difference between the two is considered (Crow and Temin, 1964). The observed lethal rates are inconsistent with the rise in mutability expected in dysgenic hybrids. Moreover, Yamaguchi and Mukai (1974) undertook a cytological examination of the chromosomes which accumulated mutations in the experiments of Mukai *et al.* (1972). They found no evidence for an abnormal rate of occurrence of chromosome rearrangements, which are known to be produced frequently in the offspring of dysgenic hybrids (Simmons and Lim, 1980; Berg *et al.*, 1980). There does not, therefore, seem to be a problem with elevated mutability in the experiments. However, the other side of the coin is that there may actually be a higher mutability in nature than is detected in experiments such as these (Berg, 1979). This would be the case if hybrid dysgenesis occurred to any great extent outside the laboratory. There is evidence that strains of the same type as the laboratory strains which produce dysgenic hybrids occur in nature, although infrequently (Engels and Preston, 1980); thus the possibility exists that hybridization between these and the more common type could occur, thereby increasing the mutation rate. This prospect makes it even more likely that what has been estimated by Mukai and Ohnishi is a minimum mutation rate; the actual rate in nature could, at least on occasion, be higher. Tajima (1978) showed that the minor viability mutation rate is also greatly enhanced in matings where hybrid dysgenesis is expected.

What about transmission ratio distortion? This is a second potentially troublesome consequence of hybrid dysgenesis. Kidwell *et al.* (1977) report that, in the gametic output of dysgenic hybrids, the laboratory chromosome is often preferentially recovered. Since viability is measured as the relative frequency of two genotypes, transmission ratio distortion could alter it spuriously. Mean viability changes would not be affected by this, nor would variance changes, as long as the level of distortion remained the same for all chromosomes throughout an experiment. Of course, if the transmission properties of the chromosomes changed, then the mean viability decline would be biased. The variance in viability would also be affected. This introduces further uncertainty in the calculation of the mutation rate for genes affecting viability.

E. Nature of Minor Viability and Fitness Mutations

The estimate of the mutation load for *Drosophila* is based on the high rate of occurrence of slightly detrimental mutations affecting viability; we might ask what sort of mutations these are, compared to lethals. One view (Mukai and Cockerham, 1977) is that they are changes in the noncoding portion of the gene, postulated by some to regulate transcription. Mutations in this portion would probably have smaller effects on viability than changes in the protein coding sequence; in fact, many of the latter might be lethal. After excluding highly repetitive sequences, the ratio of noncoding to coding DNA in *D. melanogaster* is roughly 1:20. This corresponds to the ratio of lethal to slightly detrimental mutation rates. Moreover, the lethal mutation rate is similar to that for mutations producing electrophoretically detectable charge changes in proteins (Mukai and Cockerham, 1977). The view that slightly detrimental mutations reside in the noncoding portion of the gene is therefore consistent with the statistical data on *Drosophila* gene number and DNA content.

Of course, the slightly detrimental mutations might be changes in less critical parts of the coding sequence, comparable to some of the innocuous hemoglobin mutations in the human. At the present we can only speculate about their molecular nature.

F. Conclusions About Mutation Rates

Although hybrid dysgenesis has plagued (or made more interesting) several recent experiments, it appears that Mukai and Ohnishi were both working with stocks where this system was not active. From their work it does not appear that the "normal" mutation rate for genes affecting viability on the second chromosome of *D. melanogaster* is less than 0·12. The contribution of lethal mutations to this is small, in the order of 0·005. When extrapolated to the entire genome, the viability mutation rate is at least 0·30. The value for all mutants affecting fitness may indeed be considerably higher, as is indicated by Ohnishi's results. Thus, according to Haldane's principle, the mutation load is at least 0·30 if the mutants are recessive and at least 0·60 if the mutants are partially dominant.

Even if the load is only 0·30, this is still a large price to pay for the privilege of evolution, though perhaps not intolerable for *Drosophila*. The reproductive potential of this species is probably great enough to furnish the 40–45% reproductive excess needed to cover the cost of a mutation load of 0·30. However, we should note that the load need not be manifested by reduced viability alone. The Haldane–Muller principle concerns a reduction in overall fitness. Thus a predicted load of 0·30 is not inconsistent with the observation that 95% or more of *Drosophila* eggs hatch and reach adulthood. The reduction in fitness predicted by the load principle can occur through infertility as well as through inviability; in fact, the genetic load in natural populations of *Drosophila* may be expressed primarily in this way.

If the total mutation rate for fitness genes is normally as high as one mutant per gamete generation, the load is enormous provided gene action is simply multiplicative. If $U=1$, $\bar{W}=We^{-2U}=0.14\ W$, $L=0.86$, and if the population number is to be stable ($\bar{W}=1$), the reproductive rate of a hypothetical mutant-free individual would be $W=7.39$; a reproductive excess of 639% is required. In many mammals this would be impossible.

It is necessary, then, to ask how multiple mutants are eliminated from the population and this leads to a discussion of dominance and epistasis.

III. Dominance

Based on the preceding discussion, with independent gene action the mutation load for *Drosophila* is probably no less than 0·30. If there is much mutant expression in the heterozygous state, the minimal load is closer to twice this figure. The reason, as discussed in Section I, is that with dominance most harmful mutants are eliminated through the reduced fitness of heterozygotes. For the same number of mutant genes to be eliminated from the population there must be twice as many "genetic deaths" as when a mutation is completely recessive. Thus, the issue of dominance is an important one; even a small amount of heterozygous expression has the effect of doubling the mutation load.

There is no easy way to study the heterozygous expression of mutations affecting fitness. One problem is that the heterozygous effects of such mutations are expected to be small and this makes their measurement difficult. Even after mutations have been accumulated for several generations or induced at a high rate with X-rays or chemicals, an enormous amount of data must be collected to demonstrate any heterozygous effect. In such heroic experiments there is always the danger that some irrelevant factor, unknown to the investigator, will cause a statistically significant result, leading to an unjustified conclusion. In addition, there is the difficulty of devising a control for any experiment aimed at assessing the heterozygous effects of spontaneous mutations. Against what standard will the effects of such mutations be judged? All *Drosophila* stocks are mutable.

One possible approach is to study the effects of induced mutations. There is no guarantee that these behave in the same way as spontaneous mutants, but there is probably some similarity for a general property such as dominance. This explains, in part, why induced mutations have received so much attention in the study of heterozygous effects.

A. Induced Mutations

Although Wallace (1958, 1963) was not the first to report data concerning the heterozygous effects of induced mutations, his experiments were among the largest ever performed and stimulated much further research. Their aim was to assess the effects of irradiated second chromosomes of *D. melanogaster* on viability in the heterozygous state. The surprising conclusion was that in an otherwise homozygous genotype, radiation-induced mutations enhanced, rather than diminished, viability.

Wallace's findings contradicted the view held by many, but especially by Muller, that essentially all new mutations are harmful in the heterozygous state. It is this view that would set the mutation load at twice the total mutation rate. On the other hand, if most mutants are overdominant, as Wallace argued, mutation contributes a very small fraction to the genetic load.

Initially Wallace (1958) observed a two percent *increase* in the viability of flies heterozygous for an irradiated chromosome. The radiation exposure was only 500r, a dose expected to produce lethal mutations on about three percent of the treated chromosomes. Although a significant viability improvement was seen, statistical significance could not be demonstrated in subsequent experiments of the same plan using different X-ray doses (Wallace, 1963); however, the regression of viability on X-ray dose was positive, suggesting that the greater the number of heterozygous mutations, the greater the viability.

These results are perplexing and not easily explained by conventional genetic theory. Nevertheless, there is additional evidence to support them.

Mukai et al. (1965) and Maruyama and Crow (1975) studied the heterozygous effects of treated second chromosomes on viability. In a homozygous genetic background, viability enhancement was seen. However, this was not observed by Pandey (1975) whose experimental techniques were similar to those of Maruyama and Crow (1975).

More detailed experiments by Simmons (1976) and Temin (1978) revealed no viability enhancement. In these, mutagenized chromosomes were tested for heterozygous viability effects in different genetic backgrounds. Moreover, data on the homozygous effects of the chromosomes were obtained, permitting a separation into those with lethal or nearly lethal mutations and those with only slightly detrimental mutations or none at all. Simmons worked with irradiated second chromosomes, while Temin used chromosomes treated with EMS. The overall results were in agreement: mutagenized chromosomes reduced the viability of their heterozygous carriers in both homozygous and heterozygous backgrounds. Moreover, chromosomes with a drastic effect on viability in the homozygous state had a greater impact on heterozygous viability than normal or nearly normal chromosomes. Temin, who performed the larger experiment, found that lethal chromosomes that had been treated with 5×10^{-3} M EMS lowered heterozygous viability by about three percent in an otherwise homozygous background; in a heterozygous background the viability reduction was only one percent. Chromosomes carrying severe detrimentals lowered viability by about two percent in both backgrounds, while chromosomes with mildly detrimental mutations lowered it only one percent in the homozygous background and 0·7 percent in the heterozygous background. These data, along with the homozygous viability effects which Temin measured, permitted calculation of the average dominance per gene for each class of mutations. Lethals showed the least dominance, slight detrimentals the most. This means that relative to effects in the homozygous state, slightly detrimental mutations have the greatest effects in the heterozygous state. From the standpoint of the mutation load, these minor mutations, which occur at a high spontaneous rate, account for the largest portion of the reduction in population fitness attributable to mutation.

The viability-depressing effect of induced mutations observed by Simmons and Temin is in line with data collected earlier by Falk and others (Falk, 1961, 1967a; Falk et al., 1965; Falk and Ben-Zeev, 1966) who performed a series of radiation experiments; in these the homozygous and heterozygous effects of irradiated chromosomes on viability were determined. Although there were statistical fluctuations in the data, the results indicated that radiation-induced mutations, especially lethals, reduce heterozygous viability. Other evidence bearing on the issue was obtained by Stern et al. (1952) and Murata and Tobari (1973, 1976); the former determined the viability effects of radiation-induced lethals in the heterozygous state, while the latter monitored the elimination of radiation-induced lethals from population cages. The population studies demonstrated that the lethals do indeed reduce fitness in the heterozygous condition, with the dominance parameter being something greater than 0·02.

In assessing the impact of mutation on the population, overall fitness is, of course, the quantity of interest. However, as mentioned earlier, it is a difficult quantity to measure. Direct measurement is almost impossible, requiring careful studies of viability, fertility, fecundity, longevity, mating ability, etc., and additional information on how to put these together into a summary statistic. The matter would seem hopeless were it not for indirect techniques based on the analysis of changing mutant or chromosome frequencies in a population. These were used by Falk (1967b), who studied the rate of elimination of irradiated second chromosomes in small laboratory populations of *D. melanogaster* under conditions where recombination was suppressed and the chromosomes never became homozygous. A similar strategy was employed by Simmons et al. (1978), who analysed frequency changes of heterozygous second chromosomes treated with EMS. In both cases the induced mutations were observed to lower overall fitness in the heterozygous state. Mitchell (1977) and Mitchell and Simmons (1977) carried the analysis one step further, estimating the dominance of EMS-induced mutations with mild effects on fitness. They compared the reduction in fitness of hemizygous males carrying mutagenized X chromosomes with the corresponding reduction in heterozygous females, and then estimated the dominance of the induced mutations. The estimate was 0·25, a value which agrees reasonably well with the estimates obtained by Temin (1978) and Ohnishi (1977c) in their studies of EMS-induced detrimental mutations affecting viability.

There is another aspect to the studies of overall fitness effects which warrants attention; this has to do with the relation between viability effects and total fitness effects. Although the data do not permit a detailed examination of the relationship, it appears that the heterozygous effects on viability are quite a bit less than the effects on fitness as a whole. For instance, Temin (1978) found that lethals induced on chromosomes treated with 5×10^{-3} M EMS reduced heterozygous viability between one and three percent, but Simmons et al. (1978) observed an 11% reduction in overall fitness in their study of the heterozygous effects of lethals induced on chromosomes treated with slightly less mutagen (4×10^{-3} M EMS). The experiments are not strictly comparable, but the data certainly suggest that the effects of mutagenized chromosomes on total fitness are greater than their effects on viability alone. Falk (1967b) reached a similar conclusion in his analysis of the heterozygous effects of radiation-induced lethals.

The data of Mitchell (1977) and Mitchell and Simmons (1977) also support this view; the fitness of a male carrying an X chromosome treated with 0·001 M EMS was reduced by an average of 1·7 percent in their experiments, while the corresponding reduction in viability was only 0·5 percent.

When all the evidence is weighed, there is strong support for the view that induced mutations are partially dominant; nevertheless, the evidence for overdominance obtained by Wallace (1958, 1963), Mukai et al. (1965), and Maruyama and Crow (1975) remains. Is there any way of explaining the inconsistency?

One possibility is that the viability enhancement seen in the above mentioned experiments is attributable to some other cause than radiation-induced mutations. In each of these experiments a relatively low dose of radiation was employed, limiting the amount of genetic damage that was done. Unfortunately, there is no way of knowing the extent of the damage, since the irradiated chromosomes were not made homozygous in order to identify lethal and lesser viability-affecting mutations. We can assume that about three percent of the irradiated chromosomes in the experiments of Wallace (1958) and Mukai, Yoshikawa and Sano (1965) carried lethal mutations, and that about six percent did so in the experiments of Maruyama and Crow (1975). However, the load of slightly detrimental mutations induced in each of these experiments is a matter of some uncertainty. There is a limited body of evidence which indicates that radiation does not produce many mutations with slight effects on viability, or that if it does, their cumulative effects are very small (Friedman, 1964; Falk et al., 1965; Falk and Ben-Zeev, 1966; Simmons, 1976). If this is the case, then it is hard to see how the viability enhancement observed by Wallace and others could be attributed to radiation-induced mutations; the low doses they used would not have produced many lethal mutations and would probably not have produced enough slightly detrimental damage to be detected in a test for homozygous effects. The viability enhancement they observed in heterozygotes must then have some other explanation, but none has been proposed.

B. Spontaneous Mutations

The only direct evidence on the dominance of spontaneous mutations concerns lethals. It is almost always possible to compare the effects on viability of lethal-bearing chromosomes with nonlethal chromosomes from the same population. This is about the only circumstance involving spontaneous mutations in which a satisfactory control experiment can be carried out. Studies of minor viability-affecting mutants are not so clear-cut and, as might be expected, have produced conflicting, confusing results.

Yoshikawa and Mukai (1970) determined the heterozygous effects on viability of spontaneous lethal mutations on the *D. melanogaster* second chromosome. The lethals clearly depressed viability, regardless of the genetic background in which they were tested. When two nonallelic lethals were put in the same fly, the viability depression was about twice as great as when only one lethal was present. Ohnishi (1977c) also studied the heterozygous effects of spontaneous lethal mutations, as well as those induced with EMS. In both cases, viability was lowered between three and six percent. An earlier study by Tobari (1966) revealed a puzzling temperature effect, possibly of some significance; at 17°C and 25°C spontaneous lethal mutations reduced viability, but at 29°C they enhanced it. This needs further study.

There is at least one report that spontaneous lethals are neutral with respect to the viability of heterozygotes; this is the study of Wallace (1965), whose extensive data indicated variability in the viability effects, but on the average no viability reduction. Although these data do not support the claim that lethals are partially dominant, neither do they argue against it. On the matter of dominance, the findings of Yoshikawa and Mukai (1970) and of Ohnishi (1977c) are unambiguous: there is, on the average, a small but significant reduction in viability when one member of a pair of chromosomes carries a spontaneous lethal, and this is consistent with the data on induced lethal mutations.

Excepting the massive experiments of Temin (1978) and Ohnishi (1977c), in which EMS-induced mutations with only slight effects on viability in the homozygous state were studied, our information about the heterozygous effects of mildly detrimental induced mutations is poor. As mentioned earlier, the radiation experiments probably do not have much to offer in this regard; relative to lethals, X-rays do not seem to produce very many mutations with slight viability effects or, if they do, the effects are very slight indeed. Is the evidence with respect to spontaneous detrimentals any better? Unfortunately, if anything it is worse.

There have been several attempts to study the heterozygous effects of spontaneous detrimentals on viability (Mukai *et al.*, 1964, 1965; Mukai and Yamazaki, 1968; Mukai, 1969a; Ohnishi, 1977c; for a review see Simmons and Crow, 1977); these have made use of *D. melanogaster* second chromosomes in which such mutations have accumulated for many generations. In each experiment the strategy has been to compare the viabilities of heterozygotes for these chromosomes with those of the corresponding homozygotes. Two types of heterozygotes have been studied; in one the mutant-bearing chromosome is paired with a chromosome presumed to be free of mutations, in the other it is paired with a chromosome like itself, but with an independent mutational history. In the first case the mutations under test are all on the same chromosome, in the second they are on both members of the chromosome pair.

When the mutants are coupled on the same chromosome, there is some evidence for overdominance (Mukai *et al.*, 1964; Mukai and Yamazaki, 1968; Mukai, 1969a), whereas in the repulsion configuration, all experiments agree that the average effect is to reduce viability. There is no obvious explanation for the coupling–repulsion effect; furthermore it is physiologically unexpected, and various experiments of the coupling type have given inconsistent results (Ohnishi, 1977c). This suggests that the overdominance found in some cases is spurious. From the repulsion experiments, it would seem that spontaneous detrimental mutations reduce heterozygous viability by between 35% and 50% of the reduction in homozygous viability. This is greater than the 20–30% figure seen with EMS-induced mutations, but the latter seem to have larger homozygous effects, so the net reduction in heterozygous viability for the two types is about the same (Ohnishi, 1977b). The implication is that mild spontaneous mutants are practically additive in their effects on viability; thus their impact on the fitness of the population is twice the total mutation rate.

C. Mutations from Natural Populations

The difficulties encountered in studying the dominance properties of newly arisen spontaneous mutations are amplified for the case of mutations extracted from a natural population. The reason is that the deleterious effects of such mutations in heterozygotes are expected to be less, on the average, than the corresponding effects of newly arisen mutations. Natural selection eliminates mutants with debilitating heterozygous effects, with the result that their equilibrium frequencies in a population are low. The more dominant a mutant, the lower its equilibrium frequency, all other factors being equal. This means that, relative to an unselected sample of newly arisen mutations, those from an equilibrium population are expected to have smaller heterozygous effects. These, of course, are more difficult to measure. The situation is further complicated by the possibility that some, perhaps even a large fraction of the natural genetic variability for a trait such as viability, is attributable to overdominance. Overdominant mutations, if they occur, would be expected to accumulate in a population and attain high frequencies. Such alleles could have a large impact on any experiment designed to assess the heterozygous effects of deleterious mutants segregating at low frequency in the population.

Given these difficulties, it is not surprising that much of the evidence on the dominance of mutations from natural populations is inconsistent. Even for lethals there is disagreement. Some investigators (Dobzhansky and Spassky, 1968; Anderson, 1969; Watanabe et al., 1976) have concluded that lethal mutations extracted from natural populations are overdominant, at least in native genetic backgrounds. Others have found that lethals reduce viability in the heterozygous state (Prout, 1952; Hiraizumi and Crow, 1960; Kitagawa, 1967; Watanabe and Oshima, 1970; Yoshikawa and Mukai, 1970; Mukai and Yamaguchi, 1974). Indirect evidence analysed by Crow and Temin (1964) indicates that the overall fitness is lowered by 1–2% in lethal heterozygotes. However, this conclusion requires the assumption that there is little local inbreeding in *Drosophila* populations, which, if present, would have the same effect as dominance in hastening the elimination of lethal genes from the population. On the basis of isozyme frequency data, Smith et al. (1978) have argued that there may be just enough inbreeding in *Drosophila* populations to account for the low frequencies of individual lethal mutations. There is, however, as these authors point out, a weakness in their argument, having to do with the isozyme frequency method of investigating population structure. When data from laboratory populations are analysed, there is evidence that lethals have some dominance effect (Murata, 1970).

There is even more uncertainty about the heterozygous expression of mutations with much smaller effects on viability. Mukai and co-workers (Mukai et al., 1972; Mukai and Yamaguchi, 1974; Watanabe et al., 1976) have collected enormous amounts of data on these, showing that slightly detrimental mutants from a natural population lower heterozygous viability by about 20% of the reduction seen in homozygotes. However the standard error on this estimate is large, and results from different experiments are not in complete agreement.

It should be noted that the procedure for estimating the dominance of slightly detrimental mutations from natural populations relies on the simultaneous measurement of homozygous and heterozygous viabilities. The regression of heterozygous viability on the sum of the viabilities of homozygotes for the constituent chromosomes estimates the dominance. The viabilities are, of course, measured in standard segregation tests, where one class of offspring serves as a viability standard; another carries the mutation-bearing chromosome. Since the chromosomes come from nature, and are introduced by crosses into the cytoplasm of laboratory strains, there is always the danger that dysgenic hybrids will be produced. If, as has been mentioned previously, segregation distortion occurs in these, the viability measurements could be spurious and the dominance estimates invalid. Cockerham and Mukai (1978) have analysed the effects of segregation

D. Partial Dominance of Mutants Affecting Fitness

There is indirect, but very convincing evidence that mutants having a mild homozygous effect on viability have a significant heterozygous effect on fitness. The argument has been reviewed (Crow, 1979) and will be only briefly recounted here.

From equation (20) we recall that the decrease in homozygous viability per generation of mutation accumulation is approximately $(1-\bar{h})\Sigma u_i s_i$ and the experimental value is about 0·003 for the second chromosome. If a chromosome from a natural population is made homozygous and its effect on viability is measured in the same way—by competition for survival with a reference genotype—the expected value for the dispersed load is approximately $(1-\bar{h}')\Sigma u_i s_i p_i$. In these equations $\bar{h}$ and $\bar{h}'$ are the heterozygous effects on viability of the mutants in the test system whereas p_i is the "persistence" of the gene in nature.

By persistence we mean the average number of individuals that harbor the mutant gene before it is eliminated from the population. In a very large (i.e., deterministic) population the mean persistence is also the mean number of generations that the mutant remains before elimination.

The approximate numbers are given below.

	New Mutants $(1-\bar{h})\Sigma u_i s_i$	Natural Population $(1-\bar{h}')\Sigma u_i s_i p_i$	Ratio of Natural to New $\bar{p}$
Lethals ($s=1$)	0·005	0·25	50
Milds	0·003	0·10	~50

Note that $\bar{p}$ is weighted, appropriately, by $u_i s_i$. We should also note that $\bar{h}'$ is expected to be somewhat less than $\bar{h}$ because the mutants with the greatest heterozygous effect on *fitness* in *nature* are most rapidly eliminated from the natural population and this should be correlated with their heterozygous *viability* in the *laboratory*. For this reason we have changed the ratio in the bottom line from 33 (given in Crow, 1979) to approximately 50. The maximum error from this source is a factor of 2. (No such problem applies to the lethals, which are always classified unambiguously. The value of 0·25 is simply the Poisson-corrected frequency of chromosomes with one or more lethals.)

Mukai (1979) has given a summary similar to that in the table above. He gives somewhat higher values for the dispersed load in natural populations based on studies in North Carolina. The ratio in the right column then becomes larger, about 75. Although the exact values are uncertain and may be very different from one locality to another, the overall conclusion that the persistence is roughly the same for lethals and for milds is well established.

Consider the lethal mutants first. As was noted by Sturtevant (see Dobzhansky and Wright, 1941) the frequency of lethal mutants in the population is an order of magnitude too low to be consistent with the known mutation rate, complete recessivity, and random mating, and too low *a fortiori* if they are overdominant. If the mutants are eliminated through inbreeding, the mean persistence of a recessive lethal is $1/F$, where F is Wright's inbreeding coefficient. Thus to explain the lethal frequency F must be $1/50$.

If the mutants are eliminated by heterozygotes effects on fitness, then $p_i = 1/H_i$, where H_i is the heterozygous effect of the i-th mutant on fitness. Li and Nei (1972) have shown that the expected value of p is $1/H$ even when stochastic fluctuations are taken into account, as might be anticipated intuitively. This does assume, however, that all eliminations are through heterozygous effects; that is to say, H is large enough and F small enough that the mutant never becomes homozygous. If $\bar{p} = 50$, then $\tilde{H} = 1/50$, where $\tilde{H}$ is the harmonic mean of the effects of new mutants on fitness (see Morton *et al.*, 1956).

Using the same arguments for mildly deleterious mutants, if elimination is by inbreeding $\bar{p} = \overline{(1/FS)} = 1/F\tilde{S}$, where S is the homozygous effect on fitness (in contrast to s, which measures only one component, viability) and $\tilde{S}$ is its harmonic mean. But, of course, F is the same for both drastic and mild mutants whereas $\tilde{S} \ll 1$. Since $\bar{p} \approx 50$ for both lethals and milds, the two types of mutants cannot *both* be eliminated by inbreeding.

On the other hand, the data are consistent with heterozygous elimination of both lethals and milds with $\tilde{H}$ around 0·02.

There is another possibility, namely that lethals are eliminated mainly by inbreeding with $F \approx 0·02$, while milds are eliminated mainly by heterozygous effects. That lethals are eliminated by consanguineous matings, especially sibs and half-sibs, has been suggested. The lethals cannot be eliminated by random mating in populations of small effective size; this was ruled out by tests of the lethals for allelism between and within localities (Crow and Temin, 1964). However the incestuous mating hypothesis cannot be ruled out, even though there is no evidence for it. Crow and Temin regarded it as unlikely since the persistence estimate is the same in large laboratory cages and in nature, and it is hard to imagine that incestuous matings occur with sufficient frequency in cages where there is no reason for sibships to remain together after emerging from the pupae.

E. Conclusions about Dominance

There are inconsistencies in the results of experiments designed to measure heterozygous effects directly. Such experiments involve enormous numbers of flies and the signal to noise ratio is low. These experiments are recounted in detail in Sections IIIA,B. But the conclusions of Section IIID are based on highly reproducible measurements.

We regard it as established that mildly deleterious mutants affecting *homozygous viability* have appreciable *heterozygous* effects on *fitness* and that the great majority are eliminated this way. The mean persistence of such a mutant is about 50 generations, implying a mean heterozygous effect for those mutants in an equilibrium population of 1/50 or 0·02. (This is the harmonic mean of new mutants before selection.)

We cannot rule out the possibility that lethals are eliminated, at least in part, by local consanguineous matings, although there is no evidence for this. In any case, the mutants that make the great bulk of the contribution to the mutation load are partially dominant and therefore, on the Haldane principle, the mean fitness is given by equation (8) and the load is approximately $2U$, or more accurately $1-e^{-2U}$.

But we have yet to consider epistasis as a load-reducing mechanism. We now discuss that.

IV. Epistasis

The effect of synergistic epistasis as a load-reducing mechanism was first discussed quantitatively by Kimura (1961) and was reviewed in some detail by Crow (1970). For example, if fitness decreases in proportion to the square of the number of accumulated mutants, the mutation load is reduced by half.

The relation between fitness and the number of heterozygous mutants is not known. The curve for homozygous viability *is* known (see Fig. 2). This is best fitted by a curve of the form

$$\text{Viability} = 1 - an - bn^2 \tag{26}$$

where n is the number of mutants. Reasonable values of a and b permit a load reduction of 25 to 50%, not enough to make a significant impact on solving Haldane's mutation load dilemma. We must seek elsewhere.

If we take $U=0.4$ as the total gametic mutation rate for all loci affecting fitness in *Drosophila*—from Ohnishi's study of several fitness components it could be higher—the Haldane–Muller mutation load is $1-e^{-2U}$, or 0·80, too high for comfort. And as we have just seen, synergistic epistasis of a linear-quadratic form does not give much help.

Several authors (King, 1967; Sved et al., 1967; Milkman, 1967, 1978; Wills, 1978; and several more) have suggested that something approximating truncation selection for fitness may occur in nature. It is well known by plant and animal breeders that the maximum change of gene frequency for a specified intensity of selection occurs with truncation (Haldane, 1932). Individuals below a certain level are rejected and those that are saved contribute equally to the next generation. Is nature clever enough to do this?

In an ideal model the organisms are ranked in order of "fitness potential" (Milkman, 1978). Then a specific fraction, determined by the reproductive potential of the species, survive and reproduce. Those which reproduce are those that rank highest in fitness potential. Various authors have shown, each according to his own taste in model preference, that such a system can greatly reduce the load. Crow and Kimura (1979) made what they regarded as reasonable assumptions about the distribution of fitness potential and used parameters based on *Drosophila* experiments to show that with truncation selection the load could be reduced for a mutation rate of 0·4 from 0·55 to 0·10. By eliminating mutants in groups, as illustrated by equation (17), mutants with a mean persistence of 50 generations, implying $\tilde{S}=0{\cdot}02$, actually reduce fitness by only about 0·002 if the load is equally partitioned among them.

Of course no one expects truncation to be perfect, but it does not have to be, as argued especially by Milkman (1978). For example, if the fitness function, instead of rising suddenly from 0 to 1 at the truncation point, T, rises gradually from $T-\sigma$ to $T+\sigma$, where σ is the standard deviation of the distribution of fitness potentials, this is about 85% as efficient as sharp truncation.

A species with enormous reproductive capacity, such as *Drosophila*, could tolerate a mutation load of 0·90 or more, even with considerable accidental death unrelated to fitness potential. But organisms with low fecundity, like many birds and mammals, could not. We doubt very much that load-reducing epistasis is fundamentally different in organisms that produce few progeny and take care of them or many and do not. These, we believe, are reproductive rather than evolutionary strategies. We would therefore expect to find load-reducing mechanisms even in those species, such as *Drosophila*, where a female can produce a few hundred progeny.

But here information ceases. It is clear that truncation selection, or something even roughly approximating it, based on ranking of individuals by fitness potential, can greatly reduce the mutation load. The empirical question, yet to be answered, is: does this happen in nature?

References

ABRAHAMSON, S., WÜRGLER, F. E., DEJONGH, C. and MEYER, H. U. (1980). How many loci on the X-chromosome of *Drosophila melanogaster* can mutate to recessive lethals? *Environmental Mutagenesis* **2**, 447–453.
ANDERSON, W. W. (1969). Genetics of natural populations. XLI. The selection coefficients of heterozygotes for lethal chromosomes in *Drosophila* on different genetic backgrounds. *Genetics* **62**, 827–836.
AUERBACH, C. and KILBEY, B. (1971). Mutation in eukaryotes. *Ann. Rev. Genet.* **5**, 163–218.
BERG, R. L. (1979). Global pattern of mutability in natural populations of *Drosophila melanogaster*. *Genetics* **91**, s8–s9.
BERG, R., ENGELS, W. R. and KREBER, R. A. (1980). Site-specific X-chromosome rearrangements from hybrid dysgenesis in *Drosophila melanogaster*. *Science* **210**, 427–429.
BREGLIANO, J. C., PICARD, G., BUCHETON, A., PELISSON, A., LAVIGE, J. M. and L'HERITIER, P. (1980). Hybrid dysgenesis in *Drosophilia melanogaster*. *Science* **207**, 606–611.
BRITTNACHER, J. G. (1980). Genetic variation and genetic load due to the male reproductive component of fitness in Drosophila. *Genetics* **97**, 719–730.
BUNDEGAARD, J. and CHRISTIANSEN, F. B. (1972). Dynamics of polymorphisms. I. Selection components in an experimental population of *Drosophila melanogaster*. *Genetics* **71**, 439–460.
CLAYTON, G. A. and ROBERTSON, A. (1955). Mutation and quantitative variation. *Am. Nat.* **89**, 151–158.
COCKERHAM, C. C. and MUKAI, T. (1978). Effects of marker chromosomes on relative viability. *Genetics* **90**, 827–849.
CROW, J. F. (1970). Genetic loads and the cost of natural selection. *In*: "Mathematical Topics in Population Genetics" (K. Kojima, ed.), pp. 128–177. Springer-Verlag, Heidelberg.
CROW, J. F. (1979). Minor viability mutants in *Drosophila*. *Genetics* **92**, s165–s172.
CROW, J. F. and KIMURA, M. (1979). Efficiency of truncation selection. *Proc. Natl. Acad. Sci. U.S.A.* **76**, 396–399.
CROW, J. F. and TEMIN, R. G. (1964). Evidence for the partial dominance of recessive lethal genes in natural populations of *Drosophila*. *Am. Nat.* **98**, 21–33.
DOBZHANSKY, TH. (1970). "The Genetics of the Evolutionary Process." Columbia University Press, New York.
DOBZHANSKY, TH. and SPASSKY, B. (1968). Genetics of natural populations. XL. Heterotic and deleterious effects of recessive lethals in populations of *Drosophila pseudoobscura*. *Genetics* **59**, 411–425.
DOBZHANSKY, TH. and WRIGHT, S. (1941). Genetics of natural populations. V. Relations between mutation rate and accumulation of lethals in populations of *Drosophila pseudoobscura*. *Genetics* **26**, 23–51.
ENGELS, W. R. (1979). Germ line aberrations associated with a case of hybrid dysgenesis in *Drosophila melanogaster* males. *Genet. Res.* **33**, 137–146.
ENGELS, W. R. and PRESTON, C. R. (1980). Components of hybrid dysgenesis in a wild population of *Drosophila melanogaster*. *Genetics* **95**, 111–128.
FISHER, R. A. (1958). "The Genetical Theory of Natural Selection," 2nd edition. Dover Publications, New York.
FALK, R. (1961). Are induced mutations in *Drosophila* overdominant? II. Experimental results. *Genetics* **46**, 737–757.

FALK, R. (1967a). Viability of heterozygotes for induced mutations in *Drosophila melanogaster*. III. Mutations in spermatogonia. *Mutat. Res.* **4**, 59–72.
FALK, R. (1967b). Fitness of heterozygotes for irradiated chromosomes in *Drosophila*. *Mutat. Res.* **4**, 805–819.
FALK, R. and BEN-ZEEV, N. (1966). Viability of heterozygotes for induced mutations in *Drosophila melanogaster*. II. Mean effects in irradiated autosomes. *Genetics* **53**, 65–77.
FALK, R., RAHAT, A. and BEN-ZEEV, N. (1965). Viability of heterozygotes for induced mutations in *Drosophila melanogaster*. I. Irradiated X-chromosomes. *Mutat. Res.* **2**, 438–451.
FRIEDMAN, L. D. (1964). X-ray-induced sex-linked lethal and detrimental mutations and their effects on the viability of *Drosophila melanogaster*. *Genetics* **49**, 689–699.
GREENBERG, R. and CROW, J. F. (1960). A comparison of the effect of lethal and detrimental chromosomes from *Drosophila* populations. *Genetics* **45**, 1153–1168.
HALDANE, J. B. S. (1932). A mathematical theory of natural and artificial selection. IX. Rapid selection. *Proc. Camb. Phil. Soc.* **28**, 244–248.
HALDANE, J. B. S. (1937). The effect of variation on fitness. *Am. Nat.* **71**, 337–349.
HIRAIZUMI, Y. and CROW, J. F. (1960). Heterozygous effects on viability, fertility, rate of development, and longevity of *Drosophila* chromosomes that are lethal when homozygous. *Genetics* **45**, 1071–1083.
JUDD, B. H., SHEN, M. W. and KAUFMAN, T. C. (1972). The anatomy and function of a segment of the X chromosome of *Drosophila melanogaster*. *Genetics* **71**, 139–156.
KÄFER, E. (1952). Vitalitätsmutationen ausgelöst durch Rötgenstrahlen bei *Drosophila melanogaster*. *Z. Ind. Abst. Vererb.* **84**, 508–535.
KIDWELL, M. G. (1983). Intraspecific hybrid sterility. This volume.
KIDWELL, M. G., KIDWELL, J. F. and SVED, J. A. (1977). Hybrid dysgenesis in *Drosophila melanogaster*: a syndrome of aberrant traits including mutation, sterility and male recombination. *Genetics* **86**, 813–833.
KIMURA, M. (1961). Some calculations on the mutational load. *Jap. J. Genet.* **36**, 179–190.
KIMURA, M. and CROW, J. F. (1978). Effect of overall phenotypic selection on genetic change at individual loci. *Proc. Natl. Acad. Sci. USA* **75**, 6168–6171.
KIMURA, M. and MARUYAMA, T. (1966). The mutation load with epistatic gene interactions in fitness. *Genetics* **54**, 1337–1351.
KING, J. L. (1966). The gene interaction component of the genetic load. *Genetics* **53**, 403–413.
KING, J. L. (1967). Continuously distributed factors affecting fitness. *Genetics* **55**, 483–492.
KITAGAWA, O. (1967). Interactions in fitness between lethal genes in heterozygous condition in *Drosophila melanogaster*. *Genetics* **57**, 809–820.
KNIGHT, G. R. and ROBERTSON, A. (1957). Fitness as a measurable character in *Drosophila*. *Genetics* **42**, 524–530.
LEWONTIN, R. C. (1974). "The Genetic Basis of Evolutionary Change." Columbia University Press, New York.
LI, W. and NEI, M. (1972). Total number of individuals affected by a single deleterious mutation in a finite population. *Am. J. Human Genet.* **24**, 667–679.
LIM, J. K. and SNYDER, L. A. (1974). Cytogenetic and complementation analyses of recessive lethal mutations induced in the X chromosome of *Drosophila* by three alkylating agents. *Genet. Res.* **24**, 1–10.
LIU, C. P and LIM, J. K. (1975). Complementation analysis of methyl methane-sulfonate-induced recessive lethal mutations in the zeste-white region of the X chromosome of *Drosophila melanogaster*. *Genetics* **79**, 601–611.

MARUYAMA, T. and CROW, J. F. (1975). Heterozygous effects of X-ray-induced mutations on viability of *Drosophila melanogaster*. *Mutat. Res.* **27**, 241–248.

MILKMAN, R. D. (1967). Heterosis as a major cause of heterozygosity in nature. *Genetics* **55**, 493–495.

MILKMAN, R. (1978). Selection differentials and selection coefficients. *Genetics* **88**, 391–403.

MITCHELL, J. A. (1977). Fitness effects of EMS-induced mutations on the X chromosome of *Drosophila melanogaster*. I. Viability effects and heterozygous fitness effects. *Genetics* **87**, 763–774.

MITCHELL, J. A. and SIMMONS, M. J. (1977). Fitness effects of EMS-induced mutations on the X chromosome of *Drosophila melanogaster*. II. Hemizygous fitness effects. *Genetics* **87**, 775–783.

MORTON, N. E., CROW, J. F. and MULLER, H. J. (1956). An estimate of the mutational damage in man from data on consanguineous marriages. *Proc. Natl. Acad. Sci. USA* **42**, 855–863.

MUKAI, T. (1964). The genetic structure of natural populations of *Drosophila melanogaster*. I. Spontaneous mutation rate of polygenes controlling viability. *Genetics* **50**, 1–19.

MUKAI, T. (1969a). The genetic structure of natural populations of *Drosophila melanogaster*. VI. Further studies on the optimum heterozygosity hypothesis. *Genetics* **61**, 479–495.

MUKAI, T. (1969b). The genetic structure of natural populations of *Drosophila melanogaster*. VII. Synergistic interaction of spontaneous mutant polygenes controlling viability. *Genetics* **61**, 749–761.

MUKAI, T. (1970). Spontaneous mutation rates of isozyme genes in *Drosophila melanogaster*. *Dros. Inf. Serv.* **45**, 99.

MUKAI, T. (1979). Polygenic mutations. *In*: "Quantitative Genetic Variation" (J. N. Thompson, Jr. and J. M. Thoday, eds.) pp. 177–196. Academic Press, New York and London.

MUKAI, T. (1980). The genetic structure of natural population of *Drosophila melanogaster*. XIV. Effects of the incomplete dominance of the IN(2LR)SM1(Cy) chromosome on the estimates of various genetic parameters. *Genetics* **94**, 169–184.

MUKAI, T. and COCKERHAM, C. C. (1977). Spontaneous mutation rates of isozyme genes in *Drosophila melanogaster*. *Proc. Natl. Acad. Sci. USA* **74**, 2514–2517.

MUKAI, T. and YAMAGUCHI, O. (1974). The genetic structure of natural populations of *Drosophila melanogaster*. XI. Genetic variation in a local population. *Genetics* **76**, 339–366.

MUKAI, T. and YAMAZAKI, T. (1968). The genetic structure of natural populations of *Drosophila melanogaster*. V. Coupling-repulsion effect of spontaneous mutant polygenes controlling viability. *Genetics* **59**, 513–535.

MUKAI, T., CHIGUSA, S. and YOSHIKAWA, I. (1964). The genetic structure of natural populations of *Drosophila melanogaster*. II. Overdominance of spontaneous mutant polygenes controlling viability in homozygous genetic background. *Genetics* **50**, 711–715.

MUKAI, T., CHIGUSA, S. and YOSHIKAWA, I. (1965). The genetic structure of natural populations of *Drosophila melanogaster*. III. Dominance effect of spontaneous mutant polygenes controlling viability in heterozygous genetic background. *Genetics* **52**, 493–501.

MUKAI, T., YOSHIKAWA, I. and SANO, K. (1966). The genetic structure of natural populations of *Drosophila melanogaster*. IV. Heterozygous effects of radiation-induced mutations on viability in various genetic backgrounds. *Genetics* **53**, 513–527.

MUKAI, T., CHIGUSA, S. I., METTLER, L. E. and CROW, J. F. (1972). Mutation rate and dominance of genes affecting viability in *Drosophila melanogaster*. *Genetics* **72**, 335–355.

MULLER, H. J. (1950). Our load of mutations. *Am. J. Human Genet.* **2**, 111–176.

MURATA, M. (1970). Frequency distribution of lethal chromosomes in small populations of *Drosophila melanogaster*. *Genetics* **64**, 559–571.

MURATA, M. and TOBARI, I. (1973). Changes in frequency of lethal second chromosomes in experimental populations of *Drosophila melanogaster* with radiation histories. *Jap. J. Genet.* **48**, 349–359.

MURATA, M. and TOBARI, I. (1976). Changes in frequency and allelism of recessive lethals in experimental populations of *Drosophila melanogaster* with radiation histories. *Jap. J. Genet.* **51**, 27–37.

OHNISHI, O. (1977a). Spontaneous and ethyl methanesulfonate-induced mutations controlling viability in *Drosophila melanogaster*. I. Recessive lethal mutations. *Genetics* **87**, 519–527.

OHNISHI, O. (1977b). Spontaneous and ethyl methanesulfonate-induced mutations controlling viability in *Drosophila melanogaster*. II. Homozygous effect of polygenic mutations. *Genetics* **87**, 529–545.

OHNISHI, O. (1977c). Spontaneous and ethyl methanesulfonate-induced mutations controlling viability in *Drosophila melanogaster*. III. Heterozygous effect of polygenic mutations. *Genetics* **87**, 547–556.

PANDEY, J. (1975). Further studies on heterozygous effects of radiation on viability of *Drosophilia melanogaster*. *Mutat. Res.* **27**, 249–253.

PAXMAN, G. J. (1957). A study of spontaneous mutation in *Drosophila melanogaster*. *Genetica* **29**, 39–57.

PROUT, T. (1952). Selection against heterozygotes for autosomal lethals in natural populations of *Drosophila willistoni*. *Proc. Natl. Acad. Sci. USA* **38**, 478–481.

PROUT, T. (1971a). The relation between fitness components and population prediction in *Drosophila*. I. The estimation of fitness components. *Genetics* **68**, 127–149.

PROUT, T. (1971b). The relation between fitness components and population prediction in *Drosophila*. II. Population prediction. *Genetics* **68**, 151–167.

SCHALET, A. (1960). "A Study of Spontaneous Visible Mutations in *Drosophila melanogaster*." Ph.D. Thesis, Indiana University, Bloomington.

SIMMONS, M. J. (1976). Heterozygous effects of irradiated chromosomes on viability in *Drosophila melanogaster*. *Genetics* **84**, 353–374.

SIMMONS, M. J. and CROW, J. F. (1977). Mutations affecting fitness in *Drosophila* populations. *Ann. Rev. Genet.* **11**, 49–78.

SIMMONS, M. J. and LIM, J. K. (1980). Site specificity of mutations arising in dysgenic hybrids of *Drosophila melanogaster*. *Proc. Natl. Acad. Sci. U.S.A.* **77**, 6042–6046.

SIMMONS, M. J., SHELDON, E. W. and CROW, J. F. (1978). Heterozygous effects on fitness of EMS-treated chromosomes in *Drosophila melanogaster*. *Genetics* **88**, 575–590.

SIMMONS, M. J., PRESTON, C. R. and ENGELS, W. R. (1980). Pleiotropic effects on fitness of mutations affecting viability in *Drosophila melanogaster*. *Genetics* **94**, 467–475.

SMITH, D. B., LANGLEY, C. H. and JOHNSON, F. M. (1978). Variance component analysis of allozyme frequency data from eastern populations of *Drosophila melanogaster*. *Genetics* **88**, 121–137.

STERN, C., CARSON, G., KINST, M., NOVITSKI, E. and UPHOFF, D. (1952). The viability of heterozygotes for lethals. *Genetics* **37**, 413–439.

SVED, J. A. (1971). An estimate of heterosis in *Drosophila melanogaster*. *Genet. Res.* **18**, 97–105.

SVED, J. A. (1979). The "hybrid dysgenesis" syndrome in *Drosophila melanogaster*. *BioScience* **29**, 659–664.

SVED, J. A. and AYALA, F. J. (1970). A population cage test for heterosis in *Drosophila pseudoobscura*. *Genetics* **66**, 97–113.

SVED, J. A., REED, T. E. and BODMER, W. F. (1967). The number of balanced polymorphisms that can be maintained in a natural population. *Genetics* **55**, 469–481.

TEMIN, R. G. (1966). Homozygous viability and fertility loads in *Drosophila melanogaster*. *Genetics* **53**, 27–46.

TEMIN, R. G. (1978) Partial dominance of EMS-induced mutations affecting viability in *Drosophila melanogaster*. *Genetics* **89**, 315–340.

TEMIN, R. G., MEYER, H. U., DAWSON, P. S. and CROW, J. F. (1969). The influence of epistasis on homozygous viability depression in *Drosophila melanogaster*. *Genetics* **61**, 497–519.

TOBARI, I. (1966). Effects of temperature on the viability of heterozygotes of lethal chromosomes in *Drosophila melanogaster*. *Genetics* **53**, 249–259.

TAJIMA, F. (1978). "Mutator-induced Mutation Affecting Viability in *Drosophila melanogaster*." M.Sc. Dissertation, Kyushu University, Fukuoka, Japan.

WALLACE, B. (1958). The average effect of radiation-induced mutations on viability in *Drosophila melanogaster*. *Evolution* **12**, 532–552.

WALLACE, B. (1963). Further data on the overdominance of induced mutations. *Genetics* **48**, 633–651.

WALLACE, B. (1965). The viability effects of spontaneous mutations in *Drosophila melanogaster*. *Am. Nat.* **99**, 335–348.

WALLACE, B. (1968). Mutation rates for autosomal lethals in *Drosophila melanogaster*. *Genetics* **60**, 389–393.

WALLACE, B. (1970). "Genetic load: its biological and conceptual aspects." Prentice-Hall, Englewood Cliffs, N.J.

WATANABE, T. K. and OSHIMA, C. (1970). Persistence of lethal genes in Japanese natural populations of *Drosophila melanogaster*. *Genetics* **64**, 93–106.

WATANABE, T. K., YAMAGUCHI, O. and MUKAI, T. (1976). The genetic variability of third chromosomes in a local population of *Drosophila melanogaster*. *Genetics* **82**, 63–82.

WILLS, C. (1978). Rank-order selection is capable of maintaining all genetic polymorphisms. *Genetics* **89**, 403–417.

YAMAGUCHI, O. and MUKAI, T. (1974). Variations of spontaneous occurrence rates of chromosomal aberrations in the second chromosomes of *Drosophila melanogaster*. *Genetics* **78**, 1209–1221.

YOSHIKAWA, I. and MUKAI, T. (1970). Heterozygous effects on viability of spontaneous lethal genes in *Drosophila melanogaster*. *Jap. J. Genet.* **45**, 443–455.

20. Factors Affecting Mutation Rates in Natural Populations

R. C. WOODRUFF
Department of Biological Sciences
Bowling Green State University
Bowling Green, Ohio, U.S.A.

BARTON E. SLATKO
Department of Biology
Williams College
Williamstown, Massachusetts, U.S.A.

and

JAMES N. THOMPSON, JR.
Department of Zoology
University of Oklahoma
Norman, Oklahoma, U.S.A.

I. Introduction	37
II. Measurement of Mutation Rates	39
A. Methods to Determine Lethal and Visible Mutation Rates . . .	47
B. Methods to Determine the Frequency of Chromosome Breakage . .	50
C. Precautions in Measuring the Rates of Mutation and Chromosome Breakage in Natural Population Lines	52
III. Factors Affecting Mutation Potential	54
A. Extrinsic Factors	54
B. Intrinsic Factors	60
IV. Genetic Mutator Factors	63
A. Characteristics of Some Well-known Mutator Systems from Natural Populations	64
B. Characteristics of Other Mutator Systems from Natural Populations .	101
V. Effect of Mobile DNA Elements on Spontaneous Mutation Rates . .	102
VI. Conclusions	104
Acknowledgements	107
References	107

I. Introduction*

Whether viewed from the perspective of an individual or a population, the primary goal of genetic transmission is continuity. Continuity, however, has two somewhat antithetical requirements: uniformity and variation. Although faithful reproduction of DNA is needed to insure reasonable

NOTE: For additional references to this chapter, please see page 429.

control of the growth and function of a complex living system, the organism cannot be viewed in isolation from the environment to which it is adapted. Indeed, in the face of changing environmental stresses, too accurate a replication contains the seeds of its own failure. It is for this reason that the concept of mutation has always been at the core of evolutionary theory.

In order to appreciate this balance between stability and mutability in natural populations, several questions need to be answered. First, what kinds and amounts of genetic variation are normally found in natural populations of different species? Second, how is such variation maintained? Does it have a function and, if so, what is the relationship between benefit and liability in different types of environments and under different genetic systems? Next, at what rates do new genetic changes occur in natural populations? Finally, what extrinsic and intrinsic factors influence mutation rates?

Numerous studies have provided information on the types and amounts of variation in natural populations of *Drosophila* (see, for example, reviews by Spencer, 1947; Dobzhansky, 1951; Wallace, 1968a; Lewontin, 1974; Spiess, 1977; Dobzhansky *et al.*, 1977; Futuyma, 1979). It is clear that natural populations carry mutations that affect visible and behavioral phenotypes, viability, fertility, enzyme activity, electrophoretic mobility, and changes in chromosome structure. In many populations, these variants are common, although there are exceptions such as the apparent lack of chromosome rearrangements in some species of *Drosophila* (Carson, 1965; Ashburner and Lemeunier, 1976). In addition, we are beginning to appreciate the role that variation plays in helping populations adapt to spatially or temporally heterogeneous environments. The levels and significance of various classes of genetic variation are reviewed elsewhere in this volume.

Much less is known, however, about the rate at which new mutations occur in natural populations and about the factors that influence mutation rate. This is due, at least in part, to the difficulty of assessing rates of spontaneous mutation and chromosome breakage in nature. For example, the observation that mutant alleles and chromosome aberrations are found at high levels does not tell us directly about the rate at which these changes occur. Populations that carry high levels of genetic variation can do so either because they have high spontaneous mutation rates or because the resulting mutant alleles or chromosome rearrangements are maintained by selection or drift. It is, therefore, important to obtain direct measures of mutation rates in order to evaluate the impact that mutation may have upon population structure and adaptability. In this chapter we shall first survey and evaluate some of the methods for measuring mutation frequencies in

Drosophila and then we shall discuss some of the factors that may influence rates of mutation in natural populations.

II. Measurement of Mutation Rates

Spontaneous mutation rate can be defined as the probability with which a heritable change in the genetic material occurs. Such changes can be identified as transmitted mutant phenotypes other than those caused by segregation, recombination, or dominant lethals (Rieger *et al.*, 1976). The relationship between observed and actual spontaneous mutation rates is not, however, a simple one. The observed mutation rate is probably somewhat smaller than the actual mutation rate, because observed mutation depends not only upon the occurrence of an error, but also upon the probability of its repair by the cell and the probability that it will produce a detectable phenotype if repaired incorrectly.

A sample of spontaneous gene and chromosome mutation frequencies for various types of characters is given in Table I. Although they fall within a fairly narrow range, it is clear that there is variation both among loci and among strains of *Drosophila*. Pooling these estimates, the average mutation rate per base pair per replication in *D. melanogaster* is found to be about 7.0×10^{-11} (Drake, 1969; Propping, 1972), which is similar to estimated mutation rates for other organisms from bacteria to humans.

Spontaneous mutation is commonly attributed to errors during replication, such as those introduced by tautomerization of nucleotide bases, and to the action of mutation-inducing and mutation-preventing compounds produced during metabolism (see, for example, Auerbach, 1976). This cannot, however, be the entire answer. Mutations do occur in non-replicating genomes; and sensitivity to mutation-inducing compounds and the efficiency of repair will in turn be dependent upon the physiological conditions of the individual, its temperature, age, and other internal and external factors. In addition, not all spontaneous mutations are point mutations. Slizynski (1938) found that 9 of the 19 spontaneous X-linked recessive lethals analysed cytologically were tiny chromosome deficiencies, and some spontaneous mutations in prokaryotes and eukaryotes are caused by insertions or deletions of DNA sequences (Bender, personal communication; Bingham and Judd, 1981; Bingham *et al.*, 1982; Carlos and Miller, 1980; Collins and Rubin, 1982; Gehring and Paro, 1980; Goldberg *et al.*, 1982; Karess and Rubin, 1982; Levis and Rubin, 1982; Levis *et al.*, 1982; Rubin *et al.*, 1982; Snyder *et al.*, 1982; Spradling and Rubin, 1981; Zachar and Bingham, 1982; see also *Cold Spring Harbor Symp. Quant. Biol.* Vol. XLV, 1981, for reviews of this topic.) Besides physiological factors, genetic factors also play an important role. In many populations there are mutator factors which can induce gene mutations, chromosome breakage, sterility,

TABLE I. Estimates of the spontaneous frequencies of mutation and chromosome breakage in *Drosophila melanogaster*. Data is only from apparent non-mutator lines and does not include crosses with repair defective females or males. Crosses showing known hybrid dysgenic activity are also excluded. A complete bibliography of references and unpublished reports used in this table may be obtained from the authors (some references are from Crow and Temin, 1964; Graf, 1972; Schalet and Sankaranarayanan, 1976; Würgler et al., 1982; Lee et al., 1982 and Valencia et al., 1983).

A. Recessive sex-linked lethal mutation frequencies in males of long term laboratory stocks (123 references and three unpublished reports).

Males used	Mutation frequency	Range of frequencies	Number of reports or runs
Canton-S	1,261/910,328 (0·14%)	0–0·51%	more than 298
Oregon-K	596/349,389 (0·17%)	0–0·46%	44
Oregon-R	211/168,316 (0·13%)	0·04–0·28%	26
Berlin-K	472/244,815 (0·19%)	0·04–0·49%	49
Berlin	94/37,046 (0·25%)	0·18–0·80%	9
Swedish-b	49/13,891 (0·35%)	0·14–0·57%	4
R(1)2	92/32,628 (0·28%)	0–1·62%	10
Karsnaes-60	8/11,624 (0·07%)	0·05–0·09%	2
Florida-4	9/3,891 (0·23%)	0·11–0·40%	4
Florida-5	24/3,532 (0·68%)	0·56–1·07%	2
D-32	0/3,035		2
Hikone-R	6/4,285 (0·14%)		1
S150	3/4,085 (0·07%)	0·07–0·09%	2
In(1)asc	4/4,233 (0·09%)		1
OK/Y-bb	14/6,652 (0·21%)	0·22–0·27%	2
w	5/3,216 (0·16%)		2
y	6/3,683 (0·16%)		1
y w/y⁺ Y	0/636		1
$sc^8\ w^a\ bb$	5/538 (0·93%)		1
y In(1)dl49 v	14/4,420 (0·32%)		1
Basc	22/15,228 (0·14%)	0·09–0·28%	4
In(1)EN	11/2,643 (0·42%)		1
Total	2,906/1,828,114 (0·16%)	0–1·62%	more than 445

B. Recessive sex-linked lethal mutations in females of long-term laboratory stocks (Abrahamson et al., 1980; Auerbach, 1941; Markowitz, 1970; Muller et al., 1963; Ostertag and Haake, 1966; Parker, 1960; Shakarnis, 1969, 1970; Wallace, 1970).

Females used	Mutation frequency	Number of runs or reports
Oregon-R	0/9,217	3
Canton-S	26/14,197 (0·18%)	5
Basc	16/9,466 (0·17%)	1
y dow/Basc (Df)	367/246,157 (0·15%)	
D-32	2/3,574 (0·06%)	1
$y\ In(1)dl49v/y\ w^{m4}\ In(1)ras^2$	14/4,420 (0·32%)	1
Florida 4	0/796	1

Females used	Mutation frequency	Number of runs or reports
Florida 5	3/3,559 (0·08%)	2
sc⁸ wᵃ bb	0/843	1
$In(1)sc^8, y^{3P}/X \cdot Y^S, In(1)dl-49, l(1)Jl\ v\ f\ B$	80/26,402 (0·30%)	
$y\ In(1)w^{m4}/X \cdot Y^L, y\ In(1)dl-49\ v\ f\ B$	24/9,983 (0·24%)	
Total	**532/328,614 (0·16%)**	more than 17

C. Recessive sex-linked lethal mutations in males of newly collected natural population lines (mutation frequencies were determined from males collected from nature or from isofemale lines derived from nature; hybrid dysgenesis is therefore not occurring in these crosses)

Natural population	Mutation frequency
Eleven populations from U.S.S.R. collected and scored from 1937–1964 (Berg, personal communication)	99/32,852 (0·30%)
Carpenter, Wake County, North Carolina, USA (Langley and Ito, 1976)	104/40,830 (0·26%)
Grand Rapids, Wood County, Ohio, USA (Woodruff et al., 1979)	31/10,496 (0·30%)
Chateau Tahbilk, Australia (20 isofemale lines)	100/33,117 (0·33%)
Waterville, Wood County, Ohio, USA	6/2,333 (0·26%)
Bowling Green, Wood County, Ohio, USA	0/2,265
OK1, Oklahoma City, Oklahoma, USA	4/3,743 (0·11%)
W8D, Okefenokee Swamp, Georgia, USA	10/2,152 (0·46%)
TW3, Varna, New York, USA	0/2,370
BG80A, Bowling Green, Ohio, USA	0/2,265
WV80A, Waterville, Ohio, USA (Woodruff and Thompson, unpublished observations)	6/2,333 (0·26%)
Eight natural population lines from USA (Wallace, 1970)	In males = 33/13,605 (0·24%) In females = 42/13,489 (0·31%)
Spencer and Stern (1948):	
Pittsford, New York	5/2,218 (0·23%)
Webster, New York	3/1,913 (0·16%)
New Wilmington, Pennsylvania	0/727
Sukhumi, USSR	24/2,039 (1·18%)
Akhalicikh, USSR	10/1,300 (0·77%)
Wooster, Ohio, USA	8/1,266 (0·63%)
Vladikavkag, USSR	16/5,412 (0·30%)
Formosa	8/2,054 (0·39%)
Nalchik, USSR	18/5,169 (0·35%)
Vitebsk, USSR	1/402 (0·25%)
Kuibyshev, USSR	1/506 (0·20%)
Merv, USSR	2/1,424 (0·14%)
Samarkand, USSR	5/4,416 (0·11%)
Kutaisi, USSR	6/460 (1·30%)
Oni, USSR	4/378 (1·06%)
Uman, USSR	10/838 (1·19%)
Nikitski Sad, USSR	10/1,206 (0·83%)

Natural population	Mutation frequency
Sukhumi, USSR	14/3,599 (0·39%)
Serpukhov, USSR	6/1,013 (0·59%)
Kashira, USSR	7/2,081 (0·34%)
Delizhan, USSR	1/712 (0·14%)
Gelendzhik, USSR (Dubinin, 1946)	0/576
Lines derived from a Florida, USA stock (Dubinin, 1946)	302/118,846 (0·25%)
Timofeeff-Ressovsky, 1940, 8 lines (in Dubinin, 1946)	95/68,221 (0·14%)
Florida-inbred	23/2,108 (1·09%)
Wooster, Ohio	8/1,266 (0·63%)
Formosa	8/2,054 (0·39%)
Swedish-b	3/1,627 (0·18%)
California-c	2/708 (0·28%)
Huntsville, Texas	0/938
Urbana, Illinois	1/1,016 (0·10%)
Canton, Ohio	0/911
Amherst, Massachusetts	1/572 (0·17%)
Woodbury, New Jersey	1/1,159 (0·09%)
Tuscaloosa, Alabama	1/545 (0·18%)
Lausanne	2/955 (0·18%)
Seto, Japan	0/1,236
Kyoto, Japan (Demerec, 1937)	1/875 (0·11%)
Blacksburg, Virginia	6/6,318 (0·10%)
Canonburg, Pennsylvania	6/2,571 (0·23%)
Austin, Texas	6/3,323 (0·18%)
Chapel Hill, North Carolina	5/3,456 (0·15%)
Amherst, Massachusetts (Levine and Ives, 1953)	31/13,656 (0·23%)
Total	**1,096/433,920 (0·25%)**

D. Recessive autosomal lethal mutations (*=natural population lines; others are laboratory stocks)

Second chromosome:
In males:

14/2,026 (0·69%)	(Muller et al., 1950)
69/17,572 (0·39%)	(Muller, 1928; in Dubinin, 1946)
22/3,454 (0·64%)	(Wallace, 1968)
195/46,395 (0·42%)	(Purdom et al., 1968)
46/2,071 (2·22%)	(Ives, 1950)
56/14,087 (0·40%)	(Dubinin, 1946)*
20/3,055 (0·66%)	(Berg, 1937–1940; in Dubinin, 1946)*
55/1,626 (3·38%)	(Olenov, 1937–1938; in Dubinin, 1946)*
6/412 (1·46%)	(Racine et al., 1980)
73/6,346 (1·15%)	(Altenburg and Altenburg, 1952; Altenburg et al., 1952a; Altenburg et al., 1952b)
34/999 (3·40%)	(Abrahamson and Himoe, 1963)
8/1,852 (0·43%)	(Fahmy and Fahmy, 1970)
9/499 (1·80%)	(Abrahamson et al., 1966)
50/7,281 (0·69%)	(Watanabe and Lee, 1977)
657/107,675 (0·61%)	

In females:
  20/2,337 (0·86%)   (Wallace, 1968)
  19/3,159 (0·60%)   (Myer and Muller, 1961)
  0/864   (Markowitz, 1970)
  7/668 (1·05%)   (Ratanayake, 1968)
  2/1,022 (0·20%)   (Ives, 1945)
  91/17,879 (0·51%)   (Quoted in Crow and Temin, 1964)

139/25,929 (0·54%)

Third chromosome:
In males:
  16/2,797 (0·57%)   (Wallace, 1968)
In females:
  22/3,067 (0·72%)   (Wallace, 1968)

E. Examples of spontaneous forward and reverse visible mutation frequencies at specific loci.

1. X-linked forward mutations (Abrahamson et al., 1980; Fahmy and Fahmy, 1959; Gerasimova, 1981; Glass and Ritterhof, 1956; Green, 1978b; Inagaki et al., 1974; Ives, 1950, 1959b; Jollos, 1937; Muller, 1954b; Muller et al., 1950; Schalet, personal communication; Valencia, personal communication; Woodruff et al., 1979) (numbers in parentheses are clustered events).

Locus	Frequency in males ($\times 10^{-5}$)	Frequency in females ($\times 10^{-5}$)	Frequency in males and females ($\times 10^{-5}$)
achaete (*ac*, 1–0·0)			0/120,000
carmine (*cm*, 1–18·9)	4/570,149 (0·70)	3/1,790,864 (0·17)	0/120,000
carnation (*car*, 1–62·5)	1/570,149 (0·18)	6/1,740,864 (0·34)	0/120,000
cut (*ct*, 1–20·0)	7+(2)+(10)/570,149 (1·58)	14/1,800,864 (0·77)	1/120,000 (0·83)
dusky (*dy*, 1–36·2)	0/570,149	1/440,414 (0·22)	0/120,000
echinus (*ec*, 1–5·5)	1/552,756 (0·18)	0/340,000	—
forked (*f*, 1–56·7)	3/647,299 (0·46)	2/1,138,173 (0·17)	0/120,000
garnet (*g*, 1–44·4)	5/570,149 (0·88)	1/1,790,864 (0·05)	1/120,000 (0·83)
lozenge (*lz*, 1–27·7)	2/130,149 (1·53)	0/702,898	—
miniature (*m*, 1–36·1)	0/67,289	1/100,444 (0·99)	—
Notch (*N*, 1–3·0)	7/490,000 (1·43)	17+(3)+(2)/340,000 (5·59)	—
outstretched small eye (*os*, 1–59·2)	1/510,000 (0·20)	0/340,000	—
prune (*pn*, 1–0·8)	1/570,149 (0·18)	0/1,790,864	0/120,000
raspberry (*ras*, 1–32·8)	2/527,393 (0·38)	0/1,823,959	0/120,000

Locus	Frequency in males ($\times 10^{-5}$)	Frequency in females ($\times 10^{-5}$)	Frequency in males and females ($\times 10^{-5}$)
ruby (rb, 1–7·5)	0/60,149	2/1,450,864 (0·14)	0/120,000
rudimentary (r, 1–54·5)	1/60,149 (1·66)	0/702,898	—
sable (s, 1–43·0)	0/42,756	0/702,898	—
singed (sn, 1–21·0)	2/612,905 (0·33)	0/1,125,993	0/120,000
vermilion (v, 1–33·0)	0/80,149	0/752,898	0/120,000
white (w, 1–1·5)	10/668,631 (1·50)	5/1,820,864 (0·28)	0/120,000
yellow (y, 1–0·0)	12+(3)/650,111 (2·0)	1+(8)/373,095 (0·53)	0/120,000
Average mutation rate for 20 loci in males, 20 loci in females, and 13 loci in males and females ($\times 10^{-5}$)	62/8,520,630 (0·73)	56/21,069,718 (0·27)	2/1,680,000 (0·12)
Average mutation rate for both sexes ($\times 10^{-5}$)		120/31,270,348 (0·38)	

2. Second chromosome forward mutations (Carlson and Southin, 1962; Fujikawa and Inagaki, 1979; Fujikawa et al., 1975; Fujikawa, 1980; Inagaki et al., 1974; Glass and Ritterhoff, 1956; Jenkins, 1967; Miyamoto and Nakao, 1978; Smith and Corwin, 1971; Snyder and Oster, 1964; Southin, 1966)

Locus	Frequency in males ($\times 10^{-5}$)	Frequency in females ($\times 10^{-5}$)
brown (bw, 2–104·5)	3/102,759 (2·91)	—
cinnabar (cn, 2–57·5)	—	0/100,414
dumpy (dp, 2–13·0)	110/638,234 (17·23)	1/148,120 (0·67)
purple (pr, 2–54·5)	—	1/100,414 (1·0)
star (s, 2–1·3)	—	1/100,414 (1·0)
Average mutation rate for 3 loci in males and 4 loci in females ($\times 10^{-5}$)	113/740,993 (15·25)	3/449,362 (0·66)

3. Third chromosome forward mutations (Alexander, 1960; Glass and Ritterhoff, 1956; Ives, 1959b)

Locus	Frequency in males ($\times 10^{-5}$)	Frequency in females ($\times 10^{-5}$)
blistery (by, 3–48·7)	—	0/100,414
curled (cu, 3–50·0)	0/150,673	—
ebony (e, 3–70·7)	3/253,432 (0·12)	—
hairy (h, 3–26·5)	0/150,673	—
pink (p, 3–48·0)	0/150,673	—

Locus	Frequency in males ($\times 10^{-5}$)	Frequency in females ($\times 10^{-5}$)
roughoid (ru, 3–0·0)	0/99,703	—
scarlet (st, 3–44·0)	0/150,673	—
stripe (sr, 3–62·0)	0/150,673	—
thread (th, 3–43·2)	0/150,673	—
Average mutation rate for 8 loci in males and 1 locus in females ($\times 10^{-5}$)	3/1,257,173 (0·24)	0/100,414

4. Fourth chromosome forward mutations (Glass and Ritterhoff, 1956)

Locus	Frequency in males ($\times 10^{-5}$)	Frequency in females ($\times 10^{-5}$)
eyeless (ey, 4–2·0)	6/102,759 (5·83)	0/100,414
Average forward mutation rate for:	31 loci in males ($\times 10^{-5}$) 184/10,621,555 (1·73) 30 loci in males ($\times 10^{-5}$) (dp not included) 74/9,983,321 (0·74)	25 loci in females ($\times 10^{-5}$) 59/21,719,908 (0·27) 24 loci in females ($\times 10^{-5}$) (dp not included) 58/21,571,788 (0·27)
Average forward mutation rate for 56 loci in males and females ($\times 10^{-5}$):	243/32,341,463 (0·75) (dp not included) = 132/31,555,109 (0·42)	

5. Forward mutation of genes with multiple loci (Fahmy and Fahmy, 1966, 1969; Glass, 1955; Grace *et al.*, 1968; Huang and Baker, 1976; Stromnaes and Kvelland, 1963).

Minutes (M) (or haplo-4) 639/588,642 (108·55)
($\times 10^{-5}$)

6. Reverse mutations (Banerjee *et al.*, 1978; Bowman, 1965, 1969; Ballantyne and Chovnick, 1971; Johnston and Winchester, 1934; Lefevre and Green, 1959; Lewis, 1959; Lifschytz and Falk, 1969; Green, 1957, 1961; Altenburg and Browning, 1962; Woodruff, 1975; Woodruff *et al.*, 1972)

Allele	Reversion frequency ($\times 10^{-5}$)
Beadex-2 (Bx^2, 1–59·4)	0/37,995
carnation (car, 1–62·5)	0/1,222,190
cut (ct, 1–20·0)	0/713,001
echinus (ec, 1–5·5)	0/713,001
forked (f, 1–56·7)	7/1,270,301 (0·55)

Allele		Reversion frequency ($\times 10^{-5}$)
forked-36a (f^{36a})		0/185,586
forked-3N (f^{3N})	In females:	115/1,431,477 (8·03)
	In males:	2/101,304 (1·97)
garnet (g, 1–44·4)		0/713,001
miniature (m, 1–36·1)		0/768,465
scute (sc, 1–0·0)		0/913,001
vermilion (v, 1–33·0)		0/1,230,839
white (w, 1–1·5)		0/55,464
white-apricot (w^a)		1/910,184 (0·10)
white-Brownex (w^{Bwx})		0/121,305
white-colored (w^{col})		0/116,842
white-eosin (w^e)		0/113,218
white-ivory (w^i)		56/1,076,000 (5·2)
white-spotted (w^{sp})		1/98,204 (1·01)
yellow (y, 1–0·0)		0/768,465
yellow-2 (y^2)		0/226,118
Killer prune (Kpn, 3–102·9)		$1·7 \times 10^{-6}$
rosy-2 (ry^2, 3–52·0)		0/1,910,000
rosy-5 (ry^5)		0/1,070,000
rosy-41 (ry^{41})		0/800,000
rosy-42 (ry^{42})		0/1,376,000
Average reverse mutation rate of 21 loci in males and females ($\times 10^{-5}$)		182/17,941,961 (1·01)
Average reverse mutation rate of 19 loci in males and females ($\times 10^{-5}$) (f^{3N} and w^i not included)		9/15,333,180 (0·06)

F. Electrophoretic mutations (Tobari and Kojima, 1972) (The spontaneous frequencies reported in Mukai and Cockerham, 1977 and Voelker et al., 1980 were for mutator lines under conditions of hybrid dysgenesis. These frequencies are given in the text).

Locus	Mutation frequency ($\times 10^{-5}$)
Second chromosome:	
alcohol dehydrogenase (Adh, 2–50·1)	0/88,080
alpha-glycerophosphate dehydrogenase ($alpha$-$Gpdh$, 2–20·5)	1/88,080 (1·14)
malate dehydrogenase-1 (cytoplasmic malate dehydrogenase) (Mdh-1 or $cMdh$, 2–37·2)	0/88,080
Third chromosome:	
aldehyde oxidase ($Aldox$, 3–56·7)	0/57,952
alkaline phosphatase (Aph, 3–47·7)	0/57,952
esterase-C (Est-C, 3–51·7)	0/57,952
esterase-6 (Est-6, 3–36·0)	0/57,952
isocitrate dehydrogenase (Idh, 3–27·1)	2/57,952 (3·45)
octanol dehydrogenase (Odh, 3–49·2)	0/57,952
rosy (ry, 3–52·0) (xanthine dehydrogenase)	0/57,952

G. Heritable translocations (22 references and three unpublished reports)

Frequency	Range
16/155,168 (0·01%)	0–0·07%

H. B^S Y y^+ chromosome marker loss in males (16 references)

Marker lost	Breakage frequency	Range
B^S (Y^L)	96/311,494 (0·03%)	0–0·8%
y^+ (Y^S)	51/311,494 (0·02%)	0–0·06%

I. bw^+ Y y^+ chromosome marker loss in males (Baker, 1957)

Marker lost	Breakage frequency
bw^+ (Y^L)	8/47,358 (0·02%)
y^+ (Y^S)	19/47,358 (0·04%)

distortion of segregation, and other phenomena (see later sections for a detailed discussion). Thus, it is not surprising that progress in understanding the diverse causes and effects of spontaneous mutation in natural populations has been slow.

The availability of specialized stocks of *D. melanogaster*, however, has enabled investigators to measure mutation rates fairly precisely. In the following sections we shall summarize those mutation screens that seem most efficient and accurate for estimating mutation rates in natural populations. Special precautions that apply both to visible and lethal screens and to chromosome breakage screens will be discussed in a separate section.

Reviews and additional breeding programs can be found in Abrahamson and Lewis (1971), Auerbach (1962, 1976), Crow and Abrahamson (1971), Muller and Oster (1963), Sobels and Vogel (1976), Spencer (1947), Vogel (1975), Würgler *et al.* (1977), and Environmental Protection Agency "Gene-Tox" committee reports on recessive sex-linked lethal mutations (Lee *et al.*, 1983) and chromosome breakage (Valencia *et al.*, 1983) in *D. melanogaster*. The appropriate statistical analyses of low frequency events such as mutation rates are discussed in Berchtold (1975), Kastenbaum and Bowman (1970), Stevens (1942), Würgler *et al.* (1975), and Engels (1979d). Mutations and special stocks cited below are described in Lindsley and Grell (1968).

A. Methods to Determine Lethal and Visible Mutation Rates

1. Recessive sex-linked lethal mutations

The following sample screen for recessive sex-linked lethals is an efficient measure of mutation rate, because in two generations it allows the detection of lethal effects associated with point mutations and with chromosome aberrations in approximately 20% of the D. melanogaster genome.

Individual males newly collected from natural populations or males from isofemale lines are mated with females from a balancer X chromosome stock such as *Basc*, *FM6*, *FM7*, or *Inscy*. Single F_1 heterozygous females are then mated with balancer males and F_2 progeny are scored for the presence or absence of wild type males. The absence of wild type males among the progeny of such a cross is presumably due to a sex-linked recessive lethal mutation in the germ cells of the parental wild type male; a reduced frequency of these males could be due to semilethal mutations. It is important to keep a record of all F_1 progeny from each individual parental male in order to identify any cluster of lethals that might arise from single premeiotic events. Possible clusters would also be expected in the other mutation screens discussed below.

Since males are hemizygous for the X chromosome, this plan only detects newly arising lethals, visibles, and mutations affecting viability. Thus, it avoids the difficulties caused by variation maintained in a population. For discussions of lethals and viability mutations in nature, see Spencer (1947), Wallace (1968a), Mukai (1964), Yamaguchi (1976), Dobzhansky *et al.* (1977), and the chapter by Crow and Simmons in this volume.

2. Recessive autosomal lethal mutations

The spontaneous recessive lethal mutation rate can be determined for genes on the major autosomes by use of balancer chromosomes such as *SM1* or *CyO* for the second chromosome and *TM1* or *TM3* for the third chromosome. In combination, second and third chromosome balancers can be used to identify mutations on both major autosomes, approximately 80% of the genome. As an example of an autosomal mutation screen, single wild type males are crossed with *CyO* balancer females, a number of single F_1 *CyO*/+ males are crossed with *CyO* balancer females, F_2 *CyO*/+ males and females from the same cross are mated, and F_3 progeny are scored for the presence or absence of +/+ offspring.

The F_3 offspring of this program can also be scored for viability and for visible mutations. For recessive viability effects, the *CyO*/+ F_3 progeny or +/+ progeny from interline crosses of F_2 *CyO*/+ and *CyO*/+ flies can be used as comparative wild type controls. Special care must be taken to identify mutations that are present in original wild type male parents as heterozygotes. Such hidden mutations can be identified by their expected recovery in half of the F_1 crosses from any one parental male.

The principal advantage of this type of screen is that a variety of mutations can be detected in a large portion of the genome. It is a three generation cross, however, and there are possible problems associated with recombination in heterozygous male progeny when wild type lines are being tested (c.f., Hiraizumi, 1971), and there is the potential of not detecting pre-existing heterozygous mutations if the number of tested chromosomes per original parent male is small. One advantage of an autosomal lethal mutation screen over the recessive sex-linked lethal mutation screen is that the latter, but not the former, may underestimate the true mutation frequency if there is germinal selection against hemizygous mutants in premeiotic cells.

3. Recessive visible mutations

The frequency of spontaneous-recessive sex-linked visible mutations can be determined in one generation by the following screens. In the first, single wild type males are mated with females that carry attached-X chromosomes (e.g., $C(1)DX, y f$ or $C(1)M3, y^2$), and patroclinous F_1 males are scored for mutations induced in the gametes of the parental male. Males may also be mated with chromosomally normal females having multiple mutant loci on their X chromosomes (e.g., "maple" $= y\ ac\ sc\ pn\ w\ rb\ cm\ ct^6\ ras^2\ v\ g^2\ f\ car$). The first of these screens is for mutations anywhere on the X chromosome, while the second is a locus-specific one which can also identify small single marker deficiencies. The first screen can also be used to assay chromosome breakage in males when such breaks produce X-chromosome fragments which can cover mutant markers in female offspring. Since both of these are subjective screens, all presumptive mutations should be retested to confirm that they breed true and to test the possibility that detected mutations are autosomal dominants rather than being sex-linked.

Recessive visible mutations and small deficiencies can be detected on the major autosomes by the following F_1 test. Single wild type males are mated to females that have autosomes carrying a large number of recessive mutant loci. Sample chromosomes include "twelvepl" ($al\ dp\ b\ pr\ cn\ vg\ c\ a\ px\ bw\ mr\ sp$) for the second and "rucuca" ($ru\ h\ th\ st\ cu\ sr\ e\ ca$) for the third. Resulting progeny are scored for mutant phenotypes. In order to reduce the possibility of crossing over between the tester chromosome and the wild type chromosome, only male progeny should be retested to confirm presumptive mutations unless balancers are used. Other screens for visible mutations can be found in Muller and Oster (1963).

4. Dominant visible mutations

A useful one generation screen for dominant Minute mutations has been employed by a number of investigators using different species of *Drosophila* (these studies include those by Hinton, 1979, 1981, with *D. ananassae*; Mampell, 1945, 1946, with *D. persimilis*; Spencer, 1935, with *D. funebris*; Huang and Baker, 1976, Glass, 1955, and others—See Table I—with *D. melanogaster*). The F_1 progeny of natural population lines are screened for spontaneous Minute phenotypes characterized mainly by short and thin bristles and by delayed development, small body size, and rough eyes (Lindsley and Grell, 1968). There are about 40 loci in the *D. melanogaster* genome that can mutate to a Minute phenotype, and all Minutes are recessive lethals. Huang and Baker (1976) have observed a constant ratio in the recovery of ethylmethanesulfonate induced Minutes and recessive sex-linked mutations. However, a Minute phenotype may occur upon loss of the fourth chromosome, X-linked Minutes are limited to females, and Minutes are often sterile (Hinton, 1981; Lindsley and Grell, 1968). Finally, when possible all presumptive Minute mutants should be retested to be sure that they breed true.

B. Methods to Determine the Frequency of Chromosome Breakage

1. Cytogenetic screens

The frequency of spontaneous chromosome breakage in euchromatin and heterochromatin can be measured by mating wild type males with females that have structurally normal chromosomes, and scoring F_1 larvae for rearrangements in salivary gland cells or in neuroblast cells (Abrahamson and Lewis, 1971; Bauer *et al.*, 1938; Gatti, 1979; Lefevre, 1976; Roberts, 1976). Chromosome aberrations can be identified in neuroblast cells at metaphase, with or without colchicine treatment, and at anaphase. Chromosome breakage can also be scored in spermatocytes of wild type males (e.g., Henderson *et al.*, 1978; Yannopoulos, 1978a).

Major disadvantages of these cytogenetic screens are that putative rearrangements cannot be retested, it is a subjective screen, and, since spontaneous chromosome breaks are usually rare events, large numbers of progeny or cells must be scored to get reliable measures of chromosome breakage rates.

2. Heritable translocations

The frequency of spontaneous chromosome breakage can be determined by scoring for reciprocal translocations between genetically marked nonhomologous chromosomes. For example, wild type males are mated with bw; $st\ p^p$ females. Then, individual F_1 $+/bw$; $+$ $+/st\ p^p$ males are mated to bw;$st\ p^p$

females, and F_2 progeny are scored for pseudolinkage between *bw* and the *st p^p* loci (2;3 translocation) or between the Y chromosome and autosomes (T(Y;2), T(Y;3), or T(Y;2;3) rearrangements). All eye color combinations in this cross are easily distinguishable and presumptive translocations can be retested genetically or cytologically. For details of these crosses, see Abrahamson and Lewis (1971), Würgler *et al.* (1977), and Valencia *et al.* (1983).

The disadvantages of this screen are that it is a two generation test, spontaneous translocations occur at a low frequency (about 0.01% in unstored sperm), translocation heterozygotes often have reduced fertility, and breaks involving the Y chromosome can lead to male sterility. To increase the frequency of recovered translocations, parental females may be used that contain repair defective mutations and/or attached X–Y chromosomes. Presumptive translocations should be retested by F_2 crosses or be confirmed by cytogenetic analysis. Advantages of this test include its objectivity, easy scoreability, and extremely low spontaneous levels that make it unnecessary in some cases to run a concomitant control (for example, if large within laboratory historical control data are available).

3. Crossover suppression

Some chromosome rearrangements can be identified by their ability to "suppress" recombination in heterozygotes (for a review, see Roberts, 1970, 1976). In this test, wild type males are crossed with females that have marked, cytologically normal chromosomes (e.g., "c-scan", *sc f*; *al b sp*; *ve st ca*, which is a tester for the X, second, and third chromosomes) and single F_1 heterozygous females are backcrossed to multiply-marked males. A significant reduction or elimination of recombination in any chromosome region is an indication of a possible rearrangement, especially inversions and translocations, in the wild type genome. Potential rearrangements should be retested genetically or cytologically.

Small chromosome rearrangements, such as small deficiencies, may be undetected by this scheme, and rearrangements entirely in heterochromatin would not be identified.

4. Breakage and reattachment of autosome arms

By the use of compound autosome stocks (e.g., *C(2L)RM, b; C(2R)RM, cn*, a stock that has two attached left arms and two attached right arms of the second chromosome), one can identify breakage and reattachment of autosome arms that occurs in the gametes of wild type flies (Holm, 1976). When males carrying a compound autosome are mated with wild type females, no progeny normally survive because of aneuploidy. Survivors are only possible when a breakage and rearrangement event occurs in the wild type females producing a $C(2L)RM$ or a $C(2R)RM$ product, or by double non-disjunction.

One weakness of this screen for population studies is that the exact number of chromosomally normal gametes produced by wild type parents is unknown, because these gametes give rise to inviable progeny. An estimate of the number of gametes can, however, be made from the progeny of a sample of wild type females mated with males having normal karyotypes or by egg counts.

Advantages of this approach are that it is quick and survivors are easy to score. Disadvantages include the poor viability of the compound autosome bearing males and females.

5. Position-effect screen for chromosome breakage

Some chromosome rearrangements can be identified by their influence upon gene expression, a phenomenon called "position-effect variegation" (Spofford, 1976). For example, Dubinin and Siderov (1934) showed that translocations involving the fourth chromosome and an autosome are often associated with reduced dominance of the fourth chromosome wild type alleles of cubitus interruptus (the "Dubinin effect"). This position-effect can be used as a screen for detecting chromosome breaks (Stern and Kodani, 1955; Stern et al., 1946; Abrahamson and Lewis, 1971). As an example of such a screen, wild type males are mated to homozygous cubitus interruptus females, and the F_1 progeny are scored for a reduction in the wild type expression of the fourth longitudinal (cubitus) wing vein. All flies with reduced or missing cubitus veins should be retested cytologically for rearrangements.

C. Precautions in Measuring the Rates of Mutation and Chromosome Breakage in Natural Population Lines

As indicated earlier, mutation frequency (the proportion of a particular mutant allele in the population) and mutation rate are commonly not the same. Mutation and chromosome breakage rates in natural populations must, therefore, be estimated directly from newly occurring mutations and not depend upon indirect assessments that might be influenced by variants maintained in the population. This is particularly true for recessive visible and lethal mutations.

It has also been found that many natural populations of *D. melanogaster* show low, but non-trivial, frequencies of recombination in males following a cross to laboratory marker stocks (see Kidwell *et al.*, 1977b; Thompson and Woodruff, 1978a; and Sved, 1979, for reviews). Male recombination (MR) lines have been obtained from natural populations world-wide. Thus, it is not safe to assume that recombination will be absent in heterozygous males during mutation screens. Indeed, MR-induced recombination and chromosome breakage in males (and possibly also in females) may not even be entirely eliminated in inversion heterozygotes (Cardellino and Mukai, 1975).

In addition to cautioning us about male recombination, studies of MR lines have also shown that certain crosses between natural population lines and laboratory stocks of *D. melanogaster* (Kidwell *et al.*, 1977b; Woodruff *et al.*, 1979; Woodruff and Thompson, 1980) and crosses between some geographically separated population samples (Thompson and Woodruff, 1980) can stimulate increased mutation and chromosome breakage events. This is the reason that it may be inappropriate to measure the rate of chromosome breakage in natural population lines by loss of genetic markers on the Y chromosome (e.g., in $B^S \ Y y^+$ bearing males). The generation of such males involves crossing wild-type males with a marked-Y laboratory stock. The same hybridization problem is also true for other chromosome breakage assays, including the "transvection effect" (Lewis, 1954; Mendelson, 1976), the "Craymer effect" (Craymer, 1974; Foureman, 1979), and the loss of ring-X chromosomes (Zimmering, 1981).

Some genetic factors isolated from natural populations increase the frequency of mutation or chromosome breakage on specific chromosomes. If these are present in lines tested for spontaneous mutation, the rates for different chromosomes can be markedly out of proportion to some standard, such as the number of loci per chromosome. The *delta* factor, for example, has been reported to induce second chromosome lethal mutations, but not X-linked lethal mutations (see Ito, 1974; Minamori and Ito, 1972; Minamori *et al.*, 1973; and later section). The recovery disrupter (RD) factor induces Y chromosome specific breaks (Erickson, 1965), while the $OK1$ mutator line undergoes frequent autosomal breakage upon outcrossing, but apparently induces less frequent X or Y chromosome breakage in spermatocytes (Henderson *et al.*, 1978). An even more striking example is the π_2 natural population line of *D. melanogaster* that undergoes frequent chromosome breakage at specific sites on the X chromosome (Berg *et al.*, 1980; Engels, 1981a,b). It is also important to note that many MR lines seem to have frequent "spontaneous" mutation at specific loci, though low frequency mutations are scattered among many loci (Green, 1978b).

Finally, the rate of spontaneous mutation and chromosome breakage can be affected by physiological processes in male and female parents (Graf, 1972; Würgler, 1972). There is an increased mutation rate in the first sperm obtained from males (Ives, 1963). There are storage effects on spontaneous mutation events, and possibly on chromosome breakage, that occur in males, i.e. X-linked and second-chromosome lethal mutation frequencies are higher in sperm stored in females (Purdom et al., 1968; Rinehart, 1969) than in sperm stored in males (Kaufman, 1947). The frequency of spontaneous mutations in male gametes is also influenced by the parental female used in a screen (Hildreth and Carson, 1957). This latter observation is probably due to the efficiency of maternal repair (Würgler and Maier, 1972; Smith, 1973; Baker et al., 1976a, 1976b; Graf and Würgler, 1976, 1978; Graf et al., 1979; Nguyen et al., 1979; Zimmering, 1981).

In summary, many common problems such as selectively maintained genetic polymorphisms, mutator activity induced by crossing wild-caught lines, and physiological influences force us to think through breeding programs very carefully to insure that mutation screens are measuring what we think they are measuring.

III. Factors Affecting Mutation Potential

Factors that influence mutation frequencies in natural populations of *Drosophila* can be roughly divided into two categories. Extrinsic factors include environmental conditions such as temperature, mutagen encounters (e.g., irradiation or chemical exposure), and exposure to viral or other infectious agents. Intrinsic factors include sex, age, and the general physiological and genetic constitution of the organism. The interplay among these extrinsic and intrinsic factors gives rise to a phenotype which we shall call mutation potential. But while environmental variables, sex, and age can have an effect upon mutation, that effect is often either small or of doubtful significance in natural populations. Genotypic variation, on the other hand, directly or indirectly affects the ability to prevent or repair mutations from any source. Indeed, as we have already mentioned, genetic factors such as mutators are associated with mutations at high frequency. In the following sections we shall discuss a variety of influences upon mutation potential, but we shall emphasize certain intrinsic factors.

A. Extrinsic Factors

Three of the most widely recognized extrinsic factors affecting mutation rates are temperature, mutagens, and infective agents. The effects of each of these, however, may in turn be influenced by the genotype and physiological condition of the organism. As these extrinsic factors have been reviewed separately elsewhere in detail (see Auerbach, 1976), they will be considered only briefly here. This approach is also justifiable, since the role of these factors in natural populations is mainly conjectural.

The effect of temperature on mutation rates has been widely investigated and debated (see for example Muller, 1928; Plough and Ives, 1935; Plough, 1941; Sheldon and Barker, 1964; Lindgren, 1972). A summary of many of these investigations is presented in Table II. Although the majority of both high and low temperature treatments appear to increase mutation frequencies, temperature effects are often correlated with, or confounded by, other factors such as developmental rate and sperm storage. Thus, it is important to distinguish clearly between the effects of different temperatures on stored sperm and the effects of different temperatures on sperm development.

The possible complexities of temperature effects can be illustrated by the following examples. In his review of temperature influence upon spontaneous mutation rates in many experimental organisms, Lindgren (1972) pointed out that since biological development proceeds faster at higher temperatures, most biological systems may have a temperature optimum above which biological damage might occur. In addition, since more generations would occur per unit time in the more rapid development at high temperatures, an estimate of mutation rates per generation might underestimate the real effect upon a population. Besides confounding time and generation number, early experiments also confounded aging (storage) and developmental rate correlates. Whereas Muller (1928), Plough (1941), and others reported a higher mutation rate at higher temperatures than at lower temperatures (Fig. 1), Sheldon (1958) and Sheldon and Barker (1964) did not. Instead, they concluded that the mutation rate per cell division is the same for comparable ages at each temperature. The effect of temperature upon stored sperm, however, is quite different (higher mutation frequency at higher temperatures; Byers and Muller, 1952; Byers, 1954). Thus, Sheldon and Barker (1964) explain the apparent contradiction between their results and those of earlier studies by the fact that temperature effects upon stored sperm and the age of tested males were not sufficiently controlled. There can be little doubt that in natural populations, any temperature effect will be equally complex and be correlated with many other developmental and genetic responses to temperature stress.

A second extrinsic factor, chemical mutagens, is reviewed in this series by Lee (1976). It is known, however, that *Drosophila* have enzyme systems that are capable of metabolically altering a wide range of potential environmental mutagens. These microsomal enzyme systems are located in fat bodies, Malpighian tubules, the digestive tract, and testes. The sensitivities of these metabolic systems are genotype dependent, and modifier genes map to all chromosomes (Wilkinson and Brattstein, 1972; Sobels and Vogel, 1976;

TABLE II. Investigations on the influence of temperature on mutation rate. A modified form of Table I in Lindgren (1972). Some references in this table can be found in Lindgren (1972).

Treated stages	Treatment Time	Temperature °C	Genetical character	Results	Reference
A. Cold shock treatments (semi lethal)					
Larvae and eggs	20–40 min.	−6	Lethals in chrom. 1 and 2	Cold treatment increases spont. mut. rate 3 times	Birkina, 1938
Grown-up males, spermatozoa	2 hr.	−6	Sex-linked lethals	0·53%. Control 0·10% Low temp. shock causes increased mut. rate	Byers, 1954 Gottschewski, 1934
Male adults	1·5–2 hr.	−5·5	X-linked lethals	1·42%. Control 0·26%	Kerkis, 1941
Male adults and eggs	30–40 min.	−30 (eggs only) to −5	X-linked lethals	No effect	Rendel and Sheldon, 1956
B. Treatment of eggs					
2–3 hr. old eggs	A few min.	45–60	Visibles	Increase	Lawrow, 1935
Eggs		2·5–31·5		No increase	Mann, 1923
Eggs				Increase	Promtov, 1934
C. Treatments of larvae with heat-shocks (semi lethal)					
Male and female larvae	12–14 hr.	36–38	Sex-linked visible, lethals in chrom. 1 (X) and 2	2-fold increase	Buchmann and Timoféeff-Ressovsky, 1935, 1936
Larvae	24 hr.	38	Lethals	1·55%. Control 0·24%	Efroimson, 1930, 1932a,b; 1932b
Larvae	12 hr.	37	Visible changes	No evidence of effect	Ferry et al., 1930
Larvae, 5–6 days	10–12 hr.	37	Visibles	Very high mut. rates	Goldschmidt, 1929

	Stage	Duration	Temp.	Mutation type	Effect	Reference
	Larvae, 1–6 days	16 hr.	36	Visibles	Great effect	Grossman et al., 1933
	Larvae	12–24 hr.	36	Visibles	Very high increase	Jollos, 1930, 1934 and cited
	Male and female larvae	24 hr.	36.5	Visibles	5–6 times increase	Plough and Ives, 1932, 1934, 1935
	Larvae	12 hr.	37	Visibles	Effect	Rozitsky, 1930
	Larvae, 4 days	12–24 hr.	36.5–38	X-linked lethals	No effect	Sheldon, 1958
D.	Treatment of adults					
	Adults	40 hr.	36	X-linked lethals	Increase	Mackensen, cited by Muller, 1932
					Small but significant increase	Muller and Altenburg, 1919
	Adults	Several days	18–26	X-linked lethals and in other chrom.	Increase	Muller, 1928
	Whole development	9–30 days	18–30	Sex ratio	Increase in % females	Nash and Come, 1967
			8–31	Lethals in chrom. 2 and 3	Fast increase	Plough, 1941; Plough and Child, 1937
	Male adults	15–30 min.	38–40	X-linked lethals	No effect	Sheldon, 1958
	Male adults	Whole development	15–30	X-linked lethals	No influence	Sheldon, 1958; Sheldon and Barker, 1964
		Whole development (11–22 days)	8–31	X-linked lethals	Increase	Timoféeff-Ressovsky, 1935
E.	Other treatments					
	Spermatozoa stored in females	3 weeks	7, 27	X-linked lethals	0·168%, 0·401%	Byers and Muller, 1952
	Presyngamic sperm nuclei in females	10–20 min.	0–37	X-linked lethals	No dependence	Hildreth, 1967
	Spermatozoa		10	Rec. lethals in chrom. 2	Indications of lower effect compared to Muller's data (1958)	Purdom, 1968

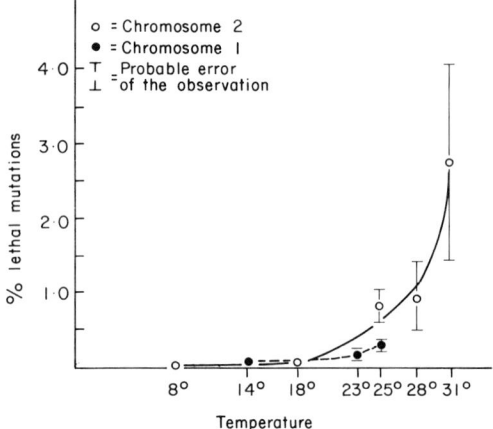

Fig. 1. Mutation and temperature (Plough, 1941).

Vogel and Sobels, 1976; Vogel, 1975, 1976; Baars et al., 1977, 1980; Magnusson and Ramel, 1978; Magnusson et al., 1979).

A forest in a thorium-rich region of Minas Gerais, Brazil, has a high naturally occurring level of radiation (from 0·05 to 3·2 mR/hour) (Penna Franca et al., 1965). *D. willistoni* and *D. nebulosa* from this forest had a significantly higher frequency of lethals plus semilethals (45·08%) when compared with a nearby control habitat (35·52%) (Cordeiro et al., 1973). The flies from the radioactive area also had a smaller decrease in egg eclosion and reproductive performance after treatment with 1,000 R of radiation, suggesting that they were more radioresistant than the control lines (Kratz, 1975).

A number of studies have reported genetic effects induced by infective agents for which *Drosophila* is the natural or artificial host. Sterility and lethality have occasionally been traced to infections by mycoplasmas, rickettsiae, spirochaetes, and microsporidia. Mutations, however, are most commonly associated with viruses; we shall concern ourselves here only with these. See Brun and Plus (1980) for a review of sigma and other *Drosophila* viruses.

Lethal CO_2 sensitivity caused by sigma virus (a rhabdovirus) has been reviewed by L'Heritier (1970). Several natural populations have been sampled for sigma virus. It was found to be present in several species of *Drosophila* (Williamson, 1961; Fleuriet, 1980). Sigma virus may be carried by gametes of either sex and may be transmitted by inoculation or transplantation of infected tissue. Some strains of sigma virus tend to reduce fecundity of host females by increasing developmental time of infected ovarian cysts. This in turn causes a build-up of infected eggs at late stages of oogenesis, blocking the ovarian tubes (Jupin et al., 1968). Sigma virus strains that affect oogenesis in this manner have not, however, been found in natural populations. They have only been observed in laboratory strains propagated through serial inoculations (L'Heritier, 1970). Sigma virus kept in this way is also reported to double the sex-linked recessive lethal mutation frequency (Baumiller, 1967; Baumiller and Lewis, 1968).

In addition, other viruses that are normally non-infectious to *Drosophila* have been shown to increase mutation frequencies. For example, Gershenson et al. (1970, 1975) found that poliomyelitis virus produces 1·24% sex-linked lethals and 2·43% second chromosome lethals, whereas Tipula iridescent virus induces 0·50% sex-linked lethals and 2·02% second chromosome lethals; similar frequencies are observed with influenza virus and with nuclear polyhedrosis virus. Burdette and Yoon (1967) reported an increase in the frequencies of sex-linked lethals, Y chromosome loss, X–Y nondisjunction, translocations, mosaics, and developmental disturbances, including melanic tumors and absent appendages, when *D. melanogster* was fed Rous sarcoma virus. Finally, Golubovsky and Plus (1982) have compared the frequencies of spontaneous recessive sex-linked lethal mutations in isogenic stocks of *D. melanogaster* in which one stock is infected by C picornavirus and the other is made virus free by outer disinfection of eggs. The frequency of lethal mutations was higher in the virus-infected flies (5/1037=0·48%) than in the virus-free flies (3/2090=0·14%), but the difference was not significant. Second chromosome lethal mutations were also recovered in the virus-infected flies at a frequency of 1·86% (5/268). Golubovsky and Plus (1982) concluded that natural viruses of *Drosophila* are partially responsible for spontaneous mutations in wild *Drosophila* populations.

The genetic impact of viruses in natural populations is clearly a field of great potential. *Drosophila* studies can be both a means of monitoring virus-induced genetic damage and a model for understanding the complexities resulting from an interaction between viruses and eukaryotic genomes in nature. One study of an unidentified infective agent presently stands out as an example of the complexities that might be uncovered. A killing agent, called Cy-killer, in some natural populations of *Drosophila* causes a reduction in the number of progeny, induction of recessive lethal mutations, and a distortion of segregation ratios when infected males and females are mated (Minamori, 1967; Minamori and Tatsukawa, 1967; Minamori et al., 1973).

First observed in a natural population from Hiroshima, Japan, Cy-killer appeared to kill selectively the $Cy/+$ progeny from matings of Cy/Pm and $+/+$, where + represents wild type chromosomes from males collected from natural populations. Later it was found that, in addition to $Cy/+$ and $Pm/+$, other heterozygotes involving laboratory and wild type chromosomes were also killed, though the $Cy/+$ individuals seem to be the most sensitive. Thus, genetic factors can modify susceptibility to the killing agent.

Although originally isolated in association with second chromosomes from nature, transmission of Cy killer can occur independently of particular chromosomes and, thus, appears to be non-chromosomal. It can be transmitted both by males and females during copulation and by larval contact with infected individuals. Infectivity appears to be temperature dependent (i.e., high temperatures increase infectivity) and age dependent (i.e., young flies appear to be noninfectious until about two days after eclosion). In addition to causing death of sensitive individuals, Cy-killer reduces developmental time in its carriers and induces recessive lethals and semi-lethals among the survivors.

B. Intrinsic Factors

A host of intrinsic factors involved in determining the ultimate physiological condition of the fly undoubtedly play roles in influencing mutation rates. Nutrition, for example, is a basic condition that could affect many aspects of mutation, including sensitivity to potential mutagens. In fact, as the nutritional content of the food supply decreases, spontaneous and induced mutation rates tend to increase (Herskowitz, 1963; Clark, 1969; deMarco *et al.*, 1975; Geer and Reno, 1976; Newberne and Zeiger, 1978). The best analysed intrinsic factors, however, are age, sex and genotype.

In laboratory populations at least, elevated mutation rates are often observed in aged flies (Muller, 1945; Lamy, 1947; Ives, 1963; Wagner and Mitchell, 1964; Clark, 1969). Except for a higher mutation rate in the first sperm collected from young males (Muller, 1954; Ives, 1963), the effect of age is generally small. Ives (1963), for example, tested 46,177 chromosomes for sex-linked recessive lethals, semi-lethals, and dominant visible mutations. Aged males showed only a small increase in mutation rate (Table III) in 1–2 day broods of males up to 18 days old. The higher mutation rate in sperm from young males would be due either to greater mutation sensitivity during larval and pupal development or to a storage effect upon sperm formed during pupation. Indeed, it is quite possible that the mutation rate *per se* is not affected by age, except in extreme instances, and that the observed age effect is simply due to the accumulation of spontaneous mutations in the aging spermatozoa (Fig. 2). Since flies in natural populations probably do not live long (c.f., an estimate of 9 days in near natural conditions in crowded population cages; Cannon, 1966), age

TABLE III. Frequencies of spontaneous recessive sex-linked lethal mutations in aged *Drosophila melanogaster* males (Ives, 1963)

Age of males	Number of lethal mutations	Total number of chromosomes scored	% Lethal mutations
1	6	4535	0·13
2–3	3	6268	0·05
4	3	4026	0·07
5	1	4024	0·02
6	3	4035	0·07
7	3	4241	0·07
8	1	4422	0·02
9–10	6	5734	0·10
11	6	3492	0·17
12	3	3667	0·09
13–18	4	2133	0·19
	39	46,577	0·08

Single males that contained an Oregon-R X chromosome and were heterozygous for the "rucuca" third chromosome (minus thread) were mated with six *Basc* females in each brood.

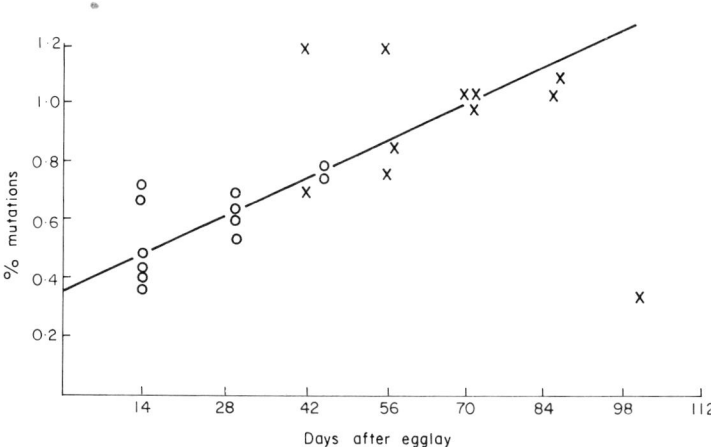

FIG. 2. Effect of age on spontaneous autosomal recessive lethal mutation rate in male *Drosophila melanogaster*. ○ = aged males only; × = aged spermatozoa (Purdom *et al.*, 1968).

might be of significance in a natural population only under unusual circumstances limiting or delaying mating.

Spontaneous and induced mutation rates in male and female germ lines may also differ, but again any differences which occur are not large (Crow and Temin, 1964; Wallace, 1968b, 1970). Nearly 12,000 lethal-free gametes were tested by Wallace (1968b) for spontaneous autosomal lethals, and a frequency of about 0·006 was obtained for each major autosome and for each sex. Wallace (1970) concluded that the sex-linked lethal mutation rate is also essentially the same in both sexes (about 0·001 out of over 18,000 gametes tested), although some variation was observed among strains isolated from different natural populations. These data contradict earlier reports suggesting lower mutation rates in females than in males (Auerbach, 1941; Muller, 1950; Glass and Ritterhoff, 1956; see also Table I).

Likely, one of the most important intrinsic factors affecting the mutation rate is genotype. This may reflect the capability of the individual to resist mutational damage (e.g., its repair capabilities) or may reflect the genomic presence of genetic elements responsible for mutation induction. A detailed discussion of the genetics and biochemistry of mutation repair in *Drosophila* is beyond the scope of this chapter. Numerous loci appear to be involved in DNA repair in *D. melanogaster* (Boyd and Setlow, 1976; Boyd *et al.*, 1976a,b, 1980, 1981; Smith, 1973, 1976; Nguyen and Boyd, 1977; Smith *et al.*, 1980). Isolation of repair defective mutants has been based upon their hypersensitivity to killing by a variety of mutagens (methylmethanesulfonate, gamma rays, nitrogen mustard). About 60 loci on all chromosomes appear to be involved (Smith and Shear, 1974; Graf and Würgler, 1976, 1978; Baker *et al.*, 1976a,b; Graf *et al.*, 1979; Baker and Smith, 1979; Nguyen *et al.*, 1979; Gatti, 1979; Boyd *et al.*, 1980; Smith *et al.*, 1980; C. Osgood and J. Boyd, personal communication).

Although most mutagen sensitive mutants do not show an increase in the spontaneous lethal mutation frequency in males (Smith, 1973; Mason, 1980), one of 13 meiotic mutants isolated from a natural population in October, 1965, in the outskirts of Rome, Italy (*mei-S282*) (Sandler *et al.*, 1968) has been observed to increase the spontaneous frequency of recessive sex-linked lethal mutations by about ten-fold in homozygous females (Carpenter and Baker, unpublished observations). This third chromosome recessive mutant reduces recombination by half and shows a high frequency of nondisjunction for all chromosomes at the first meiotic division in females (Parry, 1973). The occurrence of the lethal mutations is not associated with meiotic recombination and some mutation events seem to be premeiotic in origin.

As observed in other organisms where these parameters have been characterized, repair of genetic damage and processes of recombination and fertility are controlled by overlapping genetic functions (see reviews by Clark, 1973; Radding, 1973; Baker *et al.*, 1976b, 1978, 1980; Game *et al.*, 1980). In *Drosophila*, these mutagen sensitive mutants often fail to complement mutants characterized as having abnormal meiotic phenotypes, such as increased frequencies of nondisjunction, altered frequencies or distributions of recombination (Baker and Carpenter, 1972; Baker *et al.*, 1976b) or mitotic chromosome stability (Baker and Smith, 1979; Gatti, 1979; Gatti *et al.*, 1980). In addition, many mutagen sensitive mutants are female sterile, and it is unclear what relationship exists between the sterility and the meiotic chromosome behavior among the mutants.

Further genetic and biochemical analysis of these mutants is currently under investigation. With respect to our evaluation of mutation rate variation, however, the most important points are that at least some of these mutants can be isolated from nature and that heterogeneity in repair capabilities is, therefore, likely to influence mutation rates in some natural populations. This potential influence is made especially clear by the observation that genetic damage induced in sperm is more likely to give rise to mutations and chromosome breakage in oocytes of mutagen sensitive females than in wild-type females (Smith, 1973; Graf *et al.*, 1979; Nguyen *et al.*, 1979; Zimmering, 1981). Repair and other genetic influences, such as heterogeneity among enzymes involved in replication, may indeed make up a major proportion of the genetic component of variation observed in mutation rate differences among tested populations.

During the last ten years, a growing amount of attention has been paid to genetic factors from natural populations that dramatically increase mutation rates under certain conditions. These mutator factors will be discussed separately in the following section.

IV. Genetic Mutator Factors

In this section we shall concentrate upon genetic factors, variously called mutator genes, chromosome breakage elements, male recombination factors, or P factors that have been isolated from natural populations. At the present time, the only thing that is certain is that they often have some characteristics in common (Table IV). They increase the rate of gene mutation and/or chromosome breakage; and in many instances they are also associated with other genetic changes such as sterility, distortion of segregation, recombination in males, and nondisjunction (for reviews, see Kidwell *et al.*, 1977b; Thompson and Woodruff, 1978b; Sved, 1979; Woodruff and Thompson, 1980; Engels, 1981a). The diverse expressions of what may or may not always be a single phenomenon are sometimes referred to as "hybrid dysgenesis" (Kidwell *et al.*, 1977b; Sved, 1979; Kidwell, this volume), which in general is identified following outcrosses of individuals from the population of interest to a laboratory strain.

Strains showing some or all of the characteristics of hybrid dysgenesis have been isolated from natural populations of *D. melanogaster* worldwide and from other species, including *D. ananassae* (Hinton, 1974, 1979; Moriwaki and Tobari, 1975), *D. simulans* (Woodruff and Bortolozzi, 1976), *D. subobscura* (Philip, 1944), *D. virilis* (Kikkawa, 1935), and *D. willistoni* (Franca et al., 1968). In spite of their prevalence, our understanding of the impact of mutator factors upon the genetic structure of natural populations is only in its infancy.

Although mutator factors have been known since the early studies of Demerec and his colleagues in the 1930s, the very nature of their phenotype made them difficult to identify and trace in natural populations. The discovery of male recombination in *D. melanogaster* (Hiraizumi, 1971) provided a readily scorable trait directly correlated with mutator activity and other components of hybrid dysgenesis. Consequently, the information available about mutator factors in natural populations has increased dramatically in the last decade.

In the following survey of some of the best known mutator systems from natural populations, we have decided to use an approximately historical sequence. Not all of these systems were studied in the same detail, and direct comparisons are difficult and sometimes misleading; in a few cases it is not known when the strain under discussion was derived from nature. For these reasons, we have decided to discuss each strain separately, as in Green's (1976) review of mutable loci, to which this chapter is complementary. The generalizations that can now be drawn from these diverse mutator studies are summarized in the final section, together with a brief discussion of the possible impact of such factors upon the genetic structure of natural populations.

It should be noted that the increase in mutation rate reported for some mutator systems is low and may not differ significantly from the control rate (e.g., Neel, 1942). In addition, some control rates seem unusually high (e.g., Ives, 1945). We feel, however, that these systems should be discussed because of their historical interest and because they may add to our understanding of mutation rate variation in natural populations.

A. Characteristics of Some Well-known Mutator Systems from Natural Populations

1. Florida mutator line of Demerec

One of the first reports of mutator factors from natural populations of *Drosophila* was that of an inbred Florida stock of *D. melanogaster* obtained by Demerec (1937) from the Columbia University collection in 1928. This line had a high spontaneous mutation rate due to an autosomal mutator that was apparently also present in the Florida population from which the line had been isolated (Demerec, 1937; Plough and Holthausen, 1937; Goldschmidt, 1939).

Plough and Holthausen (1937) detected a high visible mutation rate (35 mutants among 30,000 flies scored, 0·12%) after outcrossing a wild line (Selected Line B Florida No. 7) with laboratory stocks. Unfortunately, mutator activity lasted only a few generations. Within two years, the stock had become highly sterile and mutationally stable (Goldschmidt, 1939).

Demerec (1937) screened 15 different wild stocks for frequencies of spontaneous recessive sex-linked lethal mutations. The combined frequency for all 15 stocks was 0·10% (14/13,602). Three stocks had a significantly higher mutation frequency: Formosa, Japan (0·39%, 8/2,054), Wooster, Ohio (0·63%, 8/1,266), and Florida-inbred (1·09%, 23/2,108). The lethal mutation frequency in the Florida-inbred line (the same line studied by Plough and Holthausen for visible mutation) was over 10 times as high as that of Oregon-R (0·07%, 2/3,049).

Chromosome replacement experiments indicated that the second chromosome was responsible for the increased mutation rate (Demerec, 1937), and a recessive genetic factor (mu-F) was postulated. It induced clusters of lethal and visible mutations only in germ cells, though all stages of germ cell development in males and females were affected. However, mu-F did not appear to produce chromosome aberrations, since none of the mu-F-induced lethals were associated with chromosome rearrangements.

Although there was no evidence of locus specificity among sex-linked lethals, mu-F did seem to cause visible mutations at specific loci on the X chromosome. Of 36 visible mutations that were isolated from 15,000 screened X chromosomes, for example, 24 were recovered at the yellow locus. Demerec (1937) suggested that while some of the yellow mutations may have been small, viable deficiencies, it was more likely that mu-F activity was specific to the yellow locus. In addition, some of the mutations isolated from the Florida-inbred line were genetically unstable and sterility was induced in some flies.

2. Inbred Oregon mutator line of Neel

Neel (1942) reported a high spontaneous mutation frequency in an inbred Oregon line of *D. melanogaster*. This line had been maintained in small mass culture since being obtained from C. Stern in 1938. The frequency of all types of visible mutations, as determined from inbreeding and from crosses with attached-X females, ranged from 0·01% (1/10,084) to 0·04% (13/29,264); the total frequency over a number of experiments was 0·04% (80–84/189,827). The frequency of spontaneous recessive sex-linked lethal mutations was 0·32% (24–25/7,565 total from several experiments). Both visible and lethal mutation rates, which were reported as conservative estimates, were at least two-fold higher than expected.

Although mutation frequencies fluctuated greatly in different crosses, Neel (1942) concluded from chromosome substitution experiments that the responsible factor or factors were dominant and located on the third, and possibly also the first, chromosome. Induced visible mutations were widespread in the genome, although some loci mutated at very high rates. For example, of 42 sex-linked visible mutations that were mapped, 18 were at the yellow locus.

Mutations occurred in both germinal and somatic tissues of males and females. Occasional clusters suggested a possible premeiotic origin of some germinal events. Mutator activity seemed to be limited to gene changes, since no chromosome rearrangements were recovered from the Oregon inbred line. Finally, as in many mutator systems, sterility was induced in some flies.

It is also of interest that inbreeding for more than 60 generations did not eliminate mutator expression. Mutation frequencies did, however, vary from generation to generation (e.g., recessive sex-linked lethals fluctuated between 0·10% and 1·01%). Variation may have been due to the segregation of the mutator factor or of suppressors.

3. px bl *line of Goldschmidt*

Goldschmidt (1945) in a detailed study spanning at least ten years observed that a plexus blistered stock of *D. melanogaster* (symbolized as *px bl*), which was derived in 1926 from a *px* stock of the Columbia collection, began in 1933 to undergo frequent, but sporadic, spontaneous mutation. In the original observation of genetic instability in this stock, a single pair mating gave rise to reversions of *px* and *bl*, and to visible mutations at the rudimentary, arc and silver loci. The *px* and *bl* line, which was genetically stable for long periods of time—in one case for three years—underwent additional periods of instability in which visible mutations occurred at many loci, including bran, Minutes, Lobe, Notch, ebony, dachs, dwarf and bobbed. In addition, recessive sex-linked lethal mutations and chromosome rearrangements (deficiencies, inversions, transpositions and translocations) were recovered from the *px bl* stock.

Goldschmidt (1945) also observed that the *px bl* stock and its derivatives underwent frequent spontaneous mutation and chromosome breakage after outcrossing. He also pointed out that many spontaneous visible mutations of *D. melanogaster*, including cardinal, bow, Bar, blistered, curved, bithorax, Ski, black and dachs, were recovered from hybrid crosses of laboratory mutant stocks with wild stocks, and that others, including sepia, Dichaete,

Hairless, divergent, hairy, gap, comma and fringed, were recovered from crosses of different mutant stocks. In outcrosses with *px bl* there was a tendency for specific loci to mutate at high rates. This was especially true for forward mutations at the silver, blistered and bobbed loci, and for reversions of silver and plexus. Clusters of mutational events were recovered, and abnormal sex ratios and sterility were observed in some crosses. Finally, the *px bl* stock underwent frequent simultaneous mutation at a number of loci. For example, reversions of *px* and *bl* occurred together with forward mutations at other loci. Goldschmidt (1945) attributed many of these events to chromosome rearrangements instead of point mutations.

4. hi *mutator of Ives*

In a study of the population structure of American *D. melanogaster*, Ives (1945) identified specific genetic agents that influenced the accumulation of genetic variability. In a Florida population (*Fla. 42*), which carried high levels of adverse viability genes, the spontaneous second chromosome lethal mutation frequency (6·20%) and X chromosome mutation frequency (1·21% in females and 0·82% in males) was significantly higher than that in his control lines (for example, 0·49% second chromosome lethals in Canton-S). Ives therefore concluded that the mutation rate in the Florida lines was, to some extent, under the control of mutation rate genes.

In 1950, Ives gave a more detailed report of the genetics of mutator activity in lines isolated from a Winter Park, Florida, population. Two of the ten lines from that population were found to have high spontaneous mutation rates; Florida line 73 and Florida line 30 had second chromosome lethal mutation frequencies of 3·03% (4/132) and 6·20% (8/129), respectively. A second chromosome mutator, high (*hi*), was isolated from Florida line 30. Although not mapped, line 73 may have contained a different mutator, since line 73 was homozygous for *In(2R)NS*, whereas line 30 was inversion-free (Ives, 1950).

The *hi* factor was dominant, but could be influenced by background genotype. This led Ives (1950) to suggest that dominant and recessive modifiers of *hi* are found in natural populations and that these modifiers affect the mutation rate of *hi*-containing strains. This may also be true for other mutator systems (c.f., Thompson and Woodruff, 1978a, 1980; Woodruff *et al.*, 1979).

About a ten-fold increase in recessive sex-linked and autosomal lethals and in visible mutations was found in both males and females. Although locus specificity is uncertain, *hi*-induced mutations on the X chromosome showed more locus specificity than did lethals produced by cobalt-60 gamma radiation (Ives, 1950, 1959a). Some *hi*-induced mutations seemed to be genetically unstable (Ives, 1950, 1959a), and some occurred as clusters. Mutation events were limited to germinal cells, and sterility and inviability were often associated with *hi*-bearing chromosomes. Chromosome aberrations also occurred (Ives, 1950; Hinton *et al.*, 1952). In one test of 351 mutant-bearing X chromosomes, 17 were found to contain inversions. Twelve of these inversions were analysed and 11 of them had one break in the X chromosome heterochromatin (Hinton *et al.*, 1952).

In attempting to correlate the characteristics of *hi* with those of previously reported mutators, Ives (1950) argued that although *hi* and *mu-F* (Demerec, 1937) were both isolated from Florida populations, they were not the same mutators. First, *hi* was isolated in 1940 directly from nature, while *mu-F* was isolated in 1928 from a long-term inbred line of a Florida stock. Second, *hi* was three times as active as *mu-F*. Third, *mu-F* frequently induced yellow visible mutations, whereas *hi* did not. Finally, *mu-F* did not produce chromosome rearrangements, while *hi* did.

Although Ives reported in 1959 that *hi* ceased to be useful for experimental work several years earlier, Slatko (1976b) observed that two sublines (hi^{14} and hi^{29}) still had slight mutator activity after being kept in the laboratory for over 20 years. As heterozygotes, both lines induced a 2- to 4-fold increase in sex-linked recessive lethal mutations. The *hi* lines did not, however, undergo male recombination or show segregation distortion, suggesting that *hi* and *MR* (male recombination) factors are not identical.

5. D. persimilis *mutators of Mampell*

At least two different mutators have been reported from *D. persimilis* (Mampell, 1943, 1945, 1946). Mampell (1943) first observed that a strain of unreported origin derived from a single pair mating had a high frequency of spontaneous visible mutations. Mutations occurred in somatic cells and in all germ cell stages of females and males. As many as half of the progeny of some matings were mutant. Mutator activity was associated with sterility and cytological deficiencies in some crosses, although Mampell (1943) stated that the deficiencies may have been caused by a mechanism unrelated to mutator activity.

This factor was dominant and had a dose-dependent effect. In heterozygous females the frequency of spontaneous mutations increased to 0·34% from 0·01% measured for non-mutator lines. In homozygous females the mutation frequency doubled (0·70%). In contrast to the common situation in *D. melanogaster* (c.f., Kidwell and Kidwell, 1975), there was no difference between reciprocal crosses.

A recessive second chromosome mutator was later found in the same strain (Mampell, 1945, 1946). This mutator induced a high frequency of visible mutations, particularly somatic and germinal Minute bristle mutations at a number of different loci. A large fraction of Minute mosaics were also sterile and could not be retested. In addition, the level of activity of this second chromosome mutator was different in reciprocal crosses. In crosses of mutator males with wild type females, a high frequency (6·53%, 763/11,679) of Minute progeny was observed, whereas a low frequency (0·53%, 6/7,779) was produced in the reciprocal cross (Mampell, 1945). Furthermore, few Minutes were observed in the F_2 or F_3 progeny of either cross (0 to 0·88%).

The second chromosome mutator also showed unusual patterns of transmission and interaction. It was observed to function in males and females, but only in the presence of a Y chromosome. It was affected by a cytoplasmic-mutator interaction; and it could be transmitted from fly to fly as an infective agent (Mampell, 1946). It was thought to function in the presence of a Y chromosome producing a mutator substance that was activated in female cytoplasm and transmitted through the cytoplasm of males or females. Mampell further speculated that the mutator was a virus-like particle that could become integrated into normal genes causing mutation events. The integrated mutator particles could then either replicate with the cellular genome or become dissociated and transferred to other organisms.

Support for this hypothesis came from his contention that mutator activity could be transferred by infection from *D. persimilis* to *D. melanogaster* (Mampell, 1946). Using coded cultures to remove personal bias, pairs of *D. melanogaster* adults were raised in cultures with mutator or with non-mutator *D. persimilis* adults. The resulting *D. melanogaster* progeny were mated and their offspring were scored for visible mutations. A slight increase in mutation frequency was detected in *D. melanogaster* raised with either mutator or non-mutator *D. persimilis* males. Treated lines with high induced mutation rates maintained their high rates in subsequent generations. From this Mampell concluded that the primary effect on transmission of mutability by infection was not due to the mutator, but to the presence of a Y chromosome. The data and interpretation are controversial and deserve to be repeated carefully if similar lines are again found.

6. *Chromosome breakage factor of Levitan*

After 12 years in the laboratory, a homozygous standard karyotype strain (ST_Y) of *D. robusta* suddenly began showing the ability to induce chromosome aberrations in its progeny (Levitan, 1962, 1963a). During the next 10 years, extensive studies of this line and its derivatives yielded over

2,275 new aberrations (Levitan, 1970). These aberrations included translocations, paracentric and pericentric inversions, transpositions, multiple break aberrations, deficiencies (Levitan, 1962, 1964a), and several in which a broken fragment became attached to the nucleolus (Levitan, 1970). Although the factor induces aberrations exclusively in paternal chromosomes, the distribution of breakpoints and the types of aberrations were found to fit random expectations based upon the relative sizes of the chromosomes.

Unique, newly-induced aberrations occurred only when an ST_Y female was used in a cross. Cases of multiply-affected progeny, however, showed that the ST_Y males could transmit an aberration that had occurred earlier in the ST_Y stock or in the male itself before the cross was made (Levitan and Schiller, 1963). The hypothesis that the inducing factor is cytoplasmic was supported by the findings that an ST_Y male could not transmit the inducing ability from its mother to its daughters (Levitan, 1963a) and by the fact that lines maintained their inducing ability even after the majority of the ST_Y genome had been replaced by up to eight generations of backcrossing inducer females to males from non-inducer lines (Levitan and Williamson, 1965).

Since no aberrations involved both the X and the Y chromosomes (Levitan, 1963b) and since newly-induced aberrations were not mosaic within an individual (Levitan, 1964b), it is likely that the factor breaks sperm chromosomes prior to their first post-meiotic replication. The factor increased in strength until recently (Levitan, personal communication), and it can be selected to produce "high" and "low" lines (Levitan and Schiller, 1963). The search for visible mutations associated with aberration-inducing ability was negative, however, showing that it differs in important ways from other chromosome breakage and mutator factors.

Although this factor was not obtained directly from a natural population, it or factors like it might occur naturally in this species of *Drosophila*. In addition, an understanding of its cause might focus attention upon mutator events in natural populations. From this point of view, it is interesting to note that clustering of new aberrations in the same locality (Levitan, 1962) and even in the same wild-caught female (Carson, 1958) has been reported in *D. robusta*, and the first naturally-occurring ring chromosome was reported in this species (Levitan, 1952).

7. Delta *elements of Minamori*

In 1965, the *delta* genetic system was discovered by Minamori and co-workers, who were extracting second chromosomes from a natural population of *D. melanogaster* in Japan (Minamori, 1969a, b). At least 10 of

178 tested second chromosomes were associated with distorted transmission frequencies (i.e., from crosses of bw females and $Cy/+$ males, where $+$ represents the wild chromosome from the natural population, the number of $+$ progeny was often very much reduced). A distorted sex ratio was also found among the progeny of such crosses, with males being more numerous than females.

Distortion of sex ratio appeared to occur only among the wild-type progeny from the above cross, and it was significantly correlated with the frequency of second chromosome distortion. A "reciprocal cross" effect (i.e., the cross of $Cy/+$ females and bw^D males) showed no distortion; Minamori (1970) suggested that these phenomena were, at least in part, due to a maternally inherited extrachromosomal factor which was named *delta*. *Delta* was presumed to cause distortion of sex and second chromosome ratios through differential mortality of zygotes in early development. Occasionally 95–100% one-sided recovery was observed from appropriate crosses (Minamori, 1970).

A series of crosses between laboratory stocks and strains carrying various second chromosomes from natural populations demonstrated that certain cytoplasmic-chromosome combinations allowed increased or decreased action of *delta*. One suggestion was that the strength of *delta* was due to the amount of *delta* product. This was called "retention ability." Some chromosomal combinations appeared to allow *delta* to multiply faster or in different numbers of copies than did other combinations. In addition to retention, genotype also played a vital role in determining the degree of expression of *delta*, a quantitative effect called "sensitivity." Thus, a given chromosome genotype might be sensitive to *delta* (i.e., be killed by *delta*) and also allow *delta* retention (multiplication or accumulation). Other sensitive chromosomes do not allow *delta* retention, i.e. they are sensitive to *delta* action but they cannot produce *delta* product themselves (Minamori, 1969b). For example, while the bw^D chromosome is not sensitive to *delta* action, bw^D/bw^D females appear to allow more *delta* production than either Cy/bw^D or Cy/S females (where S is a chromosome sensitive to *delta* action). In the same way, homozygotes for sensitive chromosomes are more susceptible to *delta* action, often giving rise to no progeny or to highly distorted progeny ratios (Minamori, 1969a,b, 1971, 1972a; Minamori et al., 1970).

Homozygotes for insensitive chromosomes (I) do not allow transmission of *delta* to progeny, even if the genotype inherits *delta* extrachromosomally from the mother. For example, $Cy(I)/I$ daughters do not transmit *delta* to female progeny, while $Cy(I)/S$ daughters of the same parental cross transmit *delta* from generation to generation.

From genetic analysis, it was suggested that both *delta* retention and sensitivity were largely associated with the second chromosome, with some minor contributions from other chromosomes (Minamori and Ito, 1972). This was confirmed by the mapping of a single locus responsible for these phenotypes (*Da*, 2–24·9; Minamori and Sugimoto, 1972, 1973).

The *delta* system became even more complex when it was found that several different types of *delta* elements and several different types of sensitive and insensitive second chromosomes exist both in laboratory stocks and, particularly, in natural populations (Minamori, 1971; Minamori et al., 1970). For instance, $delta^b$ is retained by an S^b (sensitive to $delta^b$) chromosome and is transmitted through females to their progeny. Much less $delta^b$ factor is transmitted to progeny through males than through females. Presumably there is too low a threshold to observe $delta^b$ activity. It was originally thought that $delta^b$ was not transmitted at all through males; however, in one experiment, after 61 successive passages through males, $delta^b$ activity was still present (Minamori, 1969b).

$Delta^r$, on the other hand, is retained by the S^r (sensitive to $delta^r$) chromosome and is fully transmitted biparentally. All descendants of S^b or S^r chromosomes accumulate $delta^b$ or $delta^r$, respectively (Minamori, 1971).

The heterogeneity observed with respect to *delta* elements reflects an underlying heterogeneity among chromosomes sensitive and insensitive to *delta*. There appear to be chromosomes that are sensitive to $delta^b$ and that allow retention of $delta^b$ (S^b). There are chromosomes insensitive to $delta^b$, but which allow retention of $delta^b$ (ID^b). There are chromosomes that are sensitive to $delta^b$, but which allow retention of *delta* for only a few generations (S^c) and chromosomes that are not sensitive to $delta^b$, but which allow retention of $delta^b$ for only a few generations (ID^c). Finally, there are insensitive chromosomes which allow no retention of *delta* (I) (Minamori et al., 1970). S^b chromosomes are not sensitive to $delta^r$, whereas S^r chromosomes are sensitive to both $delta^b$ and $delta^r$. I chromosomes appear to be insensitive to both. The observations (1) that $delta^b$ mutator action on I chromosomes, as discussed below, will induce $delta^b$ sensitivity, (2) that $delta^r$ will induce simultaneous $delta^b$ and $delta^r$ sensitivity, and (3) that ethylmethanesulfonate treatment induces both $delta^b$ and $delta^r$ sensitive chromosomes, suggests only a minor functional difference between the two. In these cases and in each case of a sensitive or insensitive chromosome tested, the phenotype maps to the *Da* locus (Minamori and Ito, 1972; Minamori and Sugimoto, 1973).

The mutational effects of *delta* elements have been investigated by Minamori and Ito (1971a,b). Their results indicate that *delta* produces a high frequency of second chromosome lethals (a 3-fold increase over spontaneous frequencies), but little increase in X-linked lethals (Ito, 1974). In addition, second chromosome lethals were found to be clustered towards the left tip and in the right arm of the chromosome (Fig. 3). The lethals appear to act at the same stage of development as does the killing action associated with distorted transmission ratios, i.e. at about hatching. The frequency of lethals and semi-lethals approaches 4% for the second chromosome.

Delta chromosomes have also been found to be associated with male recombination, as measured along the second chromosome (Minamori *et al.*, 1977). Male recombination appears to be largely premeiotic. Recombination in females is also influenced by *delta* in a way paralleling its mutational activity. Recombination distances are increased along the second chromosome, but not along the X (Minamori *et al.*, 1977; Ito, 1974).

In natural populations, *delta* is present in substantial, though somewhat varying, frequencies (10–50%), depending upon the location and season in which the collection is made (Minamori, 1970; Minamori *et al.*, 1970, 1973). Estimates from fairly large samples of second chromosomes also suggest that *delta* may in fact not function as a mutator or distorter in the natural population. It has been suggested that *delta* is suppressed and does not harm its carrier in the natural population, but does so only when crossed into a laboratory strain. A polymorphism among *delta* elements and chromosomes, described above, is also present in natural populations. For example, in one population sampled at a Hiroshima pickle factory, S^b, S^c, ID^b, ID^c, and I chromosomes were all present (Minamori *et al.*, 1970).

As an interesting aside, *delta* seems to interact with the *SD* (Segregation Distortion) (reviewed by Hartl and Hiraizumi, 1976) system in *D. melanogaster*, although *delta* itself does not have *SD* activity (Minamori,

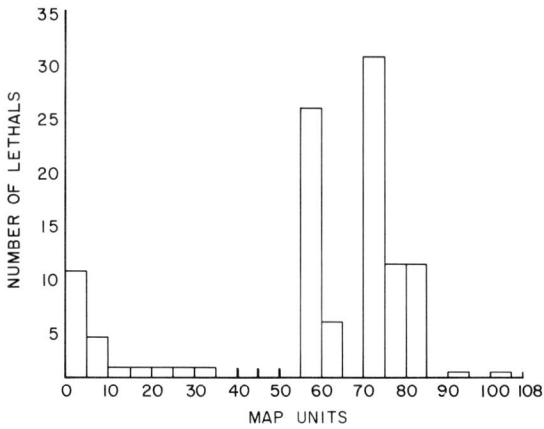

FIG. 3. Distribution of spontaneous second chromosome lethal mutations isolated in a *delta*-containing line of *Drosophila melanogaster* (Minamori and Ito, 1971b).

1971, 1972a,b). *SD* elements may need the *Da* locus in order to function. If these loci are uncoupled by recombination, *SD* no longer functions. It has been suggested that a product of *Da* is necessary for normal spermiogenesis in conjunction with *I* (Insensitive) chromosomes (Minamori, 1972b).

Finally, *delta* activity is temperature sensitive, with higher temperatures (30°C) resulting in increased activity (Minamori, 1971). There are, however, several isolated *delta* elements and chromosomes that respond to temperature treatments in the opposite way. Attempts to correlate temperature effects and seasonal or geographical frequencies of *delta* chromosomes have yielded no clear conclusions (Minamori *et al.*, 1970; Minamori, 1971).

8. MR T-007 *of Hiraizumi*

The *T-007* second chromosome line was originally isolated from a collection of about 150 *D. melanogaster* males in Harlingen, Texas (Hiraizumi, 1971). An individual male was mated to $C(1)DX, yf$; *cn bw* females, to replace the Y chromosome from the natural population with a Y chromosome from the $C(1)DX$ stock. A single progeny male from each such mating was then crossed to *cn bw* females, and the backcrosses were continued for 10 generations. In each of these matings, only one original second chromosome from the natural population (e.g., *T-007*) was maintained. All other wild chromosomes were diluted out by successive backcrosses. The *T-007* chromosome was subsequently kept over a *Cy* balancer chromosome as a *T-007/Cy* stock.

The *T-007* second chromosome is known to be associated with several of the dysgenic parameters shared by other *MR* chromosomes. These include mutator activity (Slatko and Hiraizumi, 1973), largely premeiotic male recombination (Hiraizumi *et al.*, 1973; Hiraizumi, 1979a), reduced transmission frequency of *T-007* from heterozygous males (Hiraizumi, 1971, 1977), and reduced male fertility (Hiraizumi, 1977, 1979b). A number of these parameters appear to be temperature sensitive, since a low temperature (19°C) decreases the frequency of male recombination, while at high temperature (29°C) males are almost completely sterile. Males not sterile show greatly reduced fecundity (Matthews, 1980). At 29°C almost all *T-007* females are sterile (Matthews and Gerstenberg, 1979; Matthews and Slatko, 1983) due to the gonadal dysgenesis described by Kidwell (this volume).

With respect to mutator activity, *T-007* males induce about 1·2% lethals along the X chromosome, 2·8% lethals along the second chromosome (10-fold increase over control values) and a 10- to 200-fold increase in visible mutations at a number of loci, including *sn, ras, y, pr, cn* and *bw* (Slatko and

Hiraizumi, 1973; Green and Shepherd, 1979). In addition, occasional inversions occur in the *T-007* stock giving evidence for chromosome breakage, though frequencies of new rearrangements have not been assayed directly.

Male recombination can occur on both the second and third chromosomes, though frequencies are higher on the second (Hiraizumi, 1971; Hiraizumi *et al.*, 1973; Isackson *et al.*, 1981; Slatko, 1978b) (Fig. 4). As measured along the entire second chromosome, *T-007* heterozygous males show 3·4% recombination (3·4 cM in males versus 107 cM in control females), with 0·5% recombination occurring between *pr* and *cn* (1 cM in control females). Along the third chromosome, *T-007* males show 1·3% recombination (versus 100 cM in control females). Thus, the distribution of recombination in males is different from meiotic recombination in control females, with relatively more recombination in centromeric regions (Fig. 5). Male recombination events appear to be largely premeiotic (i.e., gonial) based upon Poisson distribution analysis of clusters of recombinant genotypes (Hiraizumi *et al.*, 1973) and upon consideration of the segregation patterns of recombinant genotypes with respect to X and Y chromosomes (Hiraizumi, 1979a).

The mapping of *T-007* and its association with its various phenotypes has

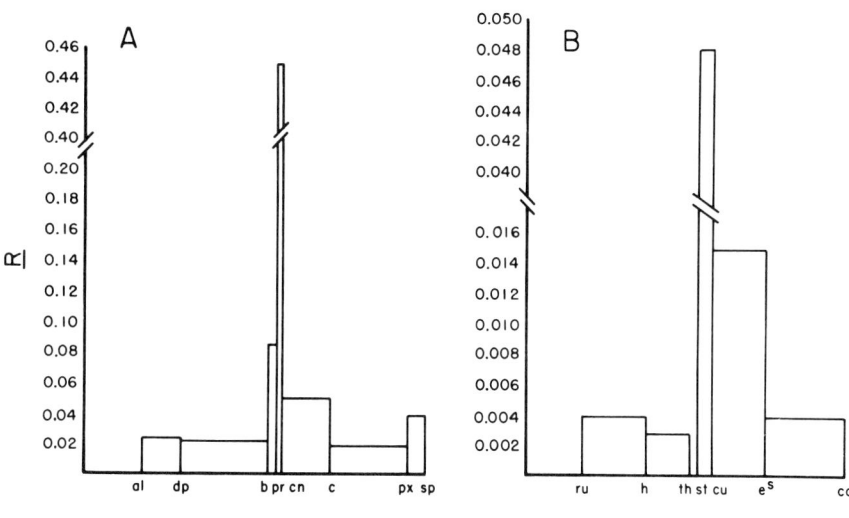

FIG. 4. Distribution of spontaneous recombination in the *T-007 MR* line of *Drosophila melanogaster*. Relative recombination frequency (R) is shown on the vertical axis, and genetic map positions are identified as mutant loci on the horizontal axis. A = second chromosome, B = third chromosome (Slatko, 1978b).

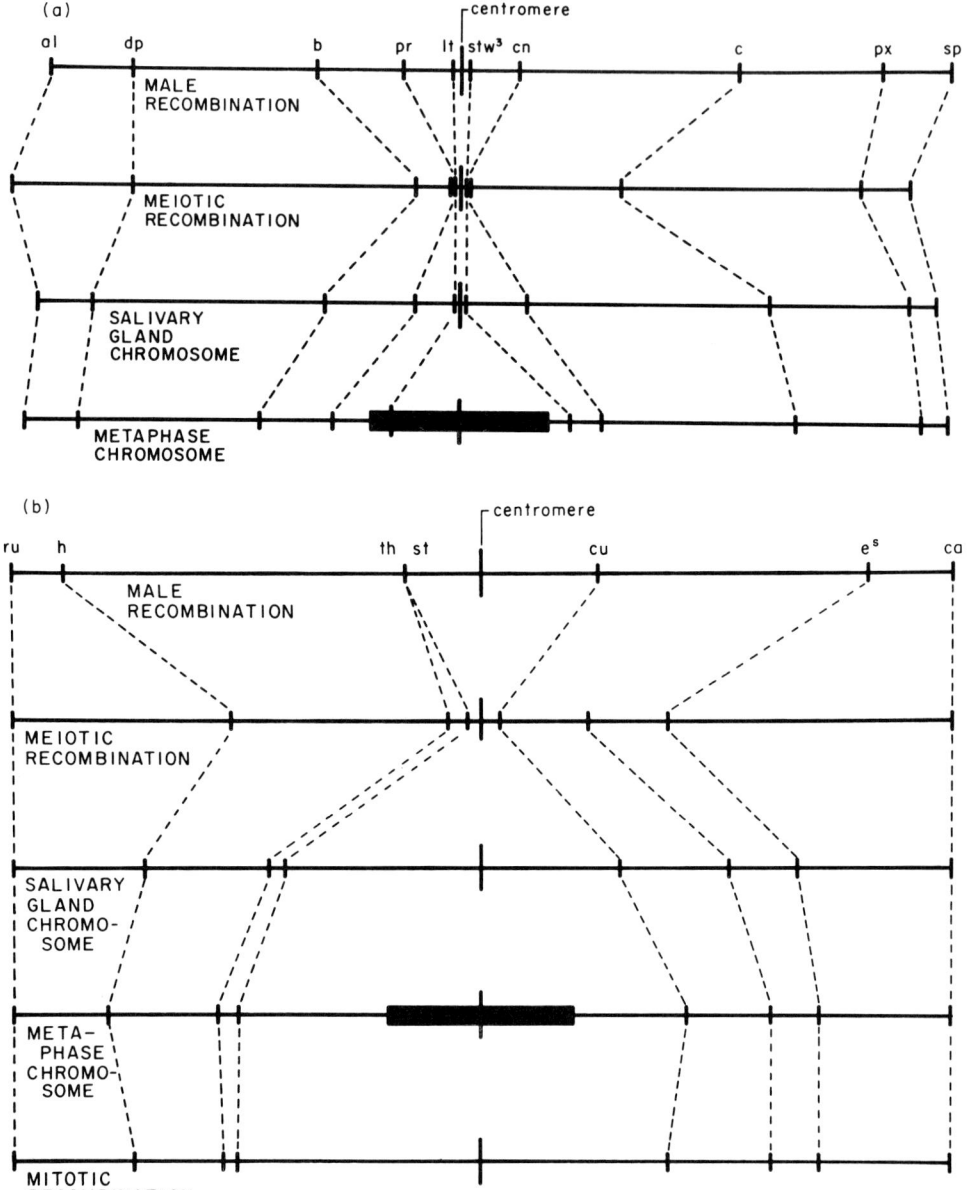

FIG. 5. A comparison of spontaneous male recombination in the *T-007 MR* line of *Drosophila melanogaster* with meiotic and mitotic recombination, and with the salivary gland and metaphase chromosomes. (a)=second chromosome, (b)=third chromosome (Slatko, 1978b).

been a laborious project. Briefly, *T-007* appears to contain several "major" genetic elements responsible for a large fraction of the male recombination and reduced transmission frequencies. These major elements cluster between the *pr* and *cn* loci (a 1 cM region including the centromere of the second chromosome) (Slatko and Hiraizumi, 1975; Slatko, 1976a, 1977). In addition to these major elements, there appear to be other minor elements widely distributed along the second chromosome which can influence frequencies of male recombination and transmission distortion (Slatko and Hiraizumi, 1975; Matthews *et al.*, 1978). Slatko and Green (1980) have located a region around the *pr* locus that is responsible for the majority of mutator activity associated with *T-007*.

Original confusion in interpreting the results of mapping experiments is now known to be due partly to the occurrence of new *MR* activity associated with both homologous and nonhomologous chromosomes kept in the presence of the *T-007* second chromosome. For example, about half of the non-recombinant *apl* (*=al dp b pr c px sp*) chromosomes isolated from heterozygous *T-007/apl* females show *MR* activity, whereas no *MR* activity was present before association with *T-007* (Slatko and Hiraizumi, 1975). Although transmission frequency is not affected by *MR* elements on a nonhomologue, male recombination frequency is affected (Matthews *et al.*, 1978). Subsequent experiments have shown that after association with *T-007*, the *apl* chromosomes can contain *MR* elements (symbolized *MR'*) in the region homologous to that of the *MR* major elements in *T-007* (Matthews *et al.*, 1978). The new *MR'* is apparently not capable of further transposition.

MR factors also appear to transpose to nonhomologous chromosomes (Slatko, 1978a). From heterozygous *T-007* females, third chromosomes originally free of *MR* activity can acquire it and can retain *MR* activity even after the removal of the normal *MR* second chromosome. This *MR* activity appears to be stable and has been mapped to the centromeric region of the third chromosome in two such lines (Slatko, 1978a). The "new" *MR* factors appear to be different from the original *MR*, however, since little transmission distortion is associated with them.

The distorted transmission frequency associated with *T-007* has also been subjected to detailed analysis. It was realized early that a negative correlation existed between the transmission of the *T-007* chromosome and the frequency of male recombination. By analysis of an extensive body of data which Hiraizumi (1979b) has accumulated, the following points are clear. First, the observed reduction in transmission of the *T-007* chromosome from heterozygous males is almost entirely due to the specific elimination of the *T-007* second chromosome. While the number of *T-007* progeny declines linearly with increasing "distortion," the number of progeny containing the homologue remains constant, except perhaps at very high distortion levels (Hiraizumi, 1977, 1979b). Second, when distortion increases, male recombination frequencies also increase. The increase appears to be due both to the number of observed recombinant progeny per male and to a decrease in the number of *T-007*-bearing progeny, which tends to inflate recombination frequencies. Third, no distortion appears to occur in heterozygous *T-007* females, and fertility of these females is essentially normal (Hiraizumi, 1977). Finally, Matthews *et al.* (1978) have shown that "minor" male recombination elements are widely distributed along the second chromosome (c.f., Slatko and Hiraizumi, 1975) and that the negative correlation between transmission and male recombination occurs in the presence or absence of the "major" element near the centromere of the second chromosome. It must be emphasized that the relationship between reduced transmission frequency and increased male recombination does not necessarily imply that they both occur at the same stage of spermatogenesis; they are defined in terms of the final products and, thus, the mechanisms of the two may be functionally, but not temporally, related.

None of the above experiments shed light on the actual mechanism of transmission distortion. From a viability study of zygotes from *T-007* fathers and an electron microscopic analysis of spermiogenesis in *T-007* males, Matthews (1981) concluded that distortion results from a combination of spermatid dysfunction and zygote inviability. About 21% of the *T-007* chromosomes expected to be recovered among progeny of heterozygous *T-007* males are lost at some point between fertilization and eclosion (representing 29% of the total distortion observed in young males). Another 52% of the expected number of *T-007* chromosomes are lost due to spermatid abortion during spermiogenesis (representing 71% of the total distortion). The extent of the reduction in number of spermatids per cyst correlates well with the degree of distortion found by Hiraizumi. Thus, it seems that the *T-007* chromosome triggers a spermiogenic event, leading to self-destructive dysfunction, while not affecting its homologue. The range of abnormalities present in spermatid cysts of *T-007* males suggests different defects in spermatogenesis may be occurring, although the question is open to future studies.

With respect to the cytogenetic effects of *T-007*, Slatko (1978b) has determined that there is an increase in recombination in heterozygous *T-007* females and that an increase over control coefficient of coincidence values occurs in the second, third and X chromosomes. Coefficient of coincidence values increases towards the centromere, as does the distribution of relative recombination in heterozygous *T-007* males. Male recombination frequencies compare more favorably with mitotic recombination, salivary gland and metaphase cytological maps than with the standard meiotic map (Figure 5).

T-007-induced male recombination does not appear to be influenced by the classical interchromosomal effect (Woodruff *et al.*, 1978), although *T-007* female values are. In addition, a 12-fold increase in second chromosome nondisjunction was found in females, but not in males. On the other hand, X–Y nondisjunction was increased four-fold in males. Thus, there appear to be both premeiotic and meiotic effects, suggesting that *T-007* may alter the preconditions (e.g., pairing) for meiotic exchange and disjunction, perhaps by functioning premeiotically.

As with most *MR* chromosomes, *T-007* is subject to a reciprocal cross effect, such that when the *T-007* chromosome is derived from a female, the dysgenic effects are largely suppressed; whereas when *T-007* is derived from a male, the dysgenic effects occur. In the case of *T-007*, it has been shown that a substantial part of the suppression is due to the influence of X chromosomes in the *T-007* stock which have the ability to suppress, to a large extent, the distortion of transmission frequency, male sterility at 29°C and, to a small extent, male recombination (Slatko and Hiraizumi, 1978; Matthews and Gerstenberg, 1979; Matthews, 1980). There appears to be no effect upon rescuing female sterility at 29°C by these X chromosomes. Mapping of X-linked suppressors from one *T-007* stock, although inconclusive, suggests that suppressing ability can reside in many regions of the X, but any given X chromosome appears to contain only one major site producing suppression (Matthews and Slatko, 1983). Cytoplasmic suppression is also involved (c.f., Kidwell, this volume; Engels and Preston, 1979), affecting female sterility and transmission distortion (Matthews and Slatko, 1983).

The suppressor X chromosomes appear to be similar to those in natural populations in Texas (Matthews and Slatko, 1983). In addition, they can rescue the male sterility which *T-007* males normally have at high temperatures and can restore *T-007* transmission to normal levels (Matthews and Slatko, 1983). They therefore appear to be modifying *MR* expression in both laboratory and natural populations.

9. Cranston, Harwich and South Amherst lines of Kidwell and Kidwell

Three natural population strains of *D. melanogaster* have been studied in detail by M. G. and J. F. Kidwell. Cranston (isolated in 1964 in Cranston, Rhode Island), Harwich (isolated in 1967 in Harwich, Massachusetts), and South Amherst (isolated in 1973 in South Amherst, Massachusetts) undergo male recombination and show sterility and frequent mutation in hybrid progeny (Kidwell, 1977, 1978, 1979; Kidwell and Kidwell, 1975,

1976; Kidwell et al., 1977a,b). Since hybridization appears to be critical to the induction of these and other associated genetic events, this set of effects has been termed "hybrid dysgenesis" (Kidwell et al., 1977b). Hybrid sterility, which has been the primary focus of the work on these lines, is reviewed in the chapter by M. G. Kidwell in this volume. We shall, therefore, summarize only the work on mutation and male recombination.

The Harwich line was used in selection experiments for high and low male recombination frequencies (Kidwell and Kidwell, 1976). Figure 6 shows that there were significant responses in both directions for third chromosome recombination, and the responses were correlated with sterility and distortion of transmission ratio. The frequency of male recombination events, some of which occurred in clusters, was affected by modifiers located in many positions in the genome. In addition, Kidwell et al. (1981) have observed that in a mixed population of P (paternally contributing) and M (maternally contributing) lines (or P and Q lines, with the Q lines being neutral strains that do not interact with either P or M) very rapid unidirectional change occurs to the P activity in the absence of artificial selection.

The Kidwells were the first to report a reciprocal cross effect associated with the activity of male recombination (MR) lines (Kidwell and Kidwell, 1975). Activity is induced when males from a natural population (P line) are crossed to a sensitive laboratory stock (M line) (cross A), but not in the reciprocal cross (B). For example, among progeny of hybrid females derived from crosses between Cranston males and *Basc* females (cross A), the

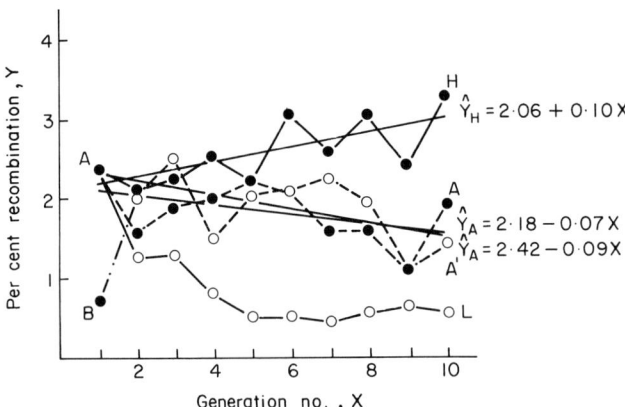

FIG. 6. The response of third chromosome spontaneous male recombination frequencies to selection in the Harwich line of *Drosophila melanogaster*. H=high selection line, L=low selection line, A and A'=unselected lines. Data points represent mean recombination events over all third chromosome intervals (Kidwell and Kidwell, 1976).

frequency of recessive sex-linked lethal mutations was much higher than that among chromosomes derived from crosses between Cranston females and *Basc* males (cross B) (Table V). The cross A frequency was also higher than that from non-hybrid parents (1/213, 0·47%). Mutation rates in subsequent generations were similar among descendants of females of both crosses. Mutations were also induced on the *Basc* chromosome in *Basc*/Cranston female progeny of cross A (12/390, 3·08%) and, but at a lower frequency, in cross B (1/459, 0·22%).

The South Amherst line also showed a high frequency of mutation after hybridization (Kidwell *et al.*, 1977a). This line contains Y and second chromosomes from the South Amherst population, X chromosomes from a *Pm/Cy Bl L* laboratory stock, and third and fourth chromosomes as a mixture from the South Amherst population and laboratory stock.

Kidwell *et al.* (1977b) suggested that the changes associated with hybrid dysgenesis are due to short term cytoplasm–chromosome interactions that show maternally influenced expression. They concluded that high frequencies of mutation and other associated traits are not common in nature but are mainly restricted to hybrids between natural population males and females from long-established laboratory stocks. Nevertheless, hybrid dysgenesis may have important implications for the increase of genetic variation and speciation in natural populations (see Kidwell, this volume).

10. OK1 of Woodruff and Thompson

In a study of spontaneous recombination in *D. melanogaster* males, Woodruff and Thompson (1977) screened 20 natural population lines captured in 1974 in Cambridge, England, and 25 lines collected in 1973 in Oklahoma City, Oklahoma. Two of the Cambridge lines and all of the Oklahoma lines showed male recombination on the second chromosome. One line from Oklahoma (Oklahoma-1, *OK1*) was chosen for detailed study because it showed an exceptionally high frequency of male recombination (3·49% on the second chromosome) in initial screens. The *OK1* line has continued to show high *MR* activity during nine years in the laboratory.

Interest in the line shifted from recombination to mutator activity when it was found that *OK1* induced frequent mutation and chromosome breakage after hybridization with other natural population lines and with laboratory stocks (Henderson *et al.*, 1978; Thompson *et al.*, 1978; Woodruff and Thompson, 1980; Thompson and Woodruff, 1980).

Male recombination in *OK1/dp b cn bw* heterozygotes is most frequent around the centromere and in the right arm ($dp–b = 0·10$ cM; $b–cn = 0·42$ cM; $cn–bw = 0·48$ cM, from 71,335 progeny scored with and without the *OK1* third chromosome; Woodruff and Thompson, 1977), with the

TABLE V. Reciprocal-cross effect on spontaneous male recombination, mutation, and chromosome breakage in F_1 progeny of *Drosophila melanogaster* natural population lines.

Line	Genetic change scored	Frequencies in lines before hybridization with laboratory stocks	Source of F_1 progeny	
			Cross A Laboratory stock females crossed with natural population males	Cross B Laboratory stock males crossed with natural population females
	Recessive sex-linked lethal mutations			
$OK1$ males[a] π^2 males[b]		4/3,743 (0·11%) Estimates as 0·2%	112/4,785 (2·34%) Estimated as 2·0% (28/70,882 = 0·04% mutations in the zeste-white region of the X chromosome)	4/3,038 (0·13%) Estimated as 3·5% (1/1,383 = 0·07% mutations in the zeste-white region of the X chromosome)
Cranston/$Basc$ females[c]		1/213 (0·47%)	Cranston X = 27/1,086 (2·49%) $Basc$ X = 12/390 (3·08%)	Cranston X = 2/968 (0·21%) $Basc$ X = 1/459 (0·22%)
	Visible mutations			
D. persimilis mutator line males[d]	Minute mutations	—	763/11,679 (6·53%)	6/7,779 (0·53%)
$OK1$ males[e]	Sex linked and dominant	0/16,833	132/27,874 (0·47%)	5/4,342 (0·12%)
Luminy and B.2 males[f]	Sex linked and dominant	—	26/42,231 (0·06%) (from R female, I male crosses)	4/62,461 (0·01%) (from I female, R male crosses)

	Male recombination	
T-007 males[g]	Second chromosome	80/3,008 (2·66%) — 1/3,944 (0·03%)
OK1 males[h]	Second chromosome	761/37,372 (2·04%) — 0/15,896
Cranston males[i]	Second chromosome	25/3,020 (0·83%) — 3/4,449 (0·07%)
Harwich males[i]	Second chromosome	42/3,731 (1·13%) — 2/1,802 (0·11%)
π^2 males[j]	Second chromosome	(a) 0·794 average frequency among 4,409 progeny — (a) 0·069 average frequency among 1,460 progeny
		(b) 0·846 average frequency among 1,537 progeny — (b) 0·043 average frequency among 9,281 progeny
		(c) 0·54% of 5,728 progeny — (c) 0·22% of 9,042 progeny
N.S.W. males[k]	Second chromosome	65/4,078 (1·60%) — 1/3,178 (0·03%)
31.1 MRF males[l]	Second chromosome	713/30,789 (2·32%) — 11/18,429 (0·06%)

	Chromosome breakage	
OK1 males[m]	In spermatocytes	0/80 — 0/192
	Loss of y^+ or B^s in $B^s Y y^+$ males	— 35/20,978 (0·17%) — 9/22,131 (0·04%)
W8D males[m]	Loss of y^+ or B^s in $B^s Y y^+$ males	31/19,915 (0·16%) — 17/19,921 (0·09%)
31.1 MRF males[n]	In spermatocytes	0/225 — 562/2,077 (27·1%) — 0/234
	In salivary glands	0/237 — 41/836 (4·9%) — 0/260
π^2 males[o]	Y-autosome, X-autosome and 2;3 translocations	0/304 — 1 (2;3)/1,504 (0·07%) — 1 (2;3)/770 (0·13%)
	X–Y translocations	0/2,875 — 1·63 average frequency among 3,384 progeny — 0·66 average frequency among 2,229 progeny
		0/9,276 — 0·45% in 5,728 progeny — 0·18% in 9,042 progeny

Footnotes to table on following page.

[a] Woodruff and Thompson (unpublished observations). F_1 males were from crosses of $OK1$ males with Canton-S or $C(1)DX,yf$ females in cross A and from $OK1$ females with Canton-S males in cross B. [b] Simmons and Lim (1980). Cross A = π^2 males with $C(1)DX,yf$ females. Cross B = π^2 females with yw males. [c] Kidwell et al. (1977a). Cross A = $Basc$ females with Cranston males. Cross B = reciprocal of cross A. [d] Mampell (1945, 1946). [e] Woodruff et al. (1979) and published results. Cross A = $OK1$ males with $C(1)DX,yf$ females. [f] Picard et al. (1978). Data are summed from a number of crosses of Inducer (I) lines with Reactive (R) lines. [g] Slatko (unpublished observations) Cross A = T-$007/Cy$ males with $al\ dp\ b\ pr\ c\ px\ sp$ females. Cross B = reciprocal of cross A. [h] Woodruff and Thompson (1977) and Henderson et al. (1978). Cross A = $OK1$ males $dp\ b\ cn\ bw$; ve females. Cross B = reciprocal of cross A. [i] Kidwell and Kidwell (1975). Cross A = Cranston or Harwich males with $al\ cl\ b\ sp^2$ females. [j] Engels (1979a,b). Cross A = π^2 males with $cn\ bw$, $C(1)DX,yf$; $cn\ bw$ or $C(1)DX,yf$; $cn\ bw$; e females. Cross B = π^2 females with $cn\ bw$ or $cn\ bw$; e males. [k] Sved (1976). Cross A = N.S.W. males with $al\ cn\ bw$ females. Cross B = reciprocal of cross A. [l] Yannopoulos (1978a, 1979), Yannopoulos and Pelecanos (1977). Cross A = in most crosses, 31.1 $MRF/Cy\ L^4$ males with $dp\ b\ cn\ bw$; ve females. Cross B = reciprocal of cross A. [m] Henderson et al. (1978), Brodberg and Woodruff, unpublished observations. In the spermatocyte breakage test, Cross A = $OK1$ males with $In(2R)bw^{VDe2}$ or $dp\ b\ cn\ bw$; ve females. Cross B = $OK1$ females with $In(2R)bw^{VDe2}$ males. The F_1 progeny of cross B gave F_2 male progeny that did undergo breakage (22/150 = 14·7%). In the chromosome marker loss test, cross A = $OK1$ or $W8D$ males with $C(1)DX,yw\ f/B^SYy^+$ females. Cross B = $OK1$ or $W8D$ females with Urbana-X^+/B^SYy^+ males. [n] Yannopoulos (1978a), Yannopoulos and Zacharapoulou (1980). In the spermatocyte breakage test, cross A = 31.1 $MRF/Cy\ L^4$ males were crossed with $dp\ b\ cn\ bw$, $dp\ b\ cn\ bw$; ve or Cy/Sp females. Cross B = 31.1 $MRF/Cy\ L^4$ females that contained cytoplasm from a Cy/Pm stock with $dp\ b\ cn\ bw$; ve males 23·8% (62/261) chromosome breakage was observed in spermatocytes. In the salivary gland polytene chromosome tests, cross A = 31.1 $MRF/Cy\ L^4$ males $dp\ b\ cn\ bw$ or $dp\ b\ cn\ bw$; ve females. Cross B = reciprocal of Cross A. [o] Engels (1979b). Cross A = π^2 males with $C(1)DX,yf$ females. Cross B = π^2 females with males from $C(1)DX,yw f$ stock.

male-derived map corresponding most closely to mitotic recombination maps (Thompson and Woodruff, 1978a). Male recombination also occurs on the third chromosome. Recombinant progeny sometimes occur in clusters from single *OK1* males, suggesting a premeiotic origin for some male recombination events (Woodruff and Thompson, 1977). Since recombination takes place in the presence of any combination of *OK1* second and third chromosome, *MR* activity seems to be due to dominant elements on both the second and third chromosomes.

Induced recombination in *OK1* males appears to be restricted to germinal cells. Thompson *et al.* (1978) screened for mitotic recombination in *OK1* males by scoring the frequency of marked clones in the wings and in the eyes of heterozygous males. There was no increase in the frequency of multiple wing hair clones or brown eye color clones in the presence of *OK1* chromosomes. In contrast, Hiraizumi *et al.* (1973) observed mosaic eyes in individuals from their crosses with the *MR* line *T-007*. In addition, unlike *T-007* and some other *MR* lines, *OK1* does not show distorted transmission ratios for second chromosomes (Woodruff and Thompson, 1977). These differences underscore the fact that, though similar in many respects, *MR* lines from different natural populations cannot be assumed to function identically.

Chromosome breakage has been analysed in detail in *OK1*. Frequent chromosome breakage occurs in spermatocytes (Henderson *et al.*, 1978), but not in neuroblast cells (Woodruff and Thompson, 1980). The absence of neuroblast chromosome breakage is consistent with the normal levels of somatic recombination which is generally due to somatic chromosome breakage. Germinal breakage events were only found in hybrid progeny from crosses between *OK1* males and laboratory stock females; no breakage events were observed in the F_1 progeny of the reciprocal cross or in the *OK1* base stock. Chromosome bridges and fragments were detected at anaphase I and II, but were absent in metaphase stages of spermatogenesis (Fig. 7). Henderson *et al.* (1978) concluded that if these breakage events were causing recombination in males (e.g., by breakage and reunion), then either some *MR* events occur at meiosis or the breakage observed at anaphase of spermatogenesis is triggered before meiosis. Frequent breakage has also been found in spermatocytes of a line of *D. melanogaster* from southern Greece (Yannopoulos, 1978a; Yannopoulos and Zacharopoulou, 1980; Stamitis *et al.*, 1981), although no chromosome breakage was detected in a small sample of spermatocytes of *T-007* males (Hiraizumi, 1979a). In addition, there is a significant increase in breakage of marked-Y chromosomes in $OK1/B^S Y y^+$ males derived from crosses of $C(1)DX, y\,w\,f; B^S Y y^+$ females with *OK1* males, but not in the same males derived from crosses of *OK1* females with $+/B^S Y y^+$ males (35/21,002, 0·17% partial marker

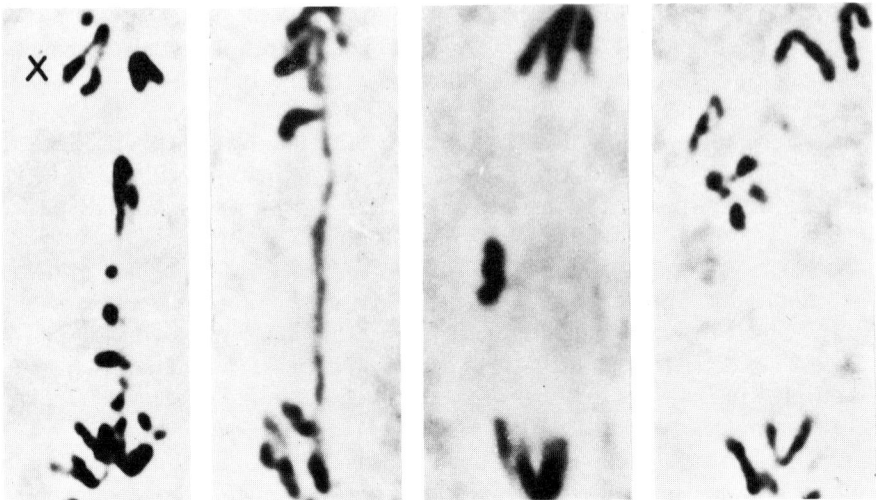

FIG. 7. Spontaneous chromosome breakage in spermatocytes of outcrossed *OK1*, a *MR* strain of *Drosophila melanogaster* (Henderson *et al.*, 1978).

loss in the former and 9/22,348, 0·04% in the latter males) (Brodberg and Woodruff, unpublished observations). A similar reciprocal cross effect was observed for the *T-007 MR* line and for a Wild-8D (*W8D*) line that was collected in Okefenokee Swamp, Georgia, in 1966 by Dr. B. Wallace.

Henderson *et al.* (1978) and Thompson and Woodruff (1978b) suggested that chromosome breakage is the cause or is at least a central element of male recombination and other dysgenic events in hybrid progeny of *OK1* flies. Others have proposed a similar mechanism for male recombination (Hiraizumi, 1971; Hiraizumi *et al.*, 1973; Slatko and Hiraizumi, 1973; Voelker, 1974; Cardellino and Mukai, 1975; Yamaguchi, 1976; Yannopoulos, 1978a).

Male recombination and its associated dysgenic events in *OK1* males are different in reciprocal crosses. When *OK1* males are mated with marked laboratory stock females, the F_1 progeny males undergo recombination. Conversely, in the reciprocal cross between *OK1* females and marked laboratory stock males, F_1 males produce no recombinant progeny (Woodruff and Thompson, 1977). The F_2, F_3 and subsequent generations do, however, undergo recombination. These results are probably due to a chromosome–cytoplasm interaction or, in some but not all cases, to an X chromosome suppressor of *MR* activity (Kidwell, 1978; Matthews and Slatko, 1983).

Reciprocal crosses also affect mutator activity, chromosome breakage and fertility in *OK1*. For example, *OK1* shows high mutator activity in crosses between *OK1* males and laboratory stock or other wild population line females (Table V; Woodruff et al., 1979; Thompson and Woodruff, 1980). While no visible mutations were observed in over 13,000 progeny scored from the *OK1* base population, very high frequencies of X-linked recessive and visible mutations were observed after outcrossing to other lines. The frequency of autosomal recessive mutations is also high in outcrossed *OK1* males (Woodruff and Thompson, 1975). Finally, the frequency of spontaneous recessive sex-linked lethal mutations in hybrid *OK1* males (112/4785, 2·34%) is 20-fold higher than the frequency in the *OK1* base stock (4/3743, 0·11%) (Woodruff, unpublished observations).

The observation that mutator activity in the *OK1* and some other *MR* lines (Woodruff et al., 1979) does not occur in the base population, but is only induced upon hybridization with other strains, has important implications for understanding mutator activity in natural populations. The "hybrid release" hypothesis of Woodruff and Thompson (1980; Thompson and Woodruff, 1980; Woodruff et al., 1979) suggests that hybridization, directly or indirectly, acts as a release mechanism for suppressed mutators in natural populations of *D. melanogaster*. This breakdown in mutation suppression could lead, for example, to an explosive increase in genetic variation through induced mutations or novel chromosome aberrations.

The mechanism of *MR* mutator activity in *OK1* and in other male recombination lines is not entirely understood. One thing that is obvious from the above discussion is that male recombination, and its correlated events, occur by a mechanism very different from female recombination (see also Woodruff et al., 1978). An interesting observation is that *MR* activity can apparently be transmitted by non-genetic means, suggesting that the ultimate cause may involve a transmissible component (Voelker, 1974; Waddle and Oster, 1974; Roberts, 1976; Sochacka and Woodruff, 1976; Yamaguchi, 1976; Green, 1977a; Woodruff and Thompson, 1977; Hellack et al., 1978; Slatko, 1978b). There are especially close parallels with the behavior of transposable elements (Thompson and Woodruff, 1980). Sochacka and Woodruff (1976) and Lyman and Woodruff (unpublished observations) have been able to induce low frequencies of male recombination in laboratory stocks by injections of extracts from *OK1* flies. Similar results have been obtained with extracts from other *MR* lines and from some laboratory stocks (Reddi et al., 1965; Gerasimova, personal communication; Slatko, unpublished observations). Low levels of male recombination have been induced by feeding laboratory strains with homogenates of *MR* lines and laboratory wild type strains (Hellack et al., 1978; Slatko, unpublished observations), and sterility has been induced by injections and feeding of extracts of lines that show dysgenic traits (Colgan and Angus, 1978; Angus and Raisbeck, 1979).

In contrast to the results with *OK1* and other lines, however, there are reports of failure to induce *MR* activity in injection and feeding experiments (Singer *et al.*, 1967; Saraswathy, 1970; Waddle and Oster, 1974; Sved *et al.*, 1978). The conflicting results of these experiments may reflect a basic genetic difference in strains, or a difference in controls, methodology or interpretation.

11. AW, JH, OYW *and Raleigh lines of Yamaguchi and Mukai*

A number of authors have identified unique inversions from various natural populations of *D. melanogaster* (e.g., Mukai and Yamaguchi, 1974; Voelker, 1974; Ashburner and Lemeunier, 1976; Stalker, 1976), and Yamaguchi and Mukai (1974) have suggested that these rearrangements might have been induced by mutators. To test this hypothesis, Yamaguchi and Mukai (1974) measured the frequency of newly-induced spontaneous second chromosome aberrations in five strains from two natural populations (*CH*, *PQ* and *RT* from Madison, Wisconsin, 1967; *AW* and *JH* from Erie, Pennsylvania, 1954). Stocks from these natural populations were formed from single males and kept by single pair matings before screening.

No chromosome aberrations were detected in 22,058 chromosome generations in the three Madison Wisconsin lines; but 112 different chromosome rearrangements were observed in 90,996 chromosome generations in the Erie, Pennsylvania lines. In the *AW* line, 19 second chromosome paracentric inversions and one 2;3 translocation were recovered in 45,990 chromosome generations (0·043%). In the *JH* line, 83 paracentric and 6 pericentric second chromosome inversions, two 2;3 translocations and one second chromosome transposition were observed out of 45,006 chromosome generations (0·204). Deletions were also induced in the mutator-free homologues (Cardellino and Mukai, 1975). All aberrations observed in the *AW* and *JH* lines were unique, and the distribution of breakpoints was not random. For example, there was a significantly higher number of breaks in 2L of the *JH* lines. Male recombination was identified in some *JH* lines.

Subsequently, Mukai and Cockerham (1977), Voelker *et al.* (1980) and Scobie and Schaffer (1982a,b) have extended the analysis to include frequencies of mutations at specific enzyme and visible loci. Several lines are associated with locus specific mutation induction, as well as male recombination, distorted transmission ratios and high frequencies of genetic instability at two second chromosome loci, dumpy and vestigial. The male recombination factors appear to map to several positions on the second chromosome. It is of interest that these chromosome lines, which have been kept for 240 generations in the laboratory after removal from a cage population started in 1954, still retain mutator activity.

The Reedy Creek Park population of *D. melanogaster*, near Raleigh, North Carolina, was reported to contain unique inversions and a high lethal and detrimental load (Chigusa *et al.*, 1969; Mukai *et al.*, 1971; Mukai and Yamaguchi, 1974) and was predicted to contain mutator factors (Chigusa *et al.*, 1969). To determine if this prediction was correct, Cardellino and Mukai (1975) measured the frequencies of viability mutations, lethals, chromosome breakage events, and male recombination in 140 inversion-free second chromosome lines (*RBR* lines) that had been isolated from the Raleigh population in 1970.

In the first generation, 90 of the 140 second chromosome lines from the Raleigh population were lethal-free (Yamaguchi *et al.*, 1976), whereas at generation 40, 80·7% (113/140 lines) carried second chromosome lethal mutations. The minimum lethal mutation rate per second chromosome was about 3·1%, which is five times as high as that in mutator-free lines (c.f., 0·63%, Mukai, 1964; Mukai *et al.*, 1972; and 0·59%–0·69%, Wallace, 1968b). In fact, Cardellino and Mukai (1975) believe that the recessive lethal mutation rate in the Raleigh lines was more than ten times higher than the normal rate. In addition, the total genetic variance in the 140 lines after 30 generations was larger than predicted due to a high mutation rate (c.f., Mukai *et al.*, 1974).

The Raleigh population also underwent frequent chromosome breakage (Cardellino and Mukai, 1975; Yamaguchi *et al.*, 1976). After 40 generations, 63 new and unique aberrations were recovered in 42 of the 140 original inversion-free lines (40 paracentric and 15 pericentric inversions, two 2;3 translocations, and 6 transpositions; 0·0112 aberrations per second chromosome per generation; Yamaguchi *et al.*, 1976). All aberrations were germinal in origin, and 90% of those carrying unique inversions were homozygous lethal.

Cardellino and Mukai (1975) and Yamaguchi *et al.* (1976) estimated that 60–70% of the tested chromosomes contained mutator factors that induced chromosome breakage, mutations and recombination in males. Although the specific positions of mutator-induced breaks were not random, the frequencies of breaks in the left and the right arms were similar. This difference from the *AW* and *JH* mutator lines was taken to mean that the mutators in the North Carolina and the Pennsylvania lines were not identical.

Recombination was detected in some males from the Raleigh population (Cardellino and Mukai, 1975). Of special interest was the observation that second chromosome recombination occurred in male $In(2LR)bw^V$ heterozygotes, the inversion being a recombination suppressor in females. Thus,

Cardellino and Mukai concluded that male recombination occurs by chromosome breakage, rather than by chiasma-type female recombination. A similar chromosome breakage mechanism has been suggested for male recombination in other lines (Slatko and Hiraizumi, 1973; Voelker, 1974; Yamaguchi, 1976; Woodruff and Thompson, 1977; Hiraizumi, 1977; Thompson and Woodruff, 1978b).

One might predict that the mutation load in natural populations would be high if the mutation and chromosome breakage observed in laboratory crosses with the Raleigh lines were also occurring in nature (Cardellino and Mukai, 1975; Yamaguchi *et al.*, 1976). To the contrary, however, the mutation rate in the Raleigh population is not unusually high (Langley and Ito, 1976). Hence, Cardellino and Mukai (1975) and Yamaguchi *et al.* (1976) speculate that mutators in natural populations are only "potential mutators." The potential mutator hypothesis states that chromosome breakage and mutations are caused by a virus that can stay and propagate in the cytoplasm only when there is a specific chromosomal gene (the potential mutator gene) in the cell. Yamaguchi (1976) supports the potential mutator hypothesis, but he adds that in natural populations there must exist some conditions that suppress the action of mutator factors.

There do seem to be times, however, when the mutators must function in the Raleigh population, as indicated by its high frequency of unique inversions. For example, 9 unique inversions were found in 135 chromosomes sampled in 1968, 3 of 146 in 1969, and 14 of 691 in 1970 (Yamaguchi *et al.*, 1976).

In order to determine the extent of mutators in other natural populations of *D. melanogaster*, Yamaguchi (1976) isolated 38 lines (called OYW lines) from five natural populations (Ishigakijama, Miyakojima, and Nago, Japan, in 1970; Yugoslavia and Taiwan in 1971) and scored them for second and third chromosome lethal mutation rates, male recombination, segregation ratio and *delta* activity. After 25 generations, 10 new aberrations were recovered in 8 of the lines (7 paracentric and 2 pericentric inversions, and one 2;3 translocation). All were of germinal origin, and there was an equal distribution of breaks in the second and third chromosomes.

Twenty-four of the OYW lines underwent recombination in males. No SD elements (Hartl and Hiraizumi, 1976) or *delta* elements were recovered, and lethal mutation rates for the major autosomes were high (0.97% for the second and third chromosomes of all OYW lines; 4.3% for OYW lines that contained mutators). Finally, Mukai and Cockerham (1977) and Voelker *et al.* (1980) have measured the spontaneous frequency of electrophoretic mutations at seven loci in hybrid progeny of AW and JH lines crossed with laboratory stocks. These frequencies, given in allele generations, are summarized in Racine *et al.* (1980) and are as follows: mitochondrial glutamate oxaloacetate transaminase (Got-2, 2–3·0)=1/431,939 (0.23×10^{-5}); alpha-glycerophosphate dehydrogenase ($alpha$-$Gpdh$, 2–20·5)=7/424,454 (1.65×10^{-5}); cytoplasmic malate dehydrogenase ($cMdh$, 2–37·2)=2/431,939 (0.46×10^{-5}); alcohol dehydrogenase (Adh, 2–50·1)=0/426,938; dipeptidase A (Dip-A, 2–55·2)=0/431,939; hexokinase C (Hex-C, 2–73·5)=5/431,939 (1.16×10^{-5}); alpha-amylase ($alpha$-Amy, 2–77·7)=1/38,565 (2.60×10^{-5}); total for all seven loci=16/3,111,598 (0.51×10^{-5}).

12. Haifa[12] of Green

MR Haifa[12] was identified in 1976 from among 17 second chromosome lines taken from a Mt. Carmel, Haifa, Israel, natural population (Green, 1977b). A single second chromosome from an isofemale line was isolated and is now kept as *Haifa[12]/Cy*. Homozygotes for *Haifa[12]* are partially female sterile and both males and females are poorly viable due to a small bristle mutation that maps just to the right of the *pr* locus (2–54·5) (Green, 1977b; Sinclair and Green, unpublished observations).

Haifa[12] induces male recombination at a frequency of approximately 0·5% along the left arm and centromeric region of the second chromosome (*al* to *cn*, 57 map units in females), and about 0·8% along the entire third chromosome (Green, 1978a; Sinclair and Green, 1979). *Haifa[12]* does not appear to be associated with distorted transmission frequencies from heterozygous males.

The mutational activity of *Haifa[12]* has been extensively analysed. Using Muller's "*Maxy*" chromosome, increased mutability was observed at the *y*, *w*, *rb*, *cm*, *ct*, *sn*, and *ras* loci, with an increase over control frequencies being as high as 250-fold in the case of the *sn* locus (Green, 1978a,b; Gerasimova, 1981). The vast majority of mutants recovered from individual male or female crosses occur as single events, although several large clusters have been observed (Green, 1977a; Gerasimova, 1981). Mutants often appear to be unstable as shown both by frequent reversions to wild type or to other alleles and by somatic instabilities (Green, 1977a,b, 1979; Golubovsky *et al.*, 1977; Gerasimova, 1981).

Both males and females show increased mutation frequencies at some loci (e.g., *sn* and *ras*) but not at others (e.g., *ct* and *v*) (Table VI). Nonetheless, a dichotomy in the mechanism of mutator activity (or repair of *MR*-induced lesions) is exemplified by the observation that *y* mutants recovered from *Haifa[12]* females are largely male lethal, while *y* mutants recovered from *Haifa[12]* males are male viable (Sinclair and Green, 1979; Green, 1979). Recently, Green and Shepherd (1979) found that *Haifa[12]* produces deficiencies at a number of autosomal loci, including *pr*, *cn* and *l(2)gl*. No tip deletions have been observed for any chromosomes.

TABLE VI. Spontaneous frequencies of visible mutations at specific sex-linked and autosomal loci in MR lines of *Drosophila melanogaster*. The MR chromosomes are paternal in origin in these crosses (Green, 1977b, 1978a, 1978b; Green and Shepherd, 1979; Gerasimova, 1981; Schalet, personal communication; Woodruff and Thompson, unpublished).

Locus and number of mutants recovered (frequency of mutation ×10⁻⁵)

Line	y	pm	w	rb	cm	ct	sn	ras	v	m	g	f	car	pr	cn	Chromosomes scored
In males																
MR-h12[a]	4 (3·3)	0	1 (0·83)	5 (5·6)	1 (0·83)	4 (4·7)	44 (51·5)	15 (17·6)	0	0	1 (0·83)	0	0	—	—	85,455
MR-h12[b]	—	—	—	—	—	—	78 (139)	5 (9)	—	—	—	—	—	—	—	(120,910 for y and w) 55,864
MR-n1	(4) (3·5)	—	—	—	(93) (235)	—	—	—	—	—	—	—	—	—	—	30–40,000
MR-s1	—	—	—	—	—	—	—	(29) (25)	—	—	—	—	—	—	—	30–40,000
MR-h12	—	—	—	—	—	—	—	—	—	—	—	—	—	6 (3·9)	10 (6·6)	152,521
MR-n1	—	—	—	—	—	—	—	—	—	—	—	—	—	4 (4·4)	72 (79·2)	90,856
MR-n1	4 (4·2)	1 (2·1)	0	0	1 (2·1)	0	11 (23)	13 (27)	0	1 (2·1)	1 (2·1)	0	0	—	—	47,354
T-007	—	—	—	—	—	—	—	—	—	—	—	—	—	6 (14·6)	18 (43·9)	(94,708 for y and w) 40,975
OK1	—	—	—	—	—	—	—	—	—	—	—	—	—	1 (3·7)	7 (26·2)	26,675
Control[c]	5 (1·0)	1 (0·2)	5 (1·0)	0	4 (0·8)	9 (1·8)	1 (0·2)	2 (0·4)	0	1 (0·2)	5 (1·0)	3 (0·6)	1 (0·2)	—	—	490,000
In females																
MR-h12	18 (25·2)	—	—	—	—	—	29 (40·1)	9 (12·6)	—	—	—	—	—	—	—	71,478
Control[c]	2 (0·6)	0	2 (0·6)	0	0	1 (0·3)	0	0	0	0	1 (0·3)	0	1	—	—	340,000

[a] MR-h12 males were crossed with females containing a Maxy X chromosome. [b] MR-12 males were crossed with attached-X females. [c] Control data from Schalet, personal communication. Also see Table I.

Recombination in heterozygous $Haifa^{12}$ females is increased over control frequencies (Sinclair and Green, 1979). On the other hand, in the presence of $c(3)G$, a meiotic mutant which vastly reduces meiotic recombination in control females (from 48% to 0·24% in the al-cn region), $Haifa^{12}$ female recombination nevertheless occurs at frequencies comparable to those in their $c(3)G^+$ male sibs, suggesting a premeiotic origin of recombinants in these females.

The mechanism of MR-induced mutator activity has been investigated from a number of viewpoints. It has been shown that $Haifa^{12}$ does not substantially increase detachment of attached-X chromosomes (Sinclair and Green, unpublished observations), and there appears to be only slight if any interaction between X-irradiation and MR-induced male recombination (Sinclair and Green, unpublished observations) or translocations (Sobels and Eeken, 1981; Slatko, unpublished observations). However, in regard to recessive lethal mutations there is a clear synergism between $Haifa^{12}$ activity and X-irradiation (Sobels and Eeken, 1981). The frequency of spontaneous visible mutations is also increased in the presence of repair defective mutations (Eeken and Sobels, 1981).

The distribution of lethals along the X chromosome, visible mutations along the X chromosome and the autosomes, and the distribution of recombination in males and females all point to the nonrandom nature of MR-induced chromosome breakage.

The majority of the mutator activity associated with the $Haifa^{12}$ MR chromosome has been mapped to a one cM region between the Tft (2–53·2) and pr (2–54·5) loci on the second chromosome (Slatko and Green, 1980). It is not currently known whether one or more "elements" responsible for mutation induction are present within this region. Overlapping deficiencies which subdivide the region show no differences in mutational activity when in combination with $Haifa^{12}$. The genetic elements responsible for the induction of male recombination also appear to be located in the Tft-pr interval of the second chromosome, and the data suggest that a correlation exists between mutational and recombinational activity among the various chromosomes generated in the mapping experiments (Slatko and Green, 1980; Green, unpublished observations). Two doses of $Haifa^{12}$, both patroclinous in origin, show no increase in mutation frequency over that of one dose (Green and Slatko, 1979). Thus, one dose of the $Haifa^{12}$ second chromosome region (either over a wild chromosome or a deficiency for the MR region) and two patroclinous doses behave similarly.

13. π_2 Strain of Engels

An inbred line (π_2) of $D.$ $melanogaster$, showing many of the characteristics of P–M hybrid dysgenesis, was derived from a natural population at the University of Wisconsin Arboretum, Madison, Wisconsin, in 1975 (Engels, 1979a,b,c, 1981a,b; Engels and Preston, 1979). Sterility and the other associated genetic changes in the π_2 line, including mutator activity, transmission ratio distortion, sex ratio distortion, male recombination, and chromosome breakage, are due to P elements on all major chromosomes and are controlled by an extrachromosomal factor (Engels, 1979c; also see Kidwell, this volume). Individuals possessing the extrachromosomal dysgenesis-inducing component are referred to as having the "M cytotype" and those lacking the component as having the "P cytotype." Although the cytotype shows limited extrachromosomal transmission, the cytotype can be switched by factors residing on all three major chromosomes (see Engels, 1981a, for a discussion of P factors and cytotype). Crosses between males of P strains (paternally contributing) from nature and females of M strains (maternally contributing and usually from laboratory stocks) cause the largest amount of dysgenic events (Engels, 1979c; Engels and Preston, 1980). This cross is generally referred to as cross A to distinguish it from the reciprocal (cross B), in which dysgenic events are usually rare or absent. Simmons et al. (1980) have observed that low, but significant, increases in recessive sex-linked lethal mutations are observed in some cross B experiments. Dysgenic events are absent from nonhybrids (Engels, 1979c). Engels (1979c) has, however, shown that maternally derived factors in presence of M cytotype can cause dysgenic events.

Hybrids derived from the π_2 line have an elevated rate of mutation for X chromosome lethal and visible mutations and show an increase in chromosome breakage involving the X and Y chromosomes (Engels, 1979b; Berg et al., 1980; Simmons and Lim, 1980; Simmons et al., 1980). X–Y translocations and large deletions of the X chromosome occurred in cross A, but there was no evidence for induction of reciprocal translocations involving the second or third chromosomes in any cross. Engels (1979b) and Berg et al. (1980) also stated that chromosome breakage did not occur at random positions on the X and Y chromosomes. Spontaneous X-chromosome rearrangements were observed at a high frequency in the π_2 line hybrids (25/272; 9·2%). These rearrangements have similar breakpoints at constriction sites and are associated in most cases with hemizygous lethality (Fig. 8). Another natural population line, Madison 75, also underwent frequent chromosome breakage (Berg et al., 1980), and the hot-spots for chromosome breakage vary in position from one P strain to the next (Engels, 1981b).

A number of the visible mutations and lethal mutations recovered from the π_2 line were genetically unstable (Engels, 1979a; Simmons et al., 1980). In particular, a singed bristle mutation (sn^w) was found to mutate at rates

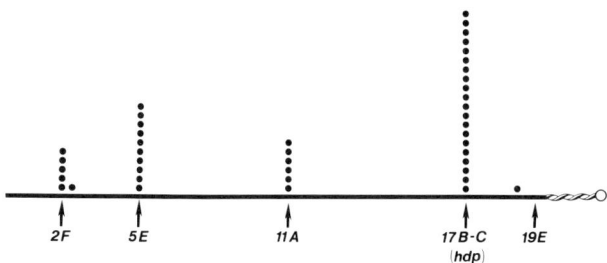

FIG. 8. Distribution of spontaneous X-chromosome breakpoints in hybrids of the π_2 line of *Drosophila melanogaster* (Engels, 1981a). Inversions and complex rearrangements were detected as males with heldup wings (*hdp*).

higher than 50%. The instability of sn^w was suppressed in flies of the *P* cytotype, indicating that the mutability in this case was under the control of a chromosomal–cytoplasm interaction. Mutation in this line is limited to germinal cells and occurs in clusters.

Male recombination and transmission of segregation ratios in crosses of the π_2 line are directly correlated and affected by temperature. Both have their highest expression at 25°C and are reduced at 17° and 29°C. In addition, there is a strong correlation between sterility and male recombination, and the causal *P* factors have widely variable distributions on the X, second and third chromosomes (Engels and Preston, 1980).

14. The 31.1 MRF Greek mutator line of Yannopoulos

31.1 MRF is a second chromosome line of *D. melanogaster* isolated in 1971 from southern Greece. It undergoes frequent chromosome breakage in spermatocytes and shows a number of the other characteristics of hybrid dysgenesis, including male recombination in the second and third chromosomes (Yannopoulos and Pelecanos, 1977; Yannopoulos, 1978a,b, 1979).

Male recombination and sterility in *31.1 MRF* is temperature-sensitive, with higher frequencies occurring at higher temperatures. For example, 0·16% recombination was observed when heterozygous males were raised at 15°C, 1·75% at 25°C, and 3·19 at 29°C (Yannopoulos and Pelecanos, 1977). Female and male sterility in *31.1 MRF* hybrids, which occurs by incomplete gonad development, is most pronounced at 28°C (Yannopoulos, 1978c; Stamatis *et al.*, 1981). Kidwell *et al.* (1977b), on the other hand, reported that male recombination in their lines was reduced at high (28°C) and at low (18°C) temperatures, again showing that all *MR* lines do not behave the same.

Since the temperature sensitive period for recombination in *31.1 MRF* males was during larval development, Yannopoulos and Pelecanos (1977) concluded that male recombination is a premeiotic event. They speculate that the temperature treatment had an indirect effect on male recombination, possibly by affecting the multiplication of a virus (Yannopoulos, 1978a). Additional support for the premeiotic origin of male recombination events comes from the observation that there is a synergistic interaction between diethyl sulfate and *31.1 MRF* induced male recombination events in premeiotic cell stages (Yannopoulos *et al.*, 1980). The transmission ratio was also temperature sensitive (at 15°C, $k=0.52$; at 25°C, $k=0.40$; at 29°C, $k=0.36$, where k is the proportion of nonrecombinant $+/marker$ progeny; Yannopoulos and Pelecanos, 1977). In other words, at higher temperatures there is a reduction in the number of *31.1 MRF*-bearing sperm produced.

Frequent chromosome breakage was seen in up to 58·6% of the meiotic anaphases in spermatocytes of outcrossed *31.1 MRF* males (Yannopoulos, 1978a) and in salivary gland chromosomes (Yannopoulos and Zacharopoulou, 1980). There was a direct correlation between the induction of male recombination and chromosome breakage. In addition, some abnormal mitotic anaphases (4/389) were found in outcrossed *31.1 MRF* males. Cytological analysis of salivary gland chromosomes showed that frequent breakage occurs in all major chromosomes in germ cells of some hybrid *31.1 MRF* males and females, and that breakage, like male recombination, is affected by a chromosomal–cytoplasmic interaction (Yannopoulos and Zacharopoulou, 1980). The types and frequencies of chromosome rearrangements recovered in hybrid flies from crosses of *31.1/Cy L^4* males with marker females (cross A) are shown in Table VII. The total frequency of breakage in three crosses was 4·9% (41 rearrangements in 836 F_2 larvae). In contrast, no breakage was observed in the F_1 larvae from crosses of *31.1/Cy L^4* males with *dp b cn bw* or *dp b cn bw; ve* females (0/237), in the F_1 larvae from crosses of *31.1/Cy L^4* females with *dp b cn bw; ve* males (0/135), or in F_2 larvae from this latter cross (F_1 *31.1/dp b cn bw; 31.1/ve* males with *dp b cn bw; ve* females, 0/125). Chromosome breakage at anaphase of spermatocytes was also observed in another natural population line (*23.5 MRF*) collected in southern Greece (Stamitis *et al.*, 1981), and both the *31.1 MRF* and *23.5 MRF* lines produce frequent chromosome deletions and duplications (Yannopoulos *et al.*, 1981).

It was suggested that the reciprocal cross influence in the *31.1 MRF* line is caused by the cytoplasmic suppressor in the *Cy L^4/Pm* laboratory stock which they used (Yannopoulos, 1978a). This would also explain the fact that the *31.1 MRF/Cy L^4* stock does not undergo chromosome breakage. A gradual acquisition of cytoplasmic resistance against the activities of *31.1 MRF* was detected when the factor was introduced into the cytoplasm of a normal strain (Yannopoulos, 1978b).

The *31.1 MRF* line is reported to act as a mutator by inducing recessive sex-linked and second chromosome lethal mutations, though no frequencies were given (Yannopoulos and Pelecanos, 1977). In addition, they hypothesized that chromosome breakage is due to a virus-like element or an episome. In support of this hypothesis, Yannopoulos (1978a, 1979) has found that all properties of *31.1 MRF* may be transmitted to other homologous chromosomes.

15. Florida lines of Voelker

Five lines of *D. melanogaster* from Jacksonville, Florida, were found to contain unique chromosome aberrations and to undergo male recombination (Voelker, 1974). Second chromosome male recombination events and unique aberrations had a similar pattern of distribution. There was an excess of both events in the centromeric region and in the right chromosome arm, though the same was not true for the third chromosome. Male recombination in the centromeric region occurred in proximal euchromatin, and recombination was not associated with the induction of deficiencies or duplications. From the latter result, Voelker (1974) concluded that male crossovers are rather precise exchange events.

TABLE VII. Occurrence of rearrangements induced in F_1 *31.1 MRF* males and females (Yannopoulos and Zacharopoulou, 1980)

Chromosome arm broken	Type of rearrangement induced					Total number of larvae tested
	Inversion	Duplication	Deficiency	Translocation	Transposition	
In males[b]						
X	3	0	0	0	0	
2L	2	4	0	0	0	
2R	7	4	1	0	0	
3L	7[a]	3	0	0	0	
3R	1	0	1	0	0	
Total	20	11	2	0	0	583
In females[b]						
X	1	1	0	1 T(1;2)	0	
2L	0	0	1	0	0	
2R	1	0	0	0	0	
3L	1	0	0	1 T(2;3)	0	
3R	0	0	0	0	1	
Total	3	1	1	2	1	253

[a] Four of the seven were pericentric inversions. [b] These rearrangements were induced in the progeny of marker females mated with *31.1 MRF* males (cross A). The reciprocal cross gave no rearrangements among 260 larvae.

16. U.S.S.R. mutator lines of Berg and Golubovsky

Berg (1966, 1972, 1974, personal communication), Golubovsky (1980) and Golubovsky et al. (1974, 1977) have detected frequent visible and lethal mutations in lines of D. melanogaster isolated from natural populations sampled in geographically distant populations from the Ukraine, Crimea, Caucasus and other locations in the U.S.S.R. from 1937 to 1976. Sex-linked recessive singed bristle (sn) and yellow (y) mutations were frequently recovered in distant populations, and many were genetically unstable. Reversions to wild type occurred at a frequency of 10^{-4} to 10^{-3}. Reversions of sn mutations occurred in clusters and in both germinal and somatic tissues. It is of particular interest that several sn males, one of which was genetically unstable, have been recovered directly from a Krasnodac, U.S.S.R., population (Green and Shepherd, 1979). This may indicate that mutators sometimes function in natural populations.

Golubovsky (1980) and Golubovsky et al. (1974; personal communication) have concluded from this long-term study of natural populations of D. melanogaster that mutation rates fluctuate with time; a 3 to 5-fold difference in lethal mutation rate was observed at different sampling times (Fig. 9). In addition, Berg (1966, 1972, 1974) has observed large variations in the frequency of visible mutations with time in U.S.S.R. natural populations of D. melanogaster, with some visible mutations increasing in frequency at the same time in a number of widely separate populations. For example, the frequency of abnormal abdomen mutations increased from a low of 0·4% in 1937 in the Crimea and 0·08% in the Transcaucasus to 11·07% and 13·76%, respectively in 1971. Although the reason for these fluctuation is unknown, Golubovsky (1980) suggests that interactions between eukaryotic and viral genomes may be one factor that affects mutation rate in nature. Spencer (1935) and Alexander (1949) have also observed periodic changes in mutation rates in natural populations of *Drosophila* as have Dubinin and his collaborators (see Dubinin, 1946, 1964; Spencer, 1947).

17. I–R System of Picard and associates

Mutator activity and sterility associated with a genetic system interaction called *I–R* has been observed in crosses with a number of natural population lines and with some laboratory stocks (examples of natural population lines are Luminy from southern France and B2′ from western France) (see Bregliano et al., 1980 for a review of this system). A chromosomal–cytoplasmic interaction is responsible for the genetic activity of these crosses, with the inducer *(I)* system being chromosomal and the reactive *(R)* system being a cytoplasmic state controlled by the genotype (Bucheton and Picard,

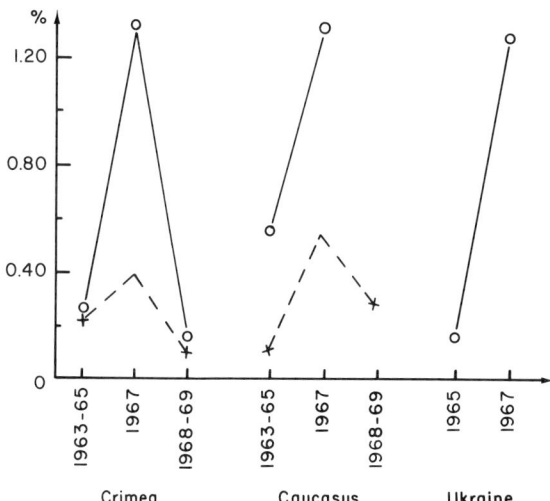

FIG. 9. Fluctuations of mutability in geographically separated populations of *Drosophila melanogaster* (Golubovsky, 1980). Lethal mutation rates in chromosome 2 (solid line) and in the X chromosome (dashed line).

1978). In addition, chromosomes that are reactive may acquire the *I* factor by a process called "chromosomal contamination" (Picard, 1976, 1978, 1979; Picard and Pelisson, 1979). The sterility associated with the *I–R* crosses is discussed by Kidwell in this volume; mutator activity will be discussed in this section.

Crosses between reactive females and inducer males give rise to X-chromosome loss in F_1 females by nondisjunction or chromosome breakage and to high rates of spontaneous mutation (Picard *et al.*, 1978). From seven such crosses, 185 X0 males derived from X-chromosome loss in females were recovered among 16,154 males (1·15%) and 31 visible mutations were recovered among 55,069 total progeny (0·06%). These frequencies can be compared to the summed results of reciprocal crosses of *I* females with *R* males and to crosses between *I* lines or between *R* lines: 17/65,989 (0·03%) for X-chromosome loss and 5/139,498 (0·004%) for visible mutations. Bregliano *et al.* (1980) also state that high frequencies of recessive sex-linked lethal mutations occur in females of *I–R* crosses. A reciprocal cross effect on chromosome loss and mutation can be shown by the following data (also see Table V). From crosses of *R* females with *I* males, 59/8,386 (0·70%) X0 males and 26/42,231 (0·06%) visible mutations were recovered, whereas in crosses of *I* females with *R* males 7/27,473 (0·03%) X0 males and 4/62,461 (0·006%) mutations were recovered. Some of the mutations occurred as clustered events, were unstable, and were associated with lethal mutations and deletions. The latter observation led Picard et al. (1978) to speculate that most of the mutability observed in I–R females was due to chromosome breakage.

Unlike strains in many natural populations, lines of the I–R system are not apparently associated with male recombination (Picard et al., 1978). From this observation and from sterility studies, Kidwell (1979) has concluded that the I–R system is different from the P–M system of some MR lines. This conclusion has been challenged (Eggleston and Kearsey, 1980).

18. D. ananassae mutators

Hinton (1979, 1981) has observed that three mutant stocks of D. ananassae, bri (bright red eyes, third chromosome marker), ca (claret eye color, second chromosome) and ca; stw (straw body color, third chromosome), undergo frequent spontaneous mutation and/or chromosome breakage that is attributable to complex mutator–suppressor systems. The bri stock was derived from the University of Texas collection and probably is of South Pacific origin, whereas the ca; stw and ca stocks were recovered from a natural population in Calcutta, India, more than 20 years ago. The ca stock was used in the synthesis of the ca; stw stock.

The bri stock mutates frequently to the dominant, visible Minute phenotype (29/6873 = 0.42% as compared to a range of 0 to 0.21% for six other mutant stocks). The spontaneous frequency of Minute mutations in D. melanogaster is 0.11% (Table I). Frequent mutation is also observed in hybrid male progeny of reciprocal crosses between bri and px (a third chromosome mutant stock with plexus wing venation that has low mutability). The increased mutation rate in the bri stock was attributed to a third-chromosome dominant mutator. The mutator functions in females and males before the completion of gametogenesis, and there is no clear evidence that the bri mutator induces chromosome breakage. Finally, there are complementary autosomal suppressors of the bri mutator activity, one present in the px stock, the other in the bri stock. The activity of these suppressors is limited to females.

Hinton (1979, 1981) has observed that the ca and ca; stw stocks also contain a mutator–suppressor system. The bri and ca mutator–suppressor systems are rather different from each other as might be expected from the geographically distant origin of the stocks. The ca and ca; stw stocks, which have a low intrastock frequency of Minute mutations (15/19,546 = 0.08%, and 12/9,287 = 0.13%, respectively) undergo frequent spontaneous mutation after hybridization with other mutant stocks. For example, in crosses of ca; stw males with bri or px females the progeny showed 0.55%

(91/16,625) and 0·61% (115/18,893) Minute mutations. The reciprocal cross, however, gave low Minute mutation frequencies (42/30,940=0·14% and 30/20,910=0·13%, respectively). Similar results are observed in crosses with the *ca* stock; progeny of *ca* male with *px* female crosses gave 1·04% (250/24,046) Minutes, whereas the reciprocal cross gave only 0·11% (26/22,857).

High frequencies of chromosome breakage are also observed in the progeny of crosses between either *ca* or *ca*; *stw* males and mutant stock females. In two sets of crosses with the *ca*; *stw* stock, ten translocations, one transposition, one inversion, and one deficiency were observed among 434 progeny (3·03%). On the other hand, only one translocation was observed in 600 progeny (0·17%) of the reciprocal cross. Five translocations were also observed among 205 progeny (2·44%) of crosses between *ca* males and marker females. All translocations involved the euchromatin of *ca* or ca; *stw* chromosomes, and breaks were of paternal origin.

From observations on backcross progeny of hybrid males and females, Hinton (1979, 1981) concluded that the mutator in the *ca* and *ca*; *stw* stocks is an extrachromosomally transmitted element whose propagation depends upon nuclear genes. The mutator functions by breaking paternal chromosomes, and its activity can be modified by a suppressor that functions in the oocyte soon after fertilization. This suppressor is chromosomally determined, but exhibits quasi-extrachromosomal transmission patterns, and it seems to function by repairing premutational lesions induced by the mutator. Neither the mutator nor the suppressor are infectious.

B. Characteristics of Other Mutator Systems from Natural Populations

There are reports of two other mutable lines of *D. melanogaster* from natural populations that should be noted. In 1937 Tiniakov (1939, 1941) isolated a mutable stock from a natural population in Khotkovo, USSR some 60 km from Moscow. From this stock, 786 mutant variants were observed among 50,437 offspring. Mutations were especially frequent at the yellow locus in males (57 mutants) and at the lozenge, forked, miniature and black loci. Somatic and germinal mutations were recovered, as were chromosome rearrangements. The cause of the high mutability was attributed to an autosomal factor.

Valadares (1937, 1940) has reported that a natural population line captured in Portugal in 1938 (see Tiniakov, 1941) underwent frequent chromosome breakage giving rise to inversions, deficiencies, translocations and duplications. In addition, 23 visible mutations and 34 gynandromorphs were recovered among 100,000 progeny.

V. Effect of Mobile DNA Elements on Spontaneous Mutation Rates

From purely genetic evidence, transposable elements (including insertion sequences, viral, and similar agents) have been repeatedly implicated as causal agents in a variety of *Dipteran* systems (see Mainx, 1964, 1966; Mainx and Doschek, 1967; and references in earlier sections). Mutator activity in the hybrid dysgenesis (*MR*) system is now being analysed extensively in this connection. Indeed, recent molecular evidence confirms that novel DNA sequences have been inserted into at least some of the genome sites where *MR*-induced mutation and chromosome breakage have occurred. This exciting finding may not only lead to a molecular explanation of such mutator systems, but may also allow the potential regulation of mutator activity (Thompson and Woodruff, 1981). In addition, it may offer a tool for cloning DNA sequences of selected interest using recombinant DNA technology (Spradling and Rubin, 1982).

The genetic basis of hybrid dysgenesis appears to be traceable to mobile chromosomal elements called "*P*" factors present in numerous locations in the genomes of the *P* strains tested so far, but absent in most *M* strains (Bingham *et al.*, 1982). Support for this hypothesis comes from many sources. Using *P* chromosomes isolated from diverse natural populations, Slatko (1978b), Matthews *et al.* (1978), and Yannopoulos (1979) have demonstrated that the genetic elements which cause male recombination in appropriate chromosomes can transpose from a *P* chromosome to an *M* chromosome homologue in the absence of exchange between closely linked genetic markers and can even transpose to nonhomologous chromosomes.

Hot-spots of mutational activity have long been associated with hybrid dysgenesis (Green, 1978; Golubovsky *et al.*, 1977; Simmons and Lim, 1980; Engels, 1981a; and Table VI). Hypermutability to other mutable or stable alleles or to wild type has led to the suggestion that the insertion of foreign DNA (and its frequent excision) plays a role in the mutational phenomena (Green, 1977a,b, 1978a,b, 1979; Golubovsky *et al.*, 1977; Thompson and Woodruff, 1978a, 1981; Engels, 1979a,b; Simmons *et al.*, 1980; Engels, 1981a). Hot-spots of chromosomal breakage have also been observed (Lim, 1979, 1980; Simmons and Lim, 1980; Berg *et al.*, 1980; Engels, 1981a, b and Figure 8). Replacement of part of a *P* chromosome with the homologous part of an *M* chromosome removes the chromosomal breakage hot-spot from that region while not affecting the hot-spots in other regions (Engels, 1981a; Raymond and Simmons, 1981). A similar result is obtained when male recombination induction is used as the assay (Hiraizumi and Slatko, unpublished observations). Novel hot-spots induced by hybrid dysgenesis in locations not present on the original P chromosome is further evidence for a genetic system involving moveable genetic sequences.

Several groups have suggested that the expression of hybrid dysgenesis is associated with the nomadic middle repetitive DNA sequences dispersed throughout the *Drosophila* genome (Bingham et al., 1982; Engels, 1981a; Thompson and Woodruff, 1981; Young and Schwartz, 1981). These families of dispersed middle repetitive DNA elements average 5000 to 6000 nucleotides in length and can wander to new chromosomal locations (Rubin et al., 1981; Spradling and Rubin, 1981; Tchurikov et al., 1981; Thompson and Woodruff, 1981; Truett et al., 1981; Young and Schwartz, 1981; and references therein). Each family of repetitive sequences may contain between approximately 10 and 100 closely related members. These families are nomadic in the sense that over 40% of the chromosomal locations of a given family may vary between pairs of strains of *D. melanogaster*, although there may be several preferred (shared) regions among several families. Comparing closely related species (e.g., *D. melanogaster* and *D. simulans*), strong sequence conservation is found for some families, while others appear to be unique to each species. In addition, the number of members in a given family may vary extensively among the two species. The sequences in *D. simulans* also locate to diverse genomic positions in different strains (Young and Schwartz, 1981).

Recent molecular evidence shows that mutants at the white, singed, lozenge, and vermillion loci contain newly inserted middle repetitive DNA elements and that revertants lose these inserts (Bregliano and Kidwell, 1982; Rubin et al., 1982). In three out of five cases, the inserted DNA sequences are members of a family of middle repetitive sequences which shares little or no homology with other nomadic DNA families and which appears to be present in P strains but absent in M strains in which the insertion mutation now resides (Bingham et al., 1982). In two of five cases, a sequence already present in both M and P strains (copia) was inserted into the locus. About 0·8 transpositions of the P factor per chromosome per generation occur to new locations in the genome in dysgenic hybrids. Rubin et al. (1981) report preliminary evidence that other middle repetitive sequences may be inserted through hybrid dysgenesis into other loci.

As an interesting aside, a recent series of observations suggests that a substantial fraction of the mutations in laboratory stocks also have insertions of DNA sequences (w^a, Bingham and Judd, 1981; ry and bx alleles, Bender, personal communication; see also references on p. 429).

As discussed by Bingham and Judd (1981), the relationship between P factors and nomadic DNA sequences provides the opportunity to clone any DNA sequence in or near a mutant induced by hybrid dysgensis (Bingham and Judd, 1981; Rubin and Spradling, 1982; Spradling and Rubin, 1982). Some mutants may, however, be difficult to obtain in this manner. Alcohol dehydrogenase mutants, for example, were not generated by hybrid dysgenesis despite extensive attempts (Bencze and Green, unpublished observations; Craymer, personal communication). A similar failure has occurred for the cyclic AMP phosphodiesterase locus (Slatko, Davis, and Kiger, unpublished observations), although only a limited number of chromosomes was examined. These failures probably reflect the low specificity at these loci required for *P* factor insertion.

Transposable element systems in other organisms provide some insight into the direction *Drosophila* studies might lead. Where molecular evidence is available, some of these elements appear to share homologies with integrated DNA copies (provirus) of RNA retrovirus sequences (Shimotohno *et al.*, 1980; McClements *et al.*, 1981). These sequences can occupy various chromosomal locations, are similar in the organization of their nucleotide sequences, and produce transcripts which, in the case of the retroviral systems, are necessary for further integration. Retroviruses use circular DNA intermediates as integrative precursors, and in *Drosophila* nuclear covalently closed circular middle repetitive DNA has been observed in standard strains and in cell cultures (Stanfield and Lengyel, 1979). Similar circles have been found in *P* strains and the distribution of sizes appears to be identical to that in wild type strains (Lee and Slatko, unpublished observations).

Recently, retroviral sequences have been observed flanking "pseudogenes" which are related, but not linked, to the cluster of functional alpha-globin genes in mice and hamsters (Leuders *et al.*, 1982). These pseudogenes are missing the intervening sequences characteristic of DNA sequences in functional globin genes and correspond to the mature messenger RNA molecule. While making a DNA copy for integration into the chromosome, a retrovirus encompassing the RNA product of a functional gene could carry with it the new pseudogene, thus depositing it into the chromosome in any location.

These interesting examples are ample evidence of the fascinating insights that molecular studies promise for the future.

VI. Conclusions

If anything is clear from this survey of influences upon mutation, it is that variation in spontaneous mutation rates provides fertile ground for future work in population genetics. The greatest promise comes from studies of genetic factors. There is no doubt that genetic factors in natural populations have the potential of inducing dramatic changes in spontaneous mutation frequencies. When and how that potential is expressed is largely still a mystery.

Part of the problem is that evolutionary geneticists have generally ignored variation in mutation rates, and studies of mutator activity have often ended as little more than a curious observation about a localized population. Investigators have focused upon different aspects of mutation, and thus conclusions and proposed mechanisms commonly differ in important ways. Indeed, whether all of these genetic systems are identical may never be known with certainty. But some reasonable generalizations can be made.

There is a strong correlation between chromosome breakage and mutation in recently studied *MR* lines, and both events show a reciprocal cross effect. In addition, mutator activity only rarely seems to occur in isolated population samples, but it is stimulated by crosses between some wild populations or between a mutator line and a laboratory stock. The idea that mutator activity is usually suppressed within populations, but can be induced under certain conditions, suggests that mutation rates can vary markedly at different times and in different parts of a population (c.f., Thompson and Woodruff, 1980; Woodruff and Thompson, 1980). Sturtevant (1937) was one of the first to suggest that genes which increase mutation rates in natural populations will be subject to selection and that any interference with the operation of this selection will increase the mutation rate.

Ives (1950), for example, believed that *hi* may function in natural populations as a dominant or a recessive, but emphasized the involvement of additional modifiers. Different rates of mutation may be due more to the action of modifiers than to the presence or absence of mutators. This prediction is important for the maintenance of mutators in natural populations, since as a dominant the mutator would constantly be active, while as a recessive it would not be selected against efficiently. Ives further argued that since selection against mutators would probably be less in larger populations, more mutators would be present there. Mutation rates in large populations should, consequently, be higher than those in small populations. He supported this prediction by citing data from Olenov *et al.* (1939), Zuitin and Pavlovetz (1940), and Zuitin (1941) in which newly derived natural population lines have higher frequencies of mutation than do standard laboratory stocks. The same observations have more recently been made on *MR* mutator activity, which is much more frequent in newly isolated wild lines than in laboratory strains (Kidwell *et al.*, 1977b; Woodruff and Thompson, 1980). An explanation of the difference between laboratory and wild strains should contribute to our understanding of the suppression and activation of mutators in nature.

The hypothesis that mutator activity can be released in crosses between flies from different wild populations has been tested by Thompson and Woodruff (1980). For example, the frequency of spontaneous recessive sex-linked lethal mutations is increased by more than ten-fold in the hybrid male progeny of crosses between some natural population lines (Woodruff and Thompson, unpublished observations). In such situations, increased mutation, chromosome breakage and sterility would be expected to contribute to greater isolation between subpopulations. Mutator effects would also be enhanced in the presence of other factors discussed in this chapter, such as defective repair enzymes in the population.

Examples of explosive increases in the frequency of mutations have been documented in several different populations (Spencer, 1935; Alexander, 1949; Berg, 1966, 1972, 1974; Golubovsky, 1980). Similarly, increased mutability has been observed after interspecific crosses of *D. persimilis* with *D. pseudoobscura* (Sturtevant, 1939) and of *D. melanogaster* with *D. simulans* (Belgovsky, 1937). In addition, Timofeeff-Ressovsky (1932) observed a sharp difference in the mutability level of two alleles of the white locus of *D. melanogaster* in strains collected from distant populations (Golubovsky, personal communication). Mutators have even been considered potential hazards for population stability (Morgan *et al.*, 1973). Further studies of localized variations in mutation rates will undoubtedly be well repaid, particularly in specialized situations such as hybrid zones (Hunt and Selander, 1973; Gould and Woodruff, 1978; Sage and Selander, 1979).

For mutation rate as for other topics in population genetics, however, it is no longer sufficient to describe phenomena in terms of population frequencies alone. For their influences to be fully understood and predicted, it is necessary to explain phenomena at a molecular level (see Green, 1977a,b). One of the most promising hypotheses to explain *MR* mutator activity is that it is caused by DNA sequences that transpose from one chromosomal site to another. If a nomadic DNA segment becomes reversibly inserted into a gene, an unstable mutation like those described for several *MR* systems may result (see Bingham *et al.*, 1982; Collins and Rubin, 1982; Karess and Rubin, 1982; Rubin *et al.*, 1982). Hybridization may, in turn, have a direct or indirect effect upon the rate of movement of these sequences (Thompson and Woodruff, 1981).

To this point in the development of population genetics, advances have occurred by making simplifying assumptions about population dynamics and the gene pool. This allows one to focus upon the principal factors influencing variation. But there is really nothing simple about natural populations. This is certainly true of mutation rates. But given the awareness of hidden complexities and the tools to analyse them, the challenges of understanding spontaneous mutation in natural populations will undoubtedly be met.

ACKNOWLEDGEMENTS

We would like to thank the following for comments on the manuscrupt: M. Ashburner, R. K. Brodberg, H. L. Carson, W. R. Engels, M. D. Golubovsky, M. M. Green, C. W. Hinton, M. G. Kidwell, M. Levitan, R. F. Lyman, J. M. Mason, K. Matthews, S. Minamori, G. Picard, N. Scobie, M. J. Simmons and G. Yannopoulos.

RCW was supported by NSF grants DEB-7923007 and DEB-8117063, and a NIEHS Research Career Development Award (K04-ES00087); BES was supported by Williams College Discretionary Funds and Research Corporation Funds; JNT was supported by NIH grant GM24809 and a Biomedical Sciences Support Grant from the University of Oklahoma.

References

NOTE: For additional references to this chapter, please see page 429.

ABRAHAMSON, S. and HIMOE, E. (1963). Induced mutation rate in sperm transmitted to sons and daughters in *Drosophila melanogaster*. *Genetics* 48, 1085–1087.

ABRAHAMSON, S. and LEWIS, E. B. (1971). The detection of mutations in *Drosophila melanogaster*. *In*: "Chemical Mutagens", (A. Hollaender, Ed.), Vol. 2, pp. 461–487. Plenum Press, New York.

ABRAHAMSON, S., MEYER, H. U., HIMOE, E. and DANIEL, G. (1966). Further evidence demonstrating germinal selection in early premeiotic germ cells of *Drosophila* males. *Genetics* 54, 687–696.

ABRAHAMSON, S., WURGLER, F. E., DEJONGH, C. and MEYER, H. V. (1980). How many loci on the X-chromosome of *Drosophila melanogaster* can mutate to recessive lethals? *Environmental Mutagenesis* 2, 447–453.

ALEXANDER, M. L. (1949). Note on gene variability in natural populations of *Drosophila*. Univ. Texas Publ. 4920, 63–69.

ALEXANDER, M. L. (1960). Radiosensitivity at specific autosomal loci in mature sperm and spermatogonial cells of *Drosophila melanogaster*. *Genetics* 45, 1019–1022.

ALTENBURG, L. S. and ALTENBURG, E. (1952). The lowering of the mutagenic effectiveness of ultraviolet by photoreactivating light in *Drosophila*. *Genetics* 37, 545–553.

ALTENBURG, E. and BROWNING, L. S. (1962). Evidence for the duplicational origin of reverse mutations at the forked locus in *Drosophila*. *Genetics* 47, 938.

ALTENBURG, E., BERGENDAHL, J. and ALTENBURG, L. S. (1952a). The non-effect of intensity on the mutagenesis of ultraviolet light within a nineteen-fold range in *Drosophila*. *Genetics* 37, 554–557.

ALTENBURG, L. S., ALTENBURG, E. and BAKER, R. N. (1952b). Evidence indicating that the mutation rate induced in *Drosophila* by low doses of ultraviolet light is an exponential function of the dose. *Genetics* 37, 558–561.

ANGUS, D. S. and RAISBECK, J. A. (1979). A transmissible factor involved in hybrid sterility in *Drosophila melanogaster*. *Genetica* 50, 81–87.

ASHBURNER, M. and LEMEUNIER, F. (1976). Relationships within the *melanogaster* species subgroup of the genus *Drosophila* (Sophophora). I. Inversion polymorphisms in *Drosophila melanogaster* and *Drosophila simulans*. *Proc. Roy. Soc. Lond. B.* **193**, 137–157.

AUERBACH, C. (1941). The effect of sex on the spontaneous mutation in *Drosophila melanogaster*. *J. Genet.* **41**, 255–265.

AUERBACH, C. (1962). "Mutation: An Introduction to Research on Mutagenesis. Part I: Methods". Oliver and Boyd, Edinburgh.

AUERBACH, C. (1976). "Mutation Research". Chapman and Hall, London.

BAARS, A. J., ZIJLSTRA, J. A., VOGEL, E. and BREIMER, D. D. (1977). The occurrence of cytochrome P-450 and aryl hydrocarbon hydroxylase activity in *Drosophila melanogaster* microsomes, and the importance of this metabolizing capacity for the screening of carcinogenic and mutagenic properties of foreign compounds. *Mutation Res.* **44**, 257–268.

BAARS, A. J., BLIJLEVEN, W. G. H., MOHN, G. R., NATARAJAN, A. T. and BREIMER, D. D. (1980). Preliminary studies on the ability of *Drosophila* microsomal preparations to activate mutagens and carcinogens. *Mutation Res.* **72**, 257–264.

BAKER, B. and CARPENTER, A. (1972). Genetic analysis of sex chromosomal meiotic mutants in *Drosophila melanogaster*. *Genetics* **71**, 255–286.

BAKER, B. and SMITH, D. (1979). The effect of mutagen sensitive mutants of *Drosophila melanogaster* in non-mutagenized cells. *Genetics* **92**, 833–844.

BAKER, B., BOYD, J., CARPENTER, A., GREEN, M., NGUYEN, T., RIPOLL, P. and SMITH, P. (1976a). Genetic controls of meiotic recombination and somatic DNA metabolism in *Drosophila melanogaster*. *Proc. Nat. Acad. Sci.* **73**, 4140–4144.

BAKER, B., CARPENTER, A., ESPOSITO, M., ESPOSITO, R. and SANDLER, L. (1976b). The genetic control of meiosis. *Ann. Rev. Genet.* **10**, 53–134.

BAKER, B., CARPENTER, A. T. C. and RIPOLL, P. (1978). The utilization during mitotic cell division of loci controlling meiotic recombination and disjunction in *Drosophila melanogaster*. *Genetics* **90**, 531–578.

BAKER, B., GATTI, M., CARPENTER, A. T. C., PIMPINELLI, S. and SMITH, D. A. (1980). Effects of recombination-deficient and repair-deficient loci on meiotic and mitotic chromosome behavior in *Drosophila melanogaster*. In: "DNA Repair and Mutagenesis in Eukaryotes", (W. M. Generoso, M. D. Shelby and F. J. de Serres, Eds.), pp. 189–208. Plenum Press, New York.

BAKER, W. K. (1957). Induced loss of a ring and telomeric chromosome in *Drosophila melanogaster*. *Genetics* **42**, 735–748.

BALLANTYNE, G. H. and CHOVNICK, A. (1971). Gene conversion in higher organisms: Non-reciprocal recombination events at rosy cistron in *Drosophila melanogaster*. *Genet. Res. Camb.* **17**, 139–149.

BANERJEE, J., HAZRA, S. K. and SEN, S. K. (1978). EMS-induced reversion studies in the white locus of *Drosophila melanogaster*. *Mutation Res.* **50**, 309–315.

BAUER, H., DEMEREC, M. and KAUFMANN, B. P. (1938). X-ray induced chromosomal alterations in *Drosophila melanogaster*. *Genetics* **23**, 610–630.

BAUMILLER, R. (1967). Virus induced point mutation. *Nature, Lond.* **214**, 806–807.

BAUMILLER, R. C. and LEWIS, K. (1968). Effect of sex and genotype on virus induced mutation: localization of mutations. *Proc. Int. Cong. Genet.* **12**, 87.

BELGOVSKY, M. L. (1937). A comparison of the frequency of induced mutations in *Drosophila simulans* and in its hybrid with *Drosophila melanogaster*. *Genetica* **19**, 370–386.

BERCHTOLD, W. (1975). Comparison of the Kastenbaum-Bowman test and Fisher's exact test. *Archiv. Genetik.* **48**, 151–157.

BERG, R. L. (1966). Studies of mutability in geographically isolated populations of *Drosophila melanogaster*. In: "Mutation in Population", (R. Honcariv, Ed.), pp. 61–74. Prague, Czechosl. Acad. Sci.

BERG, R. L. (1972). A sudden and synchronous increase in the frequency of "abnormal" abdomen (aa) in geographically isolated populations of *D. melanogaster*. *Dros. Inf. Serv.* 48, 67–69.

BERG, R. L. (1974). A simultaneous mutability rise at the singed locus in two out of three *Drosophila melanogaster* population study in 1973. *Dros. Inf. Serv.* 51, 100–102.

BERG, R., ENGELS, W. R. and KREBER, R. A. (1980). Site-specific X-chromosome rearrangements from hybrid dysgenesis in *Drosophila melanogaster*. *Science* 210, 427–429.

BINGHAM, P. M. and JUDD, B. H. (1981). A copy of the *copia* transposable element is very tightly linked to the *wa* allele at the *white* locus of *D. melanogaster*. *Cell* 25, 705–711.

BINGHAM, P. M., KIDWELL, M. G. and RUBIN, G. M. (1982). The molecular basis of P–M hybrid dysgenesis: the role of the P element, a P-strain-specific transport family. *Cell* 29, 995–1004.

BOWMAN, J. T. (1965). Spontaneous reversion of the white-ivory mutant of *Drosophila melanogaster*. *Genetics* 52, 1069–1079.

BOWMAN, J. T. (1969). Spontaneous reversion of the white-ivory mutant of *Drosophila*. *Mutation Res.* 7, 409–415.

BOYD, J. B. and SETLOW, R. B. (1976). Characterization of post-replication repair in mutagen-sensitive strains of *Drosophila melanogaster*. *Genetics* 84, 507–526.

BOYD, J. B., GOLINO, M., NGUYEN, T. and GREEN, M. M. (1976a). Isolation and characterization of X-linked mutants of *Drosophila melanogaster* which are sensitive to mutagens. *Genetics* 84, 485–506.

BOYD, J. B., GOLINO, M. and SETLOW, R. (1976b). The mei-9 mutant of *Drosophila melanogaster* increases mutagen sensitivity and decreases excision repair. *Genetics* 84, 527–544.

BOYD, J. B., HARRIS, P., OSGOOD, C. and SMITH, K. (1980). Biochemical characterization of repair deficient mutants of *Drosophila*. *In*: 'DNA Repair and Mutagenesis in Eukaryotes', (A. Hollaender *et al.*, Eds.). (In Press).

BOYD, J. B., GOLINO, M., SHAW, K. E. S., OSGOOD, C. J. and GREEN, M. M. (1981). Third-chromosome mutagen-sensitive mutants of *Drosophila melanogaster*. *Genetics* 97, 607–623.

BREGLIANO, J. C., PICARD, G., BUCHETON, A., PELISSON, A., LAVIGE, J. M. and L'HERITIER, P. (1980). Hybrid dysgenesis in *Drosophila melanogaster*. *Science* 207, 606–611.

BROWNING, L. S. (1969). The mutational spectrum produced in *Drosophila* by N-methyl-N-nitro-N-nitrosoguanidine. *Mutation Res.* 8, 157–164.

BRUN, G. and PLUS, N. (1980). The viruses of *Drosophila*. *In*: "The Genetics and Biology of *Drosophila*", (M. Ashburner and T. R. F. Wright, Eds.), Vol. 2d, pp. 625–693. Academic Press, London and New York.

BUCHETON A. and PICARD, G. (1976). Non-Mendelian female sterility in *Drosophila melanogaster*: Hereditary transmission of reactivity levels. *Heredity* 40, 207–223.

BURDETTE, W. and YOON, J. (1967). Mutations, chromosomal aberrations and tumors in insects treated with oncogenic virus. *Science* 155, 340–341.

BYERS, H. L. (1954). Thermal effects on the spontaneous mutation rate in mature spermatozoa of *Drosophila melanogaster*. Caryologia 6: Suppl. Part 1, pp. 694–696.

BYERS, H. L. and MULLER, H. J. (1952). Influence of ageing at two different temperatures on the spontaneous mutation rate in mature spermatozoa of *Drosophila melanogaster*. *Genetics* 37, 570–571.

CANNON, G. B. (1966). Intraspecies competition, viability and longevity in experimental populations. *Evolution* 20, 117–131.

CARDELLINO, R. A. and MUKAI, T. (1975). Mutator factors and genetic variance components of viability in *Drosophila melanogaster*. *Genetics* **80**, 567–583.

CARLSON, E. A. and SOUTHIN, J. L. (1962). Comparative mutagenesis of the dumpy locus in *Drosophila melanogaster*. I. X-ray treatment of mature sperm-frequency and distribution. *Genetics* **47**, 321–336.

CARSON, H. L. (1958). The population genetics of *Drosophila* robusta. *Adv. Genet.* **9**, 1–40.

CARSON, H. L. (1965). Chromosomal morphism in geographically widespread species of *Drosophila*. *In*: "The Genetics of Colonizing Species", pp. 503–531, Academic Press, New York.

CHIGUSA, S., METTLER, L. E. and MUKAI, T. (1969). Characteristics of a southern population of *Drosophila melanogaster*. *Genetics* **61**, s10.

CLARK, C. (1969). Influence of cell physiology on the realization of mutations—results and problems. *In*: "Mutation as a Cellular Process", (G. Wolstenholme and M. O'Connor, Eds.), pp. 17–28. Churchill, Ltd. Publ. Co., London.

CLARK, A. J. (1973). Recombination deficient mutants of *E. coli* and other bacteria. *Ann. Rev. Genet.* **7**, 67–86.

COLGAN, D. J. and ANGUS, D. S. (1978). Bisexual hybrid sterility in *Drosophila melanogaster*. *Genetics* **89**, 5–14.

COLLINS, M. and RUBIN, G. M. (1982). Structure of the Drosophila mutable allele, *white-crimson*, and its *white-ivory* and wild-type derivatives. *Cell* **30**, 71–79.

CORDEIRO, A. R., MARQUES, E. K. and VEIGA-NETO, A. J. (1973). Radioresistance of a natural population of *Drosophila willistoni* living in a radioactive environment. *Mutation Res.* **19**, 325–329.

CRAYMER, L. (1974) A new genetic testing procedure for potential mutagens. *Dros. Inf. Serv.* **51**, 62.

CROW, J. F. and ABRAHAMSON, S. (1971). *Drosophila* methods for mutagenicity testing. *In*: "Drugs of Abuse", (S. S. Epstein, Ed.), MIT Press, Cambridge, Massachusetts.

CROW, J. F. and TEMIN, R. G. (1964). Evidence for the partial dominance of recessive lethal genes in natural populations of *Drosophila*. *Am. Nat.* **98**, 21–33.

DEMARCO, A., BELLONI, M., COZZI, R. and OLIVIERI, G. (1975). Environmental mutagens and environmental factors that can modify their action. *Mutation Res.* **29**, 253.

DEMEREC, M. (1937). Frequency of spontaneous mutations in certain stocks of *Drosophila melanogaster*. *Genetics* **22**, 469–478.

DOBZHANSKY, TH. (1951). "Genetics and the Origin of Species". Columbia University Press, New York.

DOBZHANSKY, TH., AYALA, F. J., STEBBINS, G. L. and VALENTINE, J. W. (1977). "Evolution". W. H. Freeman and Company, San Francisco.

DRAKE, J. W. (1969). Comparative rates of spontaneous mutations. *Nature, Lond.* **221**, 1132.

DUBININ, N. P. (1946). On lethal mutations in natural populations. *Genetics* **31**, 21–38.

DUBININ, N. P. (1964). "Problems of Radiation Genetics". Oliver and Boyd, London.

DUBININ, N. P. and SIDOROV, B. N. (1934). Relation between the effect of a gene and its position in the system. *Am. Nat.* **68**, 377–381.

EEKEN, J. C. J. and SOBELS, F. H. (1981). Modification of MR mutator activity in repair-deficient strains of *Drosophila melanogaster*. *Mutation Res.* **83**, 191–200.

EGGLESTON, P. and KEARSEY, M. J. (1980). Hybrid dysgenesis in *Drosophila*: Correlation between dysgenic traits. *Heredity* **44**, 237–249.

ENGELS, W. R. (1979a). Extrachromosomal control of mutability in *Drosophila melanogaster*. *Proc. Natl. Acad. Sci. U.S.A.* **76**, 4011–4015.

ENGELS, W. R. (1979b). Germ line aberrations associated with a case of hybrid dysgenesis in *Drosophila melanogaster* males. *Genet. Res., Camb.* **33**, 137–146.

ENGELS, W. R. (1979c). Hybrid dysgenesis in *Drosophila melanogaster*: Rules of inheritance of female sterility. *Genet. Res., Camb.* 33, 219–236.
ENGELS, W. R. (1979d). The estimation of mutation rates when premeiotic events are involved. *Environmental Mutagenesis* 1, 37–43.
ENGELS, W. R. (1981a). Hybrid dysgenesis in *Drosophila* and the stochastic loss hypothesis. *Cold Spring Harbor Symp. Quant. Biol.* 45, 561–565.
ENGELS, W. R. (1981b). Identifying P factors in *Drosophila* by means of chromosome breakage hotspots. *Cell* 26, 421–428.
ENGELS, W. R. and PRESTON, C. R. (1979). Hybrid dysgenesis in *Drosophila melanogaster*: The biology of female sterility. *Genetics* 92, 161–174.
ENGELS, W. R. and PRESTON, C. R. (1980). Components of hybrid dysgenesis in a wild population of *Drosophila melanogaster*. *Genetics* 95, 111–128.
ERICKSON, J. (1965). Meiotic drive in *Drosophila* involving chromosomal breakage. *Genetics* 51, 555–571.
FAHMY, O. G. and FAHMY, M. J. (1959). Differential gene response to mutagens in *Drosophila melanogaster*. *Genetics* 44, 1149–1171.
FAHMY, O. G. and FAHMY, M. J. (1966). The nature and distribution of the mutations induced by unirradiated and irradiated heterologous deoxyribonucleic acid in *Drosophila melanogaster*. *Genetics* 54, 1123–1138.
FAHMY, O. G. and FAHMY, M. J. (1969). Cytotoxic and mutagenic activation of urethane by N-hydroxylation and O-esterification. *Chem. Biol. Interactions* 1, 257–270.
FAHMY, O. G. and FAHMY, M. J. (1970). Hydroxylamine and derivatives: cytotoxicity without mutagenicity in cellular genetic systems. *Chem. Biol. Interactions* 2, 331–348.
FLEURIET, A. (1980). Polymorphism of the hereditary sigma virus in natural populations of *Drosophila melanogaster*. *Genetics* 95, 459–465.
FOUREMAN, P. A. (1979). A translocation X;Y system for detecting meiotic nondisjunction and chromosome breakage in males of *Drosophila melanogaster*. *Environmental Health Perspectives* 31, 53–58.
FRANCA, Z. M., DACUNHA, A. B. and GARRIDO, M. C. (1968). Recombination in *Drosophila willistoni*. *Heredity* 23, 199–204.
FUJIKAWA, K. (1980). Relative sensitivity of mature *Drosophila* oocytes to the induction of *dumpy* mutations by X-rays. *Jap. J. Genet.* 55, 409–413.
FUJIKAWA, K. and INAGAKI, E. (1979). Mutagenic effectiveness of 14.1 MeV neutrons and 200 kV X-rays at the *dumpy* complex locus of *Drosophila melanogaster*. *Mutation Res.* 63, 139–146.
FUJIKAWA, K, NISHIMORI, T. and MIYAMOTO, T. (1975). Radiation-induction of fractional mutations in *Drosophila*. *Mutation Res.* 30, 283–288.
FUTUYMA, D. J. (1979). "Evolutionary Biology". Sinauer Assoc., Inc., Sunderland, MA.
GAME, J., ZAMB, T., BRAUN, R., RESNICK, M. and ROTH, R. (1980). The role of radiation (rad) genes in meiotic recombination in yeast. *Genetics* 94, 51–68.
GATTI, M. (1979). Genetic control of chromosome breakage and rejoining in *Drosophila melanogaster*: Spontaneous chromosome aberrations in X-linked mutants defective in DNA metabolism. *Proc. Natl. Acad. Sci. U.S.A.* 76, 1377–1381.
GATTI, M., PIMPINELLI, S. and BAKER, B. S. (1980). Relationship among chromatid interchanges, sister chromatid exchanges, and meiotic recombination in *Drosophila melanogaster*. *Proc. Natl. Acad. Sci. U.S.A.* 77, 1575–1579.
GEER, B. and RENO, D. (1976). The effects of dietary deficiencies for choline and nicotinic acid on the sensitivity of *Drosophila melanogaster* to mutagenic treatment. *Mutation Res.* 38, 407–408.

GEHRING, W. J. and PARO, R. (1980). Isolation of a hybrid plasmid with homologous sequences to a transposing element of *Drosophila melanogaster*. *Cell* **91**, 897–904.

GERASIMOVA, T. I. (1981). Genetic instability at the *ct* locus of *Drosophila melanogaster* induced by the *MR-h12* chromosome. *Molec. Gen. Genet.* **184**, 544–547.

GERSHENSON, S. M., ALEXANDROV, Y. and MALIUTA, S. (1970). Production of recessive lethals in *Drosophila* by viruses non-infectious for the host. *Mutation Res.* **11**, 163–173.

GERSHENSON, S. M., ALEXANDROV, Y. N. and MALIUTA, S. S. (1975). Mutagenic action of DNA and viruses in *Drosophila*. Naukova Dumka, Kiev (quoted in Golubovsky, 1980).

GLASS, B. (1955). A comparative study of induced mutation in oocytes and spermatozoa of *Drosophila melanogaster*. I. Translocations and inversions. *Genetics* **40**, 252–267.

GLASS, B. and RITTERHOF, R. (1956). Spontaneous mutation rates at specific loci in *Drosophila* males and females. *Science* **124**, 314–315.

GOLDSCHMIDT, R. (1939). Mass mutation in the Florida stock of *Drosophila melanogaster*. *Am. Nat.* **73**, 547–559.

GOLDSCHMIDT, R. (1945). A study of spontaneous mutation. *Univ. Calif. Publ. Zool.* **49**, 291–550.

GOLUBOVSKY, M. D. (1980). Mutational process and microevolution. *Genetica* **52/53**, 139–149.

GOLUBOVSKY, M. D. and PLUS, N. (1982) Mutability studies in two *Drosophila melanogaster* isogenic stocks, endemic for C Picornavirus and virus-free. *Mutation Res.* **103**, 29–32.

GOLUBOVSKY, M. D., IVANOV, YU. N., ZAKHAROV, I. K. and BERG, R. L. (1974). Investigation of synchronous and similar changes of the gene pool in geographically separated natural populations of *Drosophila melanogaster*. *Genetika* **10**, 72–83.

GOLUBOVSKY, M. D., IVANOV, YU. N. and GREEN, M. M. (1977). Genetic instability in *Drosophila melanogaster*: Putative multiple insertion mutants at the singed bristle locus. *Proc. Natl. Acad. Sci. U.S.A.* **74**, 2973–2975.

GOULD, S. J. and WOODRUFF, D. S. (1978). Natural history of Cerion VIII: Little Bahama Bank—A revision based on genetics, morphometrics and geographic distribution. *Bull. Mus. Comp. Zool.* **148**, 371–415.

GRACE, D., CARLSON, E. A. and GOODMAN, P. (1968). *Drosophila melanogaster* treated with LSD: Absence of mutation and chromosome breakage. *Science* **161**, 694–696.

GRAF, U. (1972). Spontaneous mutations in *Drosophila melanogaster*. *Humangenetik* **16**, 27–32.

GRAF, U and WÜRGLER, F. E. (1976). MMS sensitive strains in *Drosophila melanogaster*. *Mutation Res.* **34**, 251–258.

GRAF, U. and WÜRGLER, F. E. (1978). Mutagen-sensitive mutants in *Drosophila*: Relative MMS sensitivity and maternal effects. *Mutation Res.* **52**, 381–394.

GRAF, U., GREEN, M. M. and WÜRGLER, F. E. (1979). Mutagen-sensitive mutants in *Drosophila melanogaster*. Effects on premutational damage. *Mutation Res.* **63**, 101–112.

GREEN, M. M. (1957). Reverse mutation in *Drosophila* and the status of the particulate gene. *Genetics* **29**, 1–38.

GREEN, M. M. (1961). Back mutation in *Drosophila melanogaster*. I. X-ray-induced back mutations at the *yellow*, *scute* and *white* loci. *Genetics* **46**, 671–682.

GREEN, M. M. (1976). Mutable and mutator loci. *In*: "The Genetics and Biology of *Drosophila*", (M. Ashburner and E. Novitski, Eds.), Vol. 1b, pp. 929–946, Academic Press, London and New York.

GREEN, M. M. (1977a). The case for DNA insertion mutations in *Drosophila*. *In*: "DNA Insertion Elements, Plasmids and Episomes", (A. T. Bukhari, J. A. Shapiro and S. L.

Adhya, Eds.), pp. 437–443, Cold Spring Harbor Laboratory, Cold Spring Harbor, New York.
GREEN, M. M. (1977b). Genetic instability in *Drosophila melanogaster*: Denovo induction of putative insertion mutations. *Proc. Natl. Acad. Sci. U.S.A.* **74**, 3490–3493.
GREEN, M. M. (1978a). Insertion mutants and the control of gene expression in *Drosophila melanogaster*. In: "The Clonal Basis of Development", (S. Subtelny and I. M. Sussex, Eds.), Academic Press, Inc., New York and London.
GREEN, M. M. (1978b). The genetic control of mutation in *Drosophila*. *Stadler Symp.* **10**, 95–104.
GREEN, M. M. (1979). Genetic instability in *Drosophila melanogaster*: Presumptive insertion mutants at the yellow locus. *Mutation Res.* **59**, 291–293.
GREEN, M. M. and SHEPHERD, S. H. Y. (1979). Genetic instability in *Drosophila melanogaster*: The induction of specific chromosome 2 deletions by MR elements. *Genetics* **92**, 823–832.
GREEN, M. M. and SLATKO, B. (1979). Genetic instability in *Drosophila melanogaster*: Dosage and mutator activity of an MR chromosome. *Mutation Res.* **62**, 529–531.
HARTL, D. L. and HIRAIZUMI, Y. (1976). Segregation distortion. In: "The Genetics and Biology of *Drosophila*, (M. Ashburner and E. Novitski, Eds.), Vol. 1b, pp. 615–666, Academic Press, London and New York.
HELLACK, J. J., THOMPSON, J. N., JR., WOODRUFF, R. C. and HISEY, B. N. (1978). Male recombination and mosaics induced in *Drosophila melanogaster* by feeding. *Experientia* **34**, 447.
HENDERSON, S. A., WOODRUFF, R. C. and THOMPSON, J. N., JR. (1978). Spontaneous chromosome breakage at male meiosis associated with male recombination in *Drosophila melanogaster*. *Genetics* **88**, 93–107.
HERSKOWITZ, I. (1963). An influence of maternal nutrition upon the gross chromosomal mutation frequence recovered from the X-rayed sperm of *Drosophila melanogaster*. *Genetics* **48**, 703–710.
HILDRETH, P. E. and CARSON, G. L. (1957). Influences of the type of inseminated females on the lethal frequency in the X-chromosome from the male. *Proc. Natl. Acad. Sci. U.S.A.* **43**, 175–183.
HINTON, C. W. (1974). An extrachromosomal suppressor male crossing over in *Drosophila ananassae*. In: "Mechanisms in recombination", (R. F. Grell, Ed.), pp. 391–397, Plenum Press, New York.
HINTON, C. W. (1979). Two mutators and their suppressors in *Drosophila ananassae*. *Genetics* **92**, 1153–1171.
HINTON, C. W. (1981). Nucleocytoplasmic relations in a mutator-suppressor system of *Drosophila melanogaster*. *Genetics* **98**, 77–90.
HINTON, T., IVES, P. T. and EVANS, A. T. (1952). Changing the gene order and number in natural populations. *Evolution* **6**, 19–28.
HIRAIZUMI, Y. (1971). Spontaneous recombination in *Drosophila melanogaster* males. *Proc. Natl. Acad. Sci. U.S.A.* **68**, 268–270.
HIRAIZUMI, Y. (1977). The relationships among transmission frequency, male recombination and progeny production in *Drosophila melanogaster*. *Genetics* **87**, 83–93.
HIRAIZUMI, Y. (1979a). A new method to distinguish between meiotic and premeiotic recombinational events in *Drosophila melanogaster*. *Genetics* **92**, 543–554.
HIRAIZUMI, Y. (1979b). A model of the negative correlation between male recombination and transmission frequency in *Drosophila melanogaster*. *Genetics* **93**, 449–459.
HIRAIZUMI, Y., SLATKO, B., LANGLEY, C. and NILL, A. (1973). Recombination in *Drosophila melanogaster* males. *Genetics* **73**, 439–444.

HOLM, D. G. (1976). Compound autosomes. *In*: "The Genetics and Biology of *Drosophila* (M. Ashburner and E. Novitski, Eds.), Vol. 1b., pp. 529–561, Academic Press, London and New York.

HUANG, S. L. and BAKER, B. S. (1976). The mutability of *minute* loci of *Drosophila melanogaster* with ethyl methanesulfonate. *Mutation Res.* **34**, 407–414.

HUNT, W. G. and SELANDER, R. K. (1973). Biochemical genetics of hybridization in European house mice. *Heredity* **31**, 11–33.

INAGAKI, E. T., MIGAMOTO, T. and DOMOTO, T. (1974). The relationship between radiation exposure and mutation rate at the *dumpy* locus in *Drosophila*. *Jap. J. Genet.* **49**, 373–378.

ISACKSON, D. R., JOHNSON, T. K. and DENELL, R. E. (1981). Hybrid dysgenesis in *Drosophila*: The mechanism of *T-007*-induced male recombination. *Molec. Gen. Genet.* **184**, 539–543.

ISING, G. and RAMEL, C. (1976). The behaviour of a transposing element in *Drosophila melanogaster*. *In*: "The Genetics and Biology of *Drosophila*", (M. Ashburner and E. Novitski, Eds.), Vol. 1b, pp. 947–954, Academic Press, London and New York.

ITO, K. (1974). Mutagenicity of an extrachromosomal element delta for X chromosomes in *Drosophila melanogaster*. *Jap. Genet.* **49**, 25–31.

IVES, P. T. (1945). The genetic structure of American populations of *Drosophila melanogaster*. *Genetics* **30**, 167–196.

IVES, P. T. (1950). The importance of mutation rate genes in evolution. *Evolution* **4**, 236–252.

IVES, P. T. (1959a). Chromosomal distribution of mutator– and radiation–induced mutations in *Drosophila melanogaster*. *Evolution* **13**, 526–531.

IVES, P. T. (1959b). The relationship between radiation dose and dominant visible mutation rate in *Drosophila melanogaster*. *Genetics* **44**, 967–978.

IVES, P. T. (1963). Patterns of spontaneous and radiation induced mutation rates during spermiogenesis in *Drosophila melanogaster*. *Genetics* **48**, 981–996.

JENKINS, J. B. (1967). Mutagenesis at a complex locus in *Drosophila* with the monofunctional alkylating agent ethyl methanesulfonate. *Genetics* **57**, 783–793.

JOHNSTON, D. and WINCHESTER, A. M. (1934). Studies on reverse mutation in *Drosophila melanogaster*. *Am. Nat.* **68**, 351–358.

JOLLOS, V. (1937). Some attempts to test the role of cosmic radiation in the production of mutations in *Drosophila melanogaster*. *Genetics* **22**, 534–542.

JUPIN, N., PLUS, N. and FLEURIET, A. (1968). Action d'une souche de virus sigma sur la fertilite des Drosophiles femelles. *Ann. Inst. Pasteur* **114**, 577–594.

KARESS, R. E. and RUBIN, G. M. (1982). A small tandem duplication is responsible for the unstable *white-ivory* mutation in Drosophila. *Cell* **30**, 63–69.

KASTENBAUM, M. A. and BOWMAN, K. O. (1970). Tables for determining the statistical significance of mutation frequencies. *Mutation Res.* **9**, 527–549.

KAUFMAN, B. P. (1947). Spontaneous mutation rate in *Drosophila*. *Am. Nat.* **81**, 77–80.

KIDWELL, M. G. (1977). Reciprocal differences in female recombination associated with hybrid dysgenesis in *Drosophila melanogaster*. *Genet. Res., Camb.* **30**, 77–88.

KIDWELL, M. G. (1978). Hybrid dysgenesis in *Drosophila melanogaster*: No suppressor-X effect on male recombination. *J. Hered.* **69**, 417–419.

KIDWELL, M. G. (1979). Hybrid dysgenesis in *Drosophila melanogaster*: The relationship between the P-M and I-R interaction systems. *Genet. Res., Camb.* **33**, 205–217.

KIDWELL, M. G. and KIDWELL, J. F. (1975). Cytoplasm-chromosome interactions in *Drosophila melanogaster*. *Nature, Lond.* **253**, 755–756.

KIDWELL, M. G. and KIDWELL, J. F. (1976). Selection for male recombination in *Drosophila melanogaster*. *Genetics* **84**, 333–351.

KIDWELL, M. G., KIDWELL, J. F. and IVES, P. T. (1977a). Spontaneous nonreciprocal mutation and sterility in strain crosses of *Drosophila melanogaster*. *Mutation Res.* **42**, 89–98.

KIDWELL, M. G., KIDWELL, J. F. and SVED, J. A. (1977b). Hybrid dysgenesis in *Drosophila melanogaster*: A syndrome of aberrant traits including mutation, sterility and male recombination. *Genetics* **86**, 813–833.

KIDWELL, M. G., NOVY, J. B. and FEELEY, S. M. (1981). Rapid unidirectional change of hybrid dysgenesis potential in *Drosophila*. *J. Heredity* **72**, 32–38.

KIKKAWA, H. (1935). Crossing-over in the male of *Drosophila virilis*. *Cytologia* **6**, 190–194.

KRATZ, F. L. (1975). Radioresistance in natural populations of *Drosophila nebulosa* from a Brazilian area of high background radiation. *Mutation Res.* **27**, 347–355.

LAMY, R. (1947). Observed spontaneous mutation rates in relation to experimental techniques. *J. Genet.* **48**, 226–236.

LANGLEY, C. H. and ITO, K. (1976). Spontaneous mutability in *Drosophila melanogaster*, in natural and laboratory environments. *Mutation Res.* **36**, 385–386.

LEE, W. R. (1976). Chemical mutagenesis. *In*: "The Genetics and Biology of *Drosophila*", (M. Ashburner and E. Novitsky, Eds.), Vol. 1c, pp. 1299–1351. Academic Press, London and New York.

LEE, W. R., ABRAHAMSON, S., VALENCIA, R., VON HALLE, E. S., WURGLER, F. E. and ZIMMERING, S. (1983). The sex-linked recessive lethal test for mutagenesis in *Drosophila melanogaster*. A report of the U.S. EPA Gene-Tox Program. *Mutation Res.*, in press.

LEFEVRE, G., JR. (1976). A photographic representation and interpretation of the polytene chromosome of *Drosophila melanogaster* salivary glands. *In*: "The Genetics and Biology of *Drosophila*", (M. Ashburner and E. Novitski, Eds.), Vol. 1a, pp. 31–66, Academic Press, London and New York.

LEFEVRE, G., JR. and GREEN, M. M. (1959). Reverse mutation studies on the forked locus in *Drosophila melanogaster*. *Genetics* **44**, 769–776.

LEVINE, R. P. and IVES, P. T. (1953). Mutation rates and lethal gene frequencies in populations of *Drosophila melanogaster*. *Proc. Natl. Acad. Sci. U.S.A.* **39**, 817–823.

LEVITAN, M. (1952). A ring chromosome in wild-caught *Drosophila*. *Genetics* **37**, 600.

LEVITAN, M. (1962). Spontaneous chromosome aberrations in *Drosophila robusta*. *Proc. Natl. Acad. Sci. U.S.A.* **48**, 930–937.

LEVITAN, M. (1963a). A maternal factor which breaks paternal chromosomes. *Nature, Lond.* **200**, 437–438.

LEVITAN, M. (1963b). A maternal chromosome breakage factor which attacks paternal chromosomes exclusively. *Genetics* **48**, 898.

LEVITAN, M. (1964a). Spontaneous chromosome aberrations in *D. robusta* since October 1960. I. X-chromosome inversions among the first 559. *Dros. Inf. Serv.* **39**, 89–96.

LEVITAN, M. (1964b). Bilateral concordance of sperm chromosome breaks induced by a maternal factor. *Anat. Rec.* **148**, 306.

LEVITAN, M. (1970). Chromosomal breaks with attachments to the nucleolus. *Chromosoma* **31**, 452–467.

LEVITAN, M. and SCHILLER, R. (1963). Further evidence that the chromosome breakage factor in *Drosophila robusta* involves a maternal effect. *Genetics* **48**, 1231–1238.

LEVITAN, M. and WILLIAMSON, D. L. (1965). Evidence for the cytoplasmic and possibly episomal nature of a chromosome breaker. *Genetics* **52**, 456.

LEWIS, E. B. (1954). The theory and application of a new method of detecting chromosomal rearrangements in *Drosophila melanogaster*. *Am. Nat.* **88**, 225–239.

LEWIS, E. B. (1959). Germinal and somatic reversion of the ivory mutant of *Drosophila melanogaster*. *Genetics* **44**, 522.

LEWONTIN, R. C. (1974). "The Genetic Basis of Evolutionary Change". Columbia University Press, New York.

L'HERITIER, PH. (1970). *Drosophila* viruses and their roles as evolutionary factors. *Evol. Biol.* **4**, 185–208.

LIFSCHYTZ, E. and FALK, R. (1969). A genetic analysis of the Killer-prune (K-pn) locus of *Drosophila melanogaster*. *Genetics* **62**, 353–358.

LINDGREN, D. (1972). The temperature influence on the spontaneous mutation rate. I. Literature review. *Hereditas* **70**, 165–178.

LINDSLEY, D. L. (1954). Reverse mutation at the loci of white and forked. *Dros. Inf. Serv.* **28**, 131.

LINDSLEY, D. L. and GRELL, E. H. (1968). Genetic Variations of *Drosophila melanogaster*. Carnegie Institution of Washington, publ. 627.

MAGNUSSON, J. and RAMEL, C. (1978). Mutagenic effects of vinyl chloride on *Drosophila melanogaster* with and without pretreatment with sodium phenobarbiturate. *Mutation Res.* **57**, 307–312.

MAGNUSSON, J., HALLSTROM, I. and RAMEL, C. (1979). Studies on metabolic activation of vinyl chloride, in *Drosophila melanogaster* after pretreatment with phenobarbital and polychlorinated biphenyls. *Chem. Biol. Interact.* **24**, 287–298.

MAMPELL, K. (1943). High mutation frequency in *Drosophila pseudoobscura*, race B. *Proc. Natl. Acad. Sci. U.S.A.* **29**, 137–144.

MAMPELL, K. (1945). Analysis of a mutator. *Genetics* **30**, 496–505.

MAMPELL, K. (1946). Genic and nongenic transmission of mutator activity. *Genetics* **31**, 589–597.

MARKOWITZ, E. H. (1970). Gamma ray-induced mutations in *Drosophila melanogaster* oocytes: The phenomenon of dose rate. *Genetics* **64**, 313–322.

MASON, J. M. (1980). Spontaneous mutation frequencies in mutagen-sensitive mutants of *Drosophila melanogaster*. *Mutation Res.* **72**, 323–326.

MATTHEWS, K. (1980). The temperature-sensitive period of T-007-induced male sterility. *Genetics* **94**, s71–72.

MATTHEWS, K. (1981). Developmental stages of genome elimination resulting in transmission ratio distortion of the T-007 male recombination (MR) chromosomes of *Drosophila melanogaster*. *Genetics*, **97**, 95–111.

MATTHEWS, K. and GERSTENBERG, M. (1979). Non-reciprocal female sterility associated with male recombination chromosomes from Texas populations of *Drosophila melanogaster*. *Genetics* **91**, s77.

MATTHEWS, K. and SLATKO, B. (1983). X-linked suppressors of dysgenic traits associated with male recombination (MR) chromosome of *Drosophila melanogaster*. Submitted to *Genetics*.

MATTHEWS, K., SLATKO, B. E., MARTIN, D. W. and HIRAIZUMI, Y. (1978). A consideration of the negative correlation between transmission ratio and recombination frequency in a male recombination system of *Drosophila melanogaster*. *Jap. J. Genet.* **53**, 13–25.

MENDELSON, D. (1976). The improved "bithorax method" for the detection of rearrangements in *Drosophila melanogaster*. *Mutation Res.* **41**, 269–276.

MINAMORI, S. (1967). A killing agent found in natural populations of *Drosophila melanogaster*. *Jap. J. Genet.* **42**, 317–326.

MINAMORI, S. (1969a). Extrachromosome element delta in *Drosophila melanogaster*. I. Gene dependence of killing action and of multiplication. *Genetics* **62**, 583–596.

MINAMORI, S. (1969b). Extrachromosomal element delta in *Drosophila melanogaster*. II. Transmission through male parent. *Jap. J. Genet.* **44**, 347–354.
MINAMORI, S. (1970). Extrachromosomal element delta in *Drosophila melanogaster*. III. Induction of one-sided gamete recovery from male and female heterozygotes. *Genetics* **66**, 505–515.
MINAMORI, S. (1971). Extrachromosomal element delta in *Drosophila melanogaster*. V. A variant inherited biparentally at 25 C. *Jap. J. Genet.* **46**, 169–180.
MINAMORI, S. (1972a). Extrachromosomal element delta in *Drosophila melanogaster*. VIII. Inseparable association with sensitive second chromosome. *Genetics* **70**, 557–566.
MINAMORI, S. (1972b). Extrachromosomal element delta correlates with "segregation distortion" phenomena. *Dros. Inf. Serv.* **49**, 97–98.
MINAMORI, S. and ITO, K. (1971a). Effects of delta on fertility in *D. melanogaster*. *Dros. Inf. Serv.* **47**, 81–82.
MINAMORI, S. and ITO, K. (1971b). Extrachromosomal element delta in *Drosophila melanogaster*. VI. Induction of recurrent lethal mutations in definite regions of second chromosomes. *Mutation Res.* **13**, 361–369.
MINAMORI, S. and ITO, K. (1972). Extrachromosomal element delta in *Drosophila melanogaster*. VII. Relation to fertility in a second chromosome line. *Genetics* **70**, 549–556.
MINAMORI, S. and SUGIMOTO, K. (1972). Production of delta-retaining sensitive chromosomes by EMS 110 *D. melanogaster*. *Dros. Inf. Serv.* **48**, 131.
MINAMORI, S. and SUGIMOTO, K. (1973). Extrachromosomal element delta in *Drosophila melanogaster*. IX. induction of delta-retaining chromosome lines by mutation and gene mapping. *Genetics* **74**, 477–487.
MINAMORI, S. and TATSUKAWA, K. (1967). Relation of infection to population structure in *Drosophila melanogaster*. *Evolution* **32**, 337–351.
MINAMORI, Y., FUJIOKA, N., ITO, K. and IKEBUCHI, M. (1970). Extrachromosomal element delta in *Drosophila melanogaster*. IV. Variation and persistence of delta-associating second chromosomes in a natural population. *Evolution* **24**, 735–744.
MINAMORI, S., ITO, K., NAKAMURA, A., ANDO, Y. and SHIOMI, H. (1973). Increasing trend in frequencies of lethal and semilethal chromosomes in a natural population of *Drosophila melanogaster*. *Jap. J. Genet.* **48**, 41–51.
MINAMORI, S., INOUE, Y., ITO, K. and SHIMIZU, A. (1977). Extrachromosomal element delta in *Drosophila melanogaster*. X. Enhancement in female and male recombinations. *Jap. J. Genet.* **52**, 87–93.
MIYAMOTO, T. and NAKAO, Y. (1978). The frequency pattern of *dumpy* mutations induced by X-rays in the successive stages of oocytes of *Drosophila*. *Jap. J. Genet.* **53**, 175–181.
MORGAN, K., HASTINGS, P. J. and vonBORSTEL, R. C. (1973). A potential hazard: Explosive production of mutations by induction of mutators. *Environmental Health Perspectives* No. 6, 207–210.
MORIWAKI, D. and TOBARI, Y. N. (1975). Drosophila ananassae. *In*: "Handbook of Genetics", (R. C. King, Ed.), Vol. 3, pp. 513–535, Plenum Press, New York.
MUKAI, T. (1964). The genetic structure of natural populations of *Drosophila melanogaster*. I. Spontaneous mutation rate of polygenes controlling viability. *Genetics* **50**, 1–19.
MUKAI, T. and COCKERHAM, C. (1977). Spontaneous mutation rates of enzyme loci in *Drosophila melanogaster*. *Proc. Natl. Acad. Sci. U.S.A.* **74**, 2514–2517.
MUKAI, T. and YAMAGUCHI, O. (1974). The genetic structure of natural populations of *Drosophila melanogaster*. XI. Genetic variability in a local population. *Genetics* **76**, 339–366.
MUKAI, T., METTLER, L. E. and CHIGUSA, S. I. (1971). Linkage disequilibrium in a local population of *Drosophila melanogaster*. *Proc. Natl. Acad. U.S.A.* **68**, 1065–1069.

MUKAI, T., CHIGUSA, S. I., METTLER, L. E. and CROW, J. F. (1972). Mutation rate and dominance of genes affecting viability in *Drosophila melanogaster*. *Genetics* **72**, 335–355.

MUKAI, T., CARDELLINO, R. A., WATANABE, T. A. and CROW, J. F. (1974). The genetic variance for viability and its components in a local population of *Drosophila melanogaster*. *Genetics* **78**, 1195–1208.

MULLER, H. J. (1928). The measurement of mutation rates in *Drosophila*, its high variability and its dependence upon temperature. *Genetics* **13**, 279–357.

MULLER, H. J. (1945). Age in relation to the frequency of spontaneous mutants in *Drosophila*. Yearbook *Am. Phil. Soc.* for 1945, pp. 10–153.

MULLER, H. J. (1950). Our load of mutation. *Amer. J. Human Genet.* **2**, 111–176.

MULLER, H. J. (1954a). The nature of genetic effects produced by radiation. *Radiat. Biol.* **1**, 351–473.

MULLER, H. J. (1954b). A semi-automatic breeding system ("Maxy") for finding sex-linked mutations at specific "visible" loci. *Dros. Inf. Serv.* **28**, 140–141.

MULLER, H. J. and OSTER, I. I. (1963). Some mutational techniques in *Drosophila*. In: "Methodology in Basic Genetics", (W. J. Burdette, Ed.), Holden-Day, San Francisco, CA.

MULLER, H. J., VALENCIA, J. I. and VALENCIA, R. M. (1950). The frequency of spontaneous mutations at individual loci in *Drosophila*. *Genetics* **35**, 125–126.

MULLER, H. J., OSTER, I. J. and ZIMMERING, S. (1963). Are chronic and acute gamma irradiation equally mutagenic in *Drosophila*. In: "Repair from Genetic Radiation Damage", (F. H. Sobels, Ed.), Macmillan Co., New York.

MYER, H. V. and MULLER, H. J. (1961). Similarity of X-ray-induced mutation rate in gonia of *Drosophila* females and males. *Genetics* **46**, 882–883.

NEEL, J. V. (1942). A study of a case of high mutation rate in *Drosophila melanogaster*. *Genetics* **27**, 519–536.

NEWBERNE, P. M. and ZEIGER, E. (1978). Nutrition, carcinogenesis, and mutagenesis. *Adv. Modern Toxicol.* **5**, 53–84.

NGUYEN, T. D. and BOYD, J. B. (1977). The meiotic-9 (mei-9) mutants of *Drosophila melanogaster* are deficient in repair replication of DNA. *Molec. Gen. Genet.* **158**, 141–147.

NGUYEN, T. D., BOYD, J. B. and GREEN, M. M. (1979). Sensitivity of *Drosophila* mutants to chemical carcinogens. *Mutation Res.* **63**, 67–77.

OLENOV, J. M., KHARMAC, I. S., GALKOVSKAYA, K. F. and MURETOV, G. D. (1939). Factors responsible for genic composition of wild *Drosophila melanogaster* populations. *C.R. (Doklady) Acad. Sci. U.S.S.R.* **24**, 466–470.

OSTERTAG, W. and HAAKE, J. (1966). The mutagenicity in *Drosophila melanogaster* of caffeine and of other compounds which produce chromosome breakage in human cells in culture. *Z. Vererbungsl.* **98**, 299–308.

PARKER, D. R. (1960). The induction of recessive lethals in *Drosophila* oocytes. *Genetics* **45**, 135–138.

PARRY, D. M. (1973). A meiotic mutant affecting recombination in female *Drosophila melanogaster*. *Genetics* **73**, 465–486.

PENNA FRANCA, E., ALMEIDA, J. C., BECKER, J., EMMERICH, M., ROSER, F. X., KEGEL, G., HAINSBERGER, L., CULLEN, T. L., PETROW, H., DREW, R. T. and EISENBUD, M. (1965). Status of investigation in the Brazilian areas of high natural radioactivity. *Health Phys.* **11**, 699–712.

PHILIP, U. (1944). Crossing over in the males of *Drosophila subobscura*. *Nature, Lond.* **153**, 223.

PICARD, G. (1976). Non-Mendelian female sterility in *Drosophila melanogaster*: Hereditary transmission of I factor. *Genetics* **83**, 107–123.

PICARD, G. (1978). Non-Mendelian female sterility in *Drosophila melanogaster*: Further data on chromosomal contamination. *Molec. Gen. Genet.* **164**, 235–247.
PICARD, G. (1979). Non-Mendelian female sterility in *Drosophila melanogaster*: Principal characteristics of chromosomes from inducer and reactive origin after chromosomal contamination. *Genetics* **91**, 455–471.
PICARD, G. and PELISSON, A. (1979). Non-Mendelian female sterility in *Drosophila melanogaster*: Characterization of the noninducer chromosomes of inducer strains. *Genetics* **91**, 473–489.
PICARD, G., BREGLIANO, J. C., BUCHETON, A., LAVIGE, J. M., PELISSON, A. and KIDWELL, M. G. (1978). Non-Mendelian female sterility and hybrid dysgenesis in *Drosophila melanogaster*. *Genet. Res., Camb.* **32**, 275–287.
PLOUGH, H. H. (1941). Spontaneous mutability in *Drosophila*. Cold Spring Harbor. *Symp. Quant. Biol.* **9**, 127–137.
PLOUGH, H. H. and HOLTHAUSEN, C. F. (1937). A case of high mutation frequency without environmental change. *Am. Nat.* **71**, 185–187.
PLOUGH, H. H. and IVES, P. T. (1935). Induction of mutations by high temperature in *Drosophila*. *Genetics* **20**, 42–69.
PROPPING, P. (1972). Comparison of point mutation rates in different species with human mutation rates. *Humangenetik* **16**, 43–48.
PURDOM, C. E., DYER, K. and PAPWORTH, D. G. (1968). Spontaneous mutation in *Drosophila*: Studies on the rate of mutation in mature and immature germ cells. *Mutation Res.* **5**, 133–146.
RACINE, R. R., LANGLEY, C. H. and VOELKER, R. A. (1980). Enzyme mutants induced by low-dose-rate gamma-irradiation in *Drosophila*: Frequency and characterization. *Environmental Mutagenesis* **2**, 167–177.
RADDING, C. (1973). Molecular mechanisms in genetic recombination. *Ann. Rev. Genet.* **7**, 87–111.
RATNAYAKE, W. E. (1968). Tests for an effect of the Y-chromosome on the mutagenic action of formaldehyde and X-rays in *Drosophila melanogaster*. *Genet. Res., Camb.* **12**, 65–69.
REDDI, O. S., REDDY, G. M. and RAO, M. S. (1965). Induction of crossing-over in *Drosophila* males by means of ovarian extracts. *Nature, Lond.* **208**, 203.
RIEGER, R., MICHAELIS, A. and GREEN, M. M. (1976). "Glossary of Genetics and Cytogenetics". Springer-Verlag, New York.
RINEHART, R. R. (1969). Spontaneous sex-linked recessive lethal frequencies from aged and non-aged spermatozoa of *Drosophila melanogaster*. *Mutation Res.* **7**, 417–423.
ROBERTS, P. A. (1970). Screening for X-ray-induced crossover suppressors in *Drosophila melanogaster*: Prevalence and effectiveness of translocations. *Genetics* **65**, 429–448.
ROBERTS, P. A. (1976). The genetics of chromosome aberration. In: "The Genetics and Biology of *Drosophila*", (M. Ashburner and E. Novitski, Eds.), Vol. 1a, pp. 67–184, Academic Press, London and New York.
RUBIN, G. M., KIDWELL, M. G. and BINGHAM, P. M. (1982). The molecular basis of P–M hybrid dysgenesis: The nature of induced mutations. *Cell* **29**, 987–994.
SAGE, R. D. and SELANDER, R. K. (1979). Hybridization between species of the *Rana pipiens* complex in central Texas. *Evolution* **33**, 1069–1088.
SANDLER, L., LINDSLEY, D., NICOLETTI, B. and TRIPPA, G. (1968). Mutants affecting meiosis in natural populations of *Drosophila melanogaster*. *Genetics* **60**, 525–558.
SARASWATHY, T. (1970). An attempt to induce crossing-over in males of *Drosophila melanogaster* with ovarian extracts. *Experientia* **26**, 799.
SCHALET, A. P. and SANKARANARYANAN, K. (1976). Evaluation and re-evaluation of genetic radiation hazards in man. I. Interspecific comparison of estimates of mutation rates. *Mutation res.* **35**, 341–370.

SCOBIE, N. and SCHAFFER, H. (1982a). The identification of the presence of a mutator factor in a strain of *Drosophila melanogaster* by the use of mutation reversion rates and male recombination. (Submitted to *Genetics*)

SCOBIE, N. and SCHAFFER, H. (1982b). The location of a mutator factor in a strain of *Drosophila melanogaster* by the use of male recombination. (Submitted to *Genetics*)

SHAKARNIS, V. F. (1969). Induction of X-chromosome nondisjunction and recessive sex-linked lethal mutations in females of *Drosophila melanogaster* by 1,2-dichloroethane. *Sov. Genet.* 5:1666-1671 (translated from *Genetika* 5, 89–95).

SHAKARNIS, V. F. (1970). Comparative study of the action of caffeine on X-chromosome nondisjunction and recessive sex-linked lethal mutations in females of various *Drosophila melanogaster* lines. *Sov. Genet.* 6:921–924 (translated from *Genetika* 6, 83–87).

SHELDON, B. L. (1958). The effect of temperature on the mutation rate in *Drosophila melanogaster*. *Aust. J. Biol. Sci.* 11, 85–94.

SHELDON, B. L. and BARKER, J. S. F. (1964). The effect of temperature during development on mutation in *Drosophila melanogaster*. *Mutation Res.* 1, 310–317.

SIMMONS, M. J. and LIM, J. K. (1980). Site specificity of mutations arising in dysgenic hybrids of *Drosophila melanogaster*. *Proc. Natl. Acad. Sci. U.S.A.* 77, 6042–6046.

SIMMONS, M. J., JOHNSON, N. A., FAHEY, T. M., NELLETT, S. M. and RAYMOND, J. D. (1980). High mobility in male hybrids of *Drosophila melanogaster*. *Genetics* 96, 479–490.

SINCLAIR, D. and GREEN, M. M. (1979). Genetic instability in *Drosophila melanogaster*: The effect of male recombination (MR) chromosomes in females. *Molec. Gen. Genet.* 170, 219–224.

SINGER, K. M., CHOVNICK, A., SUZUKI, D. T., BAILLIE, D. and HOAR, D. (1967). Attempts to induce crossing-over in *Drosophila melanogaster* males with ovarian extracts. *Nature, Lond.* 214, 503–504.

SLATKO, B. (1976a). Genetic elements causing male recombination in *Drosophila melanogaster*. *Genetics* 83, s72 (Abs.).

SLATKO, B. (1976b). Mutator elements in natural populations of *Drosophila melanogaster*: The high mutator gene revisited. *Mutation Res.* 36, 387–390.

SLATKO, B. (1977). Genetic elements causing male recombination in *Drosophila melanogaster*. *Genetics* 86, s59 (Abs.).

SLATKO, B. (1978a). Evidence for newly induced genetic activity responsible for male recombination induction in *Drosophila melanogaster*. *Genetics* 90, 105–124.

SLATKO, B. (1978b). Parameters of male and female recombination influenced by the T-007 second chromosome in *Drosophila melanogaster*. *Genetics* 90, 257–276.

SLATKO, B. and GREEN, M. M. (1980). Genetic instability in *Drosophila melanogaster*: Mapping the mutator activity of an MR strain. *Biol. Zbl.* 99, 149–155.

SLATKO, B. and HIRAIZUMI, Y. (1973). Mutation induction in the male recombination strains of *Drosophila melanogaster*. *Genetics* 75, 643–649.

SLATKO, B. and HIRAIZUMI, Y. (1975). Elements causing male crossing-over in *Drosophila melanogaster*. *Genetics* 81, 313–324.

SLATKO, B. and HIRAIZUMI, Y. (1978). Genetic suppression of male recombination activity in *Drosophila melanogaster*. *Genetics* 88, 292 (Abs.).

SLIZYNSKI, B. M. (1938). Salivary chromosome studies of lethals in *Drosophila melanogaster*. *Genetics* 60, 389–393.

SMITH, P. D. (1973). Mutagen sensitivity of *Drosophila melanogaster*. I. Isolation and preliminary characterization of a methyl methanesulfonate-sensitive strain. *Mutation Res.* 20, 215–220.

SMITH, P. D. (1976). Mutagen sensitivity of *Drosophila melanogaster*. III. X-linked loci governing sensitivity to methylmethanesulfonate. *Molec. Gen. Genet.* 149, 73–86.

SMITH, P. D. and CORWIN, H. O. (1971). Temperature shock-induced specific visible mutations at the *dumpy* locus of *Drosophila melanogaster*. *Mutation Res.* **13**, 345–351.

SMITH, P. D. and SHEAR, C. (1974). X-ray and ultraviolet light sensitivities of a methyl methanesulfonate-sensitive strain of *Drosophila melanogaster*. *In*: "Mechanisms in Recombination", (R. F. Grell, Ed.), pp. 399–402, Plenum Press, New York.

SMITH, P. D., SNYDER, R. and DUSENBERY, R. (1980). Isolation and characterization of repair deficient mutants of *Drosophila melanogaster*. *In*: "DNA Repair and Mutagenesis in Eukaryotes" (A. Hollaender *et al.*, Eds.) (In Press).

SNYDER, L. A. and OSTER, I. I. (1964). A comparison of genetic changes induced by a monofunctional and a polyfunctional alkylating agent in *Drosophila melanogaster*. *Mutation Res.* **1**, 437–445.

SOBELS, F. H. and EEKEN, J. C. J. (1981). Influence of the MR (mutator) factor on X-ray-induced genetic damage. *Mutation Res.* **83**, 201–206.

SOBELS, F. H. and VOGEL, E. (1976). The capacity of *Drosophila* for detecting relevant genetic damage. *Mutation Res.* **41**, 95–105.

SOCHACKA, J. H. M. and WOODRUFF, R. C. (1976). Induction of male recombination in *Drosophila melanogaster* by injection of extracts of flies showing male recombination. *Nature, Lond.* **262**, 287–289.

SOUTHIN, J. L. (1966). An analysis of eight classes of somatic and gonadal mutation at the *dumpy* locus in *Drosophila melanogaster*. *Mutation Res.* **3**, 54–65.

SPENCER, W. P. (1935). The non-random nature of visible mutations in *Drosophila*. *Am. Nat.* **69**, 223–238.

SPENCER, W. P. (1947). Mutations in wild populations in *Drosophila*. *Adv. Genet.* **1**, 359–402.

SPENCER, W. P. and STERN, C. (1948). Experiments to test the validity of the linear R-dose/mutation frequency relation in *Drosophila* at low dosage. *Genetics* **33**, 43–74.

SPIESS, E. B. (1977). "Genes in Populations". John Wiley and Sons, New York.

SPOFFORD, J. B. (1976). Position-effect varigation in *Drosophila*. *In*: "The Genetics and Biology of *Drosophila*", (M. Ashburner and E. Novitski, Eds.), Vol. 1c, Academic Press, London and New York.

STALKER, H. D. (1976). Chromosome studies in wild populations of *D. melanogaster*. *Genetics* **82**, 323–347.

STAMATIS, N., YANNOPOULOS, G. and PELECANOS, M. (1981). Comparative studies of two male recombination factors (MRF) isolated from a Southern Greek *Drosophila melanogaster* population. *Genet. Res. Camb.* **38**, 125–135.

STERN, C. and KODANI, M. (1955). Studies on the position effect at the cubitus interruptus locus of *Drosophila melanogaster*. *Genetics* **40**, 343–373.

STERN, C., MACKNIGHT, R. H. and KODANI, M. (1946). The phenotypes of homozygotes and hemizygotes of position alleles and of heterozygotes between alleles in normal and translocated position. *Genetics* **31**, 598–619.

STEVENS, W. L. (1942). Accuracy of mutation rates. *J. Genet.* **43**, 301–307.

STROMNAES, O. and KVELLAND, I. (1963). The induction of Minute mutations in *Drosophila* with tritium-labelled thymidine. *Genetics* **48**, 1559–1565.

STURTEVANT, A. H. (1937). Essays on evolution I on the effects of selection on mutation rate. *Q. Rev. Biol.* **12**, 464–467.

STURTEVANT, A. H. (1939). High mutation frequency induced by hybridization. *Proc. Natl. Acad. Sci. U.S.A.* **25**, 308–310.

SVED, J. A. (1976). Hybrid dysgenesis in *Drosophila melanogaster*: A possible explanation in terms of spatial organisation of chromosomes. *Aust. J. Biol. Sci.* **29**, 375–388.

SVED, J. A. (1978). Male recombination in dysgenic hybrids of *Drosophila melanogaster*: Chromosome breakage or mitotic crossing-over? *Aust. J. Biol. Sci.* 31, 303–309.

SVED, J. A. (1979). The "hybrid dysgenesis" syndrome in *Drosophila melanogaster*. *Bioscience* 29, 659–664.

SVED, J. A., MURRAY, D. C., SCHAEFER, R. E. and KIDWELL, M. G. (1978). Male recombination is not induced to *Drosophila melanogster* by extracts of strains with male recombination potential. *Nature, Lond.* 257, 457–458.

THOMPSON, J. N., JR. and WOODRUFF, R. C. (1978a). Mutator genes—pacemaker of evolution. *Nature, Lond.* 274, 317–321.

THOMPSON, J. N., JR. and WOODRUFF, R. C. (1978b). Chromosome breakage: A possible mechanism for diverse genetic events in outbred populations. *Heredity* 40, 153–157.

THOMPSON, J. N., JR. and WOODRUFF, R. C. (1980). Increased mutation in crosses between geographically separated strains of *Drosophila melanogaster*. *Proc. Natl. Acad. Sci. U.S.A.* 77, 1059–1062.

THOMPSON, J. N., JR. and WOODRUFF, R. C. (1981). A model for spontaneous mutation in *Drosophila* caused by transposing elements. *Heredity* 47, 327–335.

THOMPSON, J. N., JR., WOODRUFF, R. C. and SCHAEFER, G. B. (1978). An assay of somatic recombination in male recombination lines of *Drosophila melanogaster*. *Genetica* 49, 77–80.

TIMOFEEFF-RESSOVSKY, N. W. (1932). Verschiedenheit der Normalen allele der white series aus swei geographisch getrennten populationen von *Drosophila melanogaster*. *Biol. Zbl.* 52, 460–476.

TINIAKOV, G. G. (1939). Highly mutable stock from a wild population of *Drosophila melanogaster*. *C.R. (Doklady) Acad. Sci. URSS* 22, 609–612.

TINIAKOV, G. G. (1941). A new case of high spontaneous mutability in *D. melanogaster*. *Dros. Inf. Serv.* 15, 40–41.

TOBARI, Y. N. and KOJIMA, K. (1972). A study of spontaneous mutation rates at ten loci detectable by starch gel electrophoresis in *Drosophila melanogaster*. *Genetics* 70, 397–403.

VALADARES, M. (1937). Declanchement d'une haute mutabilite chez une lignee pure de *Drosophila melanogaster*. *Rev. Agron. Lisbon* 25, 363–383.

VALADARES, M. (1940). Cytological evidence of spontaneous chromosome-rearrangements in *Drosophila melanogaster*: Rate of occurrence and nature of the mutants. *Dros. Inf. Serv.* 13, 76.

VALENCIA, R., ABRAHAMSON, S., LEE, W. R., VON HALLE, E. S., WOODRUFF, R. C., WÜRGLER, F. E. and ZIMMERING, S. (1983). Chromosome mutation tests for mutagenesis in *Drosophila melanogaster*. A report of the U.S. EPA Gene-Tox Program. *Mutation Res.*, in press.

VOELKER, R. A. (1974). The genetics and cytology of a mutator factor in *Drosophila melanogaster*. *Mutation Res.* 22, 265–276.

VOELKER, R. A., SCHAFFER, H. E. and MUKAI, T. (1980). Spontaneous allozyme mutations in *Drosophila melanogaster*: Rate of occurrence and nature of the mutants. *Genetics* 94, 961–968.

VOGEL, E. (1975). Some aspects of the detection of potential mutagenic agents in *Drosophila*. *Mutation Res.* 29, 241–250.

VOGEL, E. (1976). Mutagenicity of carcinogens in *Drosophila* as a function of genotype-controlled metabolism. In: "*In vitro* Metabolic Activation in Mutagenesis Testing", (F. J. DeSeves, J. R. Fouts, J. R. Bend and R. Philpott, Eds.), pp. 63–79, North Holland Publ. Co., New York.

VOGEL, E. and SOBELS, F. (1976). The function of *Drosophila* in genetic toxicology testing. In: "Chemical Mutagens", (A. Hollaender, Ed.), Vol. 4, pp. 93–142, Plenum Press, New York.

WADDLE, F. R. and OSTER, I. I. (1974). Autosomal recombination in males of *Drosophila melanogaster* caused by a transmissible factor. *J. Genet.* **61**, 177–183.
WAGNER, R. and MITCHELL, H. K. (1964). "Genetics and Metabolism", 2nd edition. John Wiley and Sons, New York.
WALLACE, B. (1968a). "Topics in Population Genetics". W. W. Norton, NY.
WALLACE, B. (1968b). Mutation rates for autosomal lethals in *Drosophila melanogaster*. *Genetics* **60**, 389–393.
WALLACE, B. (1970). Spontaneous mutation rates for sex-linked lethals in the two sexes of *Drosophila melanogaster*. *Genetics* **64**, 553–557.
WATANABE, T. K. and LEE, W. H. (1977). Sterile mutation in *Drosophila melanogaster*. *Genet. Res. Camb.* **30**, 107–113.
WILKINSON, C. and BRATTSTEIN, R. (1972). Microsomal drug metabolizing enzymes in insects. *Drug Metab. Review* **1(2)**, 153–228.
WILLIAMSON, D. (1961). Carbon dioxide sensitivity in *Drosophila affinis* and *Drosophila athabasca*. *Genetics* **46**, 1053–1060.
WOODRUFF, R. C. (1975). The control of mutational instability by a new mutator gene of *Drosophila melanogaster*. *Genet. Res. Camb.* **25**, 163–177.
WOODRUFF, R. C. and BORTOLOZZI, J. (1976). Spontaneous recombination in males of *Drosophila simulans*. *Heredity* **37**, 295–298.
WOODRUFF, R. C., BOWMAN, J. T. and SIMMONS, J. R. (1972). Sex-influenced reversion of the mutationally unstable mutant *forked -3N* of *Drosophila melanogaster*. *Mutation Res.* **15**, 86–89.
WOODRUFF, R. C. and THOMPSON, J. N., JR. (1975). Genetic analysis of male recombination in *Drosophila melanogaster*. *Genetics* **80**, s86.
WOODRUFF, R. C. and THOMPSON, J. N., JR. (1977). An analysis of spontaneous recombination in *Drosophila melanogaster* males: Isolation and characterization of male recombination lines. *Heredity* **38**, 291–307.
WOODRUFF, R. C. and THOMPSON, J. N., JR. (1980). Hybrid release of mutator activity and the genetic structure of natural populations. In: "Evolutionary Biology", (M. K. Hecht, W. C. Steere and B. Wallace, Eds.), Vol. 12, pp. 129–162, Plenum Pub. Co., New York.
WOODRUFF, R. C., SLATKO, B. and THOMPSON, J. N., JR. (1978). Lack of an interchromosomal effect associated with spontaneous recombination in males of *Drosophila melanogaster*. *Ohio J. Sci.* **78**, 310–317.
WOODRUFF, R. C., THOMPSON, J. N., JR. and LYMAN, R. F. (1979). Intraspecific hybridization and the release of mutator activity. *Nature, Lond.* **278**, 277–279.
WÜRGLER, F. E. (1972). Mutation mechanisms in *Drosophila*. *Humangenetik* **16**, 61–66.
WÜRGLER, F. E. and MAIER, P. (1972). Genetic control of mutation induction in *Drosophila melanogaster*. I. Sex-chromosome loss in X-rayed mature sperm. *Mutation Res.* **15**, 41–53.
WÜRGLER, F. E., GRAF, U. and BERCHTOLD, W. (1975). Statistical problems connected with the sex-linked recessive lethal test in *Drosophila melanogaster*. I. The use of the Kastenbaum-Bowman test. *Arch. Genet.* **48**, 158–178.
WÜRGLER, F. E., SOBELS, F. H. and VÖGEL, E. (1977). *Drosophila* as assay system for detecting genetic changes. In: "Handbook of Mutagenicity Test Procedures", (B. J. Kilbey, M. Legator, W. Nichols and C. Ramel, Eds.), Elsevier Sci. Pub. Co.
WÜRGLER, F. E., WOODRUFF, R. C., VALENCIA, R. M. and ZIMMERING, S. (1982). Heritable translocations in *Drosophila melanogaster*. A methodological review. *Mutation Res.*, in press.
YAMAGUCHI, O. (1976). Spontaneous chromosome mutation and screening of mutator factors in *Drosophila melanogaster*. *Mutation Res.* **34**, 389–406.

YAMAGUCHI, O. and MUKAI, T. (1974). Variation of spontaneous occurrence rates of chromosomal aberrations in the second chromosomes of *Drosophila melanogaster*. *Genetics* **78**, 1209–1221.

YAMAGUCHI, O., CARDELLINO, R. A. and MUKAI, T. (1976). High rates of occurrence of spontaneous chromosome aberrations in *Drosophila melanogaster*. *Genetics* **83**, 409–422.

YANNOPOULOS, G. (1978a). Studies on male recombination in a southern Greek *Drosophila melanogaster* population. (c) Chromosomal abnormalities at male meiosis. (d) Cytoplasmic factor responsible for the reciprocal cross effect. *Genet. Res. Camb.* **31**, 187–196.

YANNOPOULOS, G. (1978b). Progressive resistance against the male recombination factor *31.1 MRF* acquired by *Drosophila melanogaster*. *Experientia* **34**, 1000–1002.

YANNOPOULOS, G. (1978c). Studies on the sterility induced by the male recombination factor *31.1 MRF* in *Drosophila melanogaster*. *Genet. Res., Camb.* **32**, 239–247.

YANNOPOULOS, G. (1979). Ability of the male recombination factor *31.1 MRF* to be transposed to another chromosome in *Drosophila melanogaster*. *Molec. Gen. Genet.* **176**, 247–253.

YANNOPOULOS, G. and PELECANOS, M. (1977). Studies on male recombination in a southern Greek *Drosophila melanogaster* population. (a) Effect of temperature. (b) Suppression of male recombination in reciprocal crosses. *Genet. Res. Camb.* **29**, 231–238.

YANNOPOULOS, G. and ZACHAROPOULOU, A. (1980). Studies on the chromosomal rearrangements induced by the male recombination factor 31.1 MRF in *Drosophila melanogaster*. *Mutation Res.* **73**, 81–92.

YANNOPOULOS, G., PELECANOS, M. and ZACHAROPOULOU, A. (1980). Combined action of diethyl sulphate (DES) and of the male recombination factor *31.1 MRF* in *Drosophila*. *Genetics* **12**, 41–48.

YANNOPOULOS, G., STAMATIS, N., ZACHAROPOULOU, A. and PELECANOS, M. (1981). Differences in the induction of specific deletions and duplications by two male recombination factors isolated from the same *Drosophila* natural population. *Mutation Res.* **83**, 383–393.

ZIMMERING, S. (1981). Review of the current status of the *mei-9* test for chromosome loss in *Drosophila melanogaster*: An assay with radically improved detection capacity for chromosome lesions induced by methyl methanesulfonate (MMS), dimethylnitrosamine (DMN), and especially diethylnitrosamine (DEN) and procarbazine. *Mutation Res.* **83**, 69–80.

ZUITIN, A. I. (1941). Influence of the change from the natural complex of development conditions to the laboratory one on the mutation rate in *Drosophila melanogaster*. *C.R. (Doklady) Acad. Sci. U.S.S.R.* **30**, 61–63.

ZUITIN, A. I. and PAVLOVETZ, M. T. (1940). Mutation in general populations of *Drosophila melanogaster* under natural conditions. *C.R. (Doklady) Acad. Sci. U.S.S.R.* **29**, 483–486.

21. Intraspecific Hybrid Sterility

MARGARET G. KIDWELL

Division of Biology and Medicine
Brown University
Providence, Rhode Island, U.S.A.

I. Introduction	125
II. *Drosophila pseudoobscura*	127
III. The *Drosophila willistoni* Group	128
A. *D. willistoni*	128
B. *D. equinoxialis*	129
C. *D. tropicalis*	129
D. *D. paulistorum*	129
IV. *Drosophila melanogaster*	130
A. *SF* Sterility	131
B. *GD* Sterility	134
C. Comparison of *SF* and *GD* Sterility Systems	140
V. Other Species	146
VI. Discussion	147
Acknowledgements	148
References	149

I. Introduction

The study of intraspecific hybrid sterility is of particular interest because it is concerned with some of the first steps in the process of speciation. Patterson and Dobzhansky (1945) stated: "It is now clear that the genus *Drosophila*, contrary to the old idea, is quite rich in 'borderline cases' between races and species. Owing to the inherent advantages of many *Drosophila* species as laboratory materials, they offer excellent opportunities for studies on the genetics of speciation." The truth of this statement has been amply demonstrated during the intervening thirty-five years.

In the genus *Drosophila*, hybrid sterility is almost always of the genic variety (Dobzhansky, 1937). The extent of reproductive isolation below the species level varies from the normal condition of none at all to that of high degree which is typically, but not always, found between full species. It can occur between two different subspecies, populations or strains. It may affect hybrids and/or backcrosses of both sexes or it may be restricted to one sex. It may be reciprocal or nonreciprocal. Within any category, it may be partial or complete. The extent to which the degree of reproductive isolation is an indicator of the amount of general genetic divergence is, nonetheless, an unresolved question.

At a theoretical level, models for the evolution of genic post-mating reproductive isolating mechanisms have received little attention. Nei (1976) examined a model originally proposed by Dobzhansky (1937) and Muller (1942). Essentially, this involved two pairs (or sets) of "complementary, hybrid incapacitating" genes, one pair, or set, carried by each of two potentially diverging populations. Nei confirmed mathematically that, with some qualifications, the rise to high frequency or fixation of such genes could occur with selection acting on pleiotropic effects. In small populations, genetic drift could be a sufficient mechanism to achieve fixation. More recently, Templeton (1980) has proposed a mechanistic taxonomy of speciation in which speciation mechanisms are subdivided into two major categories, transilience modes and divergence modes. Transilience modes of speciation are characterized by isolating barriers dependent on a genetic discontinuity with extreme instability of the intermediate types.

In practice, the precise genetic mechanisms acting in early stages of the evolution of hybrid sterility have only infrequently been amenable to detailed study, but there is evidence in a number of cases that the system is considerably more complex than that of a simple two-locus interaction model. The available information does suggest that a wide diversity of genetic processes is involved in different instances of hybrid sterility. Many fascinating similarities can be observed, however, when independently evolved or evolving systems are compared. There is also wide variability in the extent of sexual isolation and of other indicators of genetic divergence, e.g. chromosomal, morphological and enzymatic, during the speciation process.

A brief overview is presented of the genetic mechanisms involved in those systems of intraspecific hybrid sterility which have received the most detailed study. A wide variety of incompatibility phenomena are included. These range from examples of hybrid sterility between largely allopatric subspecies and semispecies, often of ancient origin, as in *D. pseudoobscura*, which might even be interpreted as full species, to examples of intrapopulation incompatibility, as in *D. melanogaster* which may have evolved very recently. It should be made clear at the outset that, although there appear to be some similarities among the disparate groups described, there is as yet no evidence that similar genetic mechanisms are involved.

II. *Drosophila pseudoobscura*

Crosses between populations of *D. pseudoobscura* from widely separated localities produce fully fertile hybrids with one exception. A population from the high Andes above Bogotá is separated by a gap of about 2400 km from the closest point of the main distribution area of the species in Guatemala. Prakash (1972) reported that females from the Bogotá population produce completely sterile male hybrids when crossed with males from the mainland (United States and Guatemala locations). Female offspring from this cross and also progeny of both sexes from the reciprocal cross are normally fertile. Mating choice experiments provided no evidence for ethological isolation between Bogotá and mainland populations, but further study is needed to confirm this result.

Prakash (1972) argued that geographic isolation of the Bogotá population was historically of very recent origin. In contrast, Dobzhansky (1974) claimed that divergence from the main body of the species was of ancient origin. Furthermore, Ayala and Dobzhansky (1974) proposed that divergence was of sufficient magnitude for the two groups to be described as separate subspecies, *D. pseudoobscura pseudoobscura* and *D. pseudoobscura bogotana*. The two subspecies can be genetically differentiated on the basis of allelic frequencies at six diagnostic enzyme loci (Ayala and Dobzhansky, 1974). On average, 19 allelic substitutions have occurred for every 100 loci in the separate evolution of the subspecies. Singh *et al.* (1976) showed by sequential electrophoretic analysis that the Bogotá population was polymorphic for a largely unique set of xanthine dehydrogenase alleles. This result does not support Prakash's hypothesis of a recent divergence of the two subspecies. Also, *D. p. bogotana* differs from most populations of *D. p. pseudoobscura* in the configuration of the Y chromosome (Dobzhansky, 1935, 1974).

Dobzhansky (1974) analysed the genetic basis of the nonreciprocal hybrid sterility between the two subspecies. By means of backcrosses of fertile F_1 males to females of both subspecies, he was able to implicate one or several genes in both limbs of the X chromosome and in the second and third chromosomes. The eggs of hybrid females appeared to be influenced by the chromosomal complement of the mother. It was concluded that in backcross hybrids, sterility was a threshold character although it was rigidly determined in F_1 hybrids. The results of a population cage experiment, employing mixtures of the two subspecies, suggested that under competitive conditions, *D. p. bogotana* flies were less fit than those of *D. p. pseudoobscura*.

There are striking similarities between the two *D. pseudoobscura* subspecies and those of *D. willistoni* (Ayala and Dobzhansky, 1974). Also, there are many similarities between this hybrid sterility and that associated with hybrid dysgenesis in *D. melanogaster* (see below).

Dobzhansky (1974) interpreted the *D. pseudoobscura* sterility results as excluding cytoplasmic effects. However, Engels (1978), after making his own analysis of the data, concluded that there was evidence for a cytoplasmic component, transmissible for at least two generations. He further pointed out that the *D. pseudoobscura* results are qualitatively consistent with the rules of inheritance of hybrid dysgenesis in *D. melanogaster*.

III. The *Drosophila willistoni* Group

A. *D. WILLISTONI*

On the basis of partial hybrid sterility and allozyme differences, Ayala (1973) described the new subspecies *D. willistoni quechua* found in Peru, west of the Andes. Females of *D. w. quechua* crossed with males of the nominate subspecies produce nearly all sterile male hybrids but fully fertile female hybrids. The hybrid progeny of both sexes from the reciprocal cross are fully fertile.

In addition to partial reproductive isolation, the members of each pair of *D. willistoni* subspecies differ from each other in the patterns of their enzymes as assayed by gel electrophoresis. On the basis of allelic frequencies at five diagnostic enzyme loci, the probability of misclassification is 3.4×10^{-14} (Ayala, 1973). Ayala and Tracey (1973) determined that incipient sexual isolation between the two subspecies is no greater than that between geographic strains of the same subspecies.

Dobzhansky (1975) made a genetic analysis of the intersubspecific male hybrid sterility. Reminiscent of *D. p. pseudoobscura* and *D. p. bogotana*, major effects were associated with the X and second chromosomes and minor effects with the third chromosome. A stochastic effect was observed in backcross males in that both fertility and sterility occurred in individuals having the same hybrid genotype. Spermatogenesis in sterile males was observed to be normal except that binucleated spermatocytes were more frequent than in nonhybrid males. Both motile and nonmotile spermatozoa were found in the seminal vesicles of sterile males.

There are a number of similarities between the hybrid sterility encountered in *D. willistoni* and *D. pseudoobscura* subspecies hybrids. In both instances, the major nominate subspecies are widely distributed and the minor subspecies are isolated by geographic barriers and confined to very small areas. Also, in both cases, sterility is restricted to males and to only that reciprocal cross in which the female parent is from the minor subspecies. In neither species is there evidence for ethological isolation between subspecies.

B. *D. EQUINOXIALIS*

As in *D. willistoni*, the two allopatric subspecies *D. equinoxialis equinoxialis* and *D. equinoxialis caribbensis* are differentiated from one another by their enzyme patterns (Ayala, 1973). On the basis of allele frequencies at five diagnostic enzyme loci, considered together, the probability of subspecies misclassification is 8.0×10^{-11}. In this instance, intersubspecific hybrid sterility is fully reciprocal (Ayala *et al.*, 1974). However, although most crosses produce sterile male and fertile female progeny, irrespective of the direction of the cross, some populations of *D. e. equinoxialis* crossed to *D. e. caribbensis* are fully fertile in both directions (M. L. Tracey, personal communication). Allozyme frequencies in these aberrant populations are typical of their respective subspecies.

A small, but statistically significant, degree of incipient sexual isolation between different geographical strains of *D. equinoxialis* has been demonstrated (Ayala *et al.*, 1974), but differences in degree between intersubspecific and intrasubspecific isolation are not statistically significant.

C. *D. TROPICALIS*

The two subspecies *D. tropicalis tropicalis* and *D. tropicalis cubana* are partially reproductively isolated from one another in that both reciprocal intercrosses produce hybrid males that are completely sterile (Townsend, 1954). Degenerating spermatids were observed in the testes of sterile males. Not only were hybrid females fully fertile, but they were also heterotic for fecundity, i.e. they produced more progeny than nonhybrid females of the contributing subspecies. Sexual isolation between the two subspecies is minor and confined to one of the two reciprocal intercrosses. The relationship between these two subspecies closely resembles that between the two subspecies of *D. pallidipennis* (Patterson and Dobzhansky, 1945).

D. *D. PAULISTORUM*

Unlike the other subspecies of the *D. willistoni* group, discussed above, sexual isolation is marked among the six semispecies which make up the superspecies *D. paulistorum* (see Dobzhansky and Powell, 1975). Male hybrid sterility is also generally observed in the progeny of semispecies crosses, whereas hybrid females are fertile. Ehrman (1960) investigated the genetic basis of the male hybrid sterility in selected pairs of semispecies crosses. From backcross tests, she concluded that the presence of one or more semispecies foreign chromosomes in the mother is sufficient to cause sterility of all her male offspring. Thus, sterility is dependent on the genetic constitution of both the affected male and that of his mother. Cytological examination of F_1 hybrid males revealed disturbances in the first meiotic division which is apparently never completed. More extreme abnormalities were observed in the gonads of backcross males, suggesting premeiotic developmental arrest. The nonreciprocal male hybrid sterility in one semispecies cross (Santa Marta × Mesitas) appears to be due to a cytoplasm–Y chromosome interaction (Ehrman, 1963). A mycoplasma-like factor is possibly implicated in sterility production (Ehrman and Kernaghan, 1972). This factor can be transferred by injection. The sterility factor seems to be carried in the maternal cytoplasm, but its survival over more than one generation is dependent on the maternal genome (Ehrman, 1967). Further details are provided in Chapter 14, Volume 3b, of this series.

IV. *Drosophila melanogaster*

Two major types of intraspecific hybrid sterility have been discovered in this species during the last decade, both involving interstrain crosses. The two types of sterility, known as *SF* sterility and *GD* sterility, have a number of striking similarities, but they differ in their biological characteristics and they each are also manifestations of two apparently distinct cytoplasm–chromosome interaction systems (Kidwell, 1979). For several reasons, these two systems are of unusual interest for the study of incipient reproductive isolation despite the fact that the production of sterility usually seems to involve at least one laboratory strain.

In both instances, hybrid sterility is just one manifestation (albeit a potentially important one) of a more general syndrome of aberrant genetic traits called hybrid dysgenesis (Kidwell *et al.*, 1977b; Picard *et al.*, 1978). These other hybrid traits, such as chromosomal aberrations, lethal mutation and male recombination, appear to obey the same rules of inheritance as sterility (Engels, 1979a,b,c) and also generally tend to reduce the fitness of their carriers in that there is a reduction in progeny number in the following generation. Their frequent occurrence in association with sterility suggests major genetic disruptions of the hybrid genome. The discovery of these phenomena in *D. melanogaster* is allowing a depth of genetic analysis possible in few other organisms. For further information on other dysgenic traits associated with sterility, see reviews of Sved (1979), Bregliano *et al.* (1980) and also Woodruff *et al.* (this volume).

Others reasons for interest are the rapidity with which both sterility systems appear to be evolving and the discovery and description of a mode of non-Mendelian genetic transmission involving an unusual combination of chromosomal and extrachromosomal heredity and a phenomenon called "chromosomal contamination" implicating mobile genetic elements. The latter has important potential applications both for evolutionary studies and for providing new fundamental knowledge of genetic processes.

A. SF STERILITY

1. Qualitative and quantitative strain variability

A type of non-Mendelian hybrid sterility, subsequently called SF sterility (for sterilité femelle), was first described by Picard and L'Héritier (1971). SF sterility results from cytoplasm–chromosome interactions between two types of strains called inducer (I) and reactive (R). Inducer strains carry chromosomally linked factors called I factors which interact with a component of the cytoplasm carried by reactive strains. This interaction is nonreciprocal; it occurs when inducer males are crossed with reactive females to produce daughters called SF females (Bucheton, 1973). The effects of the interaction appear to be restricted to the female germ line. The eggs of SF females are characterized by a variable degree of reduced hatchability. The daughters of the reciprocal cross have normal fertility as do those from inducer × inducer and reactive × reactive strain matings (Picard and L'Héritier, 1971; Picard et al., 1972). Other aberrant traits produced by the I–R interaction include visible and lethal mutations and nondisjunction (Picard et al., 1978; Proust and Prudhommeau, 1982). For comprehensive reviews of the I–R system, see Bregliano et al. (1980) and Bregliano and Kidwell (1983).

In addition to the qualitative distinction between I and R strains, quantitative variation is observed among both I and R strains, but particularly within the R class (Bucheton et al., 1976), i.e. individual strains vary from strong to weak with respect to their reactive efficiencies. The majority of strains of this species can be unambiguously classified as I or R, but a third, neutral (N) type of strain has also been described. The evidence suggests that this third type is not strictly neutral but is weakly reactive, representing one extreme on the scale of reactivity.

2. Distribution of inducer and reactive strains

From a survey of strains from Europe, Africa and other parts of the world, Picard et al. (1976) reported that only inducer strains were found in natural populations. However, reactive, inducer and neutral types were identified in strains from laboratories in various parts of the world. Kidwell (1979) confirmed these broad observations in an independent sample of European and American strains. In a larger scale survey (M. G. Kidwell, in preparation), the relationship between laboratory age, geographic location and the status of strains in the I–R system has been investigated. A group of wild-type strains collected between 1930 and 1970 in the Americas and Europe showed a clear transition in the replacement of R by I types during this period. For a preliminary summary of this data, see Bregliano and Kidwell (1983). It is not clear whether the present day status of laboratory strains reflects their original condition in the wild or whether changes have occurred in formerly wild strains in response to a laboratory environment, but the available evidence tends to favor the first hypothesis. This question is discussed in Section IV, C, below.

3. Physiology of SF sterility

SF sterility is restricted to hybrid females; hybrid males have completely normal fertility (Picard, 1971). This female sterility has very specific characteristics (Picard et al., 1977) and is thus distinguishable from any other type of sterility, an advantage of considerable practical value. Gonadal morphology and cytology of SF females is completely normal. Low fertility results from the failure of eggs to complete embryonic development. It is completely determined by the chromosomal and cytoplasmic constitution of the hybrid mother and is not influenced by the male to which she has mated. Hatching failure may be complete or, more frequently, it may be partial, depending on a number of genetic and environmental factors. Developmental arrest occurs between the fifth and eighth cleavage divisions. Detailed cytological observations (Lavige and Lecher, 1982) showed a slowing down of the rate of mitosis in affected embryos. At the time of death, fragmentation of chromosomes into small pieces and spindle anomalies appear simultaneously. Embryos that successfully pass through this critical stage are themselves morphologically and physiologically normal.

4. Environmental dependence

Temperature is a critical factor affecting the manifestation of SF sterility (Picard et al., 1977). The effect of temperature varies, depending on the stage in the life cycle of SF females at which it is applied (Bucheton, 1979a). A temperature of 20°C, applied continuously during and after the development of SF females, produces the maximum reduction of egg hatchability which is characteristic for that particular cross. High temperature (29–30°C), applied at late oogenesis, results in an increase in hatchability. High temperature applied at any other stage of development does not result in a hatchability increase, on the contrary, it tends to reinforce *SF* sterility.

Egg hatchability is also strongly dependent on the age of the hybrid mother. Hatchability is minimum in the first eggs laid by *SF* females and steadily increases with female age. There is also an indirect effect on *SF* sterility of the reproductive age of mothers, maternal grandmothers and more distant female ancestors (Bucheton, 1979a).

5. *Hereditary transmission of* I *factors*

I factors may be linked to any of the four chromosomes, X, 2, 3 or 4, of inducer strains (Picard, 1976; Pelisson, 1977). In addition, inducer strains may be polymorphic for two kinds of chromosomes, those with and those without *I* factors (Picard, 1976; Picard and Pelisson, 1979). Those carrying *I* factors can interact with reactive cytoplasm, in female hybrids, to produce variable degrees of *SF* sterility. In males, *I* factors carried on inducer chromosomes are transmitted in a strictly Mendelian manner. However, in the germ line of heterozygous females bearing reactive cytoplasm, non-inducer chromosomes may acquire the properties of *I* factors by means of chromosomal contamination (Picard, 1976). Several lines of evidence support the hypothesis that *I* factors are mobile genetic elements whose transposition is normally repressed in inducer strains but greatly enhanced by reactivity (Bregliano and Kidwell, 1983). Mapping experiments indicated that *I* factors were located at only a few sites within three inducer chromosomes (Pelisson and Picard, 1979). However, the chromosomal location of *I* factors appears to vary from one inducer strain to another.

6. *Hereditary transmission of reactivity*

The degree of reactivity of *R* strains seems to be an expression of a cytoplasmic state. In any given generation, it does not depend directly on the chromosomes (Bucheton and Picard, 1978) but is mainly dependent on the cytoplasmic state of the mother. However, in the long run (up to 10 generations or more), the state and degree of reactivity of a strain is determined by the number and strength of maternal reactive chromosomes which act together additively. All three major chromosomes have been implicated in the determination of reactivity. Thus, reactivity is not determined by cytoplasmic inheritance *sensu strictu*, but by an unusual blend of cytoplasmic transmission and cumulative action of the chromosomes. It behaves rather like a conventional quantitative trait but involves a time lag of from one to several generations before the full effects of chromosomal factors are accumulated and expressed in the behavior of the cytoplasm.

In addition to the temporary effects of aging and heat treatment of *SF* females on their subsequent sterility (Picard *et al.*, 1977), it has been shown by Bucheton (1978, 1979b) that the effects of these two factors may be partially inherited when applied consistently to one or more generations of females of reactive strains. Aging of the mothers and more distant female ancestors increases the fertility of *SF* females. The reverse effect is also observed. The magnitude of this effect seems to be inversely proportional to the number of generations elapsing between treatment and observation. The long term effects of heat treatment on reactive strain females are more complex. Applied during late oogenesis, heat treatments tend to reduce the level of reactivity, but applied at any other stage in the life cycle, reactivity is strengthened.

7. *Chromosomal contamination*

Under certain, strictly defined, conditions, a non-Mendelian mode of genetic transmission called "chromosomal contamination" may occur (Picard, 1976). This phenomenon results in an irreversible acquisition of *I* factors by chromosomes previously lacking them. Whenever at least one inducer chromosome is present in a female germ cell, bearing reactive cytoplasm, all other chromosomes, whether homologous or not, may acquire the inducer property. This acquisition can occur at extremely high frequencies—approaching 100% under optimal conditions. Contaminated chromosomes seem to behave in an identical manner to donor inducer chromosomes. The donor chromosomes as well as noncontaminated chromosomes appear to remain unchanged (Picard, 1979). Noninducer chromosomes which occur together with inducer chromosomes in inducer stocks are relatively stable and do not undergo chromosomal contamination (Picard and Pelisson, 1979).

Chromosomal contamination seems to be a chance phenomenon occurring with a given probability that is dependent on a number of factors. Picard (1978a) found that this probability varied with the degree of reactivity of the maternal strain and the chromosomes involved (e.g., the X chromosome was contaminated less frequently than chromosome 2). Chromosomal contamination by *I* factors, like *SF* sterility and other aberrant traits, seems to be a manifestation of the *I–R* interaction.

B. *GD STERILITY*

1. *Qualitative and quantitative strain variability*

A type of hybrid sterility which affects both sexes and is closely associated with male recombination was first reported by Kidwell and Kidwell (1975).

Like *SF* sterility, this gonadal (*GD*) sterility results from cytoplasm–chromosome interactions between two types of strains, but the interaction system involved appears to be distinct from that which produces *SF* sterility (Kidwell, 1979). For *GD* sterility, the two types of interacting strains are called *P* (paternal contributor) and *M* (maternal contributor) (Kidwell, 1979; Engels, 1979b). *P* strains carry chromosomally linked factors called *P* factors which interact with the maternal cytoplasm of *M* strains to produce *GD* sterility. The reciprocal cross produces only a negligible degree of interaction. As a general rule, the progeny of intrastrain *P* and *M* matings have normal fertility. In addition to male recombination, a number of other aberrant traits have been found to be associated with *GD* sterility and the *P–M* interaction. These include lethal and visible mutation (Kidwell *et al.*, 1977a; Engels, 1979c), nondisjunction (Kidwell *et al.*, 1977b), chromosomal aberrations (Voelker, 1974; Engels, 1979a; Berg *et al.*, 1980), increased female recombination (Kidwell, 1977), transmission ratio distortion (Hiraizumi, 1971; Kidwell *et al.*, 1977b) and sex ratio distortion (Engels, 1979a). All aberrant traits appear to be restricted to the germ line.

In addition to the two main types of strains within the *P–M* system (Kidwell, 1979), *Q* strains are defined as those that do not interact with either *P* or *M* strains to produce significant frequencies of *GD* sterility in their hybrid progeny. There is some evidence to suggest that *Q* strains may be a subset of *P* strains because they have *P* cytotype (see Section 6, below) and frequently produce male recombination and lethal mutations when mated with *M* strain females (Simmons *et al.*, 1980; Engels and Preston, 1981; Kidwell *et al.*, 1981).

As with the *I–R* system, there is considerable quantitative variation in the strength of dysgenic potential among strains within both the *P* and *M* strain groups and between individuals within strains. At least some of this variation may be attributed to chromosomal polymorphism (Engels and Preston, 1980).

2. *Distribution of* P *and* M *strains*

Strains that have a laboratory age of at least 20–25 years have almost invariably been found to be of the *M* type, irrespective of whether they are wild type or carry markers or other specialized chromosomes (Kidwell *et al.*, 1977b; Kidwell, 1979). Strains caught in the wild, in the Americas, after 1955–60 seem increasingly likely to be of the *P* type and decreasingly likely to be of the *M* type, the younger their laboratory age (M. G. Kidwell, in preparation). Strains which are designated as *P* are, in fact, frequently polymorphic for *P* and *Q* chromosomes. A minority of strains collected during this recent period in the Americas appear to be completely neutral with respect to *GD* sterility. The frequency of *P* factors in present day American populations is quite high, but there is a possibility that on this continent the *M* type exists in the wild, albeit at low frequencies (Engels and Preston, 1980). The large degree of polymorphism in most natural populations sometimes makes a clear cut designation difficult.

Although the distribution of *P* and *M* types in natural population strains from other continents has not been studied to as great an extent as in North America, the limited data suggest some interesting differences from the American distribution. First, two strains collected in the U.S.S.R., in 1980, and two collected in Australia, in 1979 and 1980, have been unambiguously identified as *M* strains within a short period of their arrival in the laboratory (M. G. Kidwell and N. Plus, unpublished results). Furthermore, there is evidence that a transition from *M* to *P* started later in European than in American populations (M. G. Kidwell, in preparation). Also, based on a limited number of recent French samples, the frequency of *P* strains is markedly lower and that of *Q* strains higher, than in American present day natural populations (J. C. Bregliano, personal communication; M. G. Kidwell, unpublished results). Clearly, further detailed studies of the distribution of present day European populations are needed before any general conclusions can be drawn. However, even the limited available data on geographical differences between Europe and America suggest that laboratory age *per se* is not the only factor determining *P* or *M* status. Rather they suggest that differences may originate in natural populations before they are brought into the laboratory. The observation of recent outbursts of mutability at the singed locus (a hot spot for *P–M* induced mutation) in Russian natural populations (Golubovsky, 1980) might be interpreted in terms of the spread of *P* factors through original *M* populations. For further detailed discussion of this topic, see Section C below.

3. *Physiology and morphology of* GD *sterility*

GD sterility may affect both female and male hybrids, but sterility frequencies in females tend to be somewhat higher than those in males (Kidwell *et al.*, 1977b; Engels and Preston, 1979). External morphology of sterile individuals is normal and no major abnormalities in mating behavior have been observed (Engels and Preston, 1979; Kidwell and Novy, 1979). Sterile hybrid females lay no eggs. In most cases, dissection reveals a drastic reduction in size of both ovaries in adults (bilateral ovarian dysgenesis). Less frequently, ovaries are similarly affected unilaterally (Yannopoulos, 1978; Engels and Preston, 1979; Schaefer *et al.*, 1979). The ovarioles of dysgenic ovaries contain no vitellaria. Germaria lack any cells resembling the cystocyte clusters of normal ovaries. Morphologically, this type of sterility resembles that produced by the destruction of the germ cells or their determinants as a result of various treatments, usually UV-irradiation (e.g., Geigy, 1931). It is also similar to the manifestations of the maternally inherited, autosomal recessive mutant grandchildless (*gs*) of *Drosophila subobscura* (Fielding, 1967). *GD* sterility also seems to have many similarities with the sterility caused by *atrophie gonadique* (*ag*) (Periquet, 1976).

Unilateral ovarian dysgenesis does not necessarily result in sterility (Schaefer *et al.*, 1979). Broadhead *et al.* (1977) observed that the fecundity frequency distribution of fertile females differed markedly between reciprocal crosses. However, such an effect was not observed when space was a limiting factor (Engels and Preston, 1979). There is some evidence for a compensatory increase in egg production in females with unilateral rudimentary gonads.

Male *GD* sterility is also nonreciprocal and the testes may be either unilaterally or bilaterally reduced in size. However, a substantial fraction of males with normal testes are also sterile (Engels and Preston, 1979; Thompson *et al.*, 1980). The paragonia of sterile males are normal. A reduction in size of male gonads may first be observed at the late larval stage (Engels and Preston, 1979). For females there are conflicting reports as to whether such a size reduction can be seen in pre-adult stages. Schaefer *et al.* (1979) reported a marked reduction in the size of late third instar larval ovaries, but Engels and Preston (1979) found that the ovaries at this stage were normal in size. At least with respect to male sterility, strains seem to differ both quantitatively and qualitatively in the manifestation of gonadal abnormalities (Thompson *et al.*, 1980).

The results of recent experiments indicate that an additional type of partial sterility results from embryo death in the F_2 generation of *P–M* (and *Q–M*) dysgenic hybrids that have earlier escaped F_1 *GD* sterility (M. G. Kidwell, in preparation). The timing of this F_2 embryo lethality very roughly coincides with that of *SF* sterility but preliminary observations suggest that it does not coincide with any one precise stage of early development as does *SF* sterility. The gametes of both male and female F_1 dysgenic hybrids may contribute to F_2 embryo lethality. This type of F_2 partial sterility may be related to the reduced F_1 fecundity of dysgenic females (Broadhead *et al.*, 1977) and the sterility of F_1 males with normal gonads (Thompson *et al.*, 1980) which was mentioned earlier.

4. Environmental dependence

In both sexes, a high developmental temperature is essential for *GD* sterility to be manifested at high frequencies (Kidwell *et al.*, 1977b; Engels and

Preston, 1979). Kidwell and Novy (1979) constructed a temperature response curve for a developmental temperature range of 15–29°C. No significant frequencies of GD female sterility were observed below 24°C. A rapid rise in sterility occurred between 24 and 26°C and a maximum (often, but not always of 100%) is reached at and above 27°C. The critical temperature for rapid sterility increase seems to be slightly higher for males than for females (Kidwell and Novy, 1979).

The temperature sensitive period for GD sterility extends from late embryonic through early larval stages. The beginning of this period coincides, approximately, with the initiation of mitosis among primordial germ cells, a few hours before the hybrid egg hatches (Engels and Preston, 1979; Kidwell and Novy, 1979). After the completion of the temperature sensitive period, GD sterility is not reversible by environmental manipulation.

In addition to the effect of temperature, there is evidence that GD sterility increases with the age of the hybrids themselves (Kidwell *et al.*, 1977b). However, there is a tendency for the dysgenic hybrid daughters of older mothers to exhibit lower frequencies of ovarian dysgenesis than those of younger mothers (A. Bucheton, personal communication; M. G. Kidwell, unpublished results).

Frequencies of F_2 sterility, associated with embryo death, increase with parental age and with increases in developmental temperature. However, the temperature sensitive period occurs later in development than it does for GD sterility. Also, holding F_1 egg-laying females at higher temperatures tends to increase the frequency of F_2 embryo death (M. G. Kidwell, in preparation).

5. *Hereditary transmission of the* P *component*

From genetic experiments the chromosomal component of the P–M interaction appears to consist of polygenic P factors linked to one or more of the major chromosomes of P strains (Engels, 1979b). Each of these chromosomes, either singly or together, in M cytoplasm, has the ability to produce GD sterility, but the relative efficiency varies among different chromosomes from the same strain and among the same chromosomes from different strains. By the nature of the usual type of mating scheme, P chromosomes are generally transmitted to a dysgenic hybrid by the paternal parent, but, contrary to some previous reports, paternal transmission is not a necessary condition for the induction of dysgenic traits. By genetic manipulation, it can be arranged that some P chromosomes, together with M cytoplasm, are both transmitted maternally, and under these conditions it is found that GD sterility frequencies are similar to those with paternal transmission. The effects of multiple P chromosomes in the same dysgenic genome are not completely additive, but a model of independent action provides a useful approximation (Engels, 1979b).

Recently obtained molecular evidence indicates that P factors belong to a somewhat heterogeneous DNA sequence family of elements dispersed throughout the genome of P strains. Using the method of *in situ* hybridization, Bingham *et al.* (1982) demonstrated that P and Q strains carry 30–50 sites of homology to cloned copies of the P element, distributed among all the major chromosomes, and that the precise locations of these sites are specific to each P and Q strain examined. No sites of homology to the P element were found in most long-established M strains.

Biochemical analysis of four dysgenesis-induced insertion mutations in the white locus (Rubin *et al.*, 1982) indicated a range in size of inserted sequences between 0·5 and 1·4 kb. This suggests that P factors may be small relative to the size of other *Drosophila* mobile element families such as the *copia*-like elements (Rubin, 1983). The mobility of P factors is discussed in Subsection 7 below.

6. *Hereditary transmission of the* M *component*

The second component of the P–M interaction can be described as a susceptibility or lack of resistance to the action of chromosomally linked P factors. This susceptibility is a property of the maternally transmitted cytoplasm of M strains. P strains, characteristically, have cytoplasm that is resistant to their own chromosomal P factors. Thus, as a general rule, intrastrain GD sterility is not observed. This characteristic of susceptibility or resistance to P factors has been termed "cytotype" by Engels (1979b). Thus, individuals having P cytotype are resistant to the action of chromosomal P factors and those with M cytotype are susceptible.

Most individuals possess the same cytotype as their mothers, regardless of their chromosomal constitution (Engels, 1979b; Kidwell, 1981). Sometimes, however, in flies carrying a mixture of P and M or Q and M chromosomes, the cytotype can switch to the opposite state. The probability of switching to P increases with the number of chromosomes of P or Q strain origin and *vice versa*. A cytotypic switch by means of chromosomal substitution can be made in either direction with approximately equal efficiency. Furthermore, each of the three major chromosomes may be involved in determining cytotype. The strength and chromosomal location of cytotype determinants appears to vary among different strains (Kidwell, 1981). In summary, cytotype can be considered as a self-replicating property of the cytoplasm, but in the long run it is a function of the genomic constitution of the strain.

7. *Evidence that* P *factors are transposable elements*

The inordinately high spontaneous mutation frequencies associated with *MR* (male recombination) chromosomes (Green, 1977; Golubovsky *et al.*, 1977) and the hypermutability associated with the singed locus (Engels, 1979c) strongly suggested the implication of mobile elements in the *P–M* system of hybrid dysgenesis. Indirect evidence for the mobility of male recombination elements in the genome was provided by Slatko (1978), Matthews *et al.* (1978) and Yannopoulos (1979). In a screen for mutations in the zeste-white region of the X chromosome of *P–M* dysgenic hybrids, Simmons and Lim (1980) cytologically identified insertions in two different mutant stocks. Using *GD* sterility as the method of assay, Kidwell (1983) demonstrated the occurrence of chromosomal contamination in the *P–M* system in a manner analogous to Picard's (1976) demonstration in the *I–R* system. The transfer of *GD* sterility potential from *P* autosomes to *M* X chromosomes was demonstrated to take place at quite high frequencies under certain circumstances when these chromosomes were associated nonhomologously in the same genome for at least one generation. These putative transposition events were dependent on the presence of *M* cytotype.

Direct evidence that *P* factors are transposable elements was provided by the detailed biochemical analyses of four mutants at the white locus induced by *P–M* hybrid dysgenesis (Rubin *et al.*, 1982; Bingham *et al.*, 1982). In every case the occurrence of a mutation was accompanied by an insertion of nonwhite locus DNA into the white locus. These insertions were shown to have sequence homology with one another and with a cloned *P* factor. Unstable mutants at other X chromosome loci were also induced by *P–M* interaction, many at hot spots already recognized to be associated with this system. Sequence homology of these mutations with the cloned *P* factor was also demonstrated by *in situ* hybridization, again indicating insertion of *P* factors into the mutated loci.

C. Comparison of *SF* and *GD* Sterility Systems

A summary of terminology and some of the most striking differences between *SF* and *GD* sterilities are provided in Table I. Although both the *SF* and *GD* types of sterility are unusual exceptions to Haldane's rule (Haldane, 1922), they differ in the degree to which female sterility exceeds that of males. *SF* sterility is completely restricted to female hybrids, but significant frequencies of *GD* sterility are found in male hybrids, albeit at a lower frequency than in females. This difference in pattern seems to be carried through to other traits associated with the two systems of dysgenesis.

TABLE I. A comparison of terminology and sterility characteristics of the *I–R* and *P–M* systems of hybrid dysgenesis.

	System	
	I–R	*P–M*
Paternally contributing strains	*I* (inducer) all wild and some laboratory strains	*P* (paternal) most American wild strains
Maternally contributing strains	*R* (reactive) some laboratory strains	*M* (maternal) long-established laboratory strains
Neutral strains	*N* (possibly very weak *R*) some laboratory strains	*Q* (possibly very weak *P*) some wild strains
Type of sterility	*SF* (eggs of hybrids fail to hatch)	*GD* (hybrid gonads rudimentary)
Sex affected	Female only	Both (but ♀ > ♂)
Restrictive temperature	20°C[a]	>26°C
Temperature sensitive phase	Late oogenesis	Late embryonic and early larval stages

[a] The precise range of restrictive temperatures is not known.

There is no evidence that male recombination or any other male dysgenic traits are associated with the *I–R* system; all aberrant traits are restricted to hybrid females (Picard *et al.*, 1978). In contrast, several traits, including lethal mutations, transmission ratio distortion and increases in recombination frequency have been observed in both male and female hybrids associated with *GD* sterility in the *P–M* system.

The DNA changes resulting in both *SF* and *GD* sterilities appear to be restricted to the germ line, but the developmental interruption occurs at very different stages in the two types. Rudimentary ovaries have been observed in *GD* sterile female larvae and clearly the blockage occurs long before meiosis in F_1 hybrids, in contrast to the later time of induction of male recombination (Henderson *et al.*, 1978; but see also Sved, 1978). In contrast to *GD* sterility, *SF* sterility is not manifested at all in F_1 females themselves but in hatching deficiencies of their eggs.

All attempts to transmit *SF* fertility, *I* factor or reactivity by injection or feeding have been completely unsuccessful (J. C. Bregliano, personal communication). Similarly, using the Harwich and Hunter Valley strains, neither male recombination nor *GD* sterility have been found to be transmissible by injection or feeding (Sved *et al.*, 1978; J. A. Sved, personal communication; R. E. Schaefer, unpublished results). However, using a different wild strain, Sochacka and Woodruff (1976) concluded that male recombination could be transmitted by injection.

One of the most remarkable similarities between *SF* sterility and *GD* sterility lies in their mode of inheritance. Each type is a manifestation of a complex cytoplasm–chromosome interaction system involving the destabilization of transposable elements and is largely restricted to hybrids produced from one of two possible reciprocal strain crosses. In both cases, sterility can be produced by each of the major chromosomes (from I or P strains) in combination with the appropriate interacting cytoplasm (R or M). Also, the properties of the R and M cytoplasms are similarly inherited, being largely dependent on the state of the maternal cytoplasm but ultimately dependent on a combination of maternal and paternal genotypes acting together with the maternal cytoplasm over the course of several generations. The involvement of chromosomal contamination (Picard, 1976) in the inheritance of SF sterility is clear. Early attempts to demonstrate a similar phenomenon in the case of GD sterility have failed, but recently success was achieved using a different mating scheme and combination of strains (Kidwell, 1983). In addition, Kidwell *et al.* (1981) have demonstrated very rapid unidirectional changes to the P type in small populations of mixed P/M origin. It seems that chromosomal contamination may be at least quantitatively different and possibly qualitatively different in the two systems. Also, the independent action of P chromosomes (Engels, 1979b) clearly differs from the nonadditive way in which I chromosomes act together (Picard, 1978b). Single I chromosomes may cause more sterility than combinations (Pelisson, 1979).

Both SF and GD sterilities are highly dependent on environmental factors for their manifestation. The frequency of both types of sterility is increased by high developmental temperatures, but they differ in their patterns of response to a range in temperatures. There are also notable differences in the developmental stage at which the temperature sensitive period occurs. For SF sterility this coincides approximately with the stage of vitellogenesis of the eggs produced by hybrid females, i.e. it extends for a one to two day period prior to egg laying. For GD sterility the temperature sensitive period is considerably earlier, occurring during the development of the hybrids themselves. Starting at about the eighth hour, it may extend through the third or fourth day of development (Engels and Preston, 1979; Kidwell and Novy, 1979). Although the aging of GD sterile females *per se* does not reverse their infertility as it does for SF sterility, there appears to be a similar effect of aging of the mothers of SF and GD sterile females. There is a tendency for the daughters of older mothers to show a lower frequency of sterility than those of younger mothers in both systems.

Despite arguments to the contrary (see, for example, Eggleston and Kearsey, 1980), there are now several reasons to support the idea that the I–R and P–M systems of hybrid dysgenesis are essentially independent. First, the physiological features of dysgenic traits in the two systems differ.

For example, male hybrids are affected in the *P–M* system but not the *I–R* system; also, the types of sterility characteristic of the two systems differ markedly, as described earlier. Second, there is evidence for at least partial distributional independence (Kidwell, 1979) in that inducer strains may be *P*, *M* or *Q*. Third, recent evidence from biochemical experiments, using the method of *in situ* hybridization, indicates that there is no sequence homology between *P* and *I* factors (Bingham *et al.*, 1982). Fourth, there is evidence that *M* cytotype has no effect on the frequency of transposition of *I* factors; many *IM* strains exist in which *I* factors appear to be equally as stable as they are in *IP* strains. Because no *RP* strain has yet been identified, it has not been possible to determine directly whether the reactive state affects the transposition of *P* factors.

TABLE II. Types of sterility expected under appropriate restrictive developmental temperature regimes for strains with various combinations of interaction potential in the *I–R* and *P–M* systems.

Strain designation—female parent	Strain designation—male parent			
	RM	RP[a]	IM	IP
RM		GD[a]	SF	SF GD
RP[a]			SF[a]	SF[a]
IM		GD[a]		GD
IP				

[a] Hypothetical strain combinations and sterility expectations not yet observed.

Assuming the independence of the two systems, all *D. melanogaster* strains may be assigned a dual designation according to their demonstrated potential to produce either *SF* or *GD* sterility. Omitting neutral types, there are therefore four possible combinations of components from the two systems. These are shown in Table II along with the types of sterility expected from each cross. All strain types have been identified except for the *RP* combination. As described below, it is possible that this type is rare, or non-existent because the *I* type may have become fixed in natural populations before the *P* type evolved (Kidwell, 1979).

The distributions of laboratory and natural population strains with respect to their potential for sterility in the two systems show some general similarities in pattern (Bregliano and Kidwell, 1983) but some differences in detail. No natural population strain collected in the last decade has unambiguously been characterized as other than inducer. While *P* strains seem to predominate in present day American natural populations, there is evidence for a significant frequency of Q strains which are neutral for GD sterility and also the possibility of a low frequency of M strains. On the other hand, long-established laboratory strains have been found to be of all three types, inducer, reactive and neutral, but they are also exclusively of the M type. As pointed out by Engels and Preston (1980), the low frequency of dysgenic traits in natural populations (because of the preponderance of the IP type) does not at all imply that hybrid dysgenesis is irrelevant to evolutionary studies.

Two rather different but not necessarily mutually exclusive hypotheses have been advanced to explain the observed distribution of strains in laboratory and natural populations. The first hypothesis was suggested by Bucheton et al. (1976) with respect to the $I-R$ system and extended by Engels (1981), in the form of the "stochastic loss hypothesis", to include the $P-M$ system and other transposable element systems. In relation to the two hybrid dysgenesis systems, it proposes that the I and P properties are the general condition for wild populations and that the R and M states are found only in small isolates, mainly in laboratories in which a drastic change in environment and breeding structure has resulted in the loss of I and P transposable elements. An alternative hypothesis has been proposed by Kidwell (1979) in which the R and M states are assumed to have been the property of most natural populations until very recently. First the I and then the P condition are hypothesized to have originated or become activated and rapidly spread through the majority of natural populations within the last half century. In this view the R and M states of long established laboratory stocks reflect their original natural population state and may soon be viewed as a kind of evolutionary relic. It is not clear whether or not selection may be necessary for the rapid spread of I and P transposable elements in populations. Transposability *per se* might be a sufficient drive mechanism if its frequency were high. For further detailed discussion of this topic, see Kidwell et al. (1981) and Bregliano and Kidwell (1983).

A question of some interest is to what extent the various cases of intraspecific hybrid sterility reported in this species by different workers can be classified as belonging to the two major types of hybrid dysgenesis described above. Although it would be hazardous to assign individual cases rigidly to a specific system unless common reference stocks were used for hybridization, tentative classification can be attempted on the basis of similar properties and patterns of inheritance. It should also be re-emphasized that some crosses between laboratory and wild populations have the potential for both types of sterility although the existence of only one type may have, in fact, been tested.

Rosenfeld et al. (1971) reported nonreciprocal hybrid sterility in crosses between two laboratory stocks *abo/Cy* and *pr cn*. Picard *et al.* (1972) noted the similarities between this sterility and the type they were observing in France (*SF* sterility). The *abo/Cy* and *pr cn* stocks have subsequently been unambiguously characterized in the *I–R* system (Kidwell, unpublished results); *abo/Cy* has proved to be a strong *I* stock and *pr cn* a strong *R* stock. This result, in addition to the similarities previously described, makes it almost certain that Rosenfeld *et al.* were observing what is now called *SF* sterility.

The characteristics of the hybrid sterility reported by Kearsey *et al.* (1977) also resemble those of *SF* sterility in many ways. The general pattern of inheritance seems to be identical. Sterility is restricted to females, is readily affected by environmental factors and is independent of the genotype of the male used to fertilize the eggs of hybrid females. The second and third (and possibly X) chromosomes were found to be polymorphic for the presumptive inducer factors.

On the basis of their association with male recombination and the morphological and physiological properties of the observed sterility, it seems likely that the observations of Yannopoulos (1978), based on chromosomes extracted from a Greek population, can be identified as *GD* sterility. Matthews and Gerstenberg (1979) observed non-reciprocal female sterility in the progeny of males from a number of Texas male recombination (*MR*) strains when mated to laboratory strain females at 28·5°C. High frequencies of sterility were not found in the progeny of intrastrain matings raised at this temperature nor in any matings raised at room temperature (23–24°C). They concluded that Texas *MR* strains were therefore not essentially different from those of other workers with respect to their potential for sterility. Previous failures to demonstrate sterility in association with *MR* chromosomes could be completely accounted for by the use of only room temperatures which were permissive for sterility. These results provide further strong evidence for the frequent association of the *GD* type of sterility with male recombination. They also demonstrate how differences in test environments may lead to seemingly inconsistent results.

GD sterility closely resembles an essentially intrastrain type of sterility called *atrophie gonadique* and symbolized *ag* (Periquet, 1976). The *ag* character is exhibited by both females and males of certain strains collected in the wild and subsequently maintained in the laboratory. Periquet (1978a) claims it has some of the properties of a quantitative genetic trait and is determined by genes located on chromosomes 2 and 3. The penetrance of *atrophie gonadique* is strongly affected by temperature in a similar direction to *GD* sterility (Periquet, 1976, 1978b). In females, it is also affected by rank order of egg-laying (Periquet, 1978c). As for *GD* sterility, both male and female gonads may be considerably reduced in size unilaterally or bilaterally, with the blockage occurring during the first five hours of embryonic development (Periquet, 1976).

The type of sterility associated with the extrachromosomal element delta (Minamori and Ito, 1972) has some suggestive similarities but no clear parallels with either the GD or the SF type. The biological characteristics of this sterility are not clear from published reports. Sometimes, in affected males, the number and motility of sperm were reported to be deficient. In other instances, it was claimed that zygote mortality was responsible for sterility. It is apparent that this type of sterility is also temperature dependent, high temperatures being permissive and low temperatures restrictive as is the case for SF sterility.

V. Other Species

Some further examples of intraspecific hybrid sterility have been reported within species belonging to four other species groups. In each instance, the sterility is restricted to male hybrids, but the degree of sterility and its characteristics are highly variable.

In the *pallidipennis* group, crosses between the two allopatric subspecies *D. pallidipennis pallidipennis* and *D. p. centralis* exhibit partial hybrid sterility (Patterson and Dobzhansky, 1945). F_1 hybrid males are completely sterile, but F_1 females are normally fertile. No evidence for reciprocal cross effects has been reported. Backcross females and a few backcross males are fertile. Spermatogenesis in sterile males is grossly defective. Although meiotic pairing is normal, both first and second meiotic divisions are abortive, giving rise to spermatids that degenerate. From a number of backcross tests, it was observed that the greater the fraction of chromosomes from one parental species, the greater the chance that a male would be fertile. Patterson and Dobzhansky suggested that sterility results from an interaction between the two sets of subspecies genes. These two subspecies have only slight morphological differences and no observed sexual isolation. The only chromosomal difference is a single autosomal inversion.

A less drastic reduction in hybrid fertility occurs with respect to three subspecies of *D. macrospina* in the *funebris* group. When females of the western subspecies *D. macrospina limpiensis* are crossed to males of *D. m. macrospina* or *D. m. ohioensis*, hybrid males are sterile or semisterile (Mainland, 1942). In this case, sperm are usually produced, but they are mostly immotile. The genetic basis for this nonreciprocal sterility is not clear. Possible mechanisms are a Y-autosome dependency and X-autosome complementary factors (Mainland, 1942). Some sexual isolation and chromosomal differences (Warters, 1944) have been observed among this group of three subspecies.

Unidirectional hybrid male sterility has also been found in crosses between two strains of *D. micromelanica* in the *melanica* species group. Sturtevant and Novitski (1941) tested strains from Arizona and Texas and found that the progeny of both sexes were fertile from the cross Arizona females × Texas males. The reciprocal cross produced sterile males. The cause of this sterility was demonstrated to be a factor closely linked to white on the Texas X chromosome which interacted with the Arizona Y chromosome. No sex chromosome–autosome interactions were observed. It has thus been concluded that a single X chromosome mutational difference may be sufficient for cross-sterility to occur (Patterson and Stone, 1952).

Two strains of *D. peninsularis* in the *repleta* group exhibit some partial male sterility when crossed together (Patterson and Wheeler, 1947). The two strains, collected in Florida about thirty miles apart, showed some morphological differences including a difference in the shape of the spermathecal apparatus. There was also unilateral sexual isolation. Some hybrid males produced only immature nonmotile sperm bundles and no functional gametes. A degree of hybrid inviability was also observed.

In contrast to those described above, there are a number of other subspecies pairs of *Drosophila* which exhibit no observable degree of F_1 hybrid sterility. These include the pair of subspecies *D. mercatorum mercatorum* and *D. m. para repleta* in the *repleta* species group. In this instance, heterosis was observed in the F_1, but there was hybrid breakdown manifested as reduced viability in the F_2 and assumed to be the result of genic imbalance (Wharton, 1944). No sexual isolation was observed. Also in the same species group, the subspecies *D. fulvimacula fulvimacula* and *D. f. flavorepleta* show no hybrid sterility. However, there is an interesting unidirectional sexual isolation in that both F_1 males and females backcrossed to *fulvimacula* produce only a few offspring; production is normal from backcrosses to *flavorepleta* (Patterson, 1952).

VI. Discussion

In the years during and immediately after World War II, *Drosophila* studies at various taxonomic levels contributed substantially to evolutionary knowledge (Patterson and Stone, 1952). It is a fair question to ask what increases in knowledge have resulted from more recent studies, particularly at the intraspecific level. With respect to postmating reproductive isolation, the diversity of mechanisms involved has become increasingly apparent with the inclusion of new types such as infectious sterility, as in *D. paulistorum*. The involvement of multiple chromosomal and even extra-chromosomal factors in hybrid sterility has been shown by recent studies to occur quite frequently as part of complex interaction systems. These show various degrees of variation among different systems, but there are also many similarities.

The differences found between members of pairs of subspecies seem to parallel to some extent those found between members of pairs of full species (Ehrman, 1962). The wide variety of patterns of divergence seen between different pairs of subspecies suggests that postmating reproductive isolating mechanisms can arise independently of gross chromosomal changes, sexual isolation and allozyme differentiation rather than being a byproduct of general genetic divergence. This conclusion is supported by the observation that hybrids between some recently evolved pairs of species show no evidence of sterility under laboratory conditions. For example, the broadly sympatric Hawaiian species *D. heteroneura* and *D. silvestris* are morphologically, behaviorally and cytologically distinct, but laboratory hybrids between them are fully fertile (Carson, 1978).

As a result of recent findings, there is now the possibility of another solution to the old problem of how hybrid sterility can evolve (particularly in early stages) despite the resulting drastic lowering of fitness. While it is presently premature to do more than speculate, there seems to be the possibility that the phenomenon of chromosomal contamination in *D. melanogaster* involving the destabilization of chromosomally integrated mobile elements may provide a "drive" mechanism to spread sterility factors in populations without necessarily invoking the action of selection. In conjunction with this, the conditional nature of the two types of hybrid interaction described in this species suggest a further manner by which deleterious selection might be avoided under permissive environmental conditions. In relation to hybrid dysgenesis, the demonstration that a number of deleterious hybrid traits may accompany the induction of sterility suggests fascinating possible implications for directly or indirectly contributing towards incipient reproductive isolation as well as for increasing genetic variation in general. However, at present we are completely ignorant of the evolutionary importance of transposable element systems, relative to other genetic mechanisms, for generating diversity and promoting speciation.

ACKNOWLEDGEMENTS

I am indebted to Drs. W. R. Engels, J. F. Kidwell, J. A. Sved, M. L. Tracey and Professor J. C. Bregliano and his colleagues at the Université de Clermont-Ferrand II, who provided helpful advice on the first draft. Preparation of the manuscript was supported, in part, by grants from the NSF (DEB 76-82630) and the NIH (GM-25399).

References

AYALA, F. J. (1973). Two new subspecies of the *Drosophila willistoni* group. *Pan Pacific Entomol.* **49**, 273–279.

AYALA, F. J. and DOBZHANSKY, T. (1974). A new subspecies of *Drosophila pseudoobscura*. *Pan Pacific Entomol.* **50**, 211–219.

AYALA, F. J. and TRACEY, M. L. (1973). Enzyme variability in the *Drosophila willistoni* group. VIII. Genetic differentiation and reproductive isolation between two subspecies. *J. Hered.* **64**, 120–124.

AYALA, F. J., TRACEY, M. L., BARR, L. G. and EHRENFELD, J. G. (1974). Genetic and reproductive differentiation of the subspecies *Drosophila equinoxialis caribbensis*. *Evolution* **28**, 24–41.

BERG, R. L., ENGELS, W. R. and KREBER, R. A. (1980). Site specific X chromosomal rearrangements from hybrid dysgenesis in *Drosophila melanogaster*. *Science* **210**, 427–429.

BINGHAM, P. M., KIDWELL, M. G. and RUBIN, G. M. (1982). The molecular basis of P-M hybrid dysgenesis: the role of the P element, a P-strain specific transposon family. *Cell* **29**, 995–1004.

BREGLIANO, J. C., PICARD, G., BUCHETON, A., LAVIGE, J. M., PELISSON, A. and L'HÉRITIER, PH. (1980). Hybrid dysgenesis in *Drosophila melanogaster*. *Science* **207**, 606–611.

BREGLIANO, J. C. and KIDWELL, M. G. (1983). Hybrid dysgenesis determinants. In: "Mobile Genetic Elements" (J. Shapiro, ed) Academic Press, New York, London, San Francisco. (In press).

BROADHEAD, R. S., KIDWELL, J. F. and KIDWELL, M. G. (1977). Variation of the recombination fraction in *Drosophila melanogaster* females. *J. Hered.* **68**, 323–326.

BUCHETON, A. (1973). Contribution à l'étude de la stérilité femelle non mendélienne chez *Drosophila melanogaster*. Transmission héréditaire des degrés d'efficacité du facteur «reacteur». *C. R. Acad. Sci. Paris* **276**, 641–644.

BUCHETON, A. (1978). Non-Mendelian female sterility in *Drosophila melanogaster*: influence of ageing and thermic treatments. I. Evidence for a partly inheritable effect of these two factors. *Heredity, Lond.* **41**, 357–369.

BUCHETON, A. (1979a). Non-Mendelian female sterility in *Drosophila melanogaster*: influence of ageing and thermic treatments. II. Action of thermic treatments on the sterility of *SF* females and the reactivity of reactive females. *Biol. Cell.* **34**, 43–49.

BUCHETON, A. (1979b). Non-Mendelian female sterility in *Drosophila melanogaster*: influence of ageing and thermic treatments. III. Cumulative effects induced by these factors. *Genetics* **93**, 131–142.

BUCHETON, A. and PICARD, G. (1978). Non-Mendelian female sterility in *Drosophila melanogaster*: hereditary transmission of reactivity levels. *Heredity, Lond.* **40**, 207–223.

BUCHETON, A., LAVIGE, J. M., PICARD, G. and L'HÉRITIER, PH. (1976). Non-Mendelian female sterility in *Drosophila melanogaster*: quantitative variations in the efficiency of inducer and reactive strains. *Heredity, Lond.* **36**, 305–314.

CARSON, H. L. (1978). Speciation and sexual selection in Hawaiian *Drosophila*. In: "Ecological Genetics: The Interface" (P. F. Brussard, ed), pp. 93–107. Springer-Verlag, New York.

DOBZHANSKY, TH. (1935). The Y-chromosome of *Drosophila pseudoobscura*. *Genetics* **20**, 366–376.

DOBZHANSKY, TH. (1937). "Genetics and the Origin of Species." Columbia University Press, New York.

DOBZHANSKY, TH. (1974). Genetic analysis of hybrid sterility within the species *D. pseudoobscura*. *Hereditas* **77**, 81–88.

DOBZHANSKY, TH. (1975). Analysis of incipient reproductive isolation within a species of *Drosophila*. *Proc. Natl. Acad. Sci. U.S.A.* **72**, 3638–3641.

DOBZHANSKY, TH. and POWELL, J. R. (1975). The *willistoni* group of sibling species of *Drosophila*. *In*: "Handbook of Genetics. Invertebrates of Genetic Interest" (R. C. King, ed.) Vol. 3, pp. 589–622. Plenum Press, New York and London.

EGGLESTON, P. and KEARSEY, M. J. (1980). Hybrid dysgenesis in *Drosophila*: correlation between genetic traits. *Heredity* **44**, 237–249.

EHRMAN, L. (1960). The genetics of hybrid sterility in *Drosophila paulistorum*. *Evolution* **14**, 212–223.

EHRMAN, L. (1962). Hybrid sterility as an isolating mechanism in the genus *Drosophila*. *Q. Rev. Biol.* **37**, 279–302.

EHRMAN, L. (1963). Apparent cytoplasmic sterility in *Drosophila paulistorum*. *Proc. Natl. Acad. Sci. U.S.A.* **49**, 155–157.

EHRMAN, L. (1967). A study of infectious hybrid sterility in *Drosophila paulistorum*. *Proc. Natl. Acad. Sci. U.S.A.* **58**, 195–198.

EHRMAN, L. and KERNAGHAN, R. P. (1972). Infectious heredity in *Drosophila paulistorum*. *In*: "Ciba Symposium: Pathogenic Mycoplasmas," pp. 227–250, Associated Scientific Publishers, Amsterdam.

ENGELS, W. R. (1978). Chromosome–cytoplasm interactions in *Drosophila melanogaster*. Ph.D. thesis. University of Wisconsin, Madison.

ENGELS, W. R. (1979a). Germline aberrations associated with a case of hybrid dysgenesis in *Drosophila melanogaster* males. *Genet. Res., Camb.* **33**, 137–146.

ENGELS, W. R. (1979b). Hybrid dysgenesis in *Drosophila melanogaster*: rules of inheritance of female sterility. *Genet. Res., Camb.* **33**, 219–236.

ENGELS, W. R. (1979c). Extrachromosomal control of mutability in *Drosophila melanogaster*. *Proc. Natl. Acad. Sci. U.S.A.* **76**, 4011–4015.

ENGELS, W. R. (1981). Hybrid dysgenesis in *Drosophila* and the stochastic loss hypothesis. *Cold Spring Harb. Symp. Quant. Biol.* **45**, 561–565.

ENGELS, W. R. and PRESTON, C. R. (1979). Hybrid dysgenesis in *Drosophila melanogaster*: the biology of female and male sterility. *Genetics* **92**, 161–174.

ENGELS, W. R. and PRESTON, C. R. (1980). Components of hybrid dysgenesis in a wild population of *Drosophila melanogaster*. *Genetics* **95**, 111–128.

ENGELS, W. R. and PRESTON, C. R. (1981). Characteristics of a "neutral" strain in the *P–M* system of hybrid dysgenesis. *Dros. Inf. Serv.* **56**, 35–37.

FIELDING, C. J. (1967). Developmental genetics of the mutant *grandchildless* of *Drosophila subobscura*. *J. Embryol. Expl. Morphol.* **17**, 375–384.

GEIGY, R. (1931). Action de l'ultra-violet sur le pôle germinal dans l'oef de *Drosophila melanogaster*. *Rev. Suisse Zool.* **38**, 187–288.

GOLUBOVSKY, M. D. (1980). Mutational process and microevolution. *Genetica* **52/53**, 139–149.

GOLUBOVSKY, M. D., IVANOV, YU. N. and GREEN, M. M. (1977). Genetic instability in *Drosophila melanogaster*: putative multiple insertion mutants at the singed bristle locus. *Proc. Natl Acad. Sci. U.S.A.* **74**, 2973–2975.

GREEN, M. M. (1977). Genetic instability in *Drosophila melanogaster*: De novo induction of putative insertion mutants. *Proc. Natl. Acad. Sci. U.S.A.* **74**, 3490–3493.

HALDANE, J. B. S. (1922). Sex ratio and unisexual sterility in hybrid animals. *J. Genet.* **12**, 101–109.

HENDERSON, S. A., WOODRUFF, R. C. and THOMPSON, J. N. (1978). Spontaneous chromosome breakage at male meiosis associated with male recombination in *Drosophila melanogaster*. *Genetics* **88**, 93–107.

HIRAIZUMI, Y. (1971). Spontaneous recombination in *Drosophila melanogaster* males. *Proc. Natl. Acad. Sci. U.S.A.* **68**, 268–270.

KEARSEY, M. J., WILLIAMS, W. R., ALLEN, P. and COULTER, F. (1977). Polymorphism for chromosomes capable of inducing female sterility in *Drosophila. Heredity Lond.* **38**, 109–115.

KIDWELL, M. G. (1977). Reciprocal differences in female recombination associated with hybrid dysgenesis in *Drosophila melanogaster. Genet. Res. Camb.* **30**, 77–88.

KIDWELL, M. G. (1979). Hybrid dysgenesis in *Drosophila melanogaster*: the relationship between the *P–M* and the *I–R* interaction systems. *Genet. Res. Camb.* **33**, 205–218.

KIDWELL, M. G. (1981). Hybrid dysgenesis in *Drosophila melanogaster*: the genetics of cytotype determination in a neutral strain. *Genetics* **98**, 275–290.

KIDWELL, M. G. (1983). Hybrid dysgenesis in *Drosophila melanogaster*: factors affecting chromosal contamination in the *P-M* system. *Genetics* (in press).

KIDWELL, M. G. and KIDWELL, J. F. (1975). Cytoplasm–chromosome interactions in *Drosophila melanogaster. Nature, Lond.* **253**, 755–756.

KIDWELL, M. G. and NOVY, J. B. (1979). Hybrid dysgenesis in *Drosophila melanogaster*: sterility resulting from gonadal dysgenesis in the *P–M* system. *Genetics* **92**, 1127–1140.

KIDWELL, M. G., KIDWELL, J. F. and IVES, P. T. (1977a). Spontaneous non-reciprocal mutation and sterility in strain crosses of *Drosophila melanogaster. Mutation Res.* **42**, 89–98.

KIDWELL, M. G., KIDWELL, J. F. and SVED, J. A. (1977b). Hybrid dysgenesis in *Drosophila melanogaster*: a syndrome of aberrant traits including mutation, sterility and male recombination. *Genetics* **86**, 813–833.

KIDWELL, M. G., NOVY, J. B. and FEELEY, S. M. (1981). Rapid unidirectional change of hybrid dysgenesis potential in *Drosophila. J. Hered.* **72**, 32–38.

LAVIGE, J. M. and LECHER, P. (1982). Mitoses anormales dans les embryons à développement bloqué dans le systeme *I–R* de dysgénésie hybride chez *Drosophila melanogaster. Biol. Cell.* **44**, 9–14.

MAINLAND, G. B. (1942). Genetic relationships in the *Drosophila funebris* group. *Univ. Texas Publ.* **4228**, 74–112.

MATTHEWS, K. A. and GERSTENBERG, M. V. (1979). Non-reciprocal female sterility associated with male recombination chromosomes from Texas populations of *Drosophila melanogaster. Genetics* **93**, s77.

MATTHEWS, K. A., SLATKO, B. E., MARTIN, D. W. and HIRAIZUMI, Y. (1978). A consideration of the negative correlation between transmission ratio and recombination frequency in a male recombination system of *Drosophila melanogaster. Jap. J. Genet.* **53**, 13–25.

MINAMORI, S. and ITO, K. (1972). Extrachromosomal element delta in *Drosophila melanogaster*. VII. Relation to fertility in a second chromosome line. *Genetics* **70**, 549–556.

MULLER, H. J. (1942). Isolating mechanisms, evolution and temperature. *Biol. Symp.* **6**, 71–175.

NEI, M. (1976). Mathematical models of speciation and genetic distance. *In*: "Population Genetics and Ecology" (S. Karlin and E. Nevo, eds.), pp. 723–765. Academic Press, New York, San Francisco, London.

PATTERSON, J. T. (1952). A pair of allopatric subspecies belonging to the *repleta* species group. *Univ. Texas Publ.* **5204**, 114–118.

PATTERSON, J. T. and DOBZHANSKY, T. (1945). Incipient reproductive isolation between two subspecies of *Drosophila pallidipennis. Genetics* **30**, 429–438.

PATTERSON, J. T. and STONE, W. S. (1952). "Evolution in the Genus *Drosophila*," Macmillan, New York.

PATTERSON, J. T. and WHEELER, M. R. (1947). Two strains of *Drosophila peninsularis* with incipient reproductive isolation. *Univ. Texas Publ.* **4720**, 116–125.

PELISSON, A. (1977). Contribution à l'étude d'une stérilité femelle non mendélienne chez *Drosophila melanogaster*: mise en évidence d'un chromosome 4 inducteur. *C. R. Acad. Sci. Paris* **284**, 2399–2402.

PELISSON, A. (1979). The *I–R* system of hybrid dysgenesis in *Drosophila melanogaster*: influence on *SF* female sterility of their inducer and reactive paternal chromosomes. *Heredity* **43**, 423–428.

PELISSON, A. and PICARD, G. (1979). Non-Mendelian female sterility in *Drosophila melanogaster*: *I* factor mapping on inducer chromosomes. *Genetica* **50**, 141–148.

PERIQUET, G. (1976). Recherche d'une période thermosensible dans le développement d'une atrophie gonadique chez *Drosophila melanogaster*. *C. R. Acad. Sci. Paris* **283**, 1547–1550.

PERIQUET, G. (1978a). Penetrance of a threshold character in *Drosophila melanogaster*, "atrophie gonadique": theoretical models and biological interpretation. *Phys. Zool.* **51**, 371–377.

PERIQUET, G. (1978b). Recherche sur le déterminisme génétique du charactère atrophie gonadique chez *Drosophila melanogaster*. *Biol. Cell.* **33**, 33–38.

PERIQUET, G. (1978c). Influence of the rank of egg-laying on the transmission of the gonadal atrophy character in *Drosophila melanogaster*. *Experientia* **34**, 1438–1439.

PICARD, G. (1971). Un cas de stérilité femelle, chez *Drosophila melanogaster*, lié à un agent transmis maternellement. *C. R. Acad. Sci. Paris* **272**, 2484–2487.

PICARD, G. (1976). Non-Mendelian female sterility in *Drosophila melanogaster*: hereditary transmission of *I* factor. *Genetics* **83**, 107–123.

PICARD, G. (1978a). Non-Mendelian female sterility in *Drosophila melanogaster*: further data on chromosomal contamination. *Molec. Gen. Genet.* **164**, 235–247.

PICARD, G. (1978b). Non-Mendelian female sterility in *Drosophila melanogaster*: sterility in stocks derived from the genotypically inducer or reactive offspring of *SF* and *RSF* females. *Biol. Cell.* **31**, 245–254.

PICARD, G. (1979). Non-Mendelian female sterility in *Drosophila melanogaster*: principal characteristics of chromosomes from inducer and reactive origin after chromosomal contamination. *Genetics* **91**, 455–471.

PICARD, G. and L'HÉRITIER, P. (1971). A maternally inherited factor inducing sterility in *Drosophila melanogaster*. *Dros. Inf. Serv.* **46**, 54.

PICARD, G. and PELISSON, A. (1979). Non-Mendelian female sterility in *Drosophila melanogaster*: characterization of the noninducer chromosomes of inducer strains. *Genetics* **91**, 473–489.

PICARD, G., BUCHETON, A., LAVIGE, J. M. and FLEURIET, A. (1972.) Contribution à l'étude d'un phénomène de stérilité à déterminisme non mendélien chez *Drosophila melanogaster*. *C. R. Acad. Sci. Paris* **275**, 933–936.

PICARD, G., BUCHETON, A., LAVIGE, J. M. and PELISSON, A. (1976). Répartition géographique des trois type de souches impliquées dans un phénomène de stérilité à déterminisme non mendélien chez *Drosophila melanogaster*. *C. R. Acad. Sci. Paris* **282**, 1813–1816.

PICARD, G., LAVIGE, J. M., BUCHETON, A. and BREGLIANO, J. C. (1977). Non-Mendelian female sterility in *Drosophila melanogaster*: physiological pattern of embryo lethality. *Biol. Cell.* **29**, 89–98.

PICARD, G., BREGLIANO, J. C., BUCHETON, A., LAVIGE, J. M., PELISSON, A. and KIDWELL, M. G. (1978). Non-Mendelian female sterility and hybrid dysgenesis in *Drosophila melanogaster*. *Genet. Res., Camb.* **32**, 275–287.

PRAKASH, S. (1972). Origin of reproductive isolation in the absence of apparent genic differentiation in a geographic isolate of *Drosophila melanogaster*. *Genetics* **72**, 143–155.
PROUST, J. and PRUDHOMMEAU, C. (1982). Hybrid dysgenesis in *Drosophila melanogaster*. I. Further evidence for and characterization of the mutator effect of the inducer-reactive interaction. *Mutation Res.* **95**, 225–235.
ROSENFELD, A., CARPENTER, A. and SANDLER, L. (1971). A nonchromosomal factor causing female sterility in *Drosophila melanogaster*. *Dros. Inf. Serv.* **47**, 85.
RUBIN, G. M. (1983). *Drosophila* middle repetitive elements. *In*: "Mobile Genetic Elements" (J. M. Shapiro, ed.). Academic Press, New York, London, San Francisco.
RUBIN, G. M., KIDWELL, M. G. and BINGHAM, P. M. (1982). The molecular basis of *P-M* hybrid dysgenesis: the nature of induced mutations. *Cell* **29**, 987–994.
SCHAEFER, R. E., KIDWELL, M. G. and FAUSTO-STERLING, A. (1979). Hybrid dysgenesis in *Drosophila melanogaster*: morphological and cytological studies of ovarian dysgenesis. *Genetics* **92**, 1141–1152.
SIMMONS, M. J. and LIM, J. K. (1980). Site specificity of mutations arising in dysgenic hybrids of *Drosophila melanogaster*. *Proc. Natl. Acad. Sci. U.S.A.* **77**, 6042–6046.
SIMMONS, M. J., JOHNSON, N. A., FAHEY, T. M., NELLET, S. M. and RAYMOND, J. D. (1980). High mutability in male hybrids of *Drosophila melanogaster*. *Genetics* **96**, 479–490.
SINGH, R. S., LEWONTIN, R. C. and FELTON, A. A. (1976). Genetic heterogeneity within electrophoretic "alleles" of xanthine dehydrogenase in *Drosophila pseudoobscura*. *Genetics* **84**, 609–629.
SLATKO, B. E. (1978). Evidence for newly induced genetic activity responsible for male recombination induction in *Drosophila melanogaster*. *Genetics* **90**, 105–124.
SOCHACKA, J. H. M. and WOODRUFF, R. C. (1976). Induction of male recombination in *Drosophila melanogaster* by injection of extracts of flies showing male recombination. *Nature, Lond.* **262**, 287–289.
STURTEVANT, A. H. and NOVITSKI, E. (1941). Sterility in crosses of geographical races of *Drosophila micromelanica*. *Proc. Natl. Acad. Sci. U.S.A.* **27**, 392–394.
SVED, J. A. (1978). Male recombination in dysgenic hybrids of *Drosophila melanogaster*: chromosome breakage or mitotic crossing-over? *Austral. J. Biol. Sci.* **31**, 303–309.
SVED, J. A. (1979). The "hybrid dysgenesis" syndrome in *Drosophila melanogaster*. *BioScience* **29**, 659–664.
SVED, J. A., MURRAY, D. C., SCHAEFER, R. E. and KIDWELL, M. G. (1978). Male recombination is not induced in *Drosophila melanogaster* by extracts of strains with male recombination potential. *Nature, Lond.* **257**, 457–458.
TEMPLETON, A. R. (1980). Modes of speciation and inferences based on genetic distances. *Evolution* **34**, 712–729.
THOMPSON, J. N., JR., HENDERSON, S. A. and WOODRUFF, R. C. (1980). Sterility and testis structure in hybrids involving male recombination lines of *Drosophila melanogaster*. *Genetica* **51**, 221–226.
TOWNSEND, J. I. (1954). Cryptic subspeciation in *Drosophila* belonging to the subgenus *Sophora*. *Amer. Nat.* **88**, 339–351.
VOELKER, R. A. (1974). The genetics and cytology of a mutator factor in *Drosophila melanogaster*. *Mutation Res.* **22**, 265–276.
WARTERS, M. (1944). Chromosomal aberrations in wild populations of *Drosophila*. *Univ. Texas Publ.* **4445**, 129–174.
WHARTON, L. T. (1944). Interspecific hybridization in the *repleta* group. *Univ. Texas Publ.* **4445**, 175–193.
WOODRUFF, R. C., SLATKO, B. E. and THOMPSON, J. N., JR. (1982). Factors affecting mutation rate in natural populations. *In*: "The Genetics and Biology of *Drosophila*",

Volume 3c, Chapter 20 (M. Ashburner, H. L. Carson and J. N. Thompson, Jr., eds.). Academic Press, London and New York.

YANNOPOULOS, G. (1978). Studies on the sterility induced by the male recombination factor 31.1 MRF in *Drosophila melanogaster*. *Genet. Res. Camb.* **32**, 239–247.

YANNOPOULOS, G. (1979). Ability of the male recombination factor 31.1 MRF to be transposed to another chromosome in *Drosophila melanogaster*. *Molec. Gen. Genet.* **176**, 247–253.

22. Response to Selection

KENNETH MATHER

*Department of Genetics,
University of Birmingham,
Birmingham, England*

I. Introduction	155
II. General Considerations	157
A. Selection and Characters	157
B. Types of Selection	158
C. Kinds of Variation	162
D. Hidden Variability	163
E. Inbreeding and Outbreeding	168
F. Recombination	170
G. Effects of Mutation	171
III. Directional Selection	175
A. Early Generations	175
B. Selection Response in Later Generations	182
C. Chaeta Number and Fertility	186
D. Correlated Responses and Recombination	188
IV. Stabilizing Selection	191
A. Additive Genetic Variation	191
B. Canalization	196
V. Disruptive Selection	199
A. Types of Disruptive Selection	199
B. D^- Selection: Polymorphism	201
C. D^+ Selection: Divergence	203
D. Isolation	206
E. Conclusion	209
VI. Genetic Architecture of Characters	210
A. Balance and Natural Selection	210
B. Types of Character	212
Acknowledgements	215
References	215

I. Introduction

The very early days of genetics, even before *Drosophila melanogaster* established itself as an organism for genetical investigations, saw a number of selection experiments being undertaken. The Illinois Agricultural

Experiment Station's experiment to change the percentage of protein and oil in maize kernels had begun indeed in 1896. It is still continuing and the characters are still responding to selection (see for example Smith, 1908; Winter, 1929; Leng, 1962; Wright, 1977). Progress was greatest in the early generations, but has continued in a generally smooth way. Johannsen's experiment with seed weight in the bean *Phaseolus vulgaris* (reported in 1909) was shorter-lived. Because of the natural self-pollination of these beans, advance under selection ceased after a single generation. Castle's selection for the extent of coloured areas in hooded rats (Castle and Phillips, 1914; Castle and Wright, 1916) gave results broadly akin to those in the Illinois maize experiment. Selection experiments were in fact already showing themselves to be genetically informative.

The many advantages of *Drosophila* as material for genetical experimentation, including the unique opportunities it offers for genetical analysis, led ultimately to more selection experiments being undertaken with it than with any other higher organism. These studies have contributed in a unique way to our understanding both of the factors affecting response to selection and of the longer-term genetical consequences of the different types of selection that can be applied.

Most of the early selection experiments on *Drosophila melanogaster* concerned the manifestation of the effects of major mutant genes. A recessive gene which increased the number of dorsocentral bristles was isolated by MacDowell, who selected successfully for increase in the number of the extra bristles to which it gave rise (MacDowell, 1915, 1917, 1920). Sturtevant (1918) found that the dominant mutation Dichaete tended, *inter alia*, to reduce the number of dorsocentral bristles and was able to select for both enhancement and reduction of this effect. He was able to go further and show that genes on both the major autosomes were involved in these changes. The number of eye-facets in Bar-eyed flies was also changed in both directions, using the relevant type of selection, by Zeleny (1922).

In all of these cases the major advance under selection was achieved in the early generations of the experiment, with no indication of any marked change thereafter. Payne's (1918) selection for increase in the number of scutellar bristles stands in contrast in this respect. Individuals with five, instead of the customary four, scutellar bristles appear as rarities among flies from the wild as well as in culture. Though still rare, females with extra bristles are somewhat more common than males. Starting with such a five-bristled female mated to a normal four-bristled male, Payne obtained two daughters with an extra bristle, and these mated to normal brothers gave offspring of which some 4.4% carried more than four scutellars, some having five, but others six, bristles. In each generation thereafter, flies having the greatest number of bristles were taken as parents. Not only did the proportion of flies having extra bristles increase, but the numbers of bristles also rose. The proportion of individuals with extra bristles topped 50% in generation 6 and after generation 16 flies with the normal four scutellars were virtually absent. The average number of scutellars reached about nine (i.e., five extra bristles) in generation 29 and stayed in the neighbourhood of that value until the experiment came to an end in generation 38. Linkage tests showed genes in chromosomes X and 3 to be involved in the change under selection. The rise in the number of scutellars was not, however, obtained by a steady advance: scutellar number appears to have been virtually stable at just over six from generation 11 to 14 and again at about eight from generation 18 to 25, before advancing to its final value. Thus progress under selection would appear to have been by three steps of diminishing size, separated by periods of stability of increasing length.

It has been suggested by Wright (1977) that this appearance could be false and that a plausible interpretation might be sought in terms of a basically smooth curve of advance broken into a step-like appearance by unfortunate choices of parents whose phenotypes did not match their genotypes. However this may be in respect of Payne's results, such an interpretation would not serve for the results in a later experiment with scutellars, reported by Sismanidis (1942). As we shall see later, this experiment leaves no doubt of the step-wise response of scutellar number to selection.

Such step-wise advances under selection have proved to be a not uncommon feature in the many selection experiments which, following a pause of some twenty years or more after the early experiments of MacDowell, Sturtevant, Zeleny and Payne, have been undertaken with *Drosophila*. Yet, before we consider these later experiments and the light they throw on the action and consequences of selection, we must first turn to some general considerations relating to the nature of selective forces and the agencies that govern the responses they produce.

II. General Considerations

A. Selection and Characters

Selection will have occurred when the progeny contributed by individual parents differ in number, survival ability or fertility: that is when the individuals differ in Darwinian fitness. Many aspects of the environment (including, where relevant, the intervention of the experimenter) will make demands on the individuals which experience it and so will impose many forces of selection. Equally many, perhaps all, aspects of the phenotype will be involved in meeting these requirements and so contribute by their variation to the differences in fitness among individuals. Thus response to selection must reflect the impact of the totality of selective forces, whether natural or artificially imposed, on the totality of the variation whether genetic, environmental or developmental in origin.

Fitness itself is the only character that we can be sure will reflect the impact of the totality of selective forces on the totality of the variation. Unfortunately, fitness is a notoriously difficult character to measure. So we make do with other, more readily scorable, characters and in selection experiments seek to make one force of selection, which we impose in direct relation to that character, over-riding in the determination of the contributions the individuals will make to the next generation. But we must always recognize that, although at any given time this force of selection may be the most powerful that is in operation, and this character the one whose variation makes the greatest contribution to the differences in fitness, other forces of selection and variation in other characters must be expected to be playing their parts, albeit for the time being lesser ones. Furthermore, we must also recognize that the balance of power among the selective forces, and of importance among the characters, may change during the progress of an experiment, and change because of the very responses that have occurred to the selection we imposed. We must consequently accept a corresponding complexity in the experiments, in the results that they yield and in the interpretation that they require.

There is another complexity that must also be considered. Selection distinguishes directly among phenotypes, and among genotypes only to the extent that they are associated with differences of phenotype. Yet response to selection is revealed by the phenotypes of the individuals of the next generation, and these phenotypes are in general connected with those of their parents only by the genotypes that have been transmitted from parent to offspring. Thus the response to selection must depend not only on the numerical properties of the genetic determinants (that is, gene frequencies, linkage relations and recombination frequencies), but also on their physiological properties of action and interaction with one another and with the environment in producing the phenotype. We must expect, therefore, that the results we obtain from selection experiments will show a dependence on these physiological properties and will reveal something of the relations among the genes in respect of their actions and interactions that are favoured by the different kinds of selection we can recognize.

B. Types of Selection

Forces of selection may be classified in various ways. They may be referred to the agencies in the environment from which they stem, for example to temperature, humidity, light, the availability of nutrients, the presence of poisonous substances or of other organisms, whether symbionts, predators or parasites. Fluctuations in such agencies may be selectively important, too. Furthermore, at least certain of these agencies may have direct effects on the development of the phenotype. Response to the selections they impose may therefore involve sensitivity of development to environmental differences, termed genotype × environment interaction. Indeed, in the case of the effect of temperature on the number of sternopleural chaetae in *Drosophila melanogaster*, differences in sensitivity of this kind have been found not to be correlated with genetical differences in the gross expression of the character (Caligari and Mather, 1975). Temperature sensitivity may thus be regarded as a character in its own right, adjustable by selection independently of the chaeta number itself.

The impact on an individual of the selection stemming from some agencies (for example, temperature or other physical features of the environment) is not in general expected to be influenced by the number or properties of the other members of the population to which it belongs. On the other hand, with selection arising from other agencies (for example, the availability of nutrients or the activities of predators) the other members of the population constitute an important feature of the environment. We can, therefore, recognize a more general and more analytical classification of selective forces into unconditional and conditional or, to use Wallace's (1970) terms, into hard and soft. Conditional, or soft, selection may be further classified into density-dependent, frequency-dependent, or a combination of the two. We may recognize, too, selection that arises from competition among individuals (which must clearly have a density-dependent element in it) and selection which does not.

The recognition of these various types of selection, each with its own consequences for the size as well as the constitution of the population, is important in ecological genetics, from which indeed the classification has sprung. In the study of the consequences of selection for the organization of the genotype as well as for the expression of the character, a different classification (Mather, 1953a) has, however, proved to be more relevant. Where a population, or other group of individuals, occupies an effectively uniform environment, one phenotype could be expected to be more consonant than others with the requirements of that environment, and individuals displaying this optimum phenotype must be at a selective advantage over their fellows. Other individuals will be at a disadvantage by comparison with this optimal type and this disadvantage will be greater, the greater their phenotypes depart from the optimum.

Where the optimum expression of a character coincides with the mean expression of the population or group, individuals departing from the mean will be at a disadvantage. The pressure of selection will thus not be to change the mean but to reduce the variation of the phenotype around it. Such selection is said to be *Stabilizing* since it will tend to stabilize the population at the mean expression (Fig. 1, left). Where, however, the optimum phenotype is not at the mean of the population, but at some more extreme expression, the pressure of selection will be towards changing the mean by moving it closer to the optimum. The selection is then said to be *Directional* (Fig. 1, centre).

Directional (R) selection contrasts with stabilizing (S) selection in that it favours change in the mean expression of the phenotype. The third type of selection contrasts with stabilizing in that it favours not a reduction but an increase in the variation of the phenotype. Such selection is said to be *Disruptive* (D). It arises where two or more optima are favoured within the population (Fig. 1, right), either because the environment is not uniform, so making different demands on different parts of the population, or because of cooperative relations between two or more different types of individual in, for example, reproduction, avoiding predation or minimizing the impact of disease.

This classification has come to be generally used in the discussion of selection and selection experiments. A similar classification has been proposed by Simpson (1944) who refers to centripetal, centrifugal and linear selection which broadly correspond to stabilizing, disruptive and directional. It will, however, seldom be found in the literature with which we shall be chiefly concerned.

The three types of selection, stabilizing, directional and disruptive (S, R

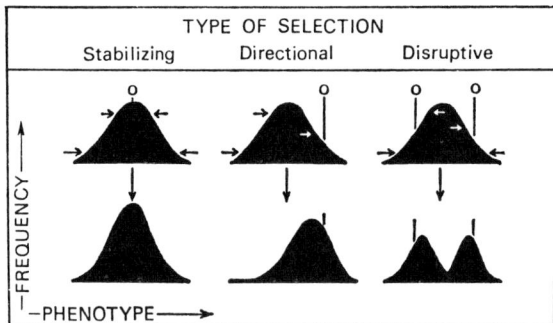

FIG. 1. The three elemental types of selection. O indicates the optimal phenotype towards which selection is acting, and the direction and force of selection is shown by the arrows in the upper row of phenotypic distributions. The lower row of distributions shows the consequences of the types of selection. (Reproduced by permission from K. Mather (1953) The genetical structure of populations. *Symp. Soc. Exp. Biol.* 7, 66–95.)

22. RESPONSE TO SELECTION

and D) will be expected to have different consequences not only for the expression of the relevant characters, but also for the organization, both mechanical and physiological, of the genotype. It must be recognized, however, that they are not mutually exclusive: rather, they are different elements or components which may be, and indeed generally are, combined in various amounts to give the selective situations that we see in nature. Only fitness is under exclusively R selection, and though in the applied genetics of plant and animal breeding desirable characters may in the long-term be under consistent R selection, albeit when a variety is developed for the market it is put under immediate S selection to stabilize its characteristics. Under natural conditions the environment is seldom wholly stable and predominantly S selection will then most commonly have an R component in it, albeit a fluctuating or even cyclical one. Equally, unless there is a gross or cataclysmic change in the environment, steady change will produce a steady, continuing R component, which will, however, always have an S component associated with it. D selection implies R selection in two or more directions simultaneously, though not of course on the same individual; and it, too, will commonly involve a component of S selection in addition. Within limits, however, we can impose a single type of selection in experiment, and so investigate the genotypic responses to it.

We should note here that selection experiments are broadly of two kinds. One of these is exemplified by experiments with cage populations kept in the laboratory using the technique first introduced by L'Héritier and Teissier (1934). In such experiments, the cage population is under pressure from the full range of whatever forces of quasi-natural selection that may develop. The generations overlap and changes are detected by taking samples from the cage at appropriate intervals. Such experiments can tell whether a given character and, in cases where the genotypes are readily distinguishable, even whether a given genetic difference is being acted on by selection. We can learn, too, what the overall consequencies of selection are, as for example the elimination of major mutant genes from the population, or the achievement of a balance between the frequencies of relatively inverted sequences in homologous chromosomes. Indeed it was by such means that Dobzhansky (1948, 1950) showed that such a balanced polymorphism is achieved when inversion sequences come from the same wild population but not when they come from different ones. In this sense such experiments provide information about the relative fitnesses of recognizably different genotypes, though they provide little, if any, information about the types of selective forces that are acting or about the pressures stemming from the individual forces.

In the second kind of selection experiment, which usually starts with a sample from a population or with the F_2 between two distinct (and preferably inbred) lines, the experimenter deliberately applies selection for some convenient character in choosing the parents for the next generation. The effect of the selection is recorded from the comparison between, on the one hand, the group of individuals from which the parents were chosen and, on the other, the progeny of these chosen parents. The process is repeated generation by generation. The experimenter can thus choose the type of selection (R, S, D or some combination of them) that he will apply and the rigour with which he will apply it. He can follow the impact of this selection not only on the expression of the character he is primarily using, but also on any other character he chooses to observe, and indeed on the organization of the relevant genotype itself. In short, it enables him to examine the consequences, long-term as well as short, of a specific type of selection. This is the type of selection experiment with which we are chiefly concerned.

C. Kinds of Variation

A number of kinds of variation of the genetic materials are recognizable at the levels of both the chromosomes and the genes. Of the possible types of variation in the number and structure of the chromosomes, only inversions have been observed in wild populations of *Drosophila* species, other than as rarities of little interest to us. Furthermore, inversions have of themselves no direct effect on the phenotype. Their responses to selection must therefore depend on the way they affect the distribution, especially the recombination, of the genes that they carry. Where inversions have been followed in selection experiments, they have been acting as tell-tales of the combinations of the genes which they contain and which, when heterozygous, they lock into super-genes. We must thus seek to elucidate the organization of genic variation if we are to understand the fate of inversions under the impact of selection.

Three kinds of genic variation are known in *Drosophila*. First there are the classical mutant genes of major, indeed often gross, effect on the phenotype—the genes on which the use of *Drosophila* as an organism for genetical experimentation was primarily built. Such mutants have arisen frequently, and even repeatedly, in experiment. With the exception of such of them as are lethal, they are readily fixed in experimental stocks. A large number of recessive mutant genes have been detected in the heterozygous condition in individuals taken from the wild (see review by Dobzhansky, 1939). Yet individuals homozygous for them are seen only with extreme rarity in wild populations. Thus, although they are easily fixed by selection in experiment, by comparison with their non-mutant alleles they must be generally disfavoured and so tend to be rapidly eliminated in the wild: a conclusion that has been amply verified by quasi-natural selection experiments with cage populations.

The second kind of variation is that in proteins, notably enzymes, first demonstrated in *Drosophila* by Lewontin and Hubby (1966). This variation, which is directly relatable to alleles at specific loci, is a regular feature of *Drosophila* populations, and is indeed to be found in virtually all the species of animals and plants in which it has been sought. It is obviously open to experimental attack at the molecular level, but little is known of its effects on the fitness of individuals. There has been a great deal of otiose argument as to whether it has any such effect at all, but cage experiments now leave little doubt that at least some protein loci are indeed under selective pressure (see for example, Minawa and Birley, 1978). The type of selection, the magnitude of its pressure, and the way it operates are, however, still far from clear.

Finally, there is the quantitative variation which is shown by virtually all characters and which is ubiquitous in populations of *Drosophila*, as in all other species of animals and plants that have been adequately examined. Its genetical element arises from genic differences characteristically of relatively small effect on the phenotype, but which act in polygenic systems, within which they are capable of reinforcing and balancing one another's effects, and which are thus capable of producing phenotypes ranging by virtually undetectable gradations between wide limits, the more central expressions being the more common. Polygenic variation is not only ubiquitous in populations: it would also appear to form the essential basis of differences between species, where these have been analysed (Mather, 1941, 1943a). The properties of quantitative variation and its underlying polygenic systems have been reviewed in the recent volume *Quantitative Genetic Variation* (Thompson and Thoday, 1979), and polygenic variation in both natural and experimental populations will be surveyed by Thompson and Thoday in a later volume of this series. It will suffice for our present purpose to note that it is the raw material both of adaptation by natural selection in the wild and of much of the improvement by artificial selection in plant and animal breeding. It is not surprising, therefore, that it has been the chief subject of the many selection experiments with which we are concerned in considering the nature and properties of response to selection.

D. Hidden Variability

Where R selection is applied to a population of individuals in respect of a character, the population is expected to respond by a change in the mean expression of the character; but it will do so only to the extent that it carries

TABLE I. Examples of wild-type characters which have been shown to respond to selection. Only a single early reference is given in each case.

Character	Species	Reference
Chaeta number		
scutellars	mel	Payne (1918)
abdominals	mel	Mather (1941)
sternopleurals	mel	Wigan (1941)
Body size		
wing length	mel	Robertson and Reeve (1952)
thorax length	mel	
Egg size	mel	Bell et al. (1955)
Fecundity	mel	
Mating choice	p/p	Koopman (1950)
	mel	Wallace (1954)
	pau	Dobzhansky and Pavlovsky (1971)
Phototaxis	rob	Carson (1958)
	mel	Hadler (1964)
Pupation site	wil	De Souza et al. (1968)
Parthenogenesis	par	Stalker (1954)
	mer	Carson (1967)

Species: mel = *melanogaster*; mer = *mercatorum*; par = *parthenogenetica*; pau = *paulistorum*; p/p = *pseudoobscura* and *persimilis*; rob = *robusta*; wil = *willistoni*.

genetic variation in respect of that character. Response to R selection is thus a test for the presence of genetic variation. A wide variety of characters, morphological, biochemical and behavioural, have been tested in *Drosophila* for heritable quantitative variation in this way, and evidence of such variation has been found in all cases (see Table I for some examples).

In some cases the individuals subjected to selection were from laboratory stocks, or descended from crosses between such stocks; but in other cases they were the immediate descendants of samples, and often very small samples, of flies taken from the wild. There can thus be no doubt that virtually all characters show heritable quantitative variation and that such variation is a regular feature of wild populations. In other words, wild populations of *Drosophila* carry the variation necessary to respond to any forces of selection imposed on them by the environments in which they come to find themselves.

The responses to R selection can be both rapid and large. This is well shown by the results of selection for the numbers of sternopleural chaetae and abdominal chaetae—characters frequently used in selection experiments because of the relative ease with which flies can be scored for them. In a selection experiment initiated by a cross between two laboratory stocks,

Mather (1941) raised the average number of chaetae jointly borne by the 4th and 5th abdominal segments from just over 41 to over 47 in eight generations of selection, and reduced it to less than 31 in the same number of generations of selection in the opposite direction (see Fig. 5). Starting with a different cross, Mather and Harrison (1949) raised the same character from about 41 to nearly 56 in twenty generations of selection (see Fig. 7). Rasmuson (1955) reports broadly similar results.

Turning to the number of sternopleural chaetae, the two sides of the fly being taken jointly, Thoday and Boam (1959) raised the average number from 20 to 45, and Barnes and Kearsey (1970) report a similar advance in mean number of sternopleurals from some 18 to 45 in eighteen generations of selection. These latter authors began their selection lines from the "Texas" cage population started by the use of 30 inseminated females taken from the wild a few years earlier. The advance to 45 took the selection line to well beyond the most extreme number of chaeta seen in the population, where indeed a fly with more than 25 or so sternopleurals had not been seen (Fig. 2).

It is a common feature of such experiments for the selection lines to transgress the limits observed for the expression of the character among the individuals of the population or F_2 from which the selection was begun. Now, it is a property of polygenic systems mediating quantitative variation that, in a randomly breeding population, by far the greater part of the genetic variability that the population carries is not expressed as differences among the phenotypes of the member individuals of the population; it is hidden in the genotypes partly by heterozygosity for alleles at a locus and partly by the effects of alleles at different loci balancing one another. Thus the effect of an increasing (or +) allele at one locus may be balanced out by that of a decreasing (or −) allele at another. The amount of variability hidden in this latter way increases with the number of loci. Where there are k loci each with two alleles (a + allele tending to raise the expression of the character relative to the mean of the population and a − allele tending to diminish it, the difference in effect between the two alleles being the same at all loci) it appears (Mather, 1960a, 1973) that, in the absence of dominance, the amount of variability expressed visibly as variation among the phenotypes (free variability) and the amount hidden by the balancing effect of non-allelic genes (homozygotic potential variability) are in the ratio $1:k-1$. Furthermore, the amount of variability hidden by the balancing action of alleles in heterozygotes (heterozygotic potential variability) is equal to the sum of the other two items. Thus, to take an example, with 10 loci contributing equally to the variability of the system and in the absence of dominance, the ratios of the three types of variability will be $1:9:10$. In other words, only 5% of the variability will be expressed in the form of

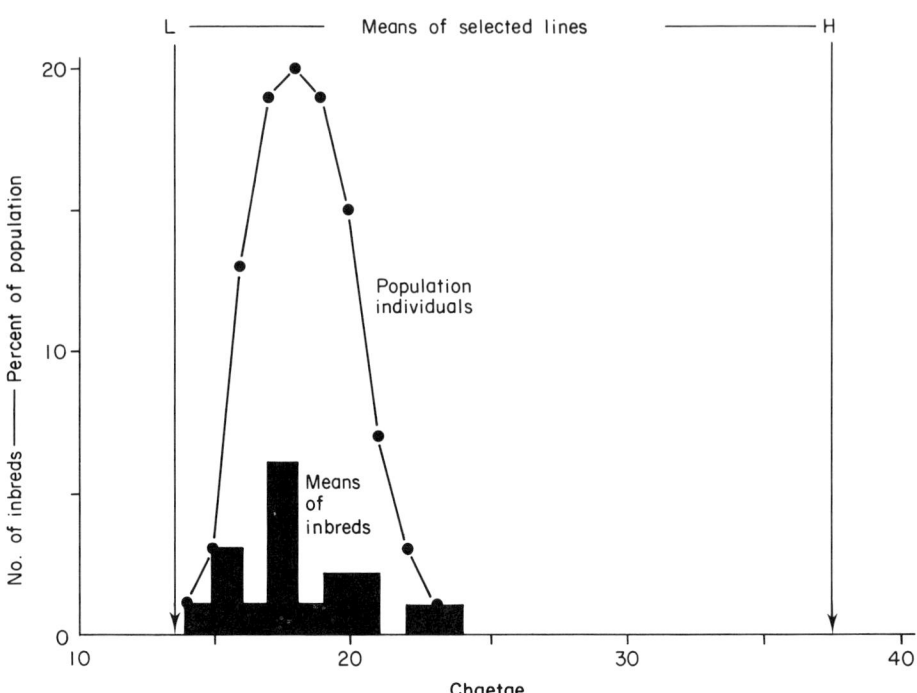

FIG. 2. The frequency distribution (in percentages) of sternopleural chaeta number among 111 females of the "Texas" population, and among the mean chaeta numbers of 18 inbred lines derived from that population. The mean chaeta numbers of the H and L lines selected from the population by Barnes and Kearsey (1970) are indicated by the arrows.

phenotypic variation: 95% will be hidden as genotypic differences which produce no variation in the phenotype. With 20 loci in such a system, the proportions of free and potential variability will be $2\frac{1}{2}\%:97\frac{1}{2}\%$ or 1:39. Nor are these assumed numbers of loci likely to be unrealistically high: the evidence indicates that the minimum number of loci mediating sternopleural chaeta number in *Drosophila* is at least 15 and may well be 20 or more (Davies, 1971; Mather, 1973). Still higher estimates of the number of genes in polygenic systems have been obtained from maize and mice (see Mather, 1979).

In the paper reporting the results of their selections for sternopleural chaeta number, Barnes and Kearsey (1970) give the overall variance of chaeta number in the females of the population from which selection was begun as 3·13. That for males was 2·94. A number of inbred lines have been made from this population and the variance between the mean chaeta numbers of the females of 18 of these lines is given by Caligari and Mather

(1980) as 5·64, after making a small correction for non-heritable variation. Since the inbred lines must be homozygous, or virtually so, the variance of their mean chaeta number will contain no heterozygous potential component and so should be twice the additive variance of the population.

The variance, 3·13, for the population females must of course be corrected for non-heritable variation, just as was the variance among the inbred lines. It is known that under the conditions in which these flies were raised, all but a small fraction of this non-heritable variation is traceable to instability of development, which can be measured by the mean square difference in chaeta number between the two sides of the fly. The progeny of single-pair matings taken from the population by Caligari and Mather gave a value of 1·52 for this mean square in females. Deducting this from 3·13 leaves 1·61 as the estimate of the genetic variation among the population females. This, however, requires a further correction since it is known that the genes affecting sternopleural chaeta number show partial dominance and that this has the effect of reducing the additive genetic variance in the populations (Caligari and Mather, 1980). Three estimates of this effect of dominance have been made by my colleagues Dr P. D. S. Caligari and Mr D. Baban who kindly allow me to quote them. They are that the additive variance is reduced by factors of 0·64, 0·56 and 0·52 relative to its value in the absence of dominance. The mean of these estimates is 0·57. The estimate for the genetic variance of the population must thus be raised to $1·61 \div 0·57 = 2·82$ to make it comparable with the genetic variance among the inbred lines, which of course is unaffected by dominance. We are thus left with genetic variances of 5·64 among the inbred lines and 2·82 in the population—a ratio of 2·00, in exact agreement with expectation. This exactness of agreement is obviously fortuitous: using the three individual estimates of the reduction due to dominance gives ratios of 2·2, 1·97 and 1·83; but it is clear that the results agree satisfactorily with expectation. We might note that the calculations have been carried out solely with females, as the hemizygous state of the X chromosome in males would complicate the expectations to the extent that any of the relevant genes are sex-linked.

Now the ratio of free to homozygotic potential variability is $1:k-1$ where k heteroallelic loci contribute equally to the variability. Thus, even though it is twice as great as that of the population, the genetic variation among the inbred lines is no more than a fraction $1/k$ of the variability available in the population as a basis for response to selection. Barnes and Kearsey have devised a way of estimating the value of k, using the F_2 and backcross generations from a cross between their line selected for high chaeta number, to which we have already referred, and a low selection line that they made simultaneously, together with the mean and additive genetic variance of the population. Using the estimate, 1·61, that we obtained above for the additive genetic variance, their method gives $k = 24.7$. This value for k is higher than the estimates obtained by Davies (1971) and Mather and Jinks (1971) using other methods, but not so much higher as to be incompatible with them when the assumptions and limitations of the various approaches are taken into account. It can, in any case, hardly be doubted that some 20, or possibly more, loci are involved, which as we have already seen means that only about 2.5% of the variability of the population is freely expressed as variation of the phenotype. Evidently the phenotypic variation shown by a population can be but a small token of the changes that selection can produce from it.

E. Inbreeding and Outbreeding

Where the potential variability is hidden by the balancing action of alleles in heterozygotes, segregation will normally ensure that half of it is freed in the next generation. Where, however, there is homozygotic potential variability, concealed by the balancing action of non-allelic genes, recombination also comes into the account. Thus where two gene-pairs, A–a and B–b are concerned (A and B being the + alleles and a and b the − alleles) and the two gene differences are of equal effect, AAbb and aaBB individuals will have the same phenotype apart from non-heritable effects. The variability hidden in their genotype difference can be freed only by crossing to give AaBb individuals (which in the absence of dominance, or with balancing dominance, will have the same phenotype as their parents) followed by interbreeding to give the range of phenotypes between AABB, at the + extreme, and aabb at the − end. Thus crossing followed by simultaneous segregation and recombination is required for the release of this variability (Fig. 3).

As we have seen, the "Texas" inbred lines, like Johannsen's (1909) dwarf beans, must have carried within their genotypes potential variability of this kind in much greater amount than the variability expressed in phenotypic variation. Furthermore, this variability must inevitably remain hidden until the lines are crossed, after which segregation and recombination can do their work. It was thus the difference in breeding systems between the naturally inbreeding beans that Johannsen used and the naturally outbreeding maize of the Illinois experiment that was responsible for the dramatic difference in their responses to selection shown by these two early experiments.

In plants, inbreeding by self-pollination, encouraged by a variety of devices, is a feature of some species, while outbreeding, ensured again by a variety of devices, is equally a feature of others. Despite their key importance for response to selection, we still know relatively little about the evolutionary origin of these various devices. It is, however, known that the efficiency of incompatibility systems, ensuring outbreeding, can be modi-

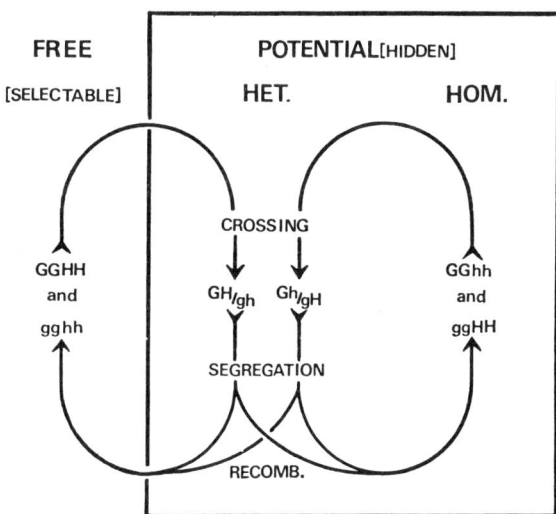

FIG. 3. The states of variability. Free variability is expressed as phenotypic differences among the genetic classes; heterozygous potential variability is concealed by the balancing effects of alleles and is thus a property of heterozygotes; and homozygotic potential is concealed by the balancing effects of non-allelic genes. Free variability is available to the action of selection, but both types of potential variability (enclosed in the box) are incapable of being acted on by selection. The arrows show the directions of flow of the variability, the rates of flow being governed by crossing, segregation and recombination. (Reproduced by permission from K. Mather (1973). *Genetical Structure of Populations*. Chapman and Hall, London.)

fied by the so-called background phenotype (see, for example, Mather, 1943b), and that the relative positions of anthers and stigmata can be changed by selection in *Nicotiana rustica* with consequences for the balance between self-pollination and outcrossing (Breese, 1959). Evidently the breeding system can be changed by selection; and in view of its central role in the control of variability, and hence of the response to selection of all the organism's characters, we can hardly doubt that it has been adjusted by natural selection in the past.

The situation is less clear in animals, most of which have a device which encourages outbreeding, namely the production of sperm and eggs by different individuals. The consequential favouring of outbreeding can evidently be circumvented by mechanical devices like the intra-uterine mating of the grass-mite *Pediculopsis* (Cooper, 1937); but a more general mechanism, capable of finer adjustment, is likely to be provided by preferential mating behaviour.

Positive assortative mating has long been known in man, and its consequences for genetic variation have been discussed by Fisher (1918) and various later writers. Negative assortative mating has been reported in the moth *Panaxia dominula* by Sheppard (1953), who attributes to it the maintenance of polymorphism for the *medionigra* gene. In *Drosophila* various mutants, such as y, w and f, have been found to affect the mating choice and success of individuals displaying them phenotypically, and in some cases their effects in mixtures of types appear to be frequency-dependent (see Wright, 1977, for a review). Results from mating-choice experiments, involving two wild-type lines Oregon and Samarkand, indicate that mating is preferentially within the line, especially in the case of Oregon (Mather and Harrison, 1949). The behaviour of certain lines, derived by selection for abdominal chaeta number from the cross between Oregon and Samarkand, implicated chromosome 2 in the determination of this discriminative mating. There was a further suggestion that one of these selection lines behaved differently from both the original parents, which would of course indicate the involvement of more than one gene-pair. More strikingly Thoday and Gibson (1962) have reported mating discrimination between lines of *Drosophila* produced by disruptive selection. Others have failed to find such discrimination in repeat experiments; but even so, it would appear that mating behaviour can be altered by selection, at least under appropriate circumstances, and so affect the organization of genetic variability for all the characters the individuals can display.

F. Recombination

The second key factor in controlling polygenic variability is recombination, which determines the redistribution of the variability in the segregating progenies of heterozygotes. Where genes are carried on different chromosomes they will recombine freely, at least in the absence of heterozygosity for interchanged segments between chromosomes. Recombination between genes carried on the same chromosome may also be affected by structural changes of the chromosomes, notably by inversions; but even apart from such structural changes, it depends on the frequency of crossing-over between the relevant loci. That this is subject to genic control has been known since Detlefsen and Roberts (1921) successfully selected for change in the frequency of recombination between loci in the X chromosome of *Drosophila*; and more recently Kidwell (1972) has shown the effects of selection on recombination between the genes Glued and Stubble in chromosome 3. Indeed there is now substantial evidence, cytological as well as genetical, from a wide variety of organisms, of the genic control of recombination and of the way in which it can be achieved by effects on the redistribution of crossing-over along the chromosome as well as by change in its overall frequency (see Mather, 1973; Catcheside, 1977).

Just as with the breeding system, therefore, we would expect the system of recombination to be subject to adjustment and readjustment by natural selection, because of its effects on the redistribution of polygenic variability and hence on response to selection in respect of all the characters on which an individual's fitness immediately depends. And in *Drosophila*, whereas we have already seen a character like sternopleural chaeta number can be governed by a polygenic system involving 20 or more loci distributed over no more than three major chromosomes, we would expect recombination to be of particular importance in governing the responses of characters to selection. Before we turn to examine the evidence from the many experiments in *Drosophila*, however, we must first consider a further agency which must affect variability but to which no reference has yet been made.

G. Effects of Mutation

The capacity of the expression of a character in selection lines greatly to transgress the limits of expression seen in the starting population or F_2 can be fully understood by reference to the large amounts of variability that lie hidden in the genotype through the balancing action of alleles in heterozygotes and, more particularly, of non-allelic genes in homozygotes. Indeed the properties of polygenic systems lead us to expect such hidden variation and such transgressive response. There is, however, a further kind of potential variability whose source is the capacity of genes to mutate and we must now assess the contribution that this makes to selective responses.

The possible contribution of mutation to change under selection was recognized by Mather (1941) and Wigan (1941) in carrying out their early experiments on selection for abdominal and sternopleural chaeta numbers respectively. They therefore selected for these characters in the inbred line, Oregon, in parallel with their main selections from the F_2's of crosses between Oregon and other lines. The results of these selections are described and discussed by Mather and Wigan (1942). Advance did indeed occur under selection but it was both much slower and much smaller than when selection was begun from F_2's. Mutation was clearly occurring among the member genes comprising the polygenic systems governing variation in these characters, but it could account for only a very small proportion of the response to selection, and hence of the variability, of the F_2's from which they also selected.

Clayton and Robertson (1955) similarly selected for abdominal chaeta number starting with an inbred line derived from the "Kaduna" population, which has been the initial material for a number of selection experiments showing rapid advances. Again, the inbred line gave a very much lower response, just as Mather and Wigan had found. Clayton and Robertson went further, however, by devising a means of estimating the increment that mutation was adding to the heritable variance in each generation, which they found to be 0·006. They also derived estimates of the increments added to the heritable variance in respect of sternopleural chaetae as well as abdominals using Mather and Wigan's results and found these to be 0·002 for sternopleurals and 0·007 for abdominals, which latter is in close agreement with the estimate from their own data. Finally they obtained estimates, by means which they do not detail, from the data of Durrant and Mather (1954), which were obtained in a different way by assaying the activity of a number of second chromosomes taken from the Oregon inbred line. These estimates were 0·006 for abdominals and 0·0025 for sternopleurals, which, since they refer solely to the effects of mutation in chromosome 2, imply higher estimates for the full set of chromosomes— perhaps as high as 0·015 for abdominals and 0·006 for sternopleurals. Even so, they are not grossly inconsistent with those from the other observations.

Averaging these various estimates gives 0·009 for the increment in the case of abdominal number and 0·004 for sternopleural number. Now, as we have seen, the "Texas" population has an additive genetic variance of 1·6 for sternopleural number which agrees well with Clayton and Robertson's figure of 1·7 for their population. Clayton and Robertson also say that a number of populations give about 5 for the additive genetic variance in respect of abdominal chaeta number. Thus the mutational increment per generation is only about 1/500 of the genetic variance of the population for the two characters; or to put the point another way, mutation would take some 500 generations to build up the genetic variance of a population to the level we see in nature.

In a general review of polygenic mutation studies, Mukai (1979) gives estimates of 0·00013 for the increment contributed per generation to the additive genetic variance of viability by mutations with minor detrimental effects. The corresponding additive genetic variance for a population is given as 0·00960. The ratio of mutational increment to population variance is thus 0·013, so indicating that in respect of this character some 70–80 generations would be needed for mutation to build up variation to the level seen in populations. This is only about one-seventh of the generations needed for the chaeta characters, which suggests, as Mukai points out, a difference in the two kinds of character.

Whichever character we consider, however, it is clear that the response of a population to directional selection must depend virtually entirely, in anything but the very long-term, on the variation that has already accumulated in it as opposed to variation newly arisen by mutation. The role of mutation must rather be that of constantly topping up the variation by making good losses due to selection or the random fixation of alleles, much as a stream is constantly topping up a reservoir from which a much greater flow of water can be obtained when circumstances demand it.

The estimates given above for the size of mutational increments were obtained by reference solely to variation expressed as differences among phenotypes, and they were compared with the additive component of the variation appearing as differences among the phenotypes of the individuals in a population. Yet by far the greater part of polygenic variability in populations is not the free variation shown by phenotypes, but is the potential variability hidden by balancing within genotypes. Equally, of course, we would expect mutation to be contributing increments to the hidden store of variability *pari passu* with its contribution to the free variation. That this is indeed the case is indicated by Mather and Wigan's (1942) selection for abdominal chaeta number in lines which started from the Oregon inbred. The inbreeding of the original Oregon line was relaxed in the selection lines, for each of which a cyclical mating system was used, similar to that adopted with the selective lines originating from the F_2's of Mather (1941). Any increment added by mutation to the store of potential variability would thus be as liable to being freed by crossing and recombination, and so becoming exposed to selection, in the lines started from the Oregon inbred as in those started from the F_2's.

Mather and Wigan's selection line for low abdominal chaeta number from the Oregon inbred was maintained continuously over the 53 generations of the experiment (Fig. 4), during which time the chaeta number, averaged over sexes, fell somewhat irregularly by 2·8 chaetae from 40·1 to 37·3. The initial high selection line was, however, terminated at generation G17, after it had shown an advance of about 2 chaetae starting at G14. A new high selection line was started from the low selection line at G17. This began to advance at G24 and continued to do so until it too was terminated at G34, by which time it had advanced by 4·7 chaetae. A third high line was begun, again from the low line, at G34 and this began to move upwards within two or three generations. It continued to do so until G44 by which time it had advanced by 3·4 chaetae. Thereafter it was stable until its termination when the experiment ended at G53.

Thus advance in the first high selection line did not start until 14 generations had elapsed. The second high line began to respond after 7 generations, and the third after only 2–3 generations of selection. Furthermore, the second line not only began its response more quickly than the first, but also moved faster once it had started. The third line began to respond sooner than the second, but did not move any faster and indeed stabilized before its response was as great as that of the second. All these

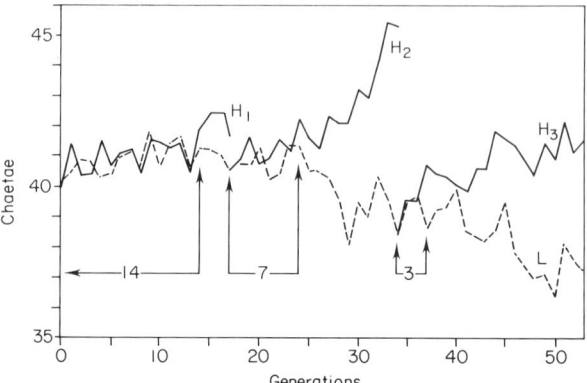

FIG. 4. The response of abdominal cheata number to selection in the inbred Oregon line (Mather and Wigan, 1942). The low selection line (L) was continuous for the 53 generations of the experiment, but three high selection lines (H1–H3) were taken off the low line at generations 0, 17 and 34 as shown. The numbers, and the arrows associated with them, show the numbers of generations that elapsed between the starts of the H lines and their first signs of response to the selection. All chaeta numbers are the means of the sexes.

features point to the accumulation through mutation in the low line of potential variability which, as a consequence of the mating system, was being steadily freed and so exposed to the action of selection. That the third high selection line did not advance for so long or so far as the second may be attributed to the erosion of the pool of potential variability in the low line by this line's own marked response to selection between G17 and G34, when these two high lines were started from it. Thus although the comparison of mutational increments and population variation is only on free variability, we can see that the mutation was contributing *pari passu* to the potential variability. We have no reason to doubt therefore that if the mutational increment to the total variability could have been assessed, and not just that to the free variation, and if it could have been compared with the total variability of the population, the picture that emerged would have been at least broadly the same as that obtained by observation of the free variation only.

One last point remains to be made before we leave our consideration of polygenic mutation. Starting with Clayton and Robertson (1955) a number of assessments have been made of the effects of radiation on the rate of this mutation. All agree in showing that, as with major gene mutations, it is stimulated by irradiation, but the estimates of the induced mutation rate per r of radiation vary by an order of magnitude and in one case by two orders (Mukai, 1979). The reasons for these differences are not clear, and in any case they need not concern us as they have little relevance to response to selection under natural conditions.

III. Directional Selection

A. Early Generations

Most of the selection experiments that have been carried out have used directional (R) selection, often in both directions simultaneously. Furthermore, a quantitative theory (well set out by Falconer, 1960) has long existed from which expectations can be derived for the response of characters showing polygenic variation to this type of selection. Where s, the selective differential, is the deviation of the mean expression of the individuals chosen as parents from the mean of the population, r, the response to selection, is the corresponding deviation of the mean expression of the offspring of these parents, and h^2, the heritability, is the ratio of the additive genetic variance of the character to the phenotypic variance of the population, the relationship expected between them is

$$r = h^2 s$$

The value of s may be found by averaging the expression of the character obtained by direct observation of the individuals taken as parents, or its average value may be found from the proportion of individuals taken as parents, using the properties of the normal distribution.

A number of selection experiments have been carried out, notably by Professor Alan Robertson and his associates, to test the validity of these expectations in practice. Clayton *et al.* (1957a) selected, in different lines, for both an increase (H) and decrease (L) in the number of abdominal chaetae, starting from a cage population. In all lines 20 females and 20 males were chosen as having the highest or lowest chaeta numbers (according to whether the selection line in question was H or L), but the intensity of selection was varied by taking the flies as the extreme members of 100, 75, 50 or 25 of each sex. Five replicate lines were used for each of the H and L selections at the greatest intensity, and three replicates each for H and L lines at the lower intensities. The chaeta numbers were scored for 200 individuals of the two sexes in each generation of each replicate of both H and L selections at each intensity. There was clear, and in some cases considerable, divergence among the replicate lines, more in fact than would perhaps be expected on the basis of sampling variation where 40 parents were used for each generation in each replicate. Too much significance must not, however, be attached to this: although 20 parents of each sex were cross-mated to give each generation, there can be no assurance that all of them contributed equally, or indeed at all, to the offspring produced. Thus the effective size of the parental groups may well have been well below 40, and sampling variance would be increased thereby.

A more striking feature of the experiment emerges, however, when we compare the advances expected with those observed in the first five generations of selection. The ratio of advance observed (O) to that expected (E) is shown in Table II for each intensity of selection of the H and L parts of the experiment. At the greater intensities of selection the average response in the H lines agrees quite well with expectation, though there is an indication that the efficacy of selection, as measured by O/E, declines with the intensity. The L selections show a similar decline of efficacy with intensity; but even at its best, in the selection of greatest intensity, the response observed is much smaller than that expected. Indeed, in the L selection of lowest intensity (20/25) such response as occurred was in the wrong direction (indicated by the $-$ sign). Perhaps this is to be attributed to the low number of individuals from which the parents were taken, and the relative low value of $O/E = 0.46$ obtained from the corresponding H selection may support this view. But even setting aside the results from this lowest intensity, the average of O/E for the three higher intensities of selection in the H lines is 0.99 (where 1.00 shows complete agreement of observation with expectation), whereas it is only 0.55 for the corresponding L lines. The ratio $0.55/0.99 = 0.56$, which compares the relative responses of L and H lines, is shown not only by the averages taken over these three intensities, but appears to a close approximation from each of the intensities taken separately. Clearly it is a feature of the experiment and hence, one assumes, stems from a property of the experimental material.

Thus, although the responses to H selection agree well with expectation, responses to L selection are consistently only just half those expected. The reason for this is not clear. Such a difference could stem from inadequacy of the scale on which the character is measured, i.e. from the differences in

TABLE II. Relation of changes under selection for increase (H) and decrease (L) observed (O) in the number of abdominal chaetae to those expected (E) (Clayton *et al.*, 1957a).

Level of selection	Expected E	Change High——Observed——Low				
		O_H	O_H/E	O_L	O_L/E	O_L/O_H
20%	2.42	2.62	1.08	1.48	0.61	0.56
27%	2.14	2.20	1.03	1.26	0.59	0.57
40%	1.68	1.46	0.87	0.79	0.47	0.54
80%	0.61	0.28	0.46	−0.08	−0.13	—

numbers of abdominal chaeta not being linearly related to the differences in the balance of the genes acting in the different individuals. Scalar problems of this kind are well known in biometrical genetics (Mather, 1946; Mather and Jinks, 1971) but doubt is cast on such an interpretation of the present case by observations in experiments, starting with other strains of *Drosophila melanogaster*, where the L selections could respond more than the H (Rasmuson, 1955); if the asymmetry was due to scalar difficulties one would expect it to be in the same direction as that oberved by Clayton *et al.* (1957a). Nor is it likely to be due to the L selection in Clayton *et al.*'s experiments being opposed by some force of natural selection which did not affect, at any rate to the same extent, their H selections. Some as yet unrecognized relationship must have been in operation, a relationship that can so distort the heritability of a character in a population as to render it different at different levels of phenotypic expression. We shall see further examples of this asymmetry of response later.

The efficacy of the quantitative theory in predicting response to selection over the short-term has been investigated in further of its aspects in a range of other experiments, again using as the character either the number of abdominal chaetae or the number of sternopleurals. The effects of assortative mating and of individual and family selection, alone or in combination, have been examined by McBride and Robertson (1963), while Sen and Robertson (1964) have looked at various ways of selecting simultaneously for two characters, as exemplified by abdominal and sternopleural chaeta numbers. Frankham *et al.* (1968a) have tested the effects of differences in the numbers of parents for each generation in different lines, and also of varying the intensity of selection, all in a large factorial experiment. The consequences of using different numbers of parents have also been investigated by Madalena and Robertson (1974), and Osman and Robertson (1968) have examined the same thing in a somewhat different way by looking at the consequences for continued L selection of introducing genes from the "Kaduna" population into a line already selected for low sternopleural chaeta number from this same population.

All the relevant experiments agree that, as theory would suggest, the use of larger numbers of parents gives greater ultimate response to selection, and they agree also in showing that replicate selection lines can diverge, considerably from one another at times, presumably as a result of sampling variation. Assortative mating increases response to selection when heritability is high, and the combination of family with individual selection is effective in raising response when heritability is low, again as theory would predict. Indeed it is a recurring theme of these various resports that the observations are in broad agreement with theoretical expectation, even though divergencies and discrepancies of a more detailed kind can arise. Sen and Robertson (1964) further note that while great differences in heritability can appear in different selection lines after 12 generations, the selection appears to have had little effect on the mean heritabilities. This is, of course, not surprising in view of the great store of hidden variability for the character that must lie awaiting release in the population.

One further test requires mention before we leave selection systems. This was undertaken, clearly for the guidance it could give in selecting domestic plants and animals, by Bell *et al.* (1955) using eight wild-type strains of *Drosophila* as initial material. They carried out family selection and recurrent cross selection (Hull, 1945) starting with a combination of all eight strains, and using an inbred tester line in the recurrent cross selection. They also carried out reciprocal recurrent selection (Comstock *et al.*, 1949) using two combinations each involving four of the strains. The characters followed were egg size, which shows a high level of additive genetic variation with little heterosis, and fecundity, which shows low heritability and marked heterosis. A combination of individual and family selection was judged to be superior for improving egg size and characters like it, whereas the two kinds of recurrent selection had the advantage, at least in the long run, in raising performance in respect of fecundity. This again accords with expectation.

Where the experiment is initiated from a population, whether brought in from the wild or made artificially by allowing a combination of strains to breed at random within itself for a number of generations, the linkage relations of the relevant genes are unlikely to have any major effect on progress under selection in the short run, especially if reasonably large numbers of parents are used in each generation: the genes will be in linkage equilibrium, or near equilibrium, at the start and only the rarer recombinations of the genes will not be available for selection. The situation is different where the selection lines are initiated from the F_2 of a cross between two unrelated lines, for the genes will not then be in, or even near, linkage equilibrium and the resolution of repulsion linkages may be the key to advance under selection even in the early generations. This is illustrated by one of Mather's early experiments (1941), to which reference has already been made (p. 165).

Directional selection was started from the F_2 of a cross between the wild-type Oregon inbred line, and a non-inbred BB stock. Selection was for increase and decrease of abdominal chaeta number in the H and L lines respectively. The progress of the two lines is illustrated in Fig. 5, where the chaeta number shown is the average of the sexes. By G2 the selection lines have slightly transgressed the chaeta number of the two parents, no doubt largely by the reassortment of the three major chromosomes as units; but the H line is stable for two generations thereafter and the L line virtually so. At

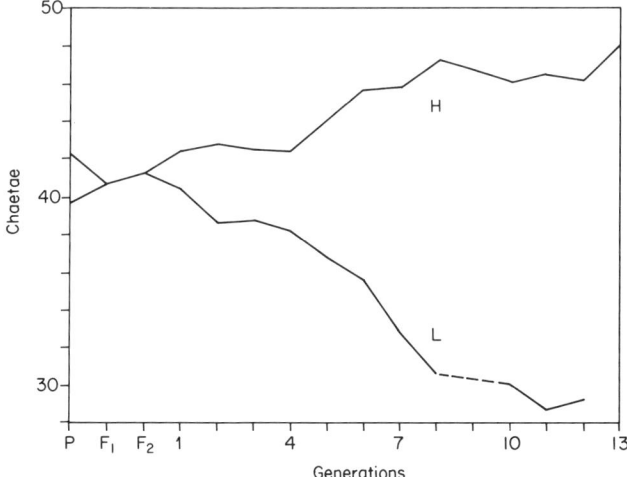

FIG. 5. The response of abdominal chaeta number to high (H) and low (L) selection starting from the F_2 of a cross between the Oregon inbred line and a BB stock (Mather, 1941). The means of the two parents, their F_1 and F_2 are also shown. All chaeta numbers are the means of the sexes.

G5, however, both H and L begin a new, more rapid and more pronounced advance under selection, continuing until G8 in H, after which this line appears to stabilize again, and showing signs of continuing until G11 in L albeit at a lower rate after G8. Something happened in both H and L at G4, which released a great deal of hitherto hidden variability. All the signs, including its simultaneous occurrence in the two lines, points to it being recombination, resulting from cross-over within one or more chromosomes.

The effects of intra-chromosome recombination on the heritable variation can be seen when a comparison is made between the components of variance in the F_2 from the original cross and the F_2 of a cross between the H and L lines after selection had been terminated at G13 and G12 respectively. The results are summarized in Table III. V_E, the non-heritable part of the variance is estimated as $\frac{1}{4} V_{P1} + \frac{1}{4} V_{P2} + \frac{1}{2} V_{F1}$, where P_1 and P_2 are the two parental lines. It is then subtracted from V_{F2} to leave V_G, the genetic variance. V_G will of course be inflated by a component arising from dominance, but comparison of the mid-parent value with the means of F_1 and F_2, as given by Mather, reveals little indication of dominance being at a level that could result in anything more than a minor disturbance. Indeed, in any case, it could not account for more than $\frac{1}{3}$ of V_G unless it were so large as to constitute overdominance. We therefore take V_G as a measure of the additive genetic variation and find that it is no more than 0·52 in the F_2 of the original cross but is 9·93 in the F_2 of the cross between H and L, that is 19

TABLE III. Components of variation for abdominal chaeta number in (a) the F_2 of the original cross and (b) the F_2 from the cross between H and L selection lines derived from it (Mather, 1941).

Cross	V_{F2}	V_E	V_G
(a)	6·93	6·41	0·52
(b)	17·47	7·54	9·93
Ratio a/b	2·52	1·17	19·10

V_E was found as $\frac{1}{4}(V_{P1}+V_{P2}+2V_{F1})$ where V_{P1}, V_{P2} and V_{F1} are the variances of the two parents of the cross and their F_1. All variances are the means of the two sexes.

times as large. Clearly linkages that were in the repulsion phase in the original cross (i.e., $+-/-+$) have been replaced by coupling linkages ($++/--$) in the cross between the selected lines. That disruptive and, by derivation, directional selection favour coupling as opposed to repulsion linkages has also been demonstrated directly, using marker genes, by Thoday (1960).

The selective consequences of intra-chromosome recombination are revealed even more strikingly by two of Sismanidis' (1942) selections for increase in the number of scutellar bristles over the normal 4. The selection lines in question, C and D, were each initiated by a single pair mating from the progeny of a female with 5 bristles, found in a *y w* laboratory stock, and one or possibly more males, all with 4 bristles, from the same stock. They were maintained thereafter by brother–sister mating, selection being practised in each generation for increase in scutellar number. The advances under selection were smaller than those found earlier by Payne, the average number in females reaching no more than 5·2 after 24 generations of selection. The average number of additional bristles was smaller in males than in females, just as Payne had found.

The progress of lines C and D is shown diagrammatically for females in Fig. 6: that for males was consistent, even though smaller, and it has been omitted in the interest of clarity. Three levels of average bristle number can be recognized, as indicated in the Figure. Both C and D moved from the initial level of 4·2 bristles (shown by the *y w* stock from which the experiment was started) to the second level at 4·5 bristles, where they remained stable, or nearly so, for some 10 generations before moving up to the third level at 5·2 bristles. D shows a slightly higher value than C at this

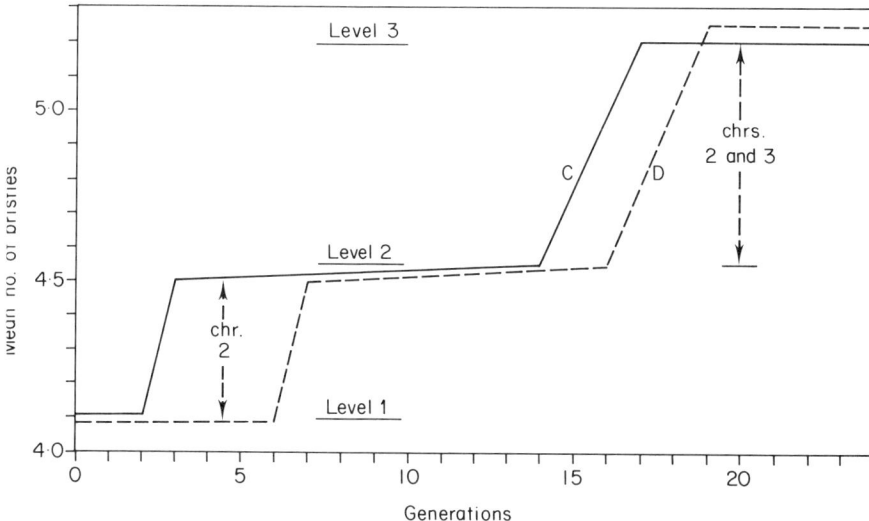

FIG. 6. Diagrammatic representation of the response of the number of scutellar bristles, in females, to selection for increased number in Sismanidis' (1942) C and D lines. The chromosomes responsible for the changes between the three levels are indicated on the diagram.

third level, but the difference is small by comparison with the inter-level steps. C moved from the first to the second level between generations G2 and G3. D did not move up until G6–G7, but when it did move the step exactly matched that of C in size. Then followed the period of near stability in each line, to be ended when C moved up to the third level at G14–G17 and D followed suit at G16–G19. Again the steps did not come at the same time, but when they did come they were of the same size.

Employing a technique used by Mather (1942) and developed further by Mather and Harrison (1949), Sismanidis assayed the contributions made by each of the three major chromosomes to these various advances under selection. The results are summarized in Table IV. The analysis is complete in line C but there is no direct measurement of the status of chromosome 3 in line D at the second level. Comparison of the effects ascribable to chromosome 2 with the overall differences between the levels allows us, however, to infer the behaviour of chromosome 3 during the two steps shown by D. The results of the analysis are clear: chromosome X was not involved in either of the two steps; the first step was traceable entirely to change in chromosome 2 in both C and D; and the second step was traceable approximately half to change in chromosome 2 and half to change in chromosome 3, again in C and D alike. Clearly the steps must have been determined by the underlying organization of the chromosomes, and in

TABLE IV. Changes in the average number of scutellar bristles of females, attributable to the effects of the three major chromosomes in Sismanidis' (1942) selection lines C and D

Change between generations	Chromosome			
	X	2	3	Sum
Line C				
0–13	0·08	0·36	0·04	0·48
13–21	−0·02	0·36	0·31	0·65
0–21	0·06	0·72	0·35	1·13
Line D				
0–14	−0·02	0·41	—	—
14–19	0·08	0·32	—	—
0–19	0·06	0·73	0·34	1·13

No assay was available for the effect of chromosome 3 in Line D at generation 14.

particular to the linkage relations of the genes they carried. The effective recombinations did not come at the same time in the two lines, and there is little reason to expect them to have done so. But when they did come they led to the same results, as indeed the production of a + + chromosome by recombination in a + −/− + heterozygote must do.

Similar evidence of dependence of response to selection on intra-chromosome recombination is not expected in the early generations of selection from populations, where, unlike F_2's, the phases of linkage are expected to be broadly in equilibrium. We should however expect it in later generations after early responses to selection had disturbed this equilibrium. Indeed McPhee and Robertson (1970) have shown by the use of inversion heterozygosity in chromosomes 2 and 3 that reduction in the rate of intra-chromosome recombination materially reduces the rate of advance of sternopleural chaetae number under selection over 17–20 generations in lines extracted from a population, within which they detected no evidence of linkage disequilibrium. We must therefore turn now to examine the consequence of selection in later generations and we must not be surprised to find them going beyond anything that simple quantitative theory would lead us to expect.

B. Selection Response in Later Generations

A great number of experiments have been carried out using directional selection. Some have lasted only a few generations, but others have been carried on for as many as 108 (Rasmuson, 1955) and 138 (Mather and Harrison, 1949). Some have started with crosses between specific lines or stocks, sometimes inbred and sometimes not (e.g. Mather, 1941; Mather and Harrison, 1949; Thoday and Boam, 1961). Others have been started from laboratory populations originating from flies caught in the wild (e.g. Reeve and Robertson, 1953; F. W. Robertson, 1955; Clayton and Robertson, 1957; Clayton et al., 1957b; Jones et al., 1968). A variety of characters have been followed, notably the number of abdominal, sterno-pleural or scutellar chaetae, wing length and thorax length, but including others such as body weight and behavioural characters like phototaxis and geotaxis (see Table I). In the short-term, as we have seen, the speed of response often, though not always, proved capable of approximate prediction from knowledge of heritabilities—but in the long-term it has invariably been found that such predictions failed. It is to the factors governing the speed and extent of responses in the long-term that we must now direct our attention. In doing so, we shall seek to elucidate and illustrate common features of these many experiments, rather than set out on the prohibitively lengthy task of reviewing them all in detail.

When starting with an F_2, even in the early stages response to selection may not progress smoothly but by a series of steps from one plateau to another, with the steps depending on the utilization of variability released by recombination consequent on crossing-over within chromosomes. Even such intra-chromosome recombination, however, may not lead to stepped responses: where there are a number of repulsion linkages to be broken, and where the number of parents and system of mating used in each generation is such that the spread of a changed chromosome through the progenies may take several generations, the steps can overlap and the overall response thus appear smooth. This would appear to have happened in Mather and Harrison's initial line selected for high abdominal chaeta number. This line rose reasonably steadily, without any obvious plateaux or steps, for 20 generations, during which time its chaeta number had risen by about 20, i.e. an average rise of 1 chaeta per generation. Selection was relaxed at G20 to avoid the loss of the line. The mean chaeta number of this massed line proceeded to fall by 17 in five generations—an average fall of 3·4 per generation. A new high selection line was taken off the unselected mass at G24 and in four generations had regained the level attained by the original selection line at G20, i.e. an average gain of over 3 chaeta per generation, so matching the rate of fall of the mass line between G20 and G25. Thus the original rate of change under selection had been replaced by a rate of change three times as high, whether in falling when selection was relaxed or rising when it was resumed. Though not individually distinguishable, the intra-chromosome recombinations upon whose occurrence the original rise had depended, had built up in the chromosomes gene combinations of more extreme effect, which further recombination could break down only as slowly as the early recombinations had built them up. The result was the trebling of the rate of change (Fig. 7).

This new selection line stabilized at G29 and did not rise further despite continued selection for 50 generations or so. A new massed line, in which selection was relaxed, was also taken off the selection line at G34. This time the chaeta number of the massed line showed no sign of falling; it was in fact quite stable. On the face of it this stability could be ascribed to the upward selection, between G24 and G29, having used up all the variability, the line having thus become homogenic. That this was not, however, the real explanation of the stability was clearly revealed by a back selection which was taken off it at G56, after the massed line had been stable for nearly 30 generations. The chaeta number of this back selection line fell sharply for some 10 generations, stabilized for a dozen generations and then fell again, even more rapidly, before stabilizing finally at a value below that of the F_2 from which the original selection had been initiated. Thus, despite its stability under both relaxed selection and continued high selection, the line was far from homogenic: in fact, it still carried enough variability to permit a fall, under back selection, greater than the rise hitherto attained under high selection. Clearly the stability depended not on it being homogenic, and so unable to respond to selection, but on a balance of the forces of selection acting on it. Before considering the nature of these forces we will look at a further example which shows us the two causes of stability, homogenicity and balanced selection at work in the same experiment.

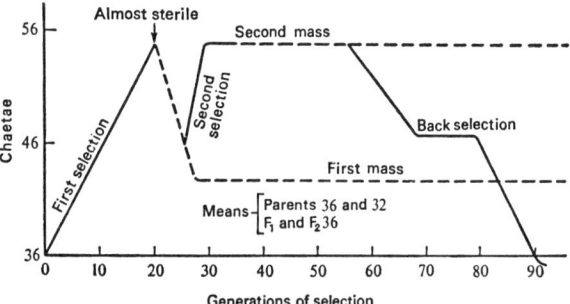

FIG. 7. Diagrammatic representation of response to selection for increase in the number of abdominal chaetae observed by Mather and Harrison (1949). Solid lines indicate selection and broken lines indicate mass culture without selection. (Reproduced by permission from K. Mather and J. L. Jinks (1971). *Biometrical Genetics* (2nd edn). Chapman and Hall, London.)

F. W. Robertson (1955) describes the behaviour of lines selected in both directions for thorax length and initiated from three populations of *D. melanogaster* each stemming from an inseminated wild female, caught in one case in Italy and in the other two in Scotland. Though differing in the details of speed and total advance of their H and L lines under selection, all three of the stocks showed the same general pattern of response. In all three cases there was fairly smooth progress of both H and L lines before response to selection ceased after 10–15 generations, the H lines showing somewhat smaller advances and ceasing to respond rather sooner than the L.

Back selection begun at G5 was effective in all H and L lines. Indeed all the lines fell back to their starting levels of thorax length at least as quickly, and commonly more quickly, than they had moved away from it under the original selection. Back selection was equally effective when started again from the H lines at G15–G17 after they had ceased to respond to forward selection: indeed, these new back selections fell more quickly than the earlier ones from the same H lines. The L lines on the other hand showed virtually no response to either back selection or relaxation of selection, when these were tested for their effects after response to forward selection had ceased.

Thus in all three cases the L line appears to have become stable because the heritable variability had been used up by response to the original selection (or at least fixed in some other way), while the H line had stabilized as a result of a balance being struck between forces of selection. That this same pattern was shown by all three stocks, stemming from flies caught as far apart as Italy and Scotland, suggests that it reflects some intrinsic feature of the genetical structure of this character, which Robertson and Reeve (1952) regard as a measure of body size. We should note in this connection that in an earlier experiment using wing length (which they also regard as a measure of body size) F. W. Robertson (1954) found the combinations of genes in a low selection line to be generally recessive to their alleles in the stock from which the selection was started, which must imply a preponderant dominance of alleles for larger size. This could provide a basis for understanding the results of selection for thorax length, since recessive genes tend to be fixed more readily than dominants. The L selection lines would thus be likely to achieve a stability arising from fixation rather than from a balance of selective forces, such as would be likely to occur in the H lines where the more dominant alleles could mark an underlying heterozygosity.

The H lines behaved in essentially the same way as Mather and Harrison's line, even to the extent of responding faster to back-selection than to the original forward selection by which they had been built up, albeit the difference in rate of response was not so dramatic as in Mather and Harrison's line. Evidently, the speed of response to selection reflects the history of earlier selection to which the character has been subject in the past.

C. CHAETA NUMBER AND FERTILITY

The frequency with which lines have been observed to stabilize, in the sense of ceasing to respond to continuing forward selection, even though they demonstrably still carry variability for the character on which the selection is based, is a feature of selection experiments. As such, it inevitably raises the question of the reason for the selective forces that result in the balance being brought about.

In Mather and Harrison's experiments this reason appears clear. The selection for the number of abdominal chaetae was observed to change other characters also. It was, in fact, observed to affect quite a number of other characters, including other kinds of chaetae, body pigmentation, mating behaviour and number of spermathecae: no doubt still more effects would have been found if the relevant characters had been examined. The effects on fertility already mentioned were, however, of particular significance. As already noted, fertility (by which they meant the number of offspring produced) declined to such an extent in the original H line that too few offspring were produced in G20 for the selection to be continued. With the relaxation of selection at G20 fertility was regained in the resulting mass line and the chaeta number fell back. There was evidently a negative correlation between chaeta number and fertility in this initial selection line. That this lowering of fertility was not simply a pleiotropic effect of raised chaeta number was, however, made clear by a low selection line started at the same time from the same F_2. This line's response to selection, as exemplified by the reduction it showed in chaeta number, was slower than that of the H line, but it nevertheless died out at G35 by reason of sterility. Loss of fertility is thus associated with response to selection in either direction, and not just simply correlated with chaeta number itself.

This relation between reduction in fertility and response to selection in either direction for abdominal chaeta number had been observed earlier by Mather (1941) and a similar relation had also been seen by Wigan when he selected for sternopleural number (1941). It was discussed by Wigan and Mather (1942), who pointed out that it could be expected to arise as a result of the intra-chromosome recombination on which the responses to selection for the chaeta character depended: the cross-overs that gave rise to the new and more extreme polygenic combinations affecting chaeta number would also frequently give rise to changes in the balance of other polygenic combinations, affecting other characters, where these were intermingled with the chaeta combinations along the chromosomes. These other characters could thus be altered by selection directed solely at chaeta number. And where the, in this sense, secondary character was fertility, it must be expected that the recombinational unbalancing of its underlying polygenic combinations would generally result in a reduction of the level. Thus in these circumstances, successful selection for chaeta number would be expected to carry with it the penalty of a reduction in fertility, no matter in which direction chaeta number was selected. A corollary of this interpretation would be, of course, that given opportunity and time for the linkages between the two polygenic systems (mediating chaeta number and fertility respectively) to be readjusted by further recombination, the newly selected chaeta numbers could be freed of their association with reduced fertility.

We know that the member genes of the polygenic systems governing abdominal and sternopleural chaeta numbers are well distributed over the lengths of all three major chromosomes in *D. melanogaster* (Davies, 1971). We do not, however, have similar detailed information about the system governing fertility, though Breese and Mather (1957, 1960) have observed that genes affecting viability (a component of fertility, in the sense that we are using the term) are widely distributed along chromosome 3, which was the only one they tested. Indeed, it is hardly to be doubted that a polygenic system mediating fertility will be as well distributed along the lengths of the chromosomes as those for the two chaeta characters.

Certainly the responses to selection in Mather and Harrison's experiment are readily interpreted on this view. It supplies a basis not only for the correlated response of reduced fertility to the selection for abdominal chaeta number but also for the decline in chaeta number in the mass line, started at G20, in which, with selection for chaeta number relaxed, fertility would take its place as the capital character and would drag chaeta number, now the subordinate character, downwards as the polygenic combinations giving higher fertility increased in frequency. Equally, with mass mating there would be opportunity for further recombination to readjust the linkage relations between the fertility and the chaeta number genes, with the result that when at G24 reselection re-established the high chaeta number, it no longer carried the same incubus of gross infertility. Thus the second selection line could be successfully established and the chaeta number maintained in the massed line taken from it without any selection for chaeta number itself being imposed, even though, as back-selection later showed, it carried the variability necessary to give rise to lower chaeta numbers. Further discussion of the correlated responses in fertility produced by Mather and Harrison's selection for abdominal chaeta number would be out of place here: a detailed account of the evidence will be found in Section 3e of their paper.

D. Correlated Responses and Recombination

Fertility was but one of the characters which showed correlated response to selection for abdominal chaeta number in Mather and Harrison's experiments. The correlated responses in the other characters can obviously be explained in the same way as those for fertility, though the data relating to them are less extensive than those for fertility, and the argument for relating the changes to recombination correspondingly less compelling. Indeed it might be thought that chaeta characters, such as abdominal, sternopleural and coxal chaeta numbers, might be correlated in ways reflecting pleiotropic action of at any rate a proportion of the genes that governed them. Clearly in such a case, selection for one of them would then inevitably produce corresponding changes in the others. Some of Mather and Harrison's results suggested that this might be the case but, even if so, the pleiotropic relations could not have been simple, and later results, especially Davies' (1971) finding that the member genes of the systems for sternopleural and abdominal chaeta numbers, though intermingled along the chromosome, are nevertheless separable from one another, leave little doubt that correlated responses for one of these characters to selection for the other must arise as a consequence of simultaneous recombination rather than from pleiotropic action.

As already noted, the relation between recombination and response to selection, whether response of the primary character or correlated responses in other characters, would be expected to appear sooner and more strikingly where selection began from an F_2 rather than from an open breeding population. Nevertheless, in the long run, recombination is a key factor even when a population provides the initial material, and correlated responses have been observed not only in sternopleural chaeta number but in the spermathecae also, in selection for abdominal chaetae from a population (Clayton et al., 1957b). Analysis revealed that a lethal chromosome 3 played a major part in this, and lethals have been found in chromosomes 2 and 3 of a number of lines, which while no longer responding to selection were still heterogenic and which variously resulted from selection for increased abdominal number (Clayton et al., 1957b; Frankham et al., 1968b), decreased numbers of sternopleurals (Madalena and Robertson, 1974), increased length of wing (Robertson and Reeve, 1952; Reeve and Robertson, 1953) and increased number of scutellar bristles (Sismanidis, 1942). Where in heterozygous condition the lethal chromosome moved the character in the direction of selection more than did the homozygous non-lethal homologue, a polymorphism should result, with the lethal homologue favoured by artificial selection and the non-lethal favoured by natural selection. Such a polymorphism could, of course, produce a line that refused to respond further to the artificial selection but was capable of responding to back-selection or even to relaxation of the artificial selection. In two cases, the effects of these lethal chromosomes on chaeta number, when heterozygous, were especially large, *viz* in Madalena and Robertson's selection for low sternopleural number and in Sismanidis's selection for increased number of scutellars.

On the face of it, such lethals could have arisen by mutation. Lethals are, however, known to arise by crossing-over in chromosome 2 of *D. melanogaster* (Misro, 1949), and while Hildreth (1956) failed to find any such arising in chromosome 3, there are some indications, even if not final evidence, that they can so arise (Breese and Mather, 1957, 1960). This interpretation is favoured by the frequency with which they have been found. Only lethal chromosomes that, when heterozygous, affect the character in the direction of selection would survive in the section line, and although dominance may be a consideration in respect of wing and thorax length, it is known to be ambidirectional for abdominal chaeta number (Breese and Mather *l.c.*) and for sternopleural number, too (Caligari and Mather, 1980). So the lethals isolated from lines selected for these characters can only be a sample of the lethals that must be assumed to have arisen. The indications are thus that the incidence of lethals is too high to be ascribed reasonably to mutation and that crossing over leading to intra-chromosome recombination must once again be implicated, the lethality being a correlated response to selection for the primary character—chaeta number or whatever it may have been.

Reference must be made to one last phenomenon before we leave response to directional selection. A number of cases have been observed, notably by Mather and Harrison (1949) and Thoday and Boam (1961), where lines responded to selection, sometimes by dramatic leaps, after having failed to do so for many generations—indeed after as many as 54 generations in one of Mather and Harrison's cases. These may, of course, be regarded as extreme cases of the short delays observed by, for example, Mather (1941) and Sismanidis (1942), which we have seen to be attributable to intra-chromosome recombination. Two further pieces of evidence reinforce this interpretation. In Mather and Harrison's case, three different lines, stable for different numbers of generations and selected in different directions, all gave their delayed responses to selection at much the same time. This would be difficult, if not impossible, reasonably to reconcile with an interpretation in terms of mutation. It is by no means so unlikely a finding if crossing-over is the origin of the changes. It is known that environmental factors affect the incidence and distribution of crossing-over, especially near the centromere, in the chromosomes of *Drosophila*, and the various selection lines would be expected to feel the effects of changes in such environmental agencies at the same time. In Thoday and Boam's case three lines responded, not at the same time, but to the same extent: all moved from having 23–24 sternopleural chaetae to having about 30. A common basis is thus indicated and it is readily found in the occurrence of a rare recombinant.

Such rare recombinants will survive and become fixed only when they produce an effect that matches the requirements of the selection that is imposed. These can hardly fail to include dominance in the direction favoured by the selection and they can also include appropriate interaction between non-allelic genes. Such non-allelic interactions were found by Mather and Harrison and were observed to be in different directions according to the selection the relevant lines were undergoing. The most striking case is, however, that of Spickett and Thoday (1966) who isolated two "genes", one in each of chromosomes 2 and 3, which both displayed dominance in the direction of selection and also a joint complementary action, such as would equally have been favoured by the selection used (see Mather, 1966, 1973).

We thus see that, no matter to which of the features of response to directional selection in *Drosophila* we turn our attention, all are interpretable in terms of crossing-over leading to intra-chromosome recombination: indeed they commonly go further and require such an interpretation. In short they all call attention to the basic importance of linkage and its resolution in determining the storage of variability and its release through crossing-over in the polygenic systems of this fly. In the short, and even medium, term the readjustment of a character by selection is dominated by the requirement for crossing-over to resolve the linkages. And so also in the longer term is the reassociation with other characters which is necessary to produce individuals that are genetically fit in the altered circumstances imposed by the selection.

It only remains to observe that, since the frequency and distribution of crossing-over within the chromosomes is itself under genetic control and hence capable of being changed by selection, the system of linkage that we see in *Drosophila* [or indeed in any other species of animals or plants, as Darlington (1939) inferred on other grounds] must itself be the product of evolution by natural selection. The system of recombination, like the breeding system, has evolved and must still be evolving under the pressure of the secondary selection arising from its effects on the variability, its conservation and its exposure to the action of primary selective forces, in respect of all the many characters on whose adjustment to the requirements of the environment the fitness of the individuals, populations and species depends.

IV. Stabilizing Selection

A. Additive Genetic Variation

In stabilizing (S) selection experiments individuals chosen to be parents of the next generation are those whose phenotypes deviate least from the mean of the population or other group from which they are taken. Deviations in either direction are penalized equally. The selection is thus for reduction in departure from the mean phenotype and hence for a lower phenotypic variance (V_P). If any such reduction is not only to be achieved but also passed on to succeeding generations, it must have a genic basis; at the same time its causation need not be confined to a reduction in the additive genetic variation (V_A) of the population or group: it might result also from a reduction in the non-heritable variation, that is from the building up of genotypes whose phenotypic expression was less subject to disturbance by environmental or developmental upsets. These would appear statistically as reductions in the environmental component of variation (V_E), from which a developmental component (V_D)* could be isolated where it was possible to compare the replicate expressions of a bilaterally symmetrical character on, for example, the two sides of a fly. Such a reduction in the non-heritable component of variation must depend on genotype × environment interaction and implies change in the physiological properties of the genotype. It will be considered in the next section, after the effect of stabilizing selection on the additive genetical component of variation has been discussed.

Reduction in the additive genetic variation may be promoted in a variety of ways. It can be the result of the selection favouring balanced combinations ($+-$ and $-+$) of the relevant genes, which give more central expressions of the character, at the expense of unbalanced combinations ($++$, $--$), associated with more extreme expressions. (The $++$, $+-$, $-+$ and $--$ are used merely as simple diagrammatic illustrations: the actual combinations will obviously be more complex and their balances more complicated.) Furthermore, where two combinations were in exact, or nearly exact, balance, their heterozygote would have virtually the same phenotype as the corresponding homozygotes ($+-/+-$, $-+/-+$) provided dominance was absent or also balanced. Where a population carried nothing but polygenic combinations so balanced, it would display no additive genetic variation. But these combinations must continually be eroded by recombinations of the relevant genes, resulting in the production of the more extreme combinations ($++$, $--$) in either a steady stream or a slight trickle according to the frequency of the recombination. Continuing

---
* Not to be confused with the component of genetic variance due to dominance, which is often also symbolized V_D.

selection would thus be necessary to maintain the near zero level of V_A, even where it has been possible to reduce it so far. Relaxation of the selection would ultimately result in the building up again of V_A, even though recombinations were rare. Mutation would, of course, also increase V_A.

Where a population had become homogenic for one or other of the balanced combinations as a result of random drift, V_A would be permanently reduced to zero (apart from the effects of such mutations as might occur), for there would then be no opportunity for recombination to disturb the balance. Inbreeding, such as occurs naturally in some plants, would have a similar effect: homozygosity of the individuals would arise, though at first without homogenicity of the population. One generation of S selection would, however, serve to reduce V_A to near zero even though heterogenicity for two or more balanced combinations was present, because of the close similarity of the phenotypes they produced. Occasional outcrossing (which also occurs in many of these plants) followed by recombination would, nevertheless, release some of the variability and result in genetic variation appearing again.

Finally, dominance could result in a reduction of V_A, even where the dominance itself was ambidirectional and balanced. Given that the dominant allele is the more common allele at each relevant locus (or at least preponderantly so when all loci are taken into account), the value of V_A is less than it would be in the absence of dominance. Caligari and Mather (1980) have shown that this relation is in fact a feature of the polygenic system mediating sternopleural chaeta number in the "Texas" cage population of *D. melanogaster*, as we saw in an earlier section.

A number of experiments have been carried out to test the effects of S selection on the variation of a number of characters, both wild-type and mutant, in *Drosophila*. In the first of them, Falconer (1957) failed to find any change in the variation shown by abdominal chaeta number; but the selection pressure he used was not great, and later experiments have regularly found the variation to be reduced. In many of these experiments the S section lines have been paralleled by others subjected to a type of disruptive selection (D^-), which may be regarded as the reverse of S selection in that the parent flies were chosen as having extreme expressions of the character, both H and L, and were mated disassortatively. Such D^- selection would, of course, be expected to increase the variation, as it had indeed been regularly observed to do.

Thoday (1959) applied S selection to sternopleural chaetae number for 44 generations and observed a decline in the phenotypic variation V_P which he traced to a reduction in V_A. The findings of Prout (1962), for the effects of S selection on time of development of the flies, were similar. Gibson and Bradley (1974) followed Thoday in using sternopleural chaeta number as the selected character, but they practised their S selection on a cage population newly established from the wild, and they did so at a fluctuating (20°/29°C) as well as at a constant (25°C) temperature. Their results show a decline in V_P and V_A under S selection in both environments. They also show a decline in V_E ($=V_P-V_A$) when averaged over the two environments (Table V). Interpretation is, however, complicated by similar declines in both V_A and V_E (the latter before G19 but the former after this generation) in control lines maintained without artificial selection. This could perhaps be taken as indicating that the population was too newly introduced to have been adjusted fully to laboratory conditions when the experiment began, with the consequence that the controls were still under quasi-natural S selection during the experiment.

There is no ambiguity, however, about the results of Scharloo et al.'s (1967) extensive experiments, which used as the character for selection the relative length of the 4th wing vein in flies carrying the gene ci^{D-G}, which shortens this vein. Being a bilateral character, this enabled them to separate the developmental component, V_D, from the rest of V_E (which, as in all the other experiments, includes non-additive genetic variation, though this is expected to be small). The decline that they found in V_P and all three of its components under S selection is set out in Table VI and illustrated in Fig. 8, where it is contrasted with the increases they observed in their D^- line.

A similar experiment was carried out by Bos and Scharloo (1973a,b, 1974) using thorax length and, in a minor way, wing length as the selected characters. Following Robertson and Reeve (1952), these characters were taken as measures of body size, itself chosen as a character differing from chaeta numbers in its biological significance, as displayed by its being subject to inbreeding depression and heterosis which the chaeta numbers

TABLE V. Effects of stabilizing selection on variation for sternopleural chaeta number in a population newly introduced into the laboratory from the wild (Gibson and Bradley, 1974)

Generation	Controls			Selection lines		
	V_A	V_E	Sum	V_A	V_E	Sum
G0	1·95	2·85	4·7	—	—	—
G19	2·80	1·20	4·0	1·10	1·80	2·9
G39	0·75	1·25	2·0	0·95	1·45	1·9

All entries are the averages from two lines kept under two different temperature regimes.

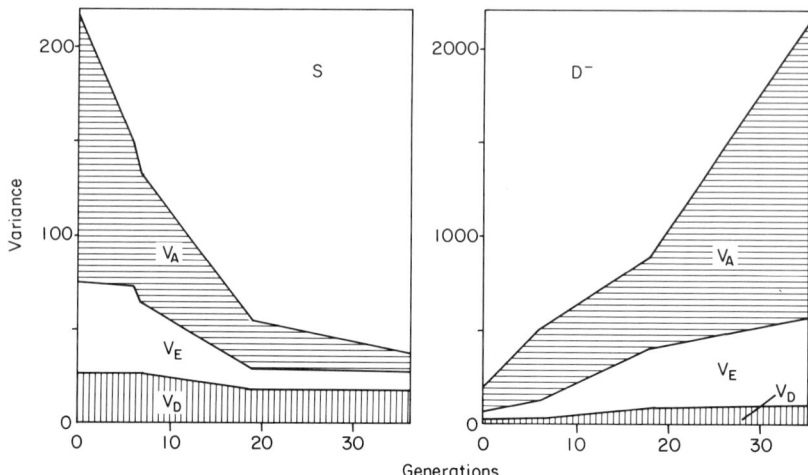

FIG. 8. Changes in the components of phenotypic variation for the relative length of the 4th wing-vein in ct^{D-G} flies as a result of stabilizing selection (S) and disruptive selection with disassortative mating (D^-), observed by Scharloo et al. (1967). The additive genetic variance (V_A), the developmental variance (V_D), and the environmental variance (V_E) with which is confounded non-additive genetic variation, all decrease under S and increase under D^- selection. Note the ten-fold difference in scale of the ordinate between the S and D^- parts of the figure.

are not. The "variance" of a culture was measured as the square of the coefficient of variation, i.e. as

$$C.V.^2 = \left(\frac{\text{standard deviation} \times 100}{\text{mean}}\right)^2$$

This gave little indication of falling under S selection. At the same time the mean thorax length fell markedly. By contrast, in the two lines they maintained under D^- selection, the variation, as measured by $C.V.^2$, rose steadily for the first 20 generations or so, after which it remained fairly stable on average, though with marked fluctuations. The mean was reasonably stable in both these D^- lines until G20 or a little later, after which it rose in one of them and fell a little in the other before re-stabilizing.

On the face of things, therefore, S selection in this experiment did not produce the expected decline in phenotypic variation. The authors attribute this to the different nature of the character, which may of course have played a part, though Kearsey and Kojima (1967) report that body size resembles chaeta characters in its genetic structure. Interpretation of Bos and Scharloo's results is not, however, unambiguous. In the first place, there must have been a clear, if unintended, element of R (directional) selection

TABLE VI. Changes in the components of variation for length of the 4th wing-vein under S and D⁻ selection (Scharloo et al., 1967).

	Base population	S seln. (G36)	D⁻ seln. (G35)
V_A	141 (1·00)	11 (0·08)	1557 (11·04)
V_D	25 (1·00)	19 (0·76)	100 (4·00)
V_E	50 (1·00)	8 (0·16)	476 (9·52)
V_P	216 (1·00)	38 (0·18)	2133 (9·88)

The bracketed figures express the values of the components after selection relative to those in the base population. V_E was found as $V_P - V_A - V_D$ and will contain variance due to non-additive genetic, as well as non-heritable, effects.

involved, as attested by the decline in the mean of the S selection lines. The response of the variation to the S element of the selection could well have been distorted by this. Furthermore, for the $C.V.^2$ to have remained broadly constant while the mean was declining requires that the standard deviation, and hence the variance proper, must also have been declining correspondingly. So the interpretation here must, in any case, depend on the justification of using $C.V.^2$ as the valid measure of variation. The justification presented is that, when measured by $C.V.^2$, the variation of males and females from the same stock is roughly equal, though no reason is advanced for expecting equal variation in the two sexes.

Thus assessment of the implication of these results must remain uncertain until this question of the metric to be used for measuring the variation has been settled. The same question arises in assessing the results from the D⁻ line, though here it is less limiting as we shall see in a later section. We may note too that Tantawy and Tayel (1970) found a decrease in the $C.V.^2$ for these same characters under S selection, no doubt because the mean was less disturbed. This would, of course, accord with expectation. Grant and Mettler (1969) also used the $C.V.$ to measure variation in an S selection experiment using an escape reaction of the flies as the character, and they found some indication of a decline in variation under the selection. Certainly Bos and Scharloo's results cannot be taken as either invalidating the expectation that variation will decline under S selection, or suggesting that any difference there may be in the biological significance of body size and chaeta number results in a difference of response to S selection.

The decline in V_A under S selection observed by Thoday (1959) for sternopleural chaeta number was investigated further by directional selection from the S selection line at generations 11, 12, 17 and 22. The S selection line responded to directional selection but did so less rapidly than the base stock (Dronfield) from which it was begun. In fact, it showed a realized heritability of 0·09 by comparison with 0·15 for the base stock. There was a decline in vigour in the S line which may well indicate a rise in homozygosity, but clearly some heterogenicity still remained in the line and the favouring of repulsion linkages must have played its part in the reduction of V_A.

A later experiment of Thoday (1960) throws further direct light on the effect of S selection on phase of linkage. This experiment started with a cross between two lines differing widely in sternopleural chaeta number and carrying certain major mutant genes, which served to mark chromosomes 2 and 3. Both S and D^- selection was practised. The S selection reduced the variance and favoured repulsion linkage of genes which had hitherto been in the coupling phase; whereas D^- selection eliminated the repulsion recombinants, just as would be expected in both cases.

B. Canalization

Not only is V_A reduced by S selection, but so too is the non-heritable variation, in so far as V_E is a measure of it. Furthermore, where V_D, which is unambiguously ascribable to developmental variation, can be estimated it also declines under S selection, though to a lesser extent then V_E (Table VI; Figure 8). Thus in addition to the additive genetic variation, the selection must be utilizing variation arising from genotype × environment interaction in the sense that it must be favouring genes which are less subject than others to disturbance from environmental differences (or developmental upsets). The character is thus becoming less variable in its expression not by the reduction of additive genetic variation, but by making it less responsive to differences in non-heritable agencies. The expression of the genotype is becoming better canalized. In the same way, the rise of V_E and V_D under D^- selection (Table VI; Fig. 8) indicates effective selection for genes less stable in their responses to differences in non-heritable agencies; the expression of the genotype is becoming less well canalized. Scharloo *et al.* (1972) give further evidence that canalizing and anti-canalizing selection are effective for the length of the 4th wing vein.

An increase in canalization has also been achieved by Waddington (1960), not by S selection of a general kind such as Scharloo *et al.* applied to the 4th wing vein, but by a direct selection for the canalization itself. He selected successfully for reduction in the difference of facet number in the eyes of a Bar-eyed stock when raised at two temperatures. Though not actually selecting for it, Caligari and Mather (1980) have shown that the variation necessary for a similar response in sternopleural chaeta number is present in the "Texas" population. They compared the chaeta numbers of the 18 inbred lines from this population at 18°C and 25°C. The temperature sensitivity, as measured by the difference between the average chaeta numbers at the two temperatures, was found to vary from 2·950 to −2·825, among the inbred lines. Furthermore, there was no evidence of a correlation between the temperature sensitivity and the overall mean chaeta of the lines. Similar results were obtained earlier by Caligari and Mather (1975) using the set of eight substitution lines based on the Samarkand and Wellington inbreds. These earlier observations also showed that the genes responsible for variation in sensitivity were preponderantly carried by chromosome 2, though some were on chromosome 3, while those for overall mean chaeta number were preponderantly on 3, though some were on 2. This does not, of course, necessarily imply that different loci were responsible for the two effects, but that the two properties are not correlated even, most probably, at the level of the locus. Sensitivity and overall chaeta number can thus be treated as distinct characters, separately adjustable by selection. To put it another way, appropriate selection could increase (or decrease) the canalization of the character round any chaeta number within the range observed.

Dr. J. M. Rendel and his associates have carried out an informative series of investigations on the canalization of the number of scutellar bristles in *D. melanogaster*, its genetical determination and its responses to selection. These have been reviewed by Rendel (1979) who gives references to the extensive original literature. He draws a distinction between, on the one hand, selection which reduces the sensitivity of the mean expression of a character to environmental influences without correspondingly reducing variation round the mean and, on the other, selection for canalization, which he sees as specifically reducing the variation round the mean. He regards stabilizing selection as being especially associated with the former. The effect of stabilizing selection in reducing V_E, which we have been discussing, brings it into this category; but the reduction of V_D, the variation due to the developmental properties of the individual fly, as observed by Scharloo *et al.* (1967), would seem to correspond to the type of selection which Rendel associates with the property of canalization. Thus the selection we have been discussing produced both effects at once, and canalization as we are using the term covers both phenomena, one of them the buffering of the phenotype against variation in external environmental agencies and the other the buffering of the phenotype against vagaries of developmental processes. This latter is the property that has been termed developmental homoeostasis or developmental stability, which can be measured in individual flies by the degree of asymmetry shown by bilateral characters and which has been found to be subject to genetical control, and hence to selective adjustment (Thoday, 1958; Mather, 1953b). Whatever the merit of the theoretical distinction between variation arising from environmental differences and that stemming from developmental instability (and it is not easy to see where the line should be drawn, especially as the former may achieve at least part of its effect through the latter), it is an operationally impossible distinction with a single as opposed to a bilateral character. Indeed we know that in respect of sternopleural chaeta number a great deal of the non-heritable variation between individuals must be no more than a reflection of the variation which can also appear as differences between the two sides of individual flies.

To return, however, to the properties of the scutellar bristle character, it is clear that the degree of canalization is under genic control, and that it can be altered by selection. It has been increased by selection in scute flies having a mean of about two bristles (Rendel and Sheldon, 1960), and it has been reduced by selection in appropriate wild-type flies (Scowcroft and Latter, 1971, quoted by Rendel, 1979). Furthermore, in this latter case, a gene on chromosome 3 appeared to play a major part in determining the degree of canalization. Much other information is to be found in Rendel's review, but to discuss it further would take us beyond our present purpose which is the consideration of responses to selection, of which the adjustment of canalization is one.

Where a meristic character is considered, particularly one like scutellar bristle number which can show only a few levels of expression and in which development is canalized in wild flies, a wide range of genotypes must all give the same phenotype. Thus although the phenotypic class may cover a wide range of genetic differences, it is impossible to select for them because they cannot be detected through their effects on the phenotype. The canalization may, however, be upset by a suitably drastic change of environment and the variation thus exposed to selection. This has been done by Waddington (1953) in respect of the posterior cross-vein in the wings of *D. melanogaster*. Normally this vein is complete in wild-type flies, but after exposure to a strong temperature shock some flies appear with broken cross-veins. If such flies are selected for use as parents and the process repeated in their progeny, Waddington found that the frequency of individuals with broken cross-veins rose. Indeed after 12 generations of such selection broken cross-veins began to appear, with increasing frequency, among flies which had received no heat shock. Eventually a line was established which when raised entirely at normal temperature was nearly uniform for broken cross-veins. The selection was made possible by the environmental intervention which over-rode the canalization mechanism and exposed the underlying polygenic variation. Whether the genetic system responsible for the canalization itself (as distinct from the variation it concealed) was affected was not made clear. Yet one can hardly doubt that, even if it had been, appropriate selection could have re-canalized the phenotype at some given level of break in the cross-vein, thus illustrating the basis of the genetic assimilation of features hitherto environmentally induced in the phenotype. This is the so-called Baldwin effect whose occurrence in nature had previously mystified biologists.

Thus the efficacy of a canalization mechanism can be changed environmentally as well as genetically, just as can the expression of any other character. Equally, as we have seen, stabilizing selection can produce its responses in several ways, by reduction in genetic variation, by the selection of genes less responsive to environment variables or the selection of genes contributing to a closer control of developmental processes. These different effects may be favoured in varying degrees by the use of different selection routines in experiment, and their contribution to the overall response to the selection may depend on different properties of the relevant genes. But this does no more than emphasize yet again that, on the one hand, a given force of selection must be expected to achieve its effect by whatever paths can contribute to it, and that, on the other, the fate of a gene or combination of genes when subjected to selection will depend on all their properties, whether of transmission from parent to offspring, or of action and interaction in the control of the development of the phenotype.

V. Disruptive Selection

A. Types of Disruptive Selection

Experiments on directional selection have been of importance especially by showing us how the mechanical properties of the genes, notably the linkages and recombinations that they show, are a key factor in determining response to selection. Those on stabilizing selection further bring out the importance of the genes' physiological properties, exemplified especially by the canalization of development, as a factor in selective response. Experiments on disruptive selection, to which we now turn, take us into a still further dimension in that they have a more immediate relevance to situations that must arise in nature.

Up to this point we have been concerned with selection in relation to a single optimum phenotype. But, because of spatial variation in the environment of a population and functional variation among its individual members, different individuals must be subject to different pressures of selection: there must in fact prospectively be more than one optimum phenotype within the population. In other words, there must be an element of disruptive selection at work. The experiments we shall discuss have concentrated on the simplest case of selection for two optima but, although experimental complexities would be encountered with more optima than this, there is no reason to believe that any new principle would be involved.

Experimentally, the disruptive element has generally been introduced by selection in opposite directions, towards increased (H) and decreased (L) expressions of the character being used. Indeed Mather's (1941) experiment (and others like it) could be regarded as bringing in disruptive selection because it involved selection, in its H and L lines respectively, towards higher and lower expression of the character (the abdominal chaeta number), though we did not consider it from this point of view. Such divergent directional selection experiments are, however, of only limited value in discussing the consequences of disruptive selection, comprising as they do two separate and distinct lines of selection, with all the matings throughout being between like individuals (H×H and L×L) and no gene-flow between the H and L lines in the experiment.

If we do away with this restriction we can recognize a wide variety of situations in respect of the matings made among the selected individuals. The mating could be assortative (D^+ selection), where in each generation flies selected for increased expression were always mated together and similarly for decreased expression. Even here, however, we could secure gene-flow between the H and L groups of the experiment by making an appropriate number of matings between H flies whose parents were themselves selected for L expression and *vice versa*, as distinct from matings between H flies all selected from the progenies of H×H matings and similarly for L flies. Equally, mating could be disassortative (D^- selection) where flies selected for increased expression were always given mates selected for decreased expression. All matings are then H×L by H×L matings. And of course we can achieve a third situation by taking all the selected H and L flies, from whatever progenies they came, mixing then together and allowing them to mate at random (D^R selection), though we should recognize that the randomness would be modified by any mating preferences that the flies might display. Experiments of all these kinds have been carried out.

The various kinds of experiment using disruptive selection mimic, and indeed can be designed to mimic, different situations found in nature. Thus D^+ selection experiments can mimic the situation where populations are separated geographically in such a way that different optima are favoured and gene-flow is precluded between them. Divergent directional selection experiments are of this kind. Where, on the other hand, the mating system is such as to ensure a given amount of gene-flow between the H and L groups, the D^+ experiment mimics the selection imposed by niche differences within the range of a population whose individuals can migrate from one niche to another. Equally D^- experiments mimic the natural situation where the individuals within a population fall into classes which are found together functionally in some way, as for example with males and females who are bound together by the need for the different, but complementary, functions they discharge in reproduction.

We would clearly expect different consequences to arise in these different situations, and it has been argued that separation of the groups under selection towards different optima (mimicked by D^+ selection) should tend to lead to isolation, while interdependence of them (mimicked by D^- selection) should lead to classical polymorphism (Mather, 1955). The subsequent experiments allow us to explore these expectations and observe the rise of the genetical mechanisms which we associate with isolation and polymorphism. Quite a number of experiments have indeed been carried out using the different kinds of disruptive selection, notably by Professors J. M. Thoday and W. Scharloo, with their respective associates. A comprehensive list of the literature is provided by Thoday (1972) in his invaluable review of the whole subject of disruptive selection.

B. D^- Selection: Polymorphism

Disruptive selection of any kind must be expected to increase phenotypic variation. We have already had occasion to note that it does so when either divergent directional selection or D^- selection is used, and indeed it has proved to be a feature of all experiments in disruptive selection (Thoday, 1972). Wherever analysis of experiments of these two kinds have been carried out, an increase in the genetical component of variation has been found, which, as we saw in earlier sections, is attributable to selection favouring coupling as opposed to repulsion linkages of the relevant genes. The non-heritable component of variation, however, does not always respond in the same way, and indeed is not expected to do so. Tables III and VI allow us to compare response in examples of divergent directional selection and D^- selection, taken respectively from Mather (1941) and Scharloo et al. (1967). In both cases V_A shows a marked increase; but the D^- selection, which is expected to favour greater genotype × environment interaction, shows V_E increased by a factor nearly as great as V_A (Table VI), whereas the divergent directional selection, which, because the selection is wholly within the H and L lines separately, would not be expected to favour increased genotype × environment interactions, has hardly changed V_E (Table III).

The effects of D^- selection on the genetical variation has been analysed further in several experiments. Thoday (1960) practised D^- selection for sternopleural chaeta number in a population started from the F_2 of a cross between two lines, both carrying marker genes and one with a high and the other a low chaeta number. This produced polymorphism for chaeta number with the H and L morphs segregating in each generation. The underlying genetic mechanism was later found to depend on a difference in chromosome 3 (see Thoday, 1972). The L morph was homozygous for a chromosome 3 giving low chaeta number, and the H morph heterozygous for this same chromosome and another which gave high number. The polymorphism thus depended on segregation of a genetic difference in one of the parents exactly as does sex-dimorphism in many animal and plant species or distyly in certain plants. Furthermore, the genetic difference was found to be compound, consisting of three loci, probably fairly closely linked, in just the same way as the super-gene responsible for the pin-thrum difference in species of *Primula* is known to be compound (Mather, 1950). And lastly, modifiers in chromosome 2, and probably the X also, had been accumulated which enhanced the innate difference in the effect of the switching gene (or super-gene) on chromosome 3—again just as the background genotype is known to determine the precise phenotypes of the different morphs in a number of polymorphisms concerned with breeding behaviour in plants (Darlington and Mather, 1949; Mather, 1973), and also of those concerned with Batesian mimicry in the butterfly, *Papillio dardanus* (Clarke and Sheppard, 1960). The D^- selection had in fact produced a genetical structure closely resembling that of natural polymorphisms.

Scharloo *et al.*'s (1967) D^R selection for the length of the 4th wing-vein in ci^{D-G} flies resulted in a situation resembling that which Thoday and his colleagues had found for sternopleural chaeta number (Scharloo, 1970a). Their D^- selection, however, had a somewhat different outcome (Scharloo, 1970b). Unlike Thoday's polymorphism and that from the D^R selection, no genetical switching element was found in it. The selection had nevertheless built up a genotype capable of giving a bimodal distribution, but one which this time appeared to be switched by environmental differences, aided by stochastic processes in development to judge by the high variation between the two wings of individual flies. The background genotype had clearly been adjusted by the selection to give prospective polymorphism, but with the switching depending on external or developmental factors, just as does the switching between the male and female morphs of the marine worm *Bonellia* or the castes in honey bees (see Mather, 1973). In other words disruptive selection where individuals of different phenotypes are bound together in some way, as here they are bound by the disassortative mating of D^- (or even the dissortative element in D^R) selection, can produce the essential genetical features of more than one kind of polymorphism found in nature.

The contrast seen between the effects of D^R and D^- selection in Scharloo *et al.*'s experiment is not characteristic of these two types of mating. D^- mating can result sometimes in a genetical switching system, as is seen in

Thoday's experiment with sternopleural chaetae and in one line of Bos and Scharloo's (1973a,b, 1974) later experiments using body size as the selected character. It can also fail to yield a genetical switch, as is seen not only in Scharloo *et al.*'s results with wing-vein length but also by another of Bos and Scharloo's lines. The chief interest of Scharloo *et al.*'s findings in this connection is that the dissortative element in the D^R mating system is evidently sufficient to make possible the production of a genetical switch: the binding between the two phenotypes does not need to be absolute for this widespread feature of natural polymorphisms to result.

In none of the experiments discussed so far were the two optima, H and L, fixed: flies with extreme expression of the character were selected as parents, irrespective of just how extreme they were. Skibinski and Thoday (1979) have recently investigated disruptive selection for sternopleural chaeta number using two optima fixed at 17–18 and 22–23 chaetae respectively, the selected parents being mated in a quasi-random system. The disruptive selection was unsuccessful in so far as no evidence of bimodality was obtained in the distribution of chaeta number even after 30 generations of selection. There was, however, evidence of divergence between the H and L groups in that $H \times H$ and $L \times L$ matings gave more extreme average chaeta numbers than did $H \times L$ and $L \times H$, but with the distributions of chaeta number overlapping.

Selection towards fixed optima implies, of course, an element of stabilizing selection round each optimum combined with the disruptive effect of selection between them. Evidently with the two optima, each 1 chaeta wide and separated only by 5 chaetae on average, the stabilizing effect could not be sufficiently disentangled from the disruptive to reduce the frequency of intermediate phenotypes to a sufficient extent over the number of generations used. Perhaps the experiment would have been more successful if the two optima had been fixed further apart, even if this had required the selection to be wholly disruptive in the early generations, with the stabilizing element entering into the selection once these optima had been attained: such a selection programme would not have been out of keeping with the magnitude of the differences seen in naturally occurring polymorphisms. Perhaps, too, the quasi-random system of mating was not the most suitable, and perhaps D^- mating would have been more effective. In the meantime the experiment can only be regarded as inconclusive.

C. D^+ Selection: Divergence

By binding the two "morphs" together in the system of mating, D^- selection makes them interdependent and so would be expected to lead to polymorphism (Mather, 1955) as indeed experiment has shown it to do. In

D^+ selection, on the other hand, like is mated to like, and simple divergence rather than polymorphism would be expected. The rate and extent of the divergence would of course further be expected to depend on the amount of gene-flow between the groups, H and L, and on the pressure of the selection that was being imposed.

We have already seen that with the extreme form of D^+ selection, divergent directional selection, divergence invariably occurs with a speed depending on the intensity of the selection. In such experiments, H flies are always mated with H flies, both coming from $H \times H$ matings and L with L both coming from $L \times L$ mating. The H and L groups are thus completely separated with no gene-flow between them. The experiments by Millicent and Thoday (1960, 1961) compared this system with others, one of which used a quasi-random mating pattern resulting in 25% gene flow per generation between the groups. In it H flies were always mated with H flies but whereas the H females were always from the progeny of $H \times H$ matings, only half the H males came from $H \times H$ progenies, the other half being H males from $L \times L$ progenies. Matings in the $L \times L$ lines were organized in the counterpart fashion. Under this mating system, giving 25% gene-flow, divergence between the H and L groups was slower than with divergent directional selection, but ultimately it was nevertheless on average as great.

A system of mating giving 50% gene flow between the H and L groups was also used by Millicent and Thoday, as it had indeed been used in an earlier experiment by Thoday and Boam (1959). In some ways this mating system resembles D^- selection, for mating was always between parents taken from the progenies of H (from $H \times L$) and L (from $H \times L$). There is, however, an important difference: under D^- selection the mating is between an H from the progeny of an $H \times H$ mating and an L fly from an $L \times L$ mating, whereas in the 50% gene flow system it is of two H flies (or two L flies) one from an $H \times H$ and the other from an $L \times L$ mating. In respect of immediate parentage these matings can be regarded as disassortative, but in respect of the flies selected for the next mating it is assortative. There is thus an element of conflict in the system, every fly having both H and L selection in its ancestry, though flies in the H group have somewhat more H than L ancestral selection, and those in the L group somewhat more L than H. Not surprisingly, therefore, though divergence was observed to occur under 50% gene flow, Millicent and Thoday found it to be slower than with 25% or 0% flow.

The divergence observed by Thoday and Boam (1959) in their earlier experiment was greater than that found by Millicent and Thoday, and indeed resulted in a polymorphism, so reflecting the disassortative element in the system. When analysed by Gibson and Thoday (1962) and Wolstenholme and Thoday (1963) the structure of this polymorphism proved to be more complex than that resulting from the wholly D^- selection of Thoday (1960), to which reference was made in the previous section. It proved to have two switching super-genes, each built up by recombination but one of them in chromosome 2 and the other in 3 (Fig. 9). The latter was confined in its operation to the H × H group of matings, which thus gave two classes of offspring, H and I (intermediate) flies. The chromosome 2 switching supergene was similarly restricted to the L × L groups of matings and gave L and I flies. Now I flies will be of lower chaeta number than H and so will be taken from H × H progenies to be used in parents in the next round of L × L matings. Similarly I flies will have a higher chaetae number than L and so will be taken from the progenies of L × L matings as the high flies to be used in the next H × H matings. The selective migration between the H × H and L × L groups of matings has resulted in a trimorphic system, where the third morph I is specialized, so to speak, in migrating between the H and L group of matings. This trimorphy neatly resolves the intrinsic conflict of the mating system, and so once again shows how the conditions under which disruptive selection operates are reflected in the genetic structure which it engenders. And it emphasizes to us, in a way all the more striking for it being so unusual, the versatility of the genotype in adjusting to the conditions imposed on it by the pattern of selection it encounters.

To return, however, to the relations between the amount of gene-flow and the intensity of selection as determinants of the outcome of the D^+ selection, all Millicent and Thoday's experiments were carried out using a selection of

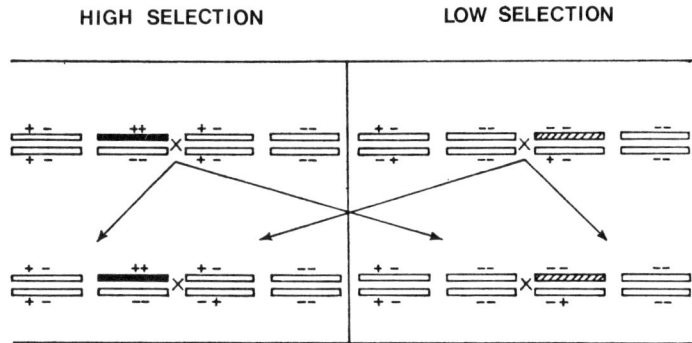

FIG. 9. The genetical structure of the polymorphism for sternopleural chaeta number established by Thoday and Boam's (1959) disruptive selection with assortative mating (D^+). The black chromosome 3 remains in the high half of the population, only its white homologue being transmitted to the low half by flies selected for low chaeta number. Similarly the hatched chromosome 2 remains in the low half of the population. (Reproduced by permission from J. M. Thoday (1965) The effects of selection for diversity. *Genetics Today* (*Proc. 11th Int. congr. Genet.*). 3, 533–540. Pergamon Press, Oxford.)

5%, i.e. 1 fly out of 20 chosen as a parent for the next generation. At this high pressure of selection, divergence was obtained even with 50% gene flow. Less powerful selection was used by Streams and Pimentel (1961, quoted by Thoday, 1972) in a somewhat different kind of experiment, though sternopleural chaeta number was again used as the character. These authors did not practise gene-flow between two groups undergoing selection in opposite directions, but fed into their selection lines flies taken at random from the population from which these selection lines had originated. They used two levels of selection, 10% (i.e., 1 fly taken from 10) and 40% (4 flies taken from 10, or 2 from 5). They also used two levels of gene flow, 50% and 20%. Little if any divergence was seen between selection line and base population with 10% selection and 50% gene-flow, though divergence was considerable at this same level of selection with 20% gene-flow. At the lower level of 40% selection, 20% gene flow still allowed some divergence, though its significance was in some doubt. Even such a low pressure of selection would, of course, give divergence with 0% gene flow. These results and those of Thoday and colleagues (summarized in Table VII), serve to emphasize that the greater the pressure of selection, the greater the gene-flow at which divergence can be obtained, as indeed would be expected.

D. Isolation

Divergent directional selection, that is D^+ selection with no gene-flow, mimics the natural situation where two populations are under selective

TABLE VII. Selection pressure, gene-flow and divergence under disruptive selection, based on results of Thoday et al. and Streams and Pimental, quoted by Thoday, 1972. (Reproduced by permission from K. Mather (1973). *Genetical Structure of Populations*. Chapman and Hall, London.)

Gene flow	Percentage of flies selected for breeding (less ← selection pressure → greater)		
	40–50%	10%	5%
50%	—	?	D
20–25%	d	D	D
0% (isolation)	D	D	D

D = marked divergence; d = slight divergence; ? = doubtful divergence.

forces while at the same time being prevented by distance or other barrier from exchanging genes. The phenotypic and genetic divergence which ensues is thus a model of allopatric speciation.

Of more interest, however, is behaviour under D^+ selection where there is gene-flow between the two groups for which selection is practised simultaneously; this system in fact mimics the natural situation where selection towards different optima is imposed by the different requirements in two ecological niches, which coexist within the territory of a population single in the sense that there is regular migration of individuals, and hence flow of genes, between the two niches. As we have just seen, experiments of this kind show that such gene flow does not prevent genetic divergence between the two groups under selection. In these experiments, however, the extent of the gene-flow was determined throughout by the pattern of mating fixed by the experimenters and imposed on the flies, which could not themselves influence the rate of gene-flow because they were offered no choice in mating.

Now, it has been known for many years that assortative mating could be promoted by selection. Thus Knight et al. (1956) raised, in the same culture container, flies of two lines which had been extracted from the same population of *D. melanogaster* and were marked by the genes ebony and vestigial respectively. The frequency of cross-mating between flies from the two lines, relative to mating within the lines, was measured by the numbers of wild-type (which must be the result of cross-mating) relative to ebony and vestigial individuals. All wild-type flies were eliminated in each generation, so penalizing parents that had cross-mated. The frequency of wild-type individuals fluctuated from one generation to another, but declined over-all, until after 10–15 generations it was well below the level in control groups, so indicating a rise in assortative mating. Wallace (1954) had reported similar results using straw and sepia as the marker genes, though even after 73 generations of the selection, which had resulted in sepia females mating with sepia males in 90% of cases, the straw females still gave no sign of preferential mating. Dobzhansky and Pavlovsky (1971) reported the achievement of marked, though not complete, isolation using a similar technique in *D. paulistorum*. In a similar way Koopman (1950) had been able to promote assortative mating in *D. pseudoobscura* and *D. persimilis* when these two species were raised together. Koopman eliminated the hybrids resulting from cross-matings in his experiments, though he noted that there was in fact no real need to do so, since they left no successful progeny under his experimental conditions. Kessler (1966) confirmed this by a somewhat different method, which allowed him further to show that selection could be effective for a reduction as well as for an increase in the incidence of assortative mating.

In all these cases, however, the selection against hybrids, and hence against parents that cross-mated, was immediate and absolute. While establishing that at any rate partial isolation between two groups, through mating preference, could arise as a result of selection, they leave open the question of whether isolation between the two phenotypic classes could arise in parallel with a divergence that disruptive selection was producing. That such a parallel promotion of divergence and isolation by mating preference can result from disruptive selection was shown by two experiments reported by Thoday and Gibson (1962) and Thoday (1965), both starting from the same wild stock. These experiments followed the same general pattern as the earlier D^+ experiments, except that the flies selected for high and low chaeta numbers were not mated in a prearranged pattern, but were all placed together in a single vial and allowed to mate as they chose. The males were then eliminated and the females separated into H and L groups, according to their chaeta numbers, for the purpose of producing progeny.

After 12 generations of the first experiment, the chaeta numbers of the progenies of the H and L mothers had diverged so far that their distributions showed no overlap. Test progenies, raised by deliberate matings between the groups showed that the hybrids had intermediate chaeta numbers that would have bridged the gap between the H and L distribution observed in generation 12. That the gap existed was thus evidence of a lack of hybrids, and therefore a lack of $H \times L$ matings among the parent flies of this generation, though these had been given a free choice. Isolation had accompanied divergence.

The second experiment gave the same result except that the distribution of chaeta number from the H and L groups separated even earlier, at generation 7. Thereafter they remained separate, apart from occasional lapses. Later tests showed that a few hybrids appeared in the L half of the experiment, but they were absent or very rare in the H half. Mating choice tests further showed that the ratio of $H \times L$ cross-matings to $H \times H$ and $L \times L$ intra-group matings was 11:78 and 42:133 with flies from the two experiments respectively (Thoday, 1965, 1972).

A number of experiments of the same kind have since been carried out in various laboratories without isolation being produced (references in Thoday, 1972), including a third experiment by Thoday and Gibson. Thoday (1972) also refers, however, to a further experiment by Beardmore and Baldawi, then in progress, which promised to produce isolation alongside divergence. The significance of the negative evidence has been discussed in detail by Thoday and Gibson (1970); but whatever the frequency of wild populations capable of responding to the selection by producing isolation in parallel with divergence, it is clear that at least some populations have the capacity to build up isolating mechanisms under disruptive selection and to do so remarkably quickly, at least where the pressure of selection is powerful.

The consequences for a population of disruptive selection, imposed on it by two habitats or niches making different selective demands, have been discussed from a theoretical point of view by Maynard Smith (1966). He assumed random mating among the adults of the population and concludes that genetical divergence will be produced provided that (i) the density-dependent factor regulating the population size operates separately in the two niches, and (ii) the selective differentials are large (at least 30% even where females choose the appropriate niche for laying). Both these conditions were satisfied by Thoday and Gibson's experiment. Such a polymorphism is likely, he concludes, to be followed by reproductive isolation between the groups in the two niches, so leading to speciation which at least many would regard as sympatric. One may question, however, whether isolation would follow divergence rather than arise in parallel with it, as Thoday and Gibson's results would indicate. The two processes would in fact seem likely to be mutually reinforcing, since even in the early stages increasing divergence would lead to increasing selection pressure against intermediate hybrids and hence increasing selection pressure towards isolation while, by reducing the number of such hybrids, increasing isolation would lead to increased pressure of selection towards divergence. Such a mutually stimulating relation between hybrid incapacity and bars to crossing must indeed be expected in general (Mather and Edwardes, 1943; Mather, 1973).

E. Conclusion

Disruptive selection, as exemplified by these experiments, can clearly have a variety of consequences, all of which are of prospective relevance to situations in nature. In all the experiments disruptive selection has been found to increase the variation of a population, and in this sense it is the converse of stabilizing selection, which tends to decrease variation. The two types of selection also contrast in their effects on the phase of linkage, D selection favouring coupling and S selection repulsion linkages. In heterogeneous habitats, which must impose some element of D selection on populations that inhabit them, we would thus expect the disruptive forces to maintain variability in the population at a level reflecting the balance between them and the inevitable stabilizing forces. Simulation experiments, reported by Maynard Smith (1979) would appear to bear out this expectation.

The way that the increased genetic variability, produced by the D selection in the experiments, is organized within the genotype depends on the type of mating, assortative or disassortative, and the level of gene-flow in the system. Disassortative mating in D^- experiments leads to overt polymorphisms basically akin in their genetical structures to the natural polymorphisms of genetical interest. The common background genotype is adjusted to facilitate the production of the opposing morphs, development being switched by segregation of a genetical difference, which has been built up by the selection, or by environmental differences or stochastic processes in development itself, just as is found in nature.

D^+ selection has regularly produced divergence between the two selected groups despite gene-flow as high as, or even higher than, that which occurs where mating is random. Again, this is relevant to situations in nature, where clines are maintained or divergence occurs over short distances despite a high rate of gene-flow consequent on migration. Furthermore, in some experiments this divergence has been accompanied by a rise in the level of genetic isolation between the groups, which throws new light on the possibility of sympatric speciation.

Following on these experiments, therefore, we can look at a number of important features of natural populations in a new way, because we can see how they could have come about, and the environmental and biological circumstances that would favour their arising. Some of these applications of experimental findings to natural populations are, however, controversial (see e.g., Mayr, 1974; and Thoday, 1974), but all have been discussed in detail by Thoday (1972).

VI. Genetic Architecture of Characters

A. Balance and Natural Selection

The many and diverse selection experiments that have been carried out with *Drosophila* establish clearly that each type of selection has its own distinctive effect on the mechanical and physiological organization of the genotype. Where, therefore, we find these characteristic types of organization in wild populations we now have a basis of experiment for seeking to interpret them as the outcome of the impact of the relevant kind of selection on the population or its ancestors. In doing so, however, two further points must be borne in mind.

In the first place, in the kind of experiment we have been discussing a powerful force for selection was deliberately imposed on the material in order that its effects should be sufficiently clearly distinguishable from any quasi-natural selection, to which the flies were simultaneously exposed, for its consequences to be identifiable with as little ambiguity as possible. Though we now know that the forces for selection in nature can be surprisingly strong, they can seldom if ever match those used in many experiments. Furthermore, the populations, and especially the numbers of parents, used in experiments are necessarily often much smaller than those commonly found in nature. This will be reflected in the response to the experimental selection being more variable, as well as more rapid, than that expected in nature, and not infrequently entailing a degree of developmental disturbance more debilitating than natural conditions would allow to persist. In other words, the experiments offer us a basis for interpreting the natural phenomena, but cannot be expected to provide us with a perfect model.

The second point is that selection can and does change the genetical structure of experimental populations in characteristic ways that we can trace by appropriate analysis to particular chromosomes, polygenic combinations and even within limits to individual loci (Thoday, 1961). Yet, by their very nature they cannot give us any confidence that, even if the type of change is characteristic of the type of selection, it will be traceable in the same measure to adjustment in the same chromosomes, the same genic combinations or the same loci. Indeed, polygenic combinations being what they are, we would expect that in general, with different starting populations, the same type of change in the genotypic organization, produced by the same type of selection, would involve different individual genes and different linkages. In other words although the overall reaction of the genotype to the selection would be the same, the detail of the individual genes, linkages and interactions, by which it was achieved, might well be different, especially where the starting populations differed in the detailed genic determination of the variation they showed. Thus, while the characteristic reorganization in both mechanical and physiological properties of the genotype must be internally consistent in any one case, it would not be expected to be consistent in its details between different selected groups.

One aspect of the lack of consistency, or correspondence, is illustrated by the fairly commonplace observation that when two distinct lines have been taken as far as they will go by directional selection in the same experiment, further progress can still be achieved by crossing the two before applying further selection. The two lines have evidently given similar responses but have done so in part by using different members of the polygenic system and by the resolution of different linkages among them.

It is, however, shown more dramatically in different ways by the observations of Dobzhansky (1948, 1950) and Wallace and Vetukhiv (1955). Dobzhansky looked at the properties of the genic combinations associated with inversion sequences in *D. pseudoobscura*, by observing their reactions to the quasi-natural selection of populations in cage experiments. Where two inversion sequences, for example AR and CH or AR and ST, were derived from the same wild population they achieved a stable polymorphic relation in the cage populations, presumably because the genic combinations they carried had been adjusted, or co-adapted, to each other by natural selection in the wild. Where, however, they came from different wild populations, no such relation was found between them: being from different populations, natural selection had had no opportunity for co-adapting them.

Wallace and Vetukhiv's observations were of a different kind, and concerned the fitness of the offspring obtained by crosses between individuals from different wild populations in three species of *Drosophila*. The F_1's from such crosses were as fit, or even fitter, than the F_1's of crosses between individuals from the same population; but in F_2 there was a marked fall in average fitness. The polygenic combinations were well adapted in respect of the immediate action of their genes, but this relation was rapidly broken by unbalance consequent on recombination. The mechanical organization of the polygenic system, though adequate and internally consistent in each population, was made inadequate by a cross-inconsistency between the populations, which revealed itself as soon as segregation and recombination took place. Again, adequate adjustment was achieved only where natural selection had had opportunity to make it so.

We may note too, that this same principle of adequate adjustment being a direct product of natural selection is illustrated by the contrast between naturally outbreeding and inbreeding species of plant. Outbreeders show inbreeding depression and heterosis as a characteristic feature of their behaviour. The inbreeders do not. Now, complete homozygotes will be relatively rare in populations of outbreeders and natural selection will have little chance of testing the adequacy of the genic combinations when in the homozygous state; but, of course, under a system of regular inbreeding, genic combinations will survive only where they are capable, when homozygous, of giving rise to fit individuals (for further discussion see Darlington and Mather, 1949; and Mather, 1973).

B. Types of Character

All the polygenic combinations of inbreeding species will be adjusted by natural selection to give adequate balance when homozygous; and equally all the combinations of outbreeding species will be adjusted to give adequate balance when heterozygous or partially so, one consequence of which is inbreeding depression and heterosis. Yet, even in outbreeding species like *Drosophila*, not all the characters show inbreeding depression and heterosis. Indeed, a distinction was long ago drawn between two classes of character, those which do show inbreeding depression and heterosis and those which do not (A. Robertson, 1955). Characters in the first class have been seen as relating more directly to fitness (e.g., fecundity, hatchability, viability) and are termed "fitness' characters. Characters in the second class, which includes the chaeta characters, were seen as "peripheral" characters whose relation to fitness was so tenuous and unimportant as to make them selectively neutral in the wild.

It has, however, been shown by Kearsey and Barnes (1970) that the fitness of individuals from the "Texas" cage population is related to their sternopleural chaetae number. Their capacity for survival under relatively crowded conditions is at a maximum where, as adults, they show the mean number of chaetae, and it falls off as the chaeta number departs increasingly from the mean, no matter in which direction (Fig. 10). Furthermore this cannot be related, as has been suggested, to the relative level of heterozygosity of the individuals, for the same differences in capacity for survival have been found where all the individuals of different chaeta numbers were taken from inbred lines (Linney *et al.*, 1971). Far from being a peripheral, selectively neutral character, sternopleural chaeta number is subject to stabilizing selection.

Now fitness is the ultimate comprehensive character, reflecting the interplay of the totality of the genotype with the totality of the selective forces, and *ex hypothesi* it must always be under wholly directional selection. This obviously does not require that every character whose expression is related to fitness, must equally be under directional selection, since, as we

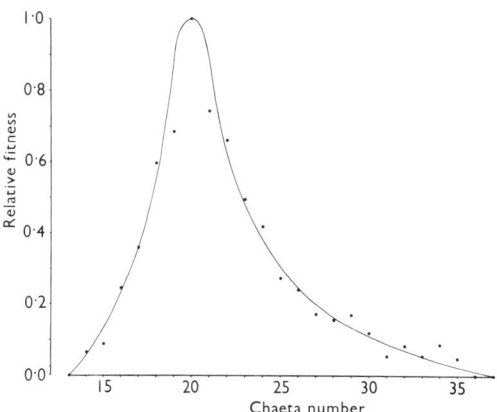

FIG. 10. Relation between sternopleural chaeta number and relative fitness in the "Texas" population. (Reproduced by permission from M. J. Kearsey and B. W. Barnes (1970) Variation for metrical characters in *Drosophila* populations, II. *Heredity*, **24**, 11–21.)

have just seen, the effect on fitness relatable to sternopleural chaeta number is distributed symmetrically round the mean of the chaeta distribution. We would, however, expect that the more closely a character is related developmentally to fitness the greater the directional element in the forces of natural selection to which it is subject. There may be a level of expression of maximal fitness with such a character, but this will be displaced from the centre of its phenotypic distribution. This is seen, for example, in the probability of survival of piglets in relation to the size of the litter (Darlington and Mather, 1949, Fig. 72). The difference between the so-called fitness and peripheral classes of character is thus not that one is subject to natural selection and the other is not, but that one will be under chiefly directional selection whereas the other will be under stabilizing selection.

As Fisher (1928, 1930) first pointed out, dominance must be regarded as the outcome of selection, the dominant gene being the favoured allele. Thus under preponderantly directional selection, dominance at the loci involved should be preponderantly in the same direction, whereas where selection is stabilizing, the commoner of the two alleles will be associated with the smaller average departure from the optimal phenotype and so will be expected to show dominance no matter whether it is the increasing or the decreasing allele. Ambidirectional, as opposed to unidirectional, dominance should thus result from stabilizing selection (Mather, 1960b; Breese and Mather, 1960; Caligari and Mather, 1980). By the same token, we should expect any interaction that may occur between non-allelic genes to be ambidirectional with stabilizing selection, but to be unidirectional and of the duplicate type with directional selection. Thus the two kinds of selection should result in the characters having different genetical architecture.

It was shown by Breese and Mather's (1960) analysis of the action of chromosome 3 in *D. melanogaster* that this expectation was realized in respect of the contrast between abdominal chaeta number, which would be expected to resemble sternopleural number in being under stabilizing selection, and viability which is a "fitness" character for which selection would be expected to be heavily directional. The chaeta character shows ambidirectional dominance (as does sternopleural number, Caligari and Mather, 1980) whereas viability not only shows unidirectional dominance towards high viability but also gives clear evidence of duplicate type interactions of non-allelic genes. Kearsey and Kojima (1967) have taken the matter further by assembling evidence that the expected difference in genetic architecture is shown by some half-dozen characters of each of the two types.

The genetic architecture of a character can thus provide us with information about the type of selection to which a character has been subject. It now appears that even the optimal expression of the characters, towards which past natural selection has been pressing it, can be estimated, at least where inbred lines are available, as with the "Texas" population (Caligari and Mather, 1980). And since the response of a character to future selection must depend on the architecture that past selection has imposed on it, we are now in a position to see more clearly what those responses are likely to be.

Acknowledgements

I am indebted for permission to reproduce figures, as follows: to the Society for Experimental Biology for Figure 1; to Chapman and Hall for Figure 3; to Professor J. L. Jinks and Chapman and Hall for Figure 7; to Professor J. M. Thoday and Pergamon Press for Figure 9; and to Drs M. J. Kearsey and B. W. Barnes and the Editor of *Heredity* for Figure 10. I am also indebted to Chapman and Hall for permission to reproduce Table VII.

References

BARNES, B. W. and KEARSEY, M. J. (1970). Variation for metrical characters in *Drosophila* populations. I. Genetic analysis. *Heredity* 25, 1–10.

BELL, A. E., MOORE, C. H. and WARREN, D. C. (1955). The evaluation of new methods for the improvement of quantitative characteristics. *Cold Spr. Harb. Symp. Quant. Biol.* 20, 197–211.

BOS, M. and SCHARLOO, W. (1973a). The effects of disruptive and stabilizing selection on body size in *Drosophila melanogaster*. I. Mean values and variances. *Genetics* 75, 679–693.

BOS, M. and SCHARLOO, W. (1973b). The effects of disruptive and stabilizing selection on body size in *Drosophila melanogaster*. II. Analysis of responses in the thorax selection lines. *Genetics* 75, 695–708.

BOS, M. and SCHARLOO, W. (1974). The effects of disruptive and stabilizing selection on body size in *Drosophila melanogaster*. III. Genetic analysis of two lines with different reactions to disruptive selection with mating of opposite extremes. *Genetica* 45, 71–90.

BREESE, E. L. (1959). Selection for differing degrees of out-breeding in *Nicotiana rustica*. *Ann. Bot.* 23, 331–344.

BREESE, E. L. and MATHER, K. (1957). The organisation of polygenic activity within a chromosome in *Drosophila*. I. Hair characteristics. *Heredity* 11, 373–395.

BREESE, E. L. and MATHER, K. (1960). The organisation of polygenic activity within a chromosome in *Drosophila*. II. Viability. *Heredity* 14, 375–399.

CALIGARI, P. D. S. and MATHER, K. (1975). Genotype-environment interaction. III. Interactions in *Drosophila melanogaster*. *Proc. Roy. Soc. Lond., B*, 191, 387–411.

CALIGARI, P. D. S. and MATHER, K. (1980). Dominance, allele frequency and selection in a population of *Drosophila melanogaster*. *Proc. Roy. Soc. Lond., B*, 208, 163–187.

CARSON, H. L. (1958). Response to selection under different conditions of recombination in *Drosophila*. *Cold Spr. Harb. Symp. Quant. Biol.* 23, 291–306.

CARSON, H. L. (1967). Selection for parthenogenesis in *Drosophila mercatorum*. *Genetics* 55, 157–171.

CASTLE, W. E. and PHILLIPS, J. C. (1914). Piebald rats and selection. An experimental test of the effectiveness of selection and the theory of gametic purity in Mendelian crosses. *Carnegie Inst. Wash.*, Publ. 195.

CASTLE, W. E. and WRIGHT, S. (1916). Studies of inheritance in guinea-pigs and rats. *Carnegie Inst. Wash.*, Publ. 226.

CATCHESIDE, D. G. (1977). "The Genetics of Recombination." Edward Arnold, London.

CLARKE, C. A. and SHEPPARD, P. M. (1960). The evolution of mimicry in the butterfly *Papilio dardanus*. *Heredity* **14**, 163–173.

CLAYTON, G. A. and ROBERTSON, A. (1955). Mutation and quantitative variation. *Amer. Nat.* **89**, 151–162.

CLAYTON, G. A. and ROBERTSON, A. (1957). An experimental check on quantitative genetical theory. II. The long-term effects of selection. *J. Genet.* **55**, 152–170.

CLAYTON, G. A., MORRIS, J. A. and ROBERTSON, A. (1957a). An experimental check on quantitative genetical theory. I. Short-term responses to selection. *J. Genet.* **55**, 131–151.

CLAYTON, G. A., KNIGHT, G. R., MORRIS, J. A. and ROBERTSON, A. (1957b). An experimental check on quantitative genetical theory. III. Correlated responses. *J. Genet.* **55**, 171–180.

COMSTOCK, R. F., ROBINSON, H. F. and HARVEY, P. H. (1949). A breeding procedure designed to make maximum use of both general and specific combining ability. *J. Amer. Soc. Agron.* **41**, 360–367.

COOPER, K. W. (1937). Reproductive behavior and haploid parthenogenesis in the grass mite, *Pediculopsis graminum* (Reut.) (Acarina, Tarsonemidae). *Proc. Nat. Acad. Sci. Wash.* **23**, 41–44.

DARLINGTON, C. D. (1939). "The Evolution of Genetic Systems." Cambridge University Press.

DARLINGTON, C. D. and MATHER, K. (1949). "The Elements of Genetics." Allen and Unwin, London.

DAVIES, R. W. (1971). The genetic relationship of two quantitative characters in *Drosophila melanogaster*. II. Location of the effects. *Genetics* **69**, 363–375.

DETLEFSEN, J. A. and ROBERTS, E. (1921). Studies on crossing over. I. The effects of selection on crossover values. *J. exp. Zool.* **32**, 333–354.

DOBZHANSKY, TH. (1939). Experimental studies on free-living populations of *Drosophila*. *Biol. Rev.* **14**, 339–368.

DOBZHANSKY, TH. (1948). Genetics of natural populations. XVIII. Experiments on chromosomes of *Drosophila pseudoobscura* from different geographical regions. *Genetics* **33**, 588–602.

DOBZHANSKY, TH. (1950). Genetics of natural populations. XIX. Origin of heterosis through natural selection in populations of *Drosophila pseudoobscura*. *Genetics* **35**, 288–302.

DOBZHANSKY, TH. and PAVLOVSKY, O. (1971). Experimentally created incipient species of *Drosophila*. *Nature, Lond.* **230**, 289–292.

DURRANT, A. and MATHER, K. (1954). Heritable variation in a long inbred line of *Drosophila*. *Genetica* **27**, 97–119.

FALCONER, D. S. (1957). Selection for phenotypic intermediates in *Drosophila*. *J. Genet.* **55**, 551–561.

FALCONER, D. S. (1960). "Introduction to Quantitative Genetics." Oliver and Boyd, Edinburgh.

FISHER, R. A. (1918). The correlation between relatives on the supposition of Mendelian inheritance. *Trans. Roy. Soc. Edin.* **52**, 399–433.

FISHER, R. A. (1928). The possible modification of the response of the wild type to recurrent mutations. *Amer. Nat.* **62**, 115–126.

FISHER, R. A. (1930). "The Genetical Theory of Natural Selection." Clarendon Press, Oxford.
FORD, E. B. (1975). "Ecological Genetics.", 4th edition. Chapman and Hall, London.
FRANKHAM, R., JONES, L. P. and BARKER, J. S. F. (1968a). The effects of population size and selection intensity in selection for a quantitative character in *Drosophila*. I. Short-term response to selection. *Genet. Res., Camb.* **12**, 237–248.
FRANKHAM, R., JONES, L. P. and BARKER, J. S. F. (1968b). The effects of population size and selection intensity in selection for a quantitative character in *Drosophila*. III. Analyses of the lines. *Genet. Res., Camb.* **12**, 267–283.
GIBSON, J. B. and BRADLEY, B. P. (1974). Stabilising selection in constant and fluctuating environments. *Heredity* **33**, 293–302.
GIBSON, J. B. and THODAY, J. M. (1962). Effects of disruptive selection. VI. A second chromosome polymorphism. *Heredity* **17**, 1–26.
GRANT, B. and METTLER, L. E. (1969). Disruptive and stabilising selection on the "escape" behavior of *Drosophila melanogaster*. *Genetics* **62**, 625–637.
HADLER, N. M. (1964). Genetic influence on phototaxis in *Drosophila melanogaster*. *Biol. Bull.* **126**, 264–273.
HILDRETH, P. E. (1956). The problem of synthetic lethals in *Drosophila melanogaster*. *Genetics* **41**, 729–742.
HULL, F. H. (1945). Recurrent selection for specific combining ability in corn. *J. Amer. Soc. Agron.* **37**, 134–145.
JOHANNSEN, W. (1909). "Elemente der exakten Erblichkeitslehre." Fischer, Jena.
JONES, L. P., FRANKHAM, R. and BARKER, J. S. F. (1968). The effects of population size and selection intensity in selection for a quantitative character in *Drosophila*. II. Long-term response to selection. *Genet. Res., Camb.* **12**, 249–266.
KEARSEY, M. J. and BARNES, B. W. (1970). Variation for metrical characters in *Drosophila* populations. II. Natural selection. *Heredity* **25**, 11–21.
KEARSEY, M. J. and KOJIMA, K. (1967). The genetic architecture of body weight and egg hatchability in *Drosophila melanogaster*. *Genetics* **56**, 23–37.
KESSLER, S. (1966). Selection for and against ethological isolation between *Drosophila pseudoobscura* and *Drosophila persimilis*. *Evolution* **20**, 634–645.
KIDWELL, M. G. (1972a). Genetic change of recombination value in *Drosophila melanogaster*. I. Artificial selection for high and low recombination and some properties of recombination-modifying genes. *Genetics* **70**, 419–432.
KIDWELL, M. G. (1972b). Genetic change of recombination value in *Drosophila melanogaster*. II. Simulated natural selection. *Genetics* **70**, 433–443.
KNIGHT, G. R., ROBERTSON, A. and WADDINGTON, C. H. (1956). Selection for sexual isolation within a species. *Evolution* **10**, 14–22.
KOOPMAN, K. F. (1950). Natural selection for reproductive isolation between *Drosophila pseudoobscura* and *Drosophila persimilis*. *Evolution* **4**, 135–148.
LENG, E. R. (1962). Results of long-term selection for chemical composition in maize and their significance in evaluating breeding systems. *Z. Pflanzenz.* **47**, 67–91.
LEWONTIN, R. C. and HUBBY, J. L. (1966). A molecular approach to the study of genic heterozygosity in natural populations. II. Amount of variation and degree of heterozygosity in natural populations of *Drosophila pseudoobscura*. *Genetics* **54**, 595–609.
L'HERITIER, P. L. and TEISSIER, G. (1934). Une expérience de sélection naturelle; Courbe d'élimination du gène "Bar" dans une population de *Drosophiles* en equilibre. *Compt. Rend. Soc. Biol.* **117**, 1049–1051.
LINNEY, R., BARNES, B. W. and KEARSEY, M. J. (1971). Variations for metrical characters in *Drosophila* populations. III. The nature of selection. *Heredity* **27**, 163–174.

McBride, G. and Robertson, A. (1963). Selection using assortative mating in *Drosophila melanogaster*. *Genet. Res., Camb.* **4**, 356–369.

MacDowell, E. C. (1915). Bristle inheritance in *Drosophila*. I. Extra bristles. *J. exp. Zool.* **19**, 61–98.

MacDowell, E. C. (1917). Bristle inheritance in *Drosophila*. II. Selection. *J. exp. Zool.* **23**, 109–146.

MacDowell, E. C. (1920). Bristle inheritance in *Drosophila*. III. Correlation. *J. exp. Zool.* **30**, 419–460.

McPhee, C. P. and Robertson, A. (1970). The effect of suppressing crossing-over on the response to selection in *Drosophila melanogaster*. *Genet. Res., Camb.* **16**, 1–16.

Madalena, F. E. and Robertson, A. (1974). Population structure in artificial selection: Studies with *Drosophila melanogaster*. *Genet. Res., Camb.* **24**, 113–126.

Mather, K. (1941). Variation and selection of polygenic characters. *J. Genet.* **41**, 159–193.

Mather, K. (1942). The balance of polygenic combinations. *J. Genet.* **43**, 309–336.

Mather, K. (1943a). Polygenic inheritance and natural selection. *Biol. Rev.* **18**, 32–64.

Mather, K. (1943b). Specific differences in *Petunia*. I. Incompatibility. *J. Genet.* **45**, 215–235.

Mather, K (1946). Dominance and heterosis. *Amer. Nat.* **80**, 91–96.

Mather, K. (1950). The genetical architecture of heterostyly in *Primula sinensis*. *Evolution* **4**, 340–352.

Mather, K. (1953a). The genetical structure of populations. *Symp. Soc. Exp. Biol.* **7**, 66–95.

Mather, K. (1953b). Genetical control of stability in development. *Heredity* **7**, 297–336.

Mather, K. (1955). Polymorphism as an outcome of disruptive selection. *Evolution* **9**, 52–61.

Mather, K. (1960a). The balance sheet of variability. *In*: "Biometrical Genetics" (O. Kempthorne, ed.), pp. 10–11. Pergamon Press, London.

Mather, K. (1960b). Evolution in polygenic systems. *In*: "Evoluzione e Genetica", 131–152. Academia Nazionale dei Lincei, Rome.

Mather, K. (1966). Variability and selection. *Proc. Roy. Soc. Lond., B,* **164**, 328–340.

Mather, K. (1973). "Genetical Structure of Populations." Chapman and Hall, London.

Mather, K. (1979). Historical overview: Quantitative variation and polygenic systems. *In*: "Quantitative Genetic Variation" (J. N. Thompson, jr. and J. M. Thoday, eds.), pp. 5–34. Academic Press, New York and London.

Mather, K. and Edwardes, P. M. J. (1943). Specific differences in *Petunia*. III. Flower colour and genetic isolation. *J. Genet.* **45**, 243–260.

Mather, K. and Harrison, B. J. (1949). The manifold effects of selection. *Heredity* **3**, 1–52 and 131–162.

Mather, K. and Jinks, J. L. (1971). "Biometrical Genetics.", 2nd edition. Chapman and Hall, London.

Mather, K. and Wigan, L. G. (1942). The selection of invisible mutations. *Proc. Roy. Soc. Lond.*, B, 50–64.

Maynard Smith, J. (1966). Sympatric speciation. *Amer. Nat.* **100**, 637–650.

Maynard Smith, J. (1979). The effects of normalizing and disruptive selection on genes for recombination. *Genet. Res., Camb.* **33**, 121–128.

Mayr, E. (1974). The definition of the term disruptive selection. *Heredity* **32**, 404–406.

Millicent, E. and Thoday, J. M. (1960). Gene flow and divergence under disruptive selection. *Science* **131**, 1311–1312.

Millicent, E. and Thoday, J. M. (1961). Effects of disruptive selection. IV. Gene-flow and divergence. *Heredity* **16**, 199–217.

MINAWA, A. and BIRLEY, A. J. (1978). The genetical response to natural selection by varied environments. I. Short-term observations. *Heredity* **40**, 39–50.

MISRO, B. (1949). Crossing over as a source of new variation. Proc. 8th Int. Congr. Genet., Hereditas, suppl. vol., 629–630.

MUKAI, T. (1979). Polygenic mutation. *In*: "Quantitative Genetic Variation." (J. N. Thompson, jr and J. M. Thoday, eds.), pp. 177–196. Academic Press, New York and London.

OSMAN, H. E. S. and ROBERTSON, A. (1968). The introduction of genetic material from inferior into superior strains. *Genet. Res., Camb.* **12**, 221–236.

PAYNE, F. (1918). An experiment to test the nature of the variations on which selection acts. *Indiana Univ. Stud.* **5**, 3–45.

PROUT, T. (1962). The effects of stabilizing selection on the time of development in *Drosophila melanogaster*. *Genet. Res., Camb.* **3**, 364–382.

RASMUSON, M. (1955). Selection for bristle numbers in some unrelated strains of *Drosophila melanogaster*. *Acta Zool.* **36**, 1–49.

REEVE, E. C. R. and ROBERTSON, F. W. (1953). Studies in quantitative inheritance. II. Analysis of a strain of *Drosophila melanogaster* selected for long wings. *J. Genet.* **51**, 276–316.

RENDEL, J. M. (1979). Canalisation and selection. *In*: "Quantitative Genetic Variation." (J. N. Thompson, jr. and J. M. Thoday, eds.), pp. 139–156. Academic Press, New York and London.

RENDEL, J. M. and SHELDON, B. L. (1960). Selection for canalization of the scute phenotype in *Drosophila melanogaster*. *Australian J. Biol. Sci.* **13**, 36–47.

ROBERTSON, A. (1955). Selection in animals: Synthesis. *Cold Spr. Harb. Symp. Quant. Biol.* **20**, 225–229.

ROBERTSON, F. W. (1954). Studies in quantitative inheritance. V. Chromosome analyses of crosses between selected and unselected lines of different body size in *Drosophila melanogaster*. *J. Genet.* **52**, 494–520.

ROBERTSON, F. W. (1955). Selection response and the properties of genetic variation. *Cold Spr. Harb. Symp. Quant. Biol.* **20**, 166–177.

ROBERTSON, F. W. and REEVE, E. C. R. (1952). Studies in quantitative inheritance. I. The effects of selection of wing and thorax length in *Drosophila melanogaster*. *J. Genet.* **50**, 414–448.

SCHARLOO, W. (1970a). Stabilizing and disruptive selection on a mutant character in *Drosophila*. II. Polymorphism caused by a genetical switch mechanism. *Genetics* **65**, 681–691.

SCHARLOO, W. (1970b). Stabilizing and disruptive selection on a mutant character in *Drosophila*. III. Polymorphism caused by a developmental switch mechanism. *Genetics* **65**, 693–705.

SCHARLOO, W., HOOGMOED, M. S. and TER KUILE, A. (1967). Stabilizing and disruptive selection on a mutant character in *Drosophila*. I. The phenotypic variance and its components. *Genetics* **56**, 709–726.

SCHARLOO, W., ZWEEP, A., SCHUITEMA, K. A. and WIJNSTRA, J. G. (1972). Stabilizing and disruptive selection on a mutant character in *Drosophila*. IV. Selection on sensitivity to temperature. *Genetics* **71**, 551–566.

SCOWCROFT, W. R. and LATTER, D. B. H. (1971). Decanalization of scutellar bristle number in *Drosophila*. *Genet. Res., Camb.* **17**, 95–101.

SEN, B. K. and ROBERTSON, A. (1964). An experimental examination of methods for the simultaneous selection of two characters, using *Drosophila melanogaster*. *Genetics* **50**, 199–209.

SHEPPARD, P. M. (1953). Polymorphism and population studies. *Symp. Soc. Exp. Biol.* **7**, 274–289.
SIMPSON, G. G. (1944). "Tempo and Mode in Evolution." Columbia Univ. Press, New York.
SISMANIDIS, A. (1942). Selection for an almost invariable character in *Drosophila. J. Genet.* **44**, 204–215.
SKIBINSKI, D. O. F. and THODAY, J. M. (1979). Disruptive selection with fixed optima. *Heredity* **42**, 327–335.
SMITH, L. H. (1908). Ten generations of corn breeding. *Ill. Agr. Expt. Sta. Bull.* **128**, 459–575.
SOUZA, H. L. DE, DA CUNHA, A. B. and DOS SANTOS, E. P. (1968). Adaptive polymorphism of behavior developed in laboratory populations of *Drosophila willistoni. Amer. Nat.* **102**, 583–586.
SPICKETT, S. G. and THODAY, J. M. (1966). Regular responses to selection. 3. Interaction between located polygenes. *Genet. Res., Camb.* **7**, 96–121.
STALKER, H. D. (1954). Parthenogenesis in *Drosophila. Genetics* **39**, 4–34.
STREAMS, F. A. and PIMENTEL, D. (1961). Effects of immigration on the evolution of populations. *Amer. Nat.* **95**, 201–210.
STURTEVANT, A. H. (1918). An analysis of the effects of selection. *Carnegie Inst. Wash.*, Publ. 264.
TANTAWY, A. O. and TAYEL, A. A. (1970). Studies on natural populations of *Drosophila*. X. Effects of disruptive and stabilizing selection on wing length and the correlated response in *Drosophila melanogaster. Genetics* **65**, 121–132.
THODAY, J. M. (1958). Homeostasis in a selection experiment. *Heredity* **12**, 401–415.
THODAY, J. M. (1959). Effects of disruptive selection. I. Genetic flexibility. *Heredity* **13**, 187–203.
THODAY, J. M. (1960). Effects of disruptive selection. III. Coupling and repulsion. *Heredity* **14**, 35–49.
THODAY, J. M. (1961). Location of polygenes. *Nature* **191**, 368–370.
THODAY, J. M. (1965). Effects of selection for genetic diversity. *Genetics Today, Proc. XIth Int. Congr. Genet.* **3**, 533–540.
THODAY, J. M. (1972). Disruptive selection. *Proc. Roy. Soc. Lond.*, B, **182**, 109–143.
THODAY, J. M. (1974). Definition of disruptive selection and of "interbreeding populations." *Heredity* **32**, 406–409.
THODAY, J. M. and BOAM, T. B. (1959). Effects of disruptive selection. II. Polymorphism and divergence without isolation. *Heredity* **13**, 205–218.
THODAY, J. M. and BOAM, T. B. (1961). Regular responses to selection. I. Description of responses. *Genet. Res., Camb.* **2**, 161–176.
THODAY, J. M. and GIBSON, J. B. (1962). Isolation by disruptive selection. *Nature, Lond.* **193**, 1164–1166.
THODAY, J. M. and GIBSON, J. B. (1970). The probability of isolation by disruptive selection. *Amer. Nat.* **104**, 219–230.
THOMPSON, J. N., JR. and THODAY, J. M. (1979). "Quantitative Genetic Variation." Academic Press, New York and London.
WADDINGTON, C. H. (1953). Genetic assimilation of an acquired character. *Evolution* **7**, 118–126.
WADDINGTON, C. H. (1960). Experiments on canalizing selection. *Genet. Res., Camb.* **1**, 140–150.
WALLACE, B. (1954). Genetic divergence of isolated populations of *Drosophila melanogaster. Proc. 9th Int. Congr. Genet., Caryologia, suppl. vol.* **2**, 761–764.

WALLACE, B. (1970). "Genetic Load." Prentice Hall, Eagleswood Cliffs, N.J.
WALLACE, B. and VETUKHIV, M. (1955). Adaptive organization of the gene pools of *Drosophila* populations. *Cold Spr. Harb. Symp. Quant. Biol.* **20**, 303–309.
WIGAN, L. G. (1941). Polygenic variability in wild *Drosophila melanogaster. Nature, Lond.* **148**, 373–374.
WIGAN, L. G. and MATHER, K. (1942). Correlated response to the selection of polygenic characters. *Ann. Eugenics* **11**, 354–364.
WINTER, F. L. (1929). Continuous selection for composition in corn. *J. Agric. Res.* **39**, 451–475.
WOLSTENHOLME, D. R. and THODAY, J. M. (1963). Effects of disruptive selection. VII. A third chromosome polymorphism. *Heredity* **18**, 413–431.
WRIGHT, S. (1977). "Evolution and the Genetics of Populations." volume III. University of Chicago Press, Chicago.
ZELENY, C. (1922). The effects of selection for eye facet number in the white bar-eye race of *Drosophila melanogaster. Genetics* **7**, 1–115.

23. Mating Behavior and Sexual Isolation in *Drosophila*

HERMAN T. SPIETH

*Department of Zoology, University of California,
Davis, California, U.S.A.*

and

JOHN M. RINGO

Department of Zoology, University of Maine, Orono, Maine, U.S.A.

I. Historical Development	224
II. Basic Nature of *Drosophila* Mating Behavior	225
III. Stimuli Involved in Mating Behavior	229
A. Auditory Stimuli	230
B. Olfactory Stimuli	233
C. Tactile Stimuli	234
D. Visual Stimuli	236
IV. Ontogeny	237
A. Male Displays and Mating Propensity	238
B. Female Behavior	239
C. Fate Mapping	242
D. Experience and Mating Preferences	243
V. Measurement of Sexual Activity	244
A. Some Commonly Used Experimental Designs	245
B. Measures of Nonrandom Mating	247
C. Models of Mating Behavior	251
VI. Adaptiveness of Mating Behavior	251
A. Courtship and Life History	251
B. Sexual Selection	254
VII. Phylogeny	257
A. Microevolutionary and Macroevolutionary Changes	257
B. Value of Behavior in Determining Phylogenetic Relationships	260
VIII. Sexual Isolation	261
A. Degree of Sexual isolation at Various Stages of Population Differentiation	261
B. Role of Males and Females	265
C. Environmental Factors Influencing Sexual Isolation	266
D. Evolutionary Origin of Sexual Isolation	268
IX. Summary	270
References	270

I. Historical Development

In the years 1900-1901 Woodworth, then on sabbatical leave at Harvard, convinced Castle that *Drosophila melanogaster* Meigen might serve as a valuable organism for the study of genetics (Davenport, 1941). This serendipitous event changed *Drosophila* from an obscure acalypterate dipteran genus into one of the best known. Fifteen years later Sturtevant described and analysed the courtship of *D. melanogaster* (Sturtevant, 1915). In 1921 he recorded the courtship of 22 species, including two species each of *Scaptomyza* and *Chymomyza* (Sturtevant, 1921). He found that the courtships are complex and species specific. Sturtevant also recorded that mutants of *D. melanogaster* are less active than the wild type and that this affects the speed of courtship.

During the years 1930-40 the concepts of the synthetic theory of evolution (Dobzhansky, 1937; Huxley, 1940; Mayr, 1942) and of modern ethology (see Eibl-Eibesfeldt, 1970) achieved prominence. Information about the species of *Drosophila* also increased dramatically. In 1900, 71 valid species were known and by 1940 the number had increased to 410 (Wheeler, 1959). A considerable number of these species were amenable to laboratory culture. Geneticists by that time had discovered numerous mutations in various species of *Drosophila*.

These developments, and especially the information provided by Sturtevant's two pioneering papers, suggested to several investigators during the early 1940's that the courtships of *Drosophila* species could be effectively described and analysed to provide valuable evolutionary and ethological data.

Studies during the 1940's primarily utilized closely related species and were directed mainly to elucidating the role of courtship in sexual isolation. Investigations during the 1950's assumed a broader scope. The courtships of numerous species were recorded, the behaviors of mutant flies were elucidated, and the nature and role of the stimuli involved in the courtships received attention. The accumulated data provided both ethological and evolutionary insights.

The 1960's saw a burgeoning of behavioral investigations. Fewer courtship rituals were described but these did include examples from the rich and unique Hawaiian fauna. Much attention was given to the behavior of mutant flies and especially to the analysis of the courtship stimuli. Sophisticated studies on the nature and role of auditory, olfactory, and visual stimuli yielded abundant and valuable data. A new line of inquiry, the study of mating speed and the analysis of its evolutionary role, was begun. A landmark review by Manning (1965), in which he emphasized both the ethological and evolutionary aspects of *Drosophila* courtships, served as a powerful stimulus for further investigations. Not surprisingly these have continued at a high rate during the ensuing years.

II. Basic Nature of *Drosophila* Mating Behavior

Mating behavior of *Drosophila* species consists of species specific fixed action patterns which are accompanied by orientation movements. Such patterns, referred to as courtship displays, are made up of a number of elements or signals, some of which are performed sequentially. The male initiates courtship by performing a fixed action pattern directed toward an individual whose size and gestalt indicate that it may be a potential mate. The response of the potential mate to the male's pattern results in information being transmitted which enables the two individuals to distinguish conspecifics from non-conspecifics, males from females and also the physiological readiness of the females to engage in copulation.

Following initiation of the courtship display, the male may terminate his actions at any point in the sequential performance of the signals or he may repeat the full pattern numerous times. If the potential mate is a non-conspecific, a conspecific male, or a previously inseminated conspecific female, then he usually terminates the action quickly. If the individual approached is a conspecific virgin female, he usually persists until either copulation occurs or one or both flies terminate the encounter.

The visually observable courtship signals and orientation movements of both sexes are performed by specific movements of the insects' bodies (Sturtevant, 1915, 1921; Spieth, 1952, 1966b; Bastock and Manning, 1955; Brown, 1964, 1966). Male actions are of two types, i.e. those involved in male-to-male encounters and those of male-to-female encounters. For the majority of species the forelegs, wings and mouthparts of the males serve as signalling structures (Table I and Figs. 1, 2a,b,c).

Interspecific diversity of the male display is great and the only parts of the body that are not used in courtship signalling by at least one or more species are the metathoracic legs. Furthermore, the males of a number of species exhibit morphological sexual dimorphisms or epigamic pigmentation of portions of their bodies. Such characters are invariably involved in the production of courtship signals.

The female's signals are more limited in number and diversity than are those of the male and are produced by the wings, legs, genitalia, and movements of the abdomen. These signals are performed in response to the male's courting overtures and are divisable into two types: rejection responses (Table II) and acceptance responses (Table III).

A number of orienting movements occur during courtships. The most frequently observed one is designated simply as orientation and occurs

TABLE I. Male courtship elements displayed by numerous species

Signaling movements

Tapping: ♂ lifts and extends one foreleg and strikes tarsus downward against potential mate (Rendel, 1945).

Wing flicking: ♂ flicks one wing vane from resting positions outward and then returns to resting state (Spieth, 1952).

Wing fluttering: In male-to-male encounters, the approached male slightly elevates wing vanes, moves them laterally a few degrees and then vibrates vanes rapidly (Spieth, 1952).

Wing semaphoring: ♂ alternately and repeatedly flicks wing vanes outward and then returns to resting position, i.e. one wing is moved outward while the other is returned to resting position (Spieth, 1966b).

Wing scissoring: ♂ repeatedly flicks both wing vanes horizontally outward and then returns to resting position (Sturtevant, 1921).

Wing vibration: ♂ extends one wing vane nearest ♀'s head and vibrates it up and down. Large interspecific variation in amount of extension, amplitude and speed of vibration, and the angle of vane with respect to the substrate, i.e. horizontal, tilted or vertical (Sturtevant, 1921). The Hawaiian species extend and vibrate both wings (Spieth, 1966b).

Wing waving: ♂ extends one wing vane and waves up and down (Sturtevant, 1921).

Leg vibration: ♂ positioned close to and directly behind tip of ♀'s abdomen, lifts and extends both forelegs forward and vibrates them against abdomen of ♀, usually her venter, but some species against sides or dorsum of ♀ abdomen. Vectors of leg movement may be vertical, horizontal, or a combination of both (Sturtevant, 1921).

Licking: ♂ positioned as for leg vibration opens labellum, extends proboscis and licks ♀ genitalia. Licking may be performed intermittently or continuously for prolonged periods (Sturtevant, 1921). Some species of the Hawaiian fauna have modified labellar lobes which grasp ♀ genitalia and others which grasp the ♀ anal papilla (Spieth, 1966b).

FIG. 1. Courtship of *D. pallidifrons* (1), *D. albomicans* (2) and *D. kepulauana* (3). (After Spieth, 1969, p. 256).

23. MATING AND SEXUAL ISOLATION IN *DROSOPHILA* 227

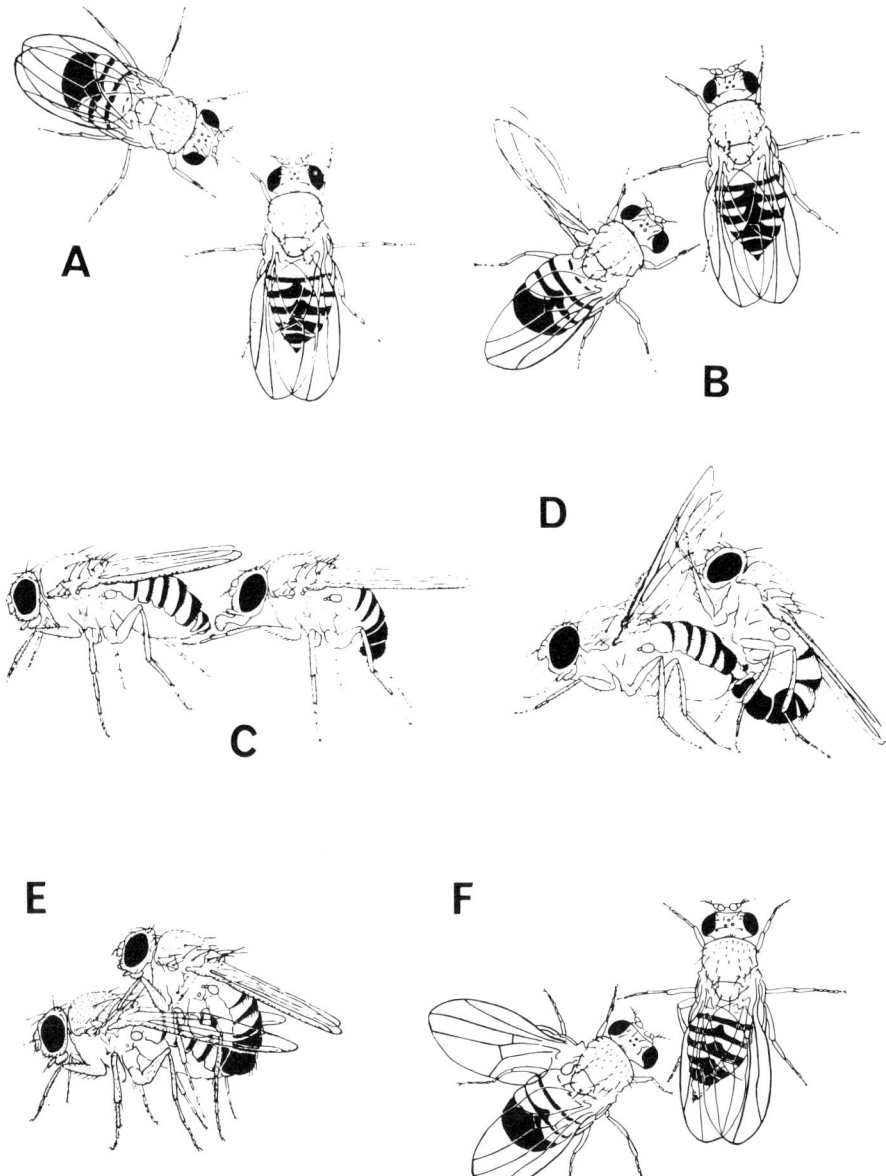

FIG. 2. Courtship behavior in *D. melanogaster*. A, orientation of the male to the female. B, wing vibration by the male. C, the male licks the female's genitalia with his proboscis. D, mounting by the male with genital contact. E, flies in copulation. F, a rejection response by the female to courtship. The female turns her abdomen towards the male and extends her ovipositor. (After Burnet and Connolly, 1974).

TABLE II. Female rejection signals

Abdomen elevation: ♀ elevates tip of abdomen high above substrate, thus inhibiting performance of licking and leg vibration signals by male (Spieth, 1952).
Abdomen depression: ♀ depresses tip of abdomen and wing tips against substrate, thus inhibiting performance of licking and leg vibration signals of males (Lutz, 1914).
Decamping: ♀ breaks contact with courting male by running, jumping, or flying (Spieth, 1947).
Flicking: Similar to ♂ flicking (Connolly and Cook, 1973).
Fluttering: Similar to ♂ fluttering.
Kicking: ♀ kicks vigorously backwards at face of male when he courts at her rear (Sturtevant, 1915).
Extrusion: ♀ extends and elongates tip of abdomen, thus exposing articulating membranes surrounding genital sclerites. Simultaneously she may direct tip of her abdomen toward face of male who usually then turns away quickly and may then engage in cleaning behavior (Rendel, 1945; Spieth, 1947).

TABLE III. Female acceptance signals

Genitalic spreading: ♀ slightly droops tip of abdomen, slightly extrudes and spreads ovipositor apart (Spieth, 1952).
Wing spreading: ♀ spreads wing vanes outward and upward, and holds them extended until the male mounts (Spieth, 1952).
Ovipositor extension: ♀ extrudes ovipositor posteriorly. Restricted to some Hawaiian species (Spieth, 1966b).

when a male visually "fixes" upon a potential mate. He slightly elevates his body, turns to face the potential mate, and approaches it. Having made physical contact with a potential mate, usually by tapping, the male of most species then quickly moves to the rear of the approached individual and positions himself close to and facing the tip of the other individual's abdomen. If the potential mate is a male, it usually spins about and faces the courting male and engages in wing flicking or wing fluttering and foreleg striking movements. Non-receptive females may also engage in similar actions but more commonly they either flee or engage in extrusion (Table II). The males of a few species (e.g., *D. subobscura*) move to the front and place themselves face to face with the potential mate.

Copulation occurs only if the female responds by performing acceptance signals. For the majority of species, the male is positioned at the rear of the female when she indicates readiness to copulate. Females of a few species (e.g., *subobscura* and members of the *nasuta* subgroup) signal acceptance when the male is positioned in front of them. The male of these species then races to the rear of the female.

A male mounts and copulates by curling the tip of the abdomen under and forward, simultaneously lunging upward and forward, thrusting his head under or between her wings if she has spread wings, grasping her body with his fore and middle legs, and then achieving intromission. Males of many species, but not all, attempt to mount and copulate with non-receptive females but they invariably are unsuccessful.

During copulation the female may walk about and even feed while the male is inert except that in most species they periodically stroke the area of genitalic contact with their hind tarsi. The Hawaiian species are exceptions: the female remains inert, often resting the venter of her thorax on the substrate, and only the male's action of stroking occurs.

Females of some species shake their bodies violently, wing flutter, or kick with the hind legs at various times during copulation. Near the end of the copulatory period, the female typically kicks vigorously rearward with her hind legs against the face and thorax of the male. These actions may influence the male's behavior, but the physical force *per se* is never effective in breaking the genitalic union and it appears that the male determines the termination of copulation. In a few species of Hawaiian flies (e.g., *D. crucigera* and its close relatives), a male near the end of a copulation releases his grasp of the female, falls backwards, and appears to enter a "trance state." The only contact between the pair is then the genitalic union. When the male withdraws and breaks from the female, he falls inertly to the substrate and remains quiescent for a short period before jumping to a standing position.

The duration of copulation is species specific but considerable interindividual variation exists and the minimum time for some species is only one-third to one-half of the maximum. The shortest time recorded is that of *D. enigma* with a mean of 5 sec (Grossfield and Rockwell, 1979); the longest is that of *D. acanthoptera* with a mean of 62 min (Spieth, 1952). *D. immigrans* has a long copulatory time, as well as great individual variation, i.e. a minimum of 14 min and a maximum of 83 min with a mean of about 50 min (Spieth, 1952). The majority of species have copulatory times of less than 10 min and closely related species almost invariably have similar mean copulatory times as well as similar minima and maxima (Sturtevant, 1942; Wheeler, 1947; Patterson, 1947; Spieth, 1952).

III. Stimuli Involved in Mating Behavior

Visual observation of the fixed action patterns displayed during courtships can provide clues as to the nature of the stimuli transmitted and received by the individuals involved, but cannot unequivocably identify the specific stimuli. By using normal flies that have had portions of their bodies surgically removed and mutant flies, investigators have successfully identified and characterized various stimuli and the roles these serve in courtships.

A courting drosophilid male provides a female with a battery of stimuli—visual, chemical (both contact and airborne), auditory, and perhaps mechanical. The stimuli function to inform the female of the species identity of the male and to stimulate the female beyond her acceptance threshold for accepting the male in copulation (Manning, 1959a,b). The female may respond with an acceptance signal or rejection. Rates and types of rejection depend on the female's age and physiological state (see Section IV). Mature virgin females vary in their acceptance threshold, but usually the male must repeat his courtship elements, often numerous times, before a female is willing to copulate. The final outcome of courtship appears to be dependent upon the physiological state of the female and the temporally summed effect of the male stimuli. If a male lacks one element of his courtship, e.g. auditory stimulus, he may still achieve copulation with a female, but he must court her for a longer time than does a normal male, on average. Thus in competition with normal males, he is seriously disadvantaged. Likewise when wild-type males are presented simultaneously with both normal mature females and mature females who are unable to perceive one element of the male's courtship signals, the males inseminate many more of the normal females than those females who have a sensory deficiency and are therefore at a mating disadvantage.

A. Auditory Stimuli

The courting males of most species engage in wing vibration, which results in the production of sounds termed courtship songs. These songs have been recorded and analysed by a number of investigators (Chang and Miller, 1978; Ewing, 1970, 1979; Ewing and Bennet-Clark, 1968; Miller et al., 1975; Shorey, 1962; von Schilcher, 1976a, 1976b; Waldron, 1964). The songs consist of two elements: the sine song and the pulse song. The sine song in *D. melanogaster* consists of a humming sound that is reminiscent of flight sounds (von Schilcher, 1976b). The pulse song is comprised of repeated sound pulses separated by time, or inter-pulse intervals (Ewing and Bennet-Clark, 1968). Pulse songs exhibit variation between and within species in the number of cycles per pulse, pulse repetition rate, and the duration of the inter-pulse interval (Fig. 3).

Immature females that are unable to engage effectively in extrusion and some inseminated females produce bursts of buzzing sounds (Ewing and Bennet-Clark, 1968). These occur without the wings being overtly extended. Presumably they are the result of wing fluttering (Table III).

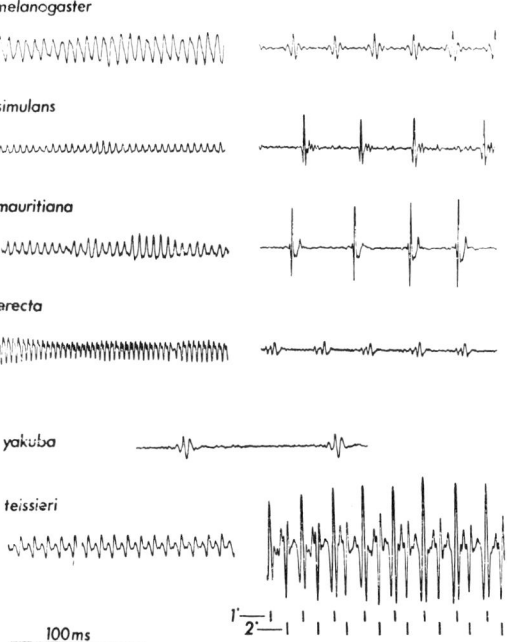

FIG. 3. Oscillograms of the courtship song of the sibling species. Sine song is shown on the left and pulse song on the right, except for *D. yakuba* which produces only pulse song. Primary (1°) and secondary (2°) pulses of *D. teissieri* are indicated by markers. (After Cowling and Burnet, 1981, p. 928).

Such sounds appear to serve as rejection signals and are similar for all species.

Both males and females are capable of receiving auditory courtship rejection signals via their antennae whose aristae serve as velocity sensitive receptors (Manning, 1967a; Bennet-Clark, 1971). The resulting movements of the sail-like arista serve to twist the large bulbous third antennal segment and thus to stimulate Johnston's organ at its base. Flies of both sexes that lack wings and aristae are "mute and deaf." Wingless *D. melanogaster* males experience mating disadvantage when tested both in the light (Ewing, 1964) and in the dark (Bastock, 1956), and aristaless *melanogaster* females are less receptive than are normal females (Manning, 1967a; Burnet et al., 1971). Morphological deficits of these types can be achieved either by surgical removal of the wings and aristae of normal flies or by mutations, e.g. vestigial, aristaless, and thread. One difficulty with mutations is that any particular mutation may, and often does, have secondary effects upon the flies. Courting males of vestigial flies persistently maintain themselves close to the tip of the female's abdomen and engage in an abnormal amount of licking and attempting to mount (Burnet and Connolly, 1974). Females with mutant aristae are inhibited from responding to the courtship signals of wild males, which is not unexpected, but aristal mutant males are also less competent than wild males when tested with wild females. Burnet *et al.* (1971) found that aristal mutant males are not as competent in maintaining orientation on the females as are wild-type males, but do spend a longer than normal time engaging in wing vibration. Apparently such secondary effects are caused either by the pleiotropic action of the mutant genes themselves (Burnet *et al.*, 1971), by genetic background, shifted by selection during maintenance over a period of time in the laboratory (Cook, 1973a,b), or by a combination of both pleiotropy and selection.

The male pulse songs are species specific. Interspecific differences involve not only the components of the song but also the volume and quality of the sound produced (Ewing and Bennet-Clark, 1968; Chang and Miller, 1978). Sympatric species, especially sympatric sibling species, typically have songs that differ significantly in mean interpulse interval, e.g. *melanogaster* and *simulans*, *pseudoobscura* and *persimilis*, while some allopatric species pairs, whether closely related or not, may have quite similar songs, e.g. *ananassae* and *athabasca*, *pseudoobscura* and *ambigua* (Ewing and Bennet-Clark, 1968; Chang and Miller, 1978).

By exposing either singly or in combination normal, wingless, and aristaless individuals to artificial songs, it has been possible to estimate the functions of the various components of the songs. Bennet-Clark and Ewing (1969) varied the pulse length and interpulse interval (i.p.i.) of the *melanogaster* male's song and concluded that the mean i.p.i.'s constitute the prime species specific portion of the song that is discriminated by the females. von Schilcher (1976a) found that an artificial "normal" song stimulated both normal and wingless males to increase their locomotor activity and to engage in homosexual courtships. Males that were exposed for 60 seconds to the song retained their increased level of activity for at least five minutes after the song had ceased. Females exposed to the same songs decreased their activities. Further study (von Schilcher, 1976b) showed that if females were prestimulated by sine song before being introduced to males then their receptivity was increased, but not if they were prestimulated by pulse song. However, if females were mixed with "deaf" (aristaless) and "dumb" (wingless) males and then stimulated with pulse song, their receptivity was increased. Thus prestimulation by pulse song is not remembered over time but that of sine song is retained. He concluded that the sine song serves as a summated stimulus and pulse song as a species specific trigger mechanism.

B. Olfactory Stimuli

The overt behavior of numerous species of *Drosophila* suggests that olfactory stimuli in the form of pheromones are involved in mating behavior. Thus non-receptive females of many species, when being courted by a male that is displaying near the tip of her abdomen, extrude their ovipositors and expose the surrounding articulating membranes. Simultaneously the female directs her abdominal tip so that it is close to and directly in front of the male's face. Typically the male turns sharply away and some males immediately vigorously clean their faces and especially their antennae. It is known that the main, if not the sole, olfactory receptors of both sexes are located on the large third antennal segment (Kellogg et al., 1962).

The males of most of the more than 100 species of the Hawaiian picture-winged species group engage in lek behavior. The male of most species stands immobile on his lek, straightens his abdomen and slightly elevates the tip, then pulsates an abdominal droplet of clear fluid for prolonged periods of time. The males of some species also periodically curl the abdominal tip down against the substrate and deposit a droplet; some additionally curl the tip against the substrate and then walk slowly back and forth and deposit trails of thin liquid films. Two such abdomen dragging species, *D. crucigera* and *D. grimshawi*, when placed in clear containers, will produce many such liquid trails. These release species specific odors that can be identified and separated by human olfaction (Ringo, 1977a). Histological study of the sex pheromone gland in *D. grimshawi* revealed that the pheromone is released into a system of canaliculi emptying into the anal region; the extruded liquid flows along the surface of the intra-anal lobes and is brushed onto the substrate by finger-like projections on the lobes' surfaces (Hodosh et al., 1979). Males of Hawaiian species which do not produce anal liquid when on their leks are never immobile; rather, they engage in highly visible body and wing movements.

Species that produce anal droplets on their leks also produce similar droplets when courting the females. If courtship occurs at the rear of the female, then wing movements accompany droplet production. If courtship occurs in front of the female, then the male of some species (e.g., species of the *adiastola* subgroup) curls his abdomen up and forward until his abdominal tip is positioned just above the vertex of his head and in close proximity to the female's face and antennae, and then engages in droplet pulsation (Spieth, 1978).

Sturtevant (1915), using clean and "scented" vials, concluded that pheromones were involved in the courtships of *D. melanogaster* but could not distinguish whether one or both sexes produced the volatile substances.

Ewing and Manning (1963) were unable to confirm Sturtevant's findings, but Shorey and Bartell (1970), using an olfactometer, showed that a volatile substance(s) produced by *D. melanogaster* females would stimulate isolated groups of males to increase their readiness to engage in sexual behavior. Females of *D. melanogaster* produce minute amounts of a volatile sex pheromone, probably in the abdomen; the pheromone, isolated by gas chromatography from this species, stimulates courtship between two wild-type males of *D. melanogaster* or *D. simulans*, but not *D. hydei* (Hall and Greenspan, 1979).

Averhoff and Richardson (1974) studied mating rates in two series of inbred lines, one series derived from standard laboratory strains (several lines per strain) and the other derived from single wild-caught females. In both series, mating between lines was initially random in multiple choice tests, becoming increasingly disassortative until generation 8, after which the tendency to mate disassortatively declined somewhat. In the mating tests among inbred lines derived from the same laboratory strain, disassortative mating did not develop, but rates of courtship and copulation were reduced, and courtship was incomplete. Using an olfactometer similar to that of Shorey and Bartell (1970), they concluded that volatile pheromones are produced by both sexes, are required for the normal initiation of courtship, and are genetically variable. Individuals were non-responsive to intraline pheromones but responsive to pheromones of other genotypes. Further study (Averhoff and Richardson, 1975) indicated that the courting male produced both a volatile and a non-volatile pheromone.

The minority male mating advantage (see Section V) depends upon volatile olfactory cues. Male-produced lipids soluble in acetone or petroleum ether are recognized by females and control the frequency-dependent mating advantage (Ehrman, 1972a). Apparently the active compounds are methyl esters of straight-chain carboxylic acids (Ehrman and Probber, 1978).

C. TACTILE STIMULI

During courtship, species specific fixed action movements result in physical contacts between males and females. The structures most widely used for such movements are the male forelegs and mouthparts. Less frequently, and typically restricted to the members of a species group, is the involvement of the male's wings (Spieth, 1978), antennae (Narda, 1966), and the female mesothoracic legs (Spieth, 1969).

It is difficult to determine whether mechanostimuli or chemostimuli, or both, are transmitted between the individuals as a result of these physical contacts. Observations and circumstantial evidence provide some clues as to the nature of the stimuli transmitted. Following visual orientation of the male upon a potential mate, the first element of the courtship is foreleg tapping by the orienting male against the body of the other fly. Observations in the field where several species are assembled on a feeding, mating, ovipositional substrate indicate that males will attempt to court any drosophilid that has a gestalt similar to his. Usually by tapping any part of the body of the approached fly, the male is able to make interspecific and conspecific discriminations. Males of *D. virilis* who have had their foretarsi surgically removed will court females of closely related species that normal males of *virilis* refuse to court (Spieth, 1952). Males of *D. sulfurigaster* in the *nasuta* subgroup, by simply tapping a female, can determine if she has any sperm stored in her genital tract. Presence of sperm in the female causes the male to desist immediately, but a few days later after the female has exhausted her stored sperm the same male will vigorously court the same female (Spieth, 1969).

Under field conditions, if a male taps a foreign female, she immediately engages in rejection responses, but if he taps a highly receptive conspecific female she may immediately give an acceptance response. It is known that drosophilid flies have chemoreceptors on their foretarsi (Hertweck, 1931) and apparently the males receive chemostimuli by tapping, while the females have chemoreceptors and mechanoreceptors on the surface of their bodies that may respond to the males' tapping actions.

The courtship displays of most males are performed at the rear of the female. The male positions himself behind the female, with his face close to the tip of her abdomen. With rare exceptions, the male then uses his forelegs to vibrate, strike, or rub against a particular portion of the female's abdomen. Such fixed action patterns are species specific. Males of many species have species specific hypertrophied setae located on the foretarsi or tibiae that are brought into contact with the female's abdomen by the foreleg movements. Typically these sexual setae are long and flexible, but for some species they are sturdy, heavily sclerotized structures, e.g. the sex combs of the *melanogaster* and *obscura* species groups and the anterior dorsal tibial spine of some Hawaiian picture-winged species such as *D. varipennis*, *hamifera*, and *paenehamifera*. Males of these latter three species vigorously move their foretarsi and the distal half of the tibia to and fro against the venter of the female. This causes the large, hook-shaped tibial spine to rake the surface of the female's venter. Drosophilid females probably receive both mechano- and chemostimuli from the foreleg action of the courting male.

When displaying at the rear of the female, the male of most species extends his proboscis, opens the labellar lobes, and touches or grasps the tip of the female's abdomen. This action may consist of repeated stabbing type movements or continuous prolonged contact during which time the labellar lobes are rapidly pulsated. Some Hawaiian species grasp a portion of the female's abdomen. Males of the modified mouthparts species group use armored labellae to grasp firmly all or a portion of the female's genitalia, while members of the *orphnopeza* subgroup have normal mouthparts but grasp the tip of their female's ovipositor. In *D. crucigera*, a member of the *grimshawi* subgroup, the anterior portion of the labellar lobes is modified for grasping the anal papillae of the female.

Probably both sexes receive chemo- and mechanostimuli from these diverse types of mouthpart–body contacts.

D. Visual Stimuli

All species of *Drosophila* can and do court during the hours of daylight, and some also will court in darkness, though none will restrict its courtships to darkness. In the field the males typically find themselves in the presence of both conspecific and non-conspecific individuals. A male visually orients upon, approaches, and taps any individual that exhibits a gestalt that is similar to its own and does not differ from the male in body size by more than a factor of two. Interspecific variation exists in the processing of the visual stimuli that the males receive. Laboratory observations by various investigators have shown that males of *D. melanogaster* will approach and court etherized or dead females. *D. subobscura* males will orient upon a moving, ovoid wax ball (Milani, 1950), whereas *D. triauraria* males refuse to court females that have had their wings removed (Grossfield, 1968).

If female *Drosophila* are etherized and decapitated they then live for several days. Such individuals will assume a normal upright position but remain immobile (Spieth, 1966c). *D. melanogaster* and *D. simulans* males will court decapitated females, frequently at the anterior end of such females rather than at the rear. *D. pseudoobscura* and *D. subobscura* males only rarely court decapitated females and typically treat them the same as inanimate objects. Other species such as *D. triauraria* will court decapitated females only if they approach directly behind the female and thus are visually unable to determine that the female lacks certain normal features (Spieth, 1966c; Grossfield, 1970).

Investigators have used various mutations of *D. melanogaster* for assessing the role of visual stimuli in courtship (Reed and Reed, 1950; Rendel, 1951; Bastock, 1956; Jacobs, 1961; Geer and Green, 1962; Connolly et al., 1969; Grossfield, 1972; Burnett and Connolly, 1973). Such mutants can be divided into two classes: (1) those which have an increased amount of black pigment in the exoskeleton, and (2) those which have reduced amounts of red (pterin) or brown (ommochrome) pigments in the primary and secondary pigment cells of the compound eye. Both classes of mutants have reduced visual acuity and some (such as *ebony*, *white*, and *vermilion: brown*) have drastically reduced acuity. Under light conditions such males display reduced ability to orient upon and maintain contact with females. Further, this disability also results in inappropriate wing vibration by the courting males. Connolly *et al.* (1969), by using phenocopy techniques, found that the inappropriate wing vibrations of *vermilion:brown* mutants were not due to pleiotropic effects of the mutations but rather that normal wing vibration is dependent upon normal orientation.

Under laboratory conditions a majority of the species that have been studied are unable to mate in darkness (light dependent species), a few species mate effectively in darkness (light independent species) and an intermediate-sized group are depressed by darkness but can achieve some copulations (facultative species) (Philip *et al.*, 1944; Wallace and Dobzhansky, 1946; Spieth and Hsu, 1950; Grossfield, 1968, 1970, 1971). In darkness orientation cannot occur and the courtships of both light independent and facultative species occur either as a result of searching movements by the males or by accidental encounters. Males of both *D. simulans* (Manning, 1959a) and *D. melanogaster* (Crossley, 1970) search for females when in the dark. Crossley (1970) observed that wild-type *D. melanogaster* males and *ebony* mutants court under darkness in a similar manner. Males of *D. auraria* (=*D. auraria* race A of early authors) can mate in both light and darkness, but white-eyed mutants, who can perceive light but lack visual acuity, refuse to mate under light but readily mate in darkness (Grossfield, 1972). Grossfield suggests that "a switch in the relative importance of different sensory stimuli occurs under different environmental conditions" and is responsible for this difference in behavior of the flies.

The light dependent species apparently either remain immobile during darkness or have a key element of their courtship which is dependent upon visual stimuli. Hawaiian species in the field remain immobile during darkness (Spieth, unpublished observations) while the *D. subobscura* female performs a dance which must be visually observed by the male. *D. grimshawi* fails to orient in the dark; this also affects non-courtship signals (Ringo, 1977a).

IV. Ontogeny

Sexual behavior first appears after eclosion, at a time ranging from hours to days, depending on the species. In males, the quality and frequency of courtship behavior, the frequency of ancillary sexual activity such as aggression or territoriality, and the competence and attractiveness of these males to females may vary with age and, in some cases, with sexual experience. In females, receptivity, sexual attractiveness to conspecific or extraspecific courting males, and the nature of rejection signals given to courting males may vary with age and with age-related physiological conditions (e.g., virgin *vs.* nonvirgin).

A. Male Displays and Mating Propensity

Males do not begin to court immediately after eclosion, but rather after a species-specific delay posteclosion (Spieth, 1958; Strömnaes and Kvelland, 1962; Lefevre and Jonsson, 1962; Fowler, 1973). The frequency and intensity of courtship changes after this delay, although this process has not been investigated extensively. Spieth (1958) observed in *D. pseudoobscura* and *D. persimilis* that males begin to court sporadically at about 12 h of age, but cannot achieve intromission until 32 h of age. Brown (1964) found virtually no differences between the interdisplay transposition probabilities in the courtship of 1 day and 3 day old males of *D. pseudoobscura* with 3 day old females, yet the older males mated at three times the rate of younger males. Part of the older males' greater success was attributed to their greater courtship time, longer courtship bouts, and shorter breaks between bouts of courtship (Brown, 1964). *D. melanogaster* males (Canton-S strain) do not show wing vibration until about 20 h of age, after which the rate of wing vibration directed towards mature virgin females rapidly approaches a plateau of 40 h (Jallon and Hotta, 1979).

Recently, the visually mediated courtship tracking ability of males has been investigated in wild-type and mutant strains of *D. melanogaster* (Cook, 1979, 1980). These studies uncovered strain-specific differences in the ability of males to track moving females and indicate that vision is required for accurate tracking and delivery of appropriately oriented wing vibration but is not required for discrimination between the female's head and her abdomen.

Relative mating success of males in *D. melanogaster* is age-dependent (Long *et al.*, 1980), as it is in *D. pseudoobscura* and *D. persimilis* (Spiess, 1970). Mating success increases from zero to a plateau value, then declines a few days after the plateau has been reached. The number of females inseminated per day in *D. melanogaster* shows the same pattern of development—increase, plateau, and decline (Kvelland, 1965). Long *et al.* (1980) found that both the duration of copulation and fecundity vary with age in the same pattern as competitive mating ability. They suggest that females may "prefer" to mate with older, and therefore more fecund, males. The age-dependent fecundity of males is related to the number of sperm available and, a close correlate of this, the amount of paragonial gland secretion in storage. The presence of motile sperm in male *Drosophila* is not necessary for mating to occur, however (Bridges, 1916; Smith, 1956; Lefevre and Jonsson, 1962).

Male mating experience, as well as age, affects male mating success. Virgin males have a significant mating advantage over nonvirgin males of the same strain and age in *D. melanogaster* (Markow *et al.*, 1978). These workers found that virgin males, in addition to being more successful at mating than nonvirgins, have higher fecundity, thus demonstrating another link between fecundity and mating success that might be explained by female preference.

In addition to courtship and copulation, males of some species show other types of sexual behavior: agonistic displays (aggression, defense, and submission), sexual advertisement, territoriality, and communal displays (Spieth, 1951, 1966b, 1968, 1974a,b; Dow and von Schilcher, 1975; Ringo, 1976; Jacobs, 1978). Ringo (1978) studied the ontogeny of sexual behavior and some nonsexual behavior in *D. grimshawi*, a lek species, from eclosion to 4 weeks posteclosion and found three patterns of development. Four lek behaviors (courtship, sexual advertisement, contact aggression, and a communal display) were virtually absent until sexual maturity (~ 2 weeks); their frequencies rose to relative maxima 1 week later. Nonsexual behaviors were less frequent 1 day posteclosion than at greater ages, and did not vary regularly with age thereafter. The frequency of noncontact aggression remained constant throughout the 4 week period. Jacobs (1978) observed *D. melanogaster* males in a 48 liter container providing a complex environment (differential lighting, food present, moving air); he found that some males had a strong tendency to persist at and defend a restricted area about 8 cm in diameter, usually on a food dish. This "evanescent and brief" territoriality (sometimes lasting several hours) was correlated with courtship; the developmental speed of both behaviors was accelerated in *ebony* and dark flies (a phenotype due to many genes on chromosome 3) which have a higher concentration of beta alanine in the hemocoel (Jacobs, 1978).

B. Female Behavior

The ontogeny of sexual receptivity and accompanying changes in rejection signals in females have been studied extensively. Sexual receptivity in virgin females is switched on at a species-specific time, which in *D. melanogaster* occurs about 48 h posteclosion (Manning, 1967b). Sexually immature females use several stereotyped actions to reject and repel males (see Section II). The switch-on of receptivity apparently involves lowering the acceptance threshold (Cook, 1973a,b). Receptivity declines in old virgins, beginning 8 days after eclosion in *D. melanogaster* (Manning, 1967b).

The switch-on of receptivity is correlated with ovarian development, which has led to the hypothesis that distension of the ovaries causes lowered acceptance threshold via neural feedback to the brain. This hypothesis has been ruled out for some species because females congenitally lacking ovaries or having underdeveloped ovaries become receptive at the usual time (Smith, 1956; Manning, 1967b). A second hypothesis is that juvenile hormone (JH), which has gonadotrophic effects in insects generally (Schneiderman, 1972), acts on the brain directly or through an intermediate neurosecretory factor to increase receptivity. Implantation of corpora allata in the pupal stage caused *D. melanogaster* females to be sexually precocious and to have faster ovarian growth within 24 h of emergence (Manning, 1966), supporting the idea that JH or a factor triggered by JH enhances or switches on receptivity. Bouletreau-Merle (1973) confirmed this hypothesis by applying JH topically to *D. melanogaster* females; she observed the precocious appearance of oocytes and early switch-on of sexual receptivity. Similarly, Ringo and Pratt (1978) induced early switch-on of sexual receptivity in *D. grimshawi* females with a topically applied JH analogue. In these experiments, certain stereotyped displays were also stimulated by the hormone analogue; oddly, both aging and the JH treatment "masculinized" the females' behavior (Ringo and Pratt, 1978; Ringo, 1978).

Females become unreceptive after copulation. Three factors inhibit receptivity: (1) the stimuli of copulation, (2) the presence of sperm in the sperm receptacles, and (3) a secretion produced in the male's paragonial gland. Copulation itself is thought to inhibit receptivity because sterile matings in *D. subobscura* are not followed by further mating for 4–24 h (Smith, 1956) and because females remate within 24 h of a sterile mating in *D. melanogaster* (Manning, 1962a,b). Of course, this "copulation effect" may be caused not by copulation itself but by a relatively shortlived (<24 h) event following copulation, such as the insemination reaction (Patterson, 1947; Patterson and Stone, 1952). The insemination reaction, which rapidly follows intra- and interspecific mating in *Drosophila* and disappears within about 12 h of intraspecific copulae, does not depend on sperm (Fowler, 1973). The presence of sperm in the sperm receptacles inhibits mating in *D. melanogaster* (Manning, 1962b, 1967b) and *D. subobscura* (Smith, 1956). Virgin *D. melanogaster* females with implanted paragonial glands show reduced receptivity, which is completely restored after 4 days, compared with fertilized females of the same strain, in which receptivity is not restored to initial levels for 9 days (Burnet *et al.*, 1973). Oviposition is also increased by the implant (Burnet and Connolly, 1974). The reduction in receptivity of *D. melanogaster* females is caused by a ninhydrin-positive (amine-containing) galactoside in paragonial secretions (Chen et al., 1977). In *D. funebris*, two ninhydrin-positive substances have been isolated from paragonia, PS-1 and PS-2; PS-1 is a peptide that lowers receptivity and PS-2 is a glycine derivative that stimulates oviposition (Baumann, 1974).

When females refuse to mate, they give several rejection signals, including fending with a midleg, kicking with the hind legs, extrusion of the vaginal plates, flicking the wings, and curling the abdomen down or to the side. In *D. melanogaster* females, 1 day old virgins reject mostly by flicking, 3–4 day old virgins reject mainly by kicking, and 4 day old fertilized females reject almost exclusively by extruding (Connolly and Cook, 1973). Virgin females with implanted paragonial glands use extrusion more than kicking and fending, so that they resemble fertilized females more than intact virgins (Burnet et al., 1973).

Remating is common in females of many species of *Drosophila* under both field and laboratory conditions (Lefevre and Jonsson, 1962; Dobzhansky and Pavlovsky, 1967; Fuerst et al., 1973; Anderson, 1974; Milkman and Zeitler, 1974; Richmond and Ehrman, 1974). Remating before the sperm supply is depleted is interesting because the sperm from the second male actively displace the sperm from the first mating; the strength and pattern of sperm displacement is strain specific (Prout and Bundgaard, 1977; Pyle and Gromko, 1978). Why do some females remate within 24 h while most wait several days to remate? Answers to this question may provide insight into mechanisms of female acceptance, generally. A correlate of the speed of remating in *D. melanogaster* is the esterase-6 genotype of the first male mate (Richmond et al., 1980). Pyle and Gromko (1981) have selected a strain of *D. melanogaster* having a decreased time to remating.

Remating may be related to the fact that oviposition is increased by paragonial secretions (Merle, 1969; Burnet et al., 1973). Consistent with this is the finding of Chiang and Hodson (1950) that there is a positive, linear relationship between fecundity of females and the number of days they were exposed to males. Receptivity and fecundity are influenced by the pars intercerebralis; destroying this portion of the brain of virgin or just-mated females inhibits both (Bouletreau-Merle, 1976).

Sexual attractiveness or "sex appeal" of a fly can be measured by the amount of courtship directed towards it by a sexually mature male. Both males and females have the same amount of sex appeal on the first day of eclosion, but males rapidly lose it by day 2 (Jallon and Hotta, 1979). Female sex appeal in *D. melanogaster* depends on pheromones that females possess in minute amounts, and which is effective over a distance of a few millimeters (Hall and Greenspan, 1979). Rejection signals given by a female affect her "sex appeal" (Cook, 1973a). Dietary factors and time since insemination also affect the amount of courtship directed towards a female

(Cook and Cook, 1975); the former may relate to ovarian action and the latter to a male pheromone in the ejaculate (Jallon et al., 1981).

C. Fate Mapping

Behavioral development can be studied at the level of its embryonic determinants, using Sturtevant's fate mapping technique (Hotta and Benzer, 1972). A fate map, or 2-dimensional representation of blastoderm sites that are fated to become larval or adult tissues, can be constructed by examining gynandromorphs, one of whose X chromosomes may carry one or more mutations under study and one or more marker mutations (e.g. yellow, Acid phosphatase-1^{null}). To date, *D. melanogaster* is the only species of *Drosophila* whose behavior has been fate mapped.

Male orientation and vibration, the first two elements of courtship, are closely correlated phenotypically and have been mapped to the head using cuticular markers (Cook, 1978; Hotta and Benzer, 1976; Nissani, 1977). Hall (1977) extended these findings to internal tissues by using a histologically identifiable enzyme variant: orientation and vibration are controlled by a portion of the dorsal brain. Orienting and wing extension are domineering over not orienting and not extending a wing, which is to say that only one side of the brain need be male in order for the gynandromorph to engage in these courtship elements. In a more detailed analysis, Hall (1979) localized the domineering focus for tapping, orientation, and wing extension to three or four small sites within a posterior part of the dorsal brain (Fig. 4). The courtship song, on the other hand, maps to a domineering focus within the mesothoracic ganglion (Schilcher and Hall, 1979); courtship song does not occur unless the wing extension focus is haplo-X but the song and its pattern are controlled within the mesothoracic ganglion. The focus for licking is in the dorsal brain, and is incompletely submissive, i.e. in gynandromorphs that lick this portion of the brain is usually haplo-X bilaterally but is sometimes haplo-X only on one side. Attempted copulation, which occurs only if the focus for orientation is male, maps to the thoracic ganglion, but not to a sharply defined area, as is the case with the other courtship elements (Hall, 1979).

Female receptivity fate maps to the head (Hotta and Benzer, 1976) and is submissive to non-receptivity (Cook, 1978). Oviposition maps to the head, close to the focus for receptivity, and is domineering to not ovipositing (Cook, 1978).

The foci of female releasers have also been fate mapped. Nissani (1977) mapped the pheromone producing foci to the tergites, sternites, and thoracic integument, in descending order of importance. In contrast, Hotta and Benzer (1976) and Hall (1977) mapped these foci to the abdomen. In

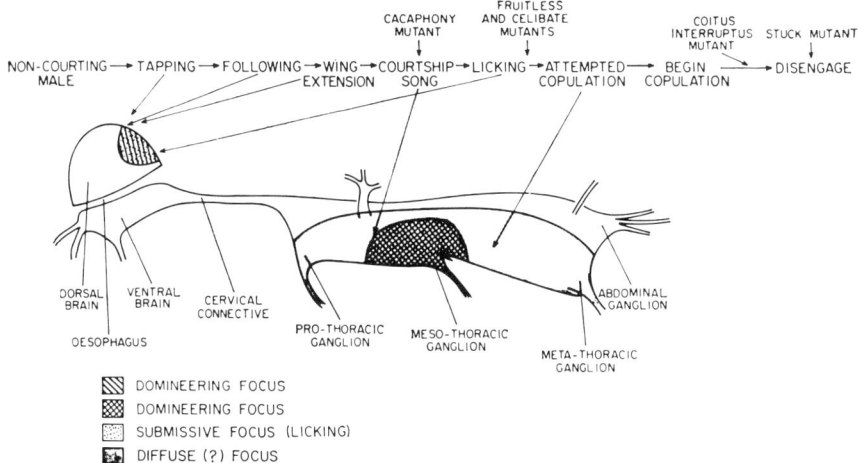

FIG. 4. Schematic representation of the courtship pathway in *D. melanogaster*, analysed with genetic variants. The parts of the pathway "controlled" by different parts of the central nervous system are shown by arrows from the behavioral step to a part of the CNS shown here in sagittal section (and the various portions of this section through an adult are shown across the bottom of the figure). (After Hall, 1979).

very careful experiments, Jallon and Hotta (1979) mapped them to the ventroposterior region of the blastoderm fate map, and suggested that the fat bodies may be the site of female pheromone synthesis.

D. EXPERIENCE AND MATING PREFERENCES

The housing conditions of adults of *D. melanogaster* affect mating frequencies (Ellis and Kessler, 1975). Given the choice between isolated or group-housed males, females tend to mate with males with similar housing experiences as their own, whereas both isolated and group-housed males mate more frequently with singly-housed females.

Pruzan and Ehrman (1974) showed that 11-day-old virgin *D. pseudoobscura* Arrowhead (AR) females mate at a higher rate with AR males than with standard (ST) males homozygous for orange (*or*); however, AR females that had mated with *or* ST males at 3 days of age mated more frequently with *or* ST males than with AR males at 11 days of age. AR females who were treated with cyclohexamide after mating with *or* ST males at 3 days of age mated more frequently with AR than with *or* ST males at 11 days of age (Pruzan *et al.*, 1977). It appeared as though the cyclohexamide-treated flies had forgotten the mating experience.

The strength of sexual isolation between sibling species or semispecies appears to be influenced by early sexual experience. Sexual isolation was increased following intraspecific mating between *D. persimilis* and *D. pseudoobscura* (Mayr and Dobzhansky, 1945; Spieth, 1958) and among the semispecies of *D. paulistorum* (O'Hara *et al.*, 1976). Sexual isolation between *D. melanogaster* and *D. simulans* was decreased after interspecific mating experience (LeMoli and Mainardi, 1972) and after passage through bispecific cultures (Eoff, 1973). The experiments cited here need to be replicated and followed up to confirm that learning of this type can take place. Experience in mixed cultures appears not to affect sexual isolation between *D. pallidosa* and *D. ananassae* (Plunka and Potter, 1977).

Individual sexually mature males of *D. melanogaster* substantially decrease rates of courtship directed towards virgin females after exposure to recently inseminated females (Siegel and Hall, 1979). Such courtship inhibition by females changes little in 2 h but decays in about 3 h. Males hemizygous for amnesiac, a sex-linked mutation selected as memory-deficient in a learning test unrelated to courtship (Quinn *et al.*, 1979), show the same conditioning effect as wild-type males, but this inhibition of courtship persists for only about 1 h (Siegel and Hall, 1979), suggesting a physiological connection between courtship inhibition by mated females and performance on an avoidance task.

V. Measurement of Sexual Activity

Drosophila is proving very useful in evolutionary and population biology for studying deviations from random mating (Petit and Ehrman, 1969; Spiess, 1970; Ehrman and Parsons, 1976). Deviations from random mating between types or populations within species can be partitioned into three orthogonal components: relative mating propensity (success) in males; relative mating propensity in females; and assortative mating. Mating between species, if it occurs, is analysed into the same three components, except that the term "assortative mating" is replaced by "sexual isolation". These three components of non-random mating can be defined and measured directly in an experiment where two types (e.g., genotypes, strains, races, or species) have an opportunity to mate. It is also possible to study the behavioral mechanisms underlying nonrandom mating by measuring attributes of courtship and mating in further experiments. The following discussion considers some widely used experimental designs for measuring rates of mating, statistical analysis of data from mating experiments, and the study of mechanisms underlying nonrandom mating.

A. Some Commonly Used Experimental Designs

Generally, virgin females and males of one or more types are confined together for a period of time. The cumulative number of matings over that period are either counted by direct observation or are determined after the experiment by dissection of the female's sperm receptacles or by observing each female's progeny. The container in which the flies are confined is usually small (1–1000 cm^3 in volume) and may be designed for easy observation of the flies (Elens and Wattiaux, 1964).

Typically, the object of a mating study is to compare rates of mating within and between pairs of types. For the sake of discussion, call the two types 1 and 2. In *no choice* experiments, type 1 males and females are placed together in one experimental unit (e.g., mating chamber), type 2 males and females are placed together in a second, and the two reciprocal combinations of 1 and 2 are placed in a third and a fourth. As few flies as a single pair may be confined in one experimental unit. This design is called "*no choice*" because only one type of each sex is present, precluding a choice between two types for either males or females. In *female choice* experiments a female of one type may accept the courtship overtures of either of two types of male: types 1 and 2 males are placed with type 1 females in one experimental unit and with type 2 females in another. In *male choice* experiments males of one type have the option of courting and attempting copulation with two types of female: type 1 and 2 females are confined with type 1 males in one experimental unit and with type 2 males in another. The minimum number of flies per experimental unit in both *female choice* and *male choice* experiments is three, one of one sex and two of the other. In *multiple choice* experiments, types 1 and 2 females are placed with types 1 and 2 males in one experimental unit. This design is called "*multiple choice*" because both males and females have the option of courting or accepting two types of the opposite sex, respectively. *No choice* and *male choice* designs do not require direct observation, whereas the other two designs may since paternity usually cannot be established conveniently after the period of confinement.

The use of the term "choice" in the names of these four basic experimental designs has been criticized on the grounds that it is anthropomorphic and may mislead the unwary reader. First, females are only "free to choose" to copulate or not to copulate with a given male (Merrell, 1950). Second, males are only "free to choose" to court or not to court a given female. Third, males apparently court indiscriminately within most species (Spieth, 1974a) and in some instances between species (Streisinger, 1948). Fourth, and most importantly, both male and female *Drosophila* may be incapable of choosing between alternative mates in the usual sense of the word. "Choice" implies a comparison between two sets of stimuli (in this case, potential mates) followed by some reaction such as approach towards one set of stimuli. There is no evidence that a male or female *Drosophila* with the exception of certain Hawaiian lek species does or even can compare two sets of sexual stimuli emanating from two directions and then react appropriately to the "more attractive" or "less noxious" of the two sets of stimuli (e.g., initiate courtship or approach and give an acceptance signal). However, "choice" is entrenched in the literature and no better term seems available. No harm will be done by talking about choice mating experiments, so long as care is taken to recognize the correct biological implications of these experiments.

The four experimental designs may or may not affect mating propensity or assortative mating in different ways, but any such effects cannot be predicted. In the few studies employing more than one design on the same material (types) the effects of design have given mixed results. Merrell (1954) crossed *D. pseudoobscura* and *D. persimilis* in *male choice*, *female choice*, and *multiple choice* experiments. Mating propensities were unaffected by design, but assortative mating was more marked in the latter two designs. Malogolowkin-Cohen *et al.* (1965) crossed six strains of *D. paulistorum* in *male choice* and *multiple choice* experiments. The isolation coefficients were comparable using the two designs.

Mating rates are commonly estimated by using equal numbers of each type in the mating experiment. However, the types may be mixed in different ratios to study the effects of relative frequency on mating rates. There is now abundant evidence that nonrandom mating sometimes depends on the relative frequencies of the types placed together. In general, frequency-dependent mating occurs wherever any of the three components of nonrandomness—female mating propensity, male mating propensity, or assortative mating—depends on the ratio of the types. However, attention has been focused on frequency-dependent mating propensities (Ehrman and Probber, 1978; Petit and Ehrman, 1969; Spiess, 1968a). When one type of male shows a greater mating advantage, the lower its relative frequency, this type is often said to have a "rare male mating advantage" or "minority male mating advantage" (Ehrman and Probber, 1978). Rare male mating advantage has been found in seven species of *Drosophila* and several other species of insects (Ayala and Campbell, 1974; Ehrman and Probber, 1978).

Frequency-dependent mating propensity, particularly in the sex having the greater variance in mating success (males, in *Drosophila*), is interesting because it provides a plausible mechanism for maintaining genetic variation in populations. Stable genetic polymorphisms can be maintained via frequency-dependent sexual selection without heterosis and its attendant genetic load (Anderson, 1969; Lewontin, 1974).

In the absence of prior knowledge about the effects of experimental design of mating behavior, the best design is the one that imitates nature most closely. Therefore, the multiple choice design would usually be preferable to the other three. It may be worthwhile to measure sexual behavior under a variety of naturalistic conditions. For example, the complex environment of an insectarium (Jacobs, 1978) might be used instead of the standard dry, plastic container.

In the laboratory the normal rearing techniques and the protocol usually followed for experiments on courtship perturb the normal ontogeny of the flies. Laboratory flies to be used for courtship experiments are typically anesthetized, segregated by sex, and allowed to mature under relatively uniform conditions, which includes an abundance of constantly available food. Usually a number of each sex is kept together, often until they reach approximately the peak of their sexual drive. Such a procedure deprives them of any opportunity to become habituated to the sexual stimuli not only of the opposite sex of their own species but also of other species that they may have been exposed to in a normal habitat. Such perturbations may contribute to the ease with which numerous investigators have observed copulations between species that rarely or never interbreed in nature, even when sympatric (Spieth, 1952).

B. Measures of Nonrandom Mating

Several methods of analysing data from mating experiments have been proposed. In this section we discuss many of these methods, including their formulation, use, strengths, and weaknesses.

The first mating index was apparently the Charles-Stalker *isolation index* (Stalker, 1942). It was designed to measure the degree of sexual isolation between two types, or the difference between the proportions of heterogamic and homogamic mating in *male choice* experiments. To discuss this and other mating indices, we follow Levene's (1949) very clear notation throughout. In a *male choice* experiment, place type 1 males with n_{11} females of type 1 and n_{12} females of type 2. After a period of time record the number of inseminated type 1 females, x_{11}, and the number of inseminated type 2 females, x_{12}. Compute the proportions of inseminated females of the two types, $p_{11} = x_{11}/n_{11}$ and $p_{12} = x_{12}/n_{12}$. The isolation index is $b_{12} = (p_{11} - p_{12})/(p_{11} + p_{12})$; $-1 \leq b_{12} \leq +1$. When equal numbers of type 1 and type 2 females are confined together, one may simply use the raw numbers of inseminated females, x_{ij}, instead of the proportions, p_{ij}. Deviations of b_{12} from zero measure the difference in rates of homogamic and heterogamic mating. Since b_{12} is subject to sampling error, one must perform a statistical test to determine whether deviations of b_{12} from zero are due to random chance. Three methods are a simple chi-square test (Levene,

1949), a test based on the normal variate (Woolf, 1968; Kaneshiro, 1976), and a modified t test (Sokal and Rohlf, 1981). For the reciprocal experiment, where type 2 males are used, we have n_{21} type 1 females and n_{22} type 2 females. Similarly, the reciprocal isolation index is $b_{21} = (p_{22} - p_{21})/(p_{22} + p_{21})$. The Charles-Stalker index was not designed for the other three experimental designs. Furthermore, it is a poor measure of assortative mating or sexual isolation as it can be strongly affected by mating propensity.

Bateman (1949) suggested averaging each pair of isolation indices to get a *joint isolation index*. Supposedly, this measures the mean level of sexual isolation. The formula is simply $b_{1 \text{ and } 2} = 1/2(b_{12} + b_{21})$. In the same paper, Bateman introduced a measure of the relative mating propensity of females or "heat." His index, which applies to *male choice* experiments, is $a_{12} = 1/2(b_{12} - b_{21})$. Levene (1949) explains why a_{12} is a poor measure of mating propensity and why both of Bateman's indices have several drawbacks.

Levene (1949) devised a *coefficient of isolation* to measure sexual isolation in a *male choice* experiment. When type 1 males are placed with types 1 and 2 females, the proportions of females still uninseminated when the experiment is terminated are q_{11} and q_{12} for type 1 and type 2 females, respectively. The coefficient of isolation for type 1 males is $K_{12} = (\log q_{11} - \log q_{12})/(\log q_{11} + \log q_{12})$. This coefficient is $+1$ for complete sexual isolation, 0 for random mating, and -1 if mating is completely heterogamic. The coefficient is much less sensitive to the proportion of females mating than the Charles-Stalker isolation index, and the standard error of the coefficient is known. Levene (1949) also gives a coefficient of *joint isolation*, the mean of two coefficients from reciprocal experiments, and a *coefficient of excess insemination*, a measure of "undirectional gene flow," along with the standard errors of these coefficients. None of these coefficients clearly measures assortative mating or mating propensities.

Merrell (1950) proposed mating indices applicable to all four designs—*no choice*, *female choice*, *male choice*, and *multiple choice*—where equal numbers of the types are used. Again, using Levene's (1949) notation, let x_{ij} be the number of matings between type i males and type j females. Merrell's measure of sexual isolation, the *isolation estimate*, is $I = (x_{12} + x_{21})/(x_{11} + x_{22})$, or the number of heterogamic matings divided by the number of homogamic matings. The index may range from zero (no heterogamic mating) to infinity (no homogamic mating); a value of 1 indicates equal rates of homogamic and heterogamic mating. His measure of relative female mating propensity or female mating ratio is $M_f = (x_{11} + x_{21})/(x_{21} + x_{22})$, or just the ratio of matings by type 1 females to those by type 2 females. Similarly, the male mating ratio is $M_m = (x_{11} + x_{12})/(x_{21} + x_{22})$. Each of the mating ratios may range between zero and infinity. Merrell (1950) recommends testing the significance of each ratio by simple chi-square tests in which the numerator and denominator are expected to be equal. The test for the equality of two percentages suggested by Sokal and Rholf (1981) is useful for power determinations. If the numbers of each sex and type (n_{ij}) are unequal, then Merrell's ratios can be computed with the mating proportions, p_{ij}, instead of the numbers of matings, x_{ij}. This is undesirable, though, as the ratios may then be biased; in this case significance testing may be carried out for a 2×2 table on the p scale as suggested by Snedecor and Cochran (1967). Unfortunately, the isolation estimate measures both assortative mating (or sexual isolation) and mating propensity. All of Merrell's measures are reasonably insensitive to the overall proportion of mated flies.

Sexual isolation can be measured by the isolation index of Malogolowkin-Cohen et al. (1965). Letting $\Sigma x_{ij} = N$, the formula for this isolation index is $I = (x_{11} + x_{22} - x_{12} - x_{21})/N$. Malogolowkin-Cohen et al. designed the index for multiple choice experiments; they give the standard deviation of I for significance testing.

Ehrman and Petit (1968) use indices of assortative mating and mating propensities developed by H. Levene. Assortative mating is measured by:

$$Z_I = \sqrt{\left(\frac{x_{11} \cdot x_{22}}{x_{12} \cdot x_{21}}\right)}.$$

Male and female mating propensities are measured, respectively, by:

$$Z_m = \sqrt{\left(\frac{x_{11} \cdot x_{12}}{x_{21} \cdot x_{22}}\right)} \text{ and } Z_f = \sqrt{\left(\frac{x_{11} \cdot x_{21}}{x_{12} \cdot x_{22}}\right)}.$$

The standard error of Z_I is approximated in large samples by:

$$\sigma(Z_I) = \tfrac{1}{2} Z_I \left(\frac{1}{x_{11}} + \frac{1}{x_{12}} + \frac{1}{x_{21}} + \frac{1}{x_{22}}\right),$$

and the standard errors of Z_m and Z_f are estimated similarly. Each index ranges between zero and infinity; a value of one indicates random mating for that index. This range is an undesirable property that can be easily remedied by a simple transformation, $Y = (Z-1)/(Z+1)$. The standard error of Y is approximated by

$$\sigma(Y) = \tfrac{1}{4}(1 - Y^2)\left(\frac{1}{x_{11}} + \frac{1}{x_{12}} + \frac{1}{x_{21}} + \frac{1}{x_{22}}\right)^{\tfrac{1}{2}}.$$

Petit and Ehrman (1969) introduced a *coefficient of mating success* to measure relative male mating propensity. Actually, it is a version of Merrell's male mating ratio, applicable to any ratio of types in an experiment or population (i.e., one of the types could be rare). If m_1=the number or proportion of type 1 males and m_2=the number or proportion of type 2 males, the coefficient of mating success is $K=m_1(x_{11}+x_{12})/m_2(x_{21}+x_{22})$. The relationship between K and Merrell's male mating ratio, M_m, is $K=m_1 M_m/m_2$. Although Petit and Ehrman do not say so, Merrell's female mating ratio could be similarly modified to account for unequal numbers of the two types.

Pruzan (1976) used a logarithmic odds ratio suggested by H. Levene to measure relative mating success of males in a *female choice* experiment. The index is $ln(x_{11}/x_{21})$, or the logarithm of the ratio of homogametic to heterogametic matings when equal numbers of the two types of male are present ($n_{11}/n_{21}=1$). Otherwise the index is $ln(x_{11}/x_{21})-ln(n_{11}/n_{21})$. The standard error of the index is $\sqrt{(1/x_{11}+1/x_{21})}$.

Except for Levene's "Z" indexes (Ehrman and Petit, 1968), all of the measures of assortative mating in the preceding discussion share a serious flaw: they confound assortative mating with male and female mating propensities. In other words, if the two types of male or the two types of female mate at different rates, then these measures may be large and statistically significant *even if the two types mate indiscriminately (nonassortatively)*. Schaffer (1968) was the first to develop a measure of true assortative mating. His *discrimination index* is based on an elegant mathematical model but cannot be calculated with a simple formula.

A simple alternative approach is to measure row and column effects as fractions and assortative mating with an index developed by Yule (1912). Measure the relative mating propensity of type 1 males, P_{m1}, with $\hat{P}_{m1}=(x_{11}+x_{12})/N$; similarly $\hat{P}_{m2}=(x_{21}+x_{22})/N$ for type 2 males. For types 1 and 2 females estimate the mating propensities by, respectively, $\hat{P}_{f1}=(x_{11}+x_{21})/N$ and $\hat{P}_{f2}=(x_{12}+x_{22})/N$. Measure assortative mating, Y, with $\hat{Y}=[(x_{11}x_{22})^{\frac{1}{2}}-(x_{12}x_{21})^{\frac{1}{2}}]/[(x_{11}x_{22})^{\frac{1}{2}}+(x_{12}x_{21})^{\frac{1}{2}}]$; recall that Y and Levene's Z are related in a simple way, $Y=(Z-1)/(Z+1)$. Significance testing for the two kinds of mating propensity and for assortative mating can be carried out with simple chi-square or G tests for row (male), column (female), or interaction (assortative mating) effects in the 2×2 table of mating data (Ehrman and Parsons, 1980; Sokal and Rohlf, 1981). The standard deviation of $\hat{Y}$, for large samples, is:

$$S_{\hat{Y}} = \tfrac{1}{4}(1-\hat{Y}^2)\left(\frac{1}{x_{11}}+\frac{1}{x_{12}}+\frac{1}{x_{21}}+\frac{1}{x_{22}}\right)^{\frac{1}{2}};$$

confidence intervals are obtained in the usual way (Sokal and Rohlf, 1981). If one of the cells of the data matrix is zero, the table should be smoothed before analysis (Bishop *et al.*, 1975), or else more data should be collected.

The relative merits of our indexes and those of Levene (Ehrman and Petit, 1968) need to be evaluated by a biometrician.

Many approaches have been used to measure frequency-dependent mating propensity (Ehrman and Petit, 1968; Petit and Ehrman, 1969; Spiess and Spiess, 1969; Ayala, 1972; Pruzan, 1976; Pot et al., 1980). Unfortunately, many workers have used biased indexes (Goux and Anxolabehere, 1980). A good, unbiased method has been developed by Adams and Duncan (1978). Their analysis is rigorous, and easy to use and interpret. Adams and Duncan (1978) show how to estimate four parameters of their mating models: Ψ (fitness differential when the two types are equally frequent), β (change in fitness with varying input frequency, δ (mean fitness differential), and P_e (equilibrium frequency for the mixture).

C. Models of Mating Behavior

There have been a few attempts to construct models of mating behavior. These models have dealt by and large either with the genetic consequences of various deviations from random mating (e.g., O'Donald, 1973, 1980) or with mating rates with one or two types present under simplified conditions (Taylor, 1975; Kence and Bryant, 1978). In no instance have parameters of these models been based upon ethological mechanisms that have been deduced or hypothesized from empirical studies. In the case of the Kence–Bryant model, at least, this may lead to difficulties (Spiess, 1982). To construct models of mating rates based on realistic assumptions about premating behavior is a challenging, open problem for future research.

VI. Adaptiveness of Mating Behavior

The complex courtships performed by *Drosophila* species involve the expenditure of considerable energy and time. To acquire evidence as to why such an "expensive" biological behavior has evolved, it is necessary to study the flies both in the laboratory and in their native habitats. In this section we consider two ways that selection has shaped courtship in *Drosophila*: (1) factors in its life history that constrain the nature of courtship and (2) sexual selection. A third way that selection affects *Drosophila* courtship is selection against interspecific hybridization; this will be taken up in Section VIII.

A. Courtship and Life History

In the field courtships and mating occur during periods of time when the flies are assembled at the food sites. *Drosophila* adults selectively consume yeasts, bacteria and fluids from decomposing plant materials. Typical substrates are rotting flowers, fruits, leaves, slime fluxes, bark wounds, rotting bark, rotting portions of cacti, and fungi. A few specialized species feed and breed on living flowers. Such food sites share in common small size, discreteness, and periodicity of occurrence (Dobzhansky and Pavan, 1950; Dobzhansky and da Cunha, 1955; Pipkin, 1965; Carson, 1971). Most species are semi-opportunistic in their feeding behavior (Patterson, 1943; Dobzhansky and da Cunha, 1955; Carson, 1971).

Typically the flies visit the food sites for a period during both the morning and the afternoon. The light intensity, temperature and humidity of the area in which a food mass is located provide the stimuli that determine the time of day when the adults seek food and how long they remain on the feeding site. There are species specific responses to these combined stimuli but usually a considerable proportion of the individuals of the species of a given area are simultaneously present on any food mass. Individuals arriving at the food site immediately begin to feed and the nutritional demands related to the production of numerous yolk-rich ova dictate that the females devote the major portion of their visit to feeding. The female may, if she finds herself on an appropriate substrate, also engage in oviposition. No species is known to oviposit into a substrate upon which it does not feed but many species have monophagic larvae, some are oligophagic and only a few have polyphagic larvae. As soon as the female is satiated, she leaves the food site so there is constant coming and leaving by the females. A few species are exceptional and their adults may remain semipermanently on the food–ovipositional sites, e.g. *D. nigrospiracula*, a desert species, and *D. melanogaster* and *D. simulans* when feeding on large food masses which accumulate during the harvesting and processing of tomatoes, fruits and grapes.

The males feed for only short periods of time. Sperm production and metabolic maintenance exert low nutritional demands. After feeding, males do not leave the site but turn to courtship. They are promiscuous, orienting upon and approaching any nearby or passing individual that possesses the gestalt of a drosophilid and that is not more than approximately twice or less than half the size of the male himself. He ignores, avoids or flees from other organisms, such as other Diptera, wasps, ants, and beetles that are also present on the food site.

The majority of females at the feeding site are mature and have previously been inseminated. A second, smaller group consists of young immature virgins. A small percentage of the females comprise a third group consisting of (a) young females that have just reached sexual maturity, (b) mature females that have exhausted their sperm supply, and (c) mature females that contain sperm but are possible candidates for multiple insemination. A sexually active male cannot visually separate any of these various types of conspecific females, nor can he separate similar appearing individuals of closely related species from his conspecifics. The male therefore randomly approaches any potential mate. If he attempts to court a foreign female, foreign male, or conspecific male, he receives rejection signals. If he courts a non-receptive female she may flee, but typically she responds vigorously with rejection responses as she simultaneously continues to feed. Such encounters are of short duration and the male usually turns away after tapping and then seeks out another individual. If he taps a female who produces the appropriate chemical stimuli and who does not respond with repelling action, he will then engage in a full courtship. A female may eventually refuse the male's overtures even after he has courted her for a period of time, while only an occasional female will engage in copulation. As a result, during each feeding period the male attempts to court many individual females and each female is approached and tested by numerous males.

Young males will engage in courtship before they are competent to achieve copulation, and mature males will readily court young females before such individuals are willing to engage in copulation. Thus the young individuals of both sexes have an opportunity to be habituated to the sexual stimuli of both their conspecifics and some non-conspecifics who are simultaneously present on the feeding site.

Drosophilid mating, except for some Hawaiian species (see below), thus occurs "at the restaurant" (Labeyrie, 1978) and appears to possess several adaptive features: (a) males are able to determine the receptivity and species identity of any female with a small expenditure of time and energy; (b) since the male stands at the rear or less frequently at the front of the female and performs his courtship display, the female is able to repel the male with a small expenditure of energy and without interruption of her feeding activity; (c) the female, even when receptive, has the opportunity to sample several males' performance (Spiess and Schwer, 1978); (d) the immatures are exposed and habituated to the stimuli of both their conspecifics and foreign individuals before they are sexually competent.

In the ancestral courtship pattern of higher Diptera, the male orients upon a female, then approaches and lunges onto her. Non-receptive females vigorously react and eventually dislodge the male. The evolution of the complex and unique behavior of *Drosophila* appears to be an evolutionary response to diurnal feeding for short periods of time on small food masses. Labeyrie (1978) alternatively suggests that the presence of a number of *Drosophila* species simultaneously feeding on such sites contributed the selection pressure that caused the development of *Drosophila* courtship patterns. Interestingly, the genus *Scaptomyza*, which evolved out of *Drosophila*, has reverted to the primitive pattern, where the male lunges onto the female. Scaptomyzids are tolerant to relatively high light intensity and low humidity, show little or no diurnality, and spend prolonged periods of time on or near their food and ovipositional substrates. The female thus has adequate time to feed, even though she must periodically cease feeding in order to dislodge males that mount her by kicking, fluttering her wings and shaking her body.

Most, if not all, of the numerous Hawaiian species exhibit exceptional mating behavior. In the field when feeding on both natural substrates of baits, the flies of both sexes are quiescent and cryptic (Spieth, 1966b). The patterns of dark maculation, the absence of agonistic behavior, any quick movements, and lack of courtship activities all contribute to crypsis.

The Hawaiian males feed, then leave and establish lek territories in the nearby vegetation. These territories are small in size and species specific in location. A male will remain on his lek for prolonged periods of time, defend it against any intruders, including insects other than drosophilids. He also steadily engages in species specific advertising behavior which may involve (a) wing waving and walking alertly about, (b) remaining immobile while a droplet of liquid is pulsated from the tip of his abdomen, or (c) walking and intermittently wiping the tip of his abdomen against the substrate to deposit a thin film of liquid (Spieth, 1968). Both field and laboratory observations indicate that females are attracted to the leks. Predation, especially by insectivorous birds which are ecologically associated with the feeding substrate sites and actively seek them out, is probably responsible for the selection pressure which led to the evolution of the lek behavior (Spieth, 1970, 1974b).

In their natural habitats the adult individuals emerge from the pupal cases in early morning. By the time their exoskeleton is hardened the environmental stress from humidity, light and temperature results in the young flies moving into secluded spots where food and close proximity to other individuals are both absent. In the afternoon these young flies will move to the nearest food site for a period of feeding and encounters with individuals of various ages and species. On subsequent days these flies will then diurnally visit feeding areas while the remainder of their lives will be spent in secluded sites, under environmental conditions that differ substantially from those of the laboratory. This is a quite different scenario from that experienced by individuals bred under laboratory conditions.

B. Sexual Selection

The foregoing discussion has dealt with features of courtship that are coadaptive with other features of *Drosophila* life history. Courtship displays are also subject to natural selection against interspecific hybridization (see

Section VIII) and sexual selection (Spieth, 1974b). Sexual selection can be subdivided as follows (Bateman, 1948; Darwin, 1871; Wilson, 1975).

1. Intrasexual selection: competition within one sex for mates.
   a. Scramble, in which competitors interfere with each other only through random encounter.
   b. Contest, in which competitors engage in ritualized competition such as territorial defense (Hutchinson, 1978).
2. Intersexual selection: mate selection (in *Drosophila*, intraspecific mate selection appears to be carried out principally by females).
   a. The selected male trait correlates positively with other components of fitness such as survival. In this case, sexual selection might develop because of the correlation.
   b. The selected male trait does not correlate with another component of fitness, or even correlates negatively.

There is ample evidence that both intra- and intersexual selection occur in *Drosophila*, although the results of many studies do not allow analysis of sexual selection into components.

Intrasexual selection acts principally in males, usually occurring as scramble or competition for females without specific behavioral responses to other males (Spieth, 1974a; Wilson, 1975). Nonaggressive competition for mates includes rate of orientation, ability to maintain orientation on rapidly moving females, ability to distinguish virgin from fecundated females (in some species), persistence, and vigor of courtship (Spiess, 1970; Spieth, 1974a). The relative ability of males to compete non-aggressively for mates is sometimes reduced in inbred strains (Connolly *et al.*, 1974; Maynard-Smith, 1956), in mutant strains (Spiess, 1970; Dow, 1976; Sciandra and Bennett, 1976; Spiess and Schwer, 1978), and in chromosomally different types (Spiess, 1970). Intrasexual selection in the form of contest (agonistic behavior) occurs in males of *D. melanogaster* (Dow and von Schilcher, 1975; Jacobs, 1978; von Skrzipek *et al.*, 1979), some species of the *virilis* group (Spieth, 1951), and many species of Hawaiian *Drosophila* (Spieth, 1966b, 1968, 1974b; Hodosh *et al.*, 1979).

Intersexual selection by females is well-documented (Petit and Ehrman, 1969; Spiess, 1970). In many cases the selected male trait correlates with other components of fitness, such as fertility or survival of offspring. Maynard-Smith (1956) found that males of inbred strains in *D. subobscura* were less vigorous courters and less fertile than outcrossed males. Females selected against inbred males because they were less effective in delivering courtship stimuli. Prakash (1967) found a correlation between mating success and fertility in *D. robusta*. Virgin males of *D. melanogaster* were more successful than males that had just mated (Markow *et al.*, 1978); in this case females were apparently selecting males on the basis of attributes that were not measured in the experiments. The number of offspring per mating is lower for once-mated than for virgin males (Markow et al., 1978).

The minority mating advantage (see Section V) results in increasing heterosis and decreasing the probability of inbreeding in a population. Johnston and Heed (1976) found that cactus breeding *D. nigrospiracula* migrated in small numbers (6·83%) from one feeding–breeding site to another; the migrant females were vigorously courted by the "resident" males and migrant males immediately courted numerous females as soon as they reached a new food mass. When copulating pairs were collected, 20% of the females and 10% of the males were migrants, thus indicating a minority effect in both males and females.

A behavioral basis for minority male mating advantage has been established for one set of genotypes in *D. melanogaster*; females tended to avoid the first male to court (Spiess and Kruckeberg, 1980).

Females of some species "hedge their bets" by engaging in multiple inseminations (Dobzhansky and Pavlovsky, 1967; Anderson, 1974; Pyle and Gromko, 1978). Females of other species such as the widespread *D. sulfurigaster* become totally unattractive to the males as soon as they have copulated, and remain so during the following 8–10 days during which production and oviposition exhaust the sperm supply. The risk of death is considerable during this time period, and thus the selection of her first mate is critical for the mature virgin female. Correlated is the fact that virgin females of *D. sulfurigaster* and other species of the *nasuta* species subgroup mate much more selectively than do the virgins of *D. pseudoobscura* and *D. melanogaster*, which engage in multiple insemination (Spieth, 1969).

There are many cases in which courtship traits are uncorrelated with other components of fitness, or may be negatively correlated. The most striking examples of this occur in Hawaiian *Drosophila*, in which many species exhibit hypertrophied and even bizarre courtship rituals reminiscent of courtship in some birds and reptiles (Spieth, 1966b, 1968, 1974a,b; Carson et al., 1970). Many of these displays (e.g., wing waving, pheromone dispersal and ritualized tactile stimulation) appear to be unrelated to male fertility or survival rate of offspring, but rather are merely stimulating to the female.

Each of the above phenomena is predicted from theories of sexual selection: see Lande (1980) for a general quantitative genetic theory on sexual selection and sexual dimorphism. The most intense intrasexual selection should occur in males whereas females should be more selective than males. First, the parental investment (Trivers, 1972) of females is higher than that of males. The metabolic investment in yolk-rich eggs is much higher than in sperm. Females must spend hours each day feeding to support the metabolic drain of egg production, and must spend time selecting oviposition sites and laying eggs. Second, the rate of gamete production is much lower in females than in males, so that it is more critical for females to ensure the survival of each offspring. Third, the storage and use of sperm by females magnifies the effect of the male's "quality." This is because females store hundreds of sperm, many of which are used to fertilize eggs over a period of days before re-mating is likely to occur (Fowler, 1973; Gromko and Pyle, 1978; Lefevre and Jonsson, 1962). The longer the delay from first to second mating in females of a species the more selective should the females of that species be. Thus, female mating propensity should be positively correlated with rate of multiple mating. In contrast, males are capable of re-mating rapidly (Spieth, 1974a), which reduces the negative effect of mating with a female having low fertility or survival ability. Fourth, repeated mating enhances the fitness of males to a greater extent than that of females (Bateman, 1948; Prout and Bundgaard, 1977; Pyle and Gromko, 1978).

Some of the male traits selected by females should be either uncorrelated with other fitness characters or negatively correlated with them (Darwin, 1871; Fisher, 1958). A number of rigorous population genetic models have made this prediction (O'Donald, 1962, 1967, 1973). These models show that sexual selection can be effective even when the sexually selected trait confers lower survival value (i.e., is opposed by selection for survival).

VII. Phylogeny

Evolutionary studies of diverse groups of animals have clearly demonstrated that behavior can be used in the same manner as morphology to draw phylogenetic inferences. The genetically programmed, stereotyped displays that comprise *Drosophila* sexual behavior have been particularly instrumental in phylogenetic analysis (Spieth, 1968, 1974a). Qualitative and quantitative differences in courtship rituals between species can reflect ecological adaptations, sexual selection, and selection for sexual isolation between these species.

On the other hand, the manner in which behavior evolves can be analysed in comparative studies of species whose phylogeny has been established from cytological and morphological data (Spieth, 1966b, 1968). In some cases, comparative studies can be followed by genetic analyses, since interspecific hybridization between closely related species is often possible (Ewing, 1969; Spieth, 1969; Schilcher and Manning, 1975).

A. MICROEVOLUTIONARY AND MACROEVOLUTIONARY CHANGES

Typically, members of small species groups and species subgroups have qualitatively similar courtship patterns and interspecific differences are quantitative (i.e., a matter of degree). The microevolution of these small, quantitative interspecific differences often resembles changes within species due to artificial selection. Between species groups and subgenera major qualitative differences are evident. The macroevolutionary changes leading to these differences are presumably the result of microevolutionary changes acting for long periods of time in different lineages.

Manning (1965) and Ewing and Manning (1967) have reviewed the microevolution of *Drosophila* behavior and insect behavior, respectively. These authors suggest that displays evolve by gradual, quantitative changes in their speed, frequency, sequence, and form or emphasis, and that new displays are unlikely to arise *de novo*. Small quantitative evolutionary changes in a display occur through gradual shifts in allele frequencies of the many genes affecting the display.

The speed of a display may change in several ways: the duration of the entire display, the duration and spacing of elements within the display, and the response latency of the display may increase or decrease. *D. simulans* males have a greater lag before initiating courtship than do males of its sibling species, *D. melanogaster*; *D. simulans* males also orient continuously for longer periods (Manning, 1959a). Similarly, the duration of courtship bouts vary among the semispecies of *D. paulistorum* (Koref-Santibañez, 1972a). Many of the attributes of courtship song—interpulse interval, interburst interval, burst length, and song duration—vary among closely related species (Ewing and Bennet-Clark, 1968; Ewing, 1970, 1979). In the case of the *funebris* species group, the songs are very complex, and some of the interspecific differences are striking (Ewing, 1979). The duration of copulation tends to vary little within species groups, but there are notable exceptions, as in the *willistoni*, *melanogaster*, and *obscura* groups (Spieth, 1952).

The frequency of displays has been used by many investigators because it is relatively easy to record. Brown (1966) found many differences in the frequencies of courtship elements in species of the *obscura* group. The frequencies of displays in the semispecies of *D. paulistorum* (Koref-Santibañez, (1972a,b) and of species in the *mesophragmatica* group (Koref-Santibañez and del Solar, 1961; Koref-Santibañez, 1963) have been analysed in relation to sexual isolation. Spieth (1969) found that the six species of the *nasuta* subgroup of the *immigrans* group differ in the frequency of many male courtship displays. Interspecific hybrid males in some cases deliver courtship displays at frequencies intermediate between the parents, and in some cases at frequencies characteristic of one parent species. Some of the interspecific differences within the *nasuta* subgroup are qualitative. Another approach to the study of frequency of displays in closely related species of

Drosophila is to use multivariate analysis to compare species. This has the advantage of simultaneous consideration of all the displays being measured and their interrelationships (Ringo and Hodosh, 1978).

A third way that sexual behavior can evolve is through changes in sequence. It is natural to consider the serial arrangement of displays used during courtship, since the sequencing of courtship elements is so striking and since it may have signal values (Brown, 1964; Manning, 1959a). To analyse sequences of displays, one may construct simple kinematic diagrams (Sustare, 1978), in which the names of displays are connected by arrows to indicate the transitions that occur between them. The thickness of each arrow may be made proportional to the relative frequency of the transition. The sequences of closely related species can in this way be compared by visual inspection and followed by statistical tests of significance. Manning (1959a) described the courtship of *D. melanogaster* and its sibling *D. simulans* in this way. Brown (1964) analysed factors affecting mating success in *D. pseudoobscura* by considering sequences of courtship, and compared courtship sequences of species in the *obscura* group (Brown, 1966). Wood *et al.* (1980) analysed courtship sequences in *D. melanogaster*, *D. simulans*, and their hybrids, using statistical tests to compare the *conditional* frequencies of transitions between behavior (i.e., conditional upon the occurrence of the first behavior in a sequential pair) (Fig. 5). Unfortunately, the quantitative analysis of courtship sequences in *Drosophila* has not been used extensively, despite the potential value in evolutionary studies.

There are often shifts in the form or emphasis of courtship displays in *Drosophila* (Spieth, 1952, 1966a). For example, the wings may be moved from the resting position to varying degrees during wing vibration (Spieth,

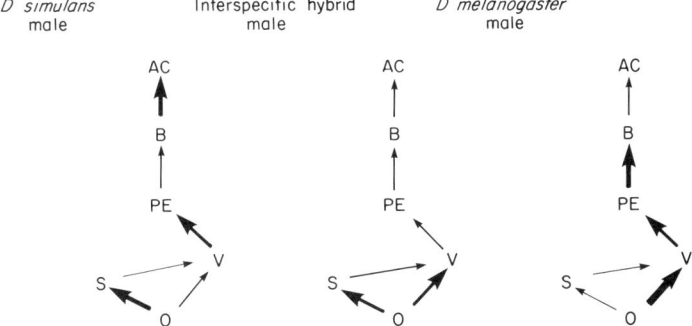

FIG. 5. Kinematic diagrams summarizing transitional frequencies. Relative widths of arrows indicate magnitude of transitional frequencies. O, orient; S, scissor; V, vibrate; PE, proboscis extend; B, bend; AC, attempt copulation. Only three pair matings are shown. Type of male in the pair mating is given; the female was *D. simulans* in each case. (After Wood *et al.*, 1980).

1952, 1969). In species whose males use their legs to stimulate the female's abdomen, as in many endemic Hawaiian species, the positioning of the legs and the direction and intensity of their movements vary among species (Spieth, 1966b). The position of the courting male relative to the female when certain displays (e.g., wing vibration) are given also varies among closely related species (Spieth, 1952, 1966b, 1969).

All four aspects of courtship—speed, frequency, sequence, and form or emphasis—rarely differ dramatically between closely related species. Usually, then, the courtship behavior of each species evolves by the gradual modification of pre-existing behavior rather than by the sudden appearance of new behavior. This is compatible with the usual model of polygenic changes in allele frequencies (Falconer, 1960; Manning, 1967c). Ewing and Manning (1967) suggested that these gradual, quantitative changes in behavior may often be the result of altered neural thresholds. Such changes could occur in sense organs, motor neurons, or the central nervous system. In addition, changes could occur in the number, size, and connectivity of the neurons involved.

Simultaneous with the evolution of male displays, changes in receptivity and in responsiveness to specific displays occur in females. Decreased receptivity of females is accompanied by an increase in the duration or intensity of male courtship in that species, and conversely (Spieth, 1951, 1968, 1974a). In endemic Hawaiian *Drosophila*, the relatively low receptivity of females may have been an important factor in the evolution of elaborate and lengthy courtship of lek species (Spieth, 1968, 1974b). Females of closely related, even crossable, species may differ considerably in their mating propensities, as may geographic races (Spieth, 1951; Spiess, 1970; Ehrman and Parsons, 1980).

B. Value of Behavior in Determining Phylogenetic Relationships

Behavioral phenotypes can be helpful in determining phylogenetic relationships among taxa, discovering new species, and separating morphologically similar species. Two approaches may be used: the measurement of sexual isolation between two species or putative species and the study of courtship displays and male–female interactions.

The observation or measurement of sexual isolation in mating experiments (see Section V) may be a valuable step in discovering new species or in separating known sibling species. For example, Sturtevant (1920) observed strong sexual isolation between the sibling species, *D. melanogaster* and *D. simulans*. Subsequent to this study, sexual isolation has been observed between many closely related species, e.g., five species of the *virilis* group

(Spieth, 1951; Stalker, 1942); *D. pseudoobscura* and *D. persimilis* (Bateman, 1949; Wallace and Dobzhansky, 1946); many species of the *quinaria* group (Sears, 1947); several species of the *willistoni* group (Burla *et al.*, 1949; Spieth, 1949); *D. affinis* and *D. algonquin* (Miller, 1950); *D. pavani* and *D. gaucha* (Koref-Santibañez and del Solar, 1961), and among the semispecies of *D. paulistorum* (Ehrman, 1965).

Male courtship behavior differs visibly and measurably between species, and so can be used to help establish evolutionary relationships. Patterns of courtship have been used to identify sibling species: *D. pullipes* and *D. grimshawi* (Ringo, 1974), *D. grimshawi* from Molokai and Oahu (Ringo and Hodosh, 1978) that probably represent sibling species or semispecies (Ohta, 1980), and *D. nasuta* and its siblings (Spieth, 1969). Courtship has been used to help characterize the *D. paulistorum* semispecies (Koref-Santibañez, 1972a,b), and to help establish phylogenics within and among species groups endemic to Hawaii (Spieth, 1968, 1978; Carson *et al.*, 1970). Most taxonomic studies of courtship displays have been purely descriptive (Spieth, 1974a), but some have used quantitative, even multivariate analysis (e.g., Koref-Santibañez, 1972a,b; Ringo and Hodosh, 1978).

VIII. Sexual Isolation

Ethological barriers to interspecific mating may comprise the most important isolating mechanism between animal species (Mayr, 1963). Certainly in *Drosophila*, interspecific mating is rare in nature, even between sympatric sibling species (Spieth, 1974a). *Drosophila* provides ample material for the experimental study of sexual isolation between populations and species. Sexual isolation between closely related species of *Drosophila* has been measured using a variety of approaches in a large number of studies (Section V), but only a fraction of these have attempted to elucidate the *mechanisms* of ethological isolation. Unfortunately, many different methods of measuring sexual isolation are not comparable and the methodology in many studies has been poor, making it difficult to draw general conclusions about the strength and relative importance of sexual isolation in evolution.

In the following discussion, we take "sexual isolation" between two species to be equivalent to positive assortative mating between two populations, as measured on some continuous scale (Section V).

A. Degree of Sexual Isolation at Various Stages of Population Differentiation

Assortative mating has been found between many laboratory stocks of the same species, stocks which differ owing to artificial selection, inbreeding, or some mutant gene or genes.

Mather and Harrison (1949) found deviations from random mating among lines selected for abdominal chaeta number in *D. melanogaster*; unfortunately, the raw data were not reported, and the analysis did not permit the inference that positive assortative mating had occurred. Koref-Santibañez and Waddington (1958) reported weak sexual isolation between a few lines of *D. melanogaster* selected for high and low abdominal chaeta number. Ehrman (1964) found weak but statistically significant sexual isolation among populations of *D. pseudoobscura* maintained at 16 and 27°C for $4\frac{1}{2}$ years. After $2\frac{1}{2}$ and 5 more years, Ehrman (1969b) retested these populations for sexual isolation, again with similar results. However, the pattern of sexual isolation was inconsistent between tests. Parsons (1965) observed positive assortative mating in lines selected for sternopleural bristle number in *D. melanogaster*. Del Solar (1966) found significant sexual isolation in *D. pseudoobscura* between lines selected either for geotaxis or phototaxis. Isolation indices did not increase materially between the first test at generation 5 and the second test at generation 11. Lines in which selection had been relaxed for several generations also continued to show strong preferences for homogamic over heterogamic matings. Burnet and Connolly (1974) observed significant sexual isolation between lines of *D. melanogaster* that Connolly (1966) had selected for high and low activity. The values were comparable to the isolation indices measured by del Solar (1966). Van Dijken and Scharloo (1979) found sexual isolation between lines of *D. melanogaster* selected for high and low locomotor activity, but after further selection this tendency to mate assortatively disappeared.

The experiments cited above involved genetically isolated selection lines. There are many more reports of disruptive selection experiments attempting to produce reproductive isolation. These experiments were designed to test sympatric speciation models. It is assumed (Grant and Mettler, 1969) that it would be much easier to produce reproductive isolation as a by-product of selection in isolated populations than in populations exchanging genes. Thoday and Gibson (1970) reviewed 22 disruptive selection experiments, arguing that most of these were not suitable; at any rate only two (Thoday and Gibson, 1962; Thoday, 1964) resulted in strong sexual isolation. In one of them (Barker and Cummins, 1969) there was evidence of postzygotic isolation between high and low lines. Grant and Mettler (1969) also failed to obtain reproductive isolation as a by-product of disruptive selection. Coyne and Grant (1972) found that some sexual isolation developed as a by-product of disruptive selection for escape response. Barker and Karlsson (1974) found significant sexual isolation developing in some lines selected disruptively for number of sternopleural bristles; high intensity of selection seemed to be an important factor in producing sexual isolation.

Assortative mating and nonrandom rates of courtship can develop as a by-product of inbreeding. Merrell (1949) found no assortative mating between two inbred strains of *D. melanogaster* (Lausanne-S and Oregon-R); a large proportion (30·4%) of the females did not mate within 60 min. Koref-Santibañez and Waddington (1958) found that four out of six highly inbred strains of *D. melanogaster* tended to mate homogamically in male choice experiments; two out of these four also mated assortatively in female choice experiments. Hoenigsberg and Koref-Santibañez (1960) found that males of inbred strains of *D. melanogaster* courted females of their own strain more than females of other strains; however, the frequencies of homogamic and heterogamic matings did not differ significantly. Outcrossed stocks in this study tended to copulate more homogamically than heterogamically, whereas males did not discriminate during courtship. Merrell (1960) reanalysed Hoenigsberg and Koref-Santibañez (1960) data and was able to explain most of the apparent "mate preferences" as simple strain differences in mating propensity. Averhoff and Richardson (1974) found that with increasing generations of inbreeding in *D. melanogaster*, strains exhibited increasing disassortative mating in multiple choice tests. This tendency reached a maximum at eight generations of inbreeding. The reluctance to mate within strains was mediated by a dual pheromone system in both sexes (Averhoff and Richardson, 1975). That is, both males and females produce volatile and nonvolatile pheromones that stimulate the opposite sex to mate if they are dissimilar, but not if they are identical. Powell and Morton (1979) observed random mating between inbred lines (up to 12 generations of sib-mating) of *D. pseudoobscura* in multiple choice tests. The disparities between the three studies cited here remain to be resolved.

Koref-Santibañez and Waddington (1958) found significant sexual isolation between the mutant yellow and mutants vermilion and white of *D. melanogaster*, in male choice tests. Rendel (1945) found that yellow males of *D. subobscura* mated more frequently with yellow females than with wild-type females, whereas wild-type males mated with both types successfully. Tan (1946) reported a similar situation with yellow stocks of *D. pseudoobscura*; Tan's data were reanalysed and discussed by Bateman (1949) and Merrell (1950). Koref-Santibañez and Waddington (1958) found significant sexual isolation between the mutant yellow and the mutants vermilion and white of *D. melanogaster*. A lack of assortative mating between different mutant stocks has been reported by several workers (Merrell, 1949; Bösiger, 1962; Geer and Green, 1962).

There have been several studies of mating between geographic populations of the same species; 18 studies, involving 21 species, were summarized by Anderson and Ehrman (1969). Unfortunately, most of these early studies used only a few strains, and employed the male choice method. Some assortative mating occurred in these experiments, although variation in mating propensity was sometimes reported as significant "one-way isolation"; of course, there is no such thing as one-way sexual isolation. Ayala (1965) found strong assortative mating in non-choice tests involving many geographic stocks of species in the *serrata* group. We are aware of seven studies of mating between geographic populations, involving multiple choice tests. Anderson and Ehrman (1969) found no evidence for assortative mating among five geographically distant populations of *D. pseudoobscura*. Ayala *et al.* (1974) observed weak assortative mating in 1 out of 4 crosses between geographic stocks of *D. equinoxialis*. Petit *et al.* (1976) crossed five French and five Japanese stocks of *D. melanogaster*, and observed weak but significant ($p < 0.05$) assortative mating in 7 out of 45 crosses and disassortative mating in one cross. Matings between French and African populations of *D. melanogaster* were not assortative, although the African males and the French females had significantly higher mating frequencies (Cohet and David, 1980). Ringo and Wood (1980) crossed nine geographic stocks of *D. simulans* and found weak but significant ($p < 0.05$) assortative mating in 2 out of 36 crosses. One of their stocks, from Tunisia, showed a significant overall tendency to mate assortatively. In contrast to these four studies, Ehrman (1972b) and Ehrman and Parsons (1980) reported substantial and widespread assortative mating between geographic stocks of *D. immigrans*; these workers suggested that random drift and adaptation to local ecological conditions may play important roles in promoting the sexual isolation in this species. Zouros and d'Entrement (1980) measured a moderate degree of sexual isolation between stocks of *D. mohavensis*, with stronger isolation observed between geographic races than within races. In all seven studies cited here, the geographic stocks were genetically heterogeneous, having been established from many wild-caught females. Ehrman and Petit (1968) studied three species in the *willistoni* group with mixed results: assortative mating occurred in 3 out of 6 crosses using *D. equinoxialis*, 7 out of 10 crosses involving *D. tropicalis*, but none of 6 involving *D. willistoni*.

Ayala *et al.* (1974) found weak assortative mating between subspecies of *D. equinoxialis* in multiple choice tests. The quantitative index of isolation used in this study is confounded to some extent with mating propensity, but the strength of assortative mating was clearly less than that found among geographic stocks of *D. immigrans* (Ehrman and Parsons, 1980). The semispecies of *D. paulistorum* showed consistently strong assortative mating in multiple choice tests (Malogolowkin-Cohen *et al.*, 1965) and in male choice tests (Carmody *et al.*, 1962). Ohta (1978) found striking differences in the mating propensities of populations of *D. grimshawi* and *D. pullipes* from different islands, as well as some assortative mating. Prakash (1972) discovered an interesting situation in *D. pseudoobscura*; a geographic isolate from Bogota, Colombia mated randomly in multiple choice tests with geographic stocks from other locations, even though Bogota females crossed to males from anywhere else have sterile sons.

Strong or complete sexual isolation has been found between almost all pairs of closely related species that have been tested. Most of these studies have used male choice tests (e.g., Stalker, 1942; Tan, 1946; Wallace and Dobzhansky, 1946; Patterson, 1947; Patterson et al., 1947; Patterson and Stone, 1949; Bateman, 1949; Kaneshiro, 1976; Ohta, 1978; Ahearn et al., 1974; Watanabe and Kawanishi, 1979), while no choice tests were used in some studies (e.g., Dreyfus, 1948; Burla et al., 1949). Merrell (1954) found a moderate degree of sexual isolation between *D. pseudoobscura* and *D. persimilis* in multiple choice tests. Dobzhansky et al. (1968) found very strong sexual isolation in 14/15 crosses among five species of the *obscura* group; in one cross (*D. imaii* × *D. pseudoobscura*) the sexual isolation was in the range of isolation indices for the *paulistorum* semispecies.

B. Role of Males and Females

The relative importance of the sexes in maintaining sexual isolation between species is an important but unresolved issue. Merrell (1949, 1954) and Bateman (1949) hypothesized that females have the primary role in determining the degree of sexual isolation between closely related species of *Drosophila*. Spieth (1952, 1974a) has maintained the opposite, that males function as the primary agents of sexual isolation in nature. Schilcher and Dow (1977) and Wood and Ringo (1980) have suggested an intermediate hypothesis, that the relative importance of males and females in sexual isolation varies from case to case. The only way to test these hypotheses is by direct observation and measurement of interspecific courtship.

There are several species of *Drosophila* in which males are relatively poor discriminators, and readily court with females of other species: *D. melanogaster* (Streisinger, 1948; Manning, 1959b; Wood and Ringo, 1980) and several species in the *obscura* group (Wallace and Dobzhansky, 1946; Mayr, 1946a,b; Streisinger, 1948). *D. subobscura* males will court moving balls of wax (Milani, 1950) and *D. melanogaster* males will court the severed heads of conspecific females (Jacobs, 1960). In a substantially greater number of species, males court homogamically more than heterogamically, in most cases breaking off courtship at the tapping stage: several species in the *virilis* group (Stalker, 1942; Spieth, 1951) and in the *willistoni* group (Spieth, 1949), many species in the *quinaria* group (Sears, 1947), *D. affinis* and *D. algonquin* (Miller, 1950), *D. pavani* and *D. gaucha* (Koref-Santibañez and del Solar, 1961), and *D. simulans* (Wood and Ringo, 1980). Schilcher and Dow (1977) specifically tested Bateman's (1949) and Merrell's (1949) hypothesis in *D. melanogaster* and five of its sibling species. They found that males, not females, of *D. melanogaster* and *D. simulans* were primarily responsible for sexual isolation; males of the other four species were not studied. Schilcher and Dow's results (1977) with *D. melanogaster* may differ from other studies because Schilcher and Dow looked for differences between males and females, not differences between species, the data from *D. melanogaster* and *D. simulans* males having been lumped together.

Males that court interspecifically waste time and energy, possibly increase the probability of predation, and even waste gametes if they manage to inseminate an allospecific female. In species that share feeding–mating–oviposition sites with other species, males courting only conspecifics would be at a selective advantage, all other things being equal. It would not be surprising to find a strong male component to sexual isolation in such species.

Intraspecifically, females dominate sexual selection (Petit and Ehrman, 1969; Spiess, 1970; Spieth, 1974a; Schilcher and Dow, 1977), but this is not to say that males always court or attempt copulation at random within species. It might be easier for sexual isolation behavior to be selected in species already having a tendency to court nonrandomly, intraspecifically. So far as we know, no one has attempted to study the relationship between strength of sexual selection and strength of sexual isolation in a group of closely related species.

C. Environmental Factors Influencing Sexual Isolation

Sexual isolation can be modified by temperature. Mayr and Dobzhansky (1945) found that mating between *D. persimilis* males and *D. pseudoobscura* females occurred more frequently at low temperatures, in male choice tests; sexual isolation was not temperature-sensitive in the reciprocal cross. In crosses between *D. americana* and *D. virilis*, lowering the temperature from 22 to 16·5°C reduced matings within *D. virilis* and abolished all matings by *D. americana*. However, the tests were carried out only for 8 days, and courtship was not observed directly. The increased sexual isolation at the lower temperature appears to be an artifact of slower development in *D. americana* (Spieth, unpublished observations).

Density or crowding can affect sexual isolation. Barker (1967) found that mass crosses gave higher percentages of heterogamic matings than single pair crosses with *D. melanogaster* and *D. simulans*. Parsons (1972) confirmed this finding using the same species. The mechanism may simply be that the presence of conspecific males stimulates *D. simulans* males to court. Levene and Dobzhansky (1945) found that varying the proportions of *D. pseudoobscura* and *D. persimilis* did not affect isolation indices.

There is even some evidence that experience influences sexual isolation. Mayr and Dobzhansky (1945) reported that *D. persimilis* males were more isolated from *D. pseudoobscura* females when they were reared with an excess of conspecific females than when they were reared with large numbers of heterospecific females. Spieth (1968) obtained similar results with *D. persimilis* females. LeMoli and Mainardi (1972) showed that male *D. melanogaster* experienced with *D. simulans* females were subsequently less sexually isolated from *D. simulans* females than males that experienced only conspecific females. Using a different experimental design, Eoff (1973) obtained reduced sexual isolation between *D. melanogaster* females and *D. simulans* males when larval, pupal and early adult stages were passed in bispecific cultures. O'Hara *et al.* (1976) claim that homogamic mating experience in female *D. paulistorum* enhanced assortative mating, but the number of intersemispecific matings was too small to draw a firm conclusion. Experience in mixed cultures does not affect sexual isolation between *D. pallidosa* and *D. ananassae* (Plunka and Potter, 1977).

The sensory basis of sexual isolation in *Drosophila* appears to be primarily chemoreceptive. Miller (1950) observed that courtship between *D. affinis* and *D. alongonquin* is rare and usually does not proceed to the tapping stage; he inferred that males do not court heterospecific females owing to stimuli acting at a distance, probably chemical ones. Airborne chemicals may be the basis for sexual isolation in the *quinaria* group as well (Sears, 1947). Contact chemoreception in males may underlie sexual isolation in species where interspecific courtship often breaks off after tapping, as in the *willistoni* group (Spieth, 1949; Koref-Santibañez, 1972b), the *virilis* group (Spieth, 1951), and the *obscura* group (Spieth, 1958).

The antennae, which contain receptors for both airborne chemicals and auditory stimuli, have been implicated in sexual isolation. Mayr (1946b) found that antennectomy reduced the isolation index in crosses between *D. pseudoobscura* and *D. persimilis*; he later confirmed that the antennae in females are critical for isolation between these species (Mayr, 1950). On the other hand, sexual isolation between *D. melanogaster* and *D. simulans* is not affected by antennectomy of either females or males, but mating discrimination is reduced by the ablation of male tarsi (Manning, 1959b).

Wing vibration may be one stimulus used in species recognition, as suggested by the distinct courtship songs of closely related species, especially sympatric ones (Ewing and Bennet-Clark, 1968; Miller *et al.*, 1975).

There is little evidence that visual displays contribute significantly to sexual isolation, although they are often important components of courtship

(Spieth, 1974a). Mayr (1946b) found that sexual isolation between *D. pseudoobscura* and *D. persimilis* is not light dependent, confirming the previous findings of Mayr and Dobzhansky (1945). Many species requiring light to mate (Grossfield, 1966) tend to use species-specific visual displays in their courtship rituals (Spieth, 1966b, 1974a). Many of the Hawaiian *Drosophila* depend heavily upon visual displays in courtship (Spieth, 1966b, 1974), but some of these species are not completely sexually isolated under laboratory conditions (Ahearn *et al.*, 1974; Kaneshiro, 1976; Ohta, 1978).

D. Evolutionary Origin of Sexual Isolation

Sexual isolation depends on a system of mate recognition in which species-specific cues or signals elicit positive responses, and stimuli from other species elicit less positive or even negative responses. Sexual isolation evolves when a population gains or loses signals, or it changes in responsiveness, either to "correct" or "incorrect" signals. In either case, the divergence of mate recognition systems in two speciating populations could either be a result of direct selection when the populations are sympatric or could be an accidental by-product of genetic divergence. There is evidence for both causes.

If there be direct selection for sexual isolation in nature between closely related species, then this isolation should be greater in regions of sympatry than elsewhere. Increased isolation in regions of sympatry or between sympatric *vs.* allopatric species was found in the semispecies of *D. paulistorum* (Ehrman, 1965) and in species of the *obscura* group (Dobzhansky *et al.*, 1968). Courtship songs, which contribute to species recognition, tend to be less similar in sibling species than in distantly related species (Ewing and Bennet-Clark, 1968).

Direct natural selection for sexual isolation can occur only if there is genetic variation upon which selection can act. Such genetic variation has been sought for and found in four species. Sexual isolation between *D. pseudoobscura* and *D. persimilis* was strengthened or weakened by artificial selection for or against sexual isolation, respectively, in both species (Koopman, 1950; Kessler, 1966). There are strain differences in rates of hybridization between *D. melanogaster* and *D. simulans* (Sturtevant, 1920; Barker, 1962, 1967; Parsons, 1972; Kawaniski and Watanabe, 1981). Artificial selection for or against sexual isolation between *D. melanogaster* and *D. simulans* was also successful, both in *D. melanogaster* females (Eoff, 1975) and in *D. simulans* males (Eoff, 1977). These results imply that these four species carry genetic variation for the tendencies to be sexually isolated from their sibling species. Not all such selection experiments have been successful (Chinnici, 1975; Fukatami and Moriwaki, 1970).

A final point in favor of the direct selection hypothesis is the ability to select for assortative mating between laboratory stocks of a single species (Crossley, 1974; Hoenigsberg et al., 1956; Knight et al., 1956; Wallace, 1954). In these experiments, pairs of genetically heterogeneous lines marked with different recessive mutations were intercrossed, and wild-type offspring were discarded. In each case, mating was not assortative at the start of the experiment, but became so as the experiment proceeded.

The second cause of sexual isolation is accidental divergence of mate recognition systems in two populations. There is evidence from laboratory experiments suggesting that sexual isolation may frequently develop as a by-product of genetic divergence owing to ecological adaptation, or to genetic drift. Assortative mating has been found among laboratory selection lines in *D. pseudoobscura* (del Solar, 1966; Ehrman, 1964, 1969b) and *D. melanogaster* (Koref-Santibañez and Waddington, 1958; Thoday and Gibson, 1962; Coyne and Grant, 1972; Barker and Karlsson, 1974; Burnet and Connolly, 1974). Of these studies only del Solar (1966) and Ehrman (1964, 1969b) measured reproductive isolation more than once; in the former case, sexual isolation increased between generation 5 and generation 11 whereas in the second case sexual isolation was inconsistent between tests. Van Dijken and Scharloo (1980) found no consistent assortative mating among lines selected for high or low adult activity in *D. melanogaster*. Increasingly strong assortative mating developed between three out of eight cage populations of *D. pseudoobscura* that were passed through four successive bottlenecks in population size (Powell, 1978). This experiment was designed specifically to test Carson's (1975) "flush–crash" speciation hypothesis. In a follow-up study, Powell and Morton (1979) found no assortative mating between several inbred lines of *D. pseudoobscura*, suggesting that the flush–crash procedure was not equivalent to a simple reduction in genetic variation. Other studies have found that assortative mating can develop as a by-product of inbreeding (Koref-Santibañez and Waddington, 1958). Averhoff and Richardson (1974) found, contrary to the other studies cited here, that inbreeding in *D. melanogaster* led to strong disassortative mating among lines.

The development of prezygotic isolation as a by-product of selection or genetic drift is consistent with a widely accepted model of allopatric speciation (Muller, 1942; Dobzhansky, 1970). In this model, two populations differentiate genetically, primarily owing to adaptation to different ecological conditions, and gradual multigenic divergence is accompanied by gradual pleiotropic decrements in reproductive compatibility. A new version of this model uses sexual selection, not natural selection, as the driving force of genetic change (Spieth, 1974b; Ringo, 1977b). To date, there has been no experimental test of the hypothesis, although detailed comparative studies of sexual behavior in Hawaiian *Drosophila* (Carson *et al.*, 1970; Spieth, 1966b, 1968) support the notion.

Another approach to the study of sexual isolation is to analyse its behavioral basis genetically, through interspecific hybridization. The little work that has been done along these lines has proved fruitful. Manning (1959b) and Wood and Ringo (1980) crossed *D. simulans* and *D. melanogaster* and found that the male hybrids courted *D. simulans* females more frequently than *D. melanogaster* females, and copulated with them more frequently. By using a mutation which rescues the otherwise lethal male hybrids between *D. melanogaster* and *D. simulans*, Kawanishi and Watanabe (unpublished observations) have studied courtship song and mating preference in reciprocal interspecific crosses. They found that the X chromosome influences the male's sexual recognition system whereas courtship song is inherited autosomally.

IX. Summary

It was quickly realized in the second decade of this century that *D. melanogaster* and other species of *Drosophila* provide excellent material for the study and elucidation of genetics. Fortuitously, *Drosophila* species engage in observable complex species specific courtship behavior. Further, these behaviors are neither depressed nor distorted by the laboratory environments. Some of the mutations that appeared in the laboratory stocks do, however, modify the courtship behaviors. Lutz and Sturtevant decided that *Drosophila* could therefore serve not only as superior organisms for the study of genetics but also for behavioral and sexual selection investigations. The diverse and numerous investigations on *Drosophila* courtships, sexual isolation and sexual selection that have been achieved since their pioneering studies have amply confirmed their judgments. In this review we have attempted to identify, elucidate and evaluate the accumulated data presented by a host of investigators who have concerned themselves with mating behaviors and/or sexual isolation in *Drosophila*. Inevitably there have been omissions, either through oversight on our part or to avoid redundancy in the review. Hopefully, it will serve as a catalyst for future investigations. We need to know much more about the role of courtships in the evolution of species.

References

During the time interval between writing the manuscript and its final assembly for publication a number of pertinent articles have appeared in print, i.e. *reviews*: Hall *et al.*, 1980a and 1980b; Jallon *et al.*, 1981; *additional papers*: Kyriacou and Hall, 1980; Markow, 1981; Markow and Hanson, 1981; Pruzan-Hotchkiss *et al.*, 1981; Pyle and Gromko, 1981;

Spiess and Carson, 1981; Spiess and Dapples, 1980; Wasserman and Koepfer, 1980; Zouros, 1981.

The data and conclusions found in these investigations have not been incorporated into the text but each is cited in the References.

ADAMS, W. T. and DUNCAN, G. T (1978). A maximum likelihood statistical method for analyzing frequency-dependent fitness experiments. *Behav. Genet.* **9**, 7–22.

AHEARN, I. N., CARSON, H. L., DOBZHANSKY, TH. and KANESHIRO, K. Y. (1974). Ethological isolation among three species of the planitibia subgroup of Hawaiian *Drosophila*. *Proc. Natl. Acad. Sci. USA* **71**, 901–903.

ANDERSON, W. W. (1969). Polymorphism resulting from the mating advantage of rare male genotypes. *Proc. Natl. Acad. Sci. USA* **64**, 190–197.

ANDERSON, W. W. (1974). Frequent multiple insemination in a natural population of *Drosophila pseudoobscura*. *Amer. Nat.* **108**, 709–711.

ANDERSON, W. W. and EHRMAN, L. (1969). Mating choice in crosses between geographic populations of *Drosophila pseudoobscura*. *Amer. Midl. Nat.* **81**, 47–53.

AVERHOFF, W. W. and RICHARDSON, R. H. (1974). Pheromonal control of mating patterns in *Drosophila melanogaster*. *Behav. Genet.* **4**, 207–225.

AVERHOFF, W. W. and RICHARDSON, R. H. (1975). Multiple pheromone system controlling mating in *Drosophila melanogaster*. *Proc. Natl. Acad. Sci. USA* **73**, 1–4.

AYALA, F. J. (1965). Sibling species of the *Drosophila serrata* group. *Evolution* **19**, 538–545.

AYALA, F. J. (1972). Frequency-dependent mating advantage in *Drosophila*. *Behav. Genet.* **2**, 85–91.

AYALA, F. J. and CAMPBELL, C. (1974). Frequency dependent selection. *Ann. Rev. Ecol. Syst.* **5**, 115–137.

AYALA, F. J., TRACEY, M. L., BARR, L. G. and EHRENFIELD, J. G. (1974). Genetic and reproductive differentiation of the subspecies, *Drosophila equinoxialis caribbensis*. *Evolution* **28**, 24–41.

BARKER, J. S. F. (1962). Sexual isolation between *Drosophila melanogaster* and *Drosophila simulans*. *Amer. Nat.* **96**, 105–115.

BARKER, J. S. F. (1967). Factors affecting sexual isolation between *Drosophila melanogaster* and *Drosophila simulans*. *Amer. Nat.* **101**, 277–287.

BARKER, J. S. F. and CUMMINS, L. J. (1969). The effect of selection for sternopleural bristle numbers on mating behavior in *Drosophila melanogaster*. *Genetics* **61**, 713–719.

BARKER, J. S. F. and KARLSSON, L. J. E. (1974). Effects of population size and selection intensity on responses to disruptive selection in *Drosophila melanogaster*. *Genetics* **78**, 715–735.

BASTOCK, M. (1956). A gene mutation that changes a behaviour pattern. *Evolution* **10**, 421–439.

BASTOCK, M. and MANNING, A. (1955). The courtship of *Drosophila melanogaster*. *Behaviour* **8**, 85–111.

BATEMAN, A. J. (1948). Intrasexual selection in *Drosophila*. *Heredity* **2**, 349–368.

BATEMAN, A. J. (1949). Analysis of data on sexual isolation. *Evolution* **3**, 174–177.

BAUMANN, H. (1974). The isolation, partial characterization, and biosynthesis of the paragonial substances, PS-1 and PS-2, of *Drosophila funebris*. *J. Insect. Physiol.* **20**, 2181–2194.

BENNET-CLARK, H. C. (1971). Acoustics of insect song. *Nature* **234**, 255–259.

BENNET-CLARK, H. C. and EWING, A. W. (1969). Pulse interval as a critical parameter in the courtship song of *Drosophila melanogaster*. *Anim. Behav.* **17**, 755–759.

BISHOP, Y. M. M., FIENBERG, S. E. and HOLLAND, P. W. (1975). "Discrete Multivariate Analysis: Theory and Practice." The MIT Press, Cambridge, Massachusetts.

BÖSIGER, E. (1962). Sur le degré d'hétérozygotie des populations naturelles de *Drosophila melanogaster* et son maintieu par la sélection sexuelle. *Bull. Biol. Fr. Belg.* **96**, 3–122.

BOULETREAU-MERLE, J. (1973). Réceptivité sexuelle et vitellogenèse chez les femmes de *Drosophila melanogaster*: effets d'une application d'hormone juvénile et de deux analogues hormonaux. *C. R. Acad. Sci. Paris* **277**, 2045–2048.

BOULETREAU-MERLE, J. (1976). Destruction de la pars intercerebralis chez *D. melanogaster*. Effet sur la fecondité, et sur la stimulation par l'accouplement. *J. Insect. Physiol.* **22**, 933–940.

BRIDGES, C. B. (1916). Non-disjunction as proof of the chromosome theory of heredity. *Genetics* **1**, 1–51.

BROWN, R. G. B. (1964). Courtship behaviour in the *Drosophila obscura* group. I. *D. pseudoobscura*. *Behaviour* **23**, 61–106.

BROWN, R. G. B. (1966). Courtship behaviour in the *Drosophila obscura* group. Part II. Comparative studies. *Behaviour* **25**, 281–322.

BURLA, H., DA CUNHA, A. B., CORDIERO, A. R., DOBZHANSKY, TH., MALOGOLOWKIN, C. and PAVAN, C. (1949). The *willistoni* group of sibling species of *Drosophila*. *Evolution* **3**, 300–314.

BURNET, B. and CONNOLLY, K. (1973). The visual component in the courtship of *Drosophila melanogaster*. *Experientia* **29**, 487–489.

BURNET, B. and CONNOLLY, K. (1974). Activity and sexual behavior in *Drosophila melanogaster*. *In*: "The Genetics of Behaviour", pp. 201–258. (J. H. F. van Abeelen, ed.) American Elsevier Publishing Co., New York.

BURNET, B., CONNOLLY, K. and DENNIS, L. (1971). The functioning and processing of auditory information in the courtship of *Drosophila melanogaster*. *Anim. Behav.* **19**, 409–415.

BURNET, B., CONNOLLY, K., KEARNEY, M. and COOK, R. (1973). Effects of male paragonial gland secretion on sexual receptivity and courtship behavior of female *Drosophila melanogaster*. *J. Insect. Physiol.* **19**, 2421–2431.

BURNET, B., EASTWOOD, L. and CONNOLLY, K. (1978). Genetic analysis of courtship song in *Drosophila melanogaster*. *Heredity* **39**, 425–426.

CARMODY, G., DIAS COLLAZO, A., DOBZHANSKY, TH., EHRMAN, L., SILAGI, I. S., TIDWELL, T. and ULLRICH, R. (1962). Mating preferences and sexual isolation within and between the incipient species of *Drosophila paulistorum*. *Amer. Midl. Nat.* **68**, 67–82.

CARSON, H. L. (1971). The ecology of *Drosophila* breeding sites. Harold L. Lyon Arboretum Lecture No. 2, Univ. of Hawaii, Honolulu, Hawaii.

CARSON, H. L. (1975). Genetics of speciation. *Amer. Nat.* **109**, 83–92.

CARSON, H. L. (1978). Speciation and sexual selection in Hawaiian *Drosophila*. *In*: P. F. Brussard (ed.), *Ecological Genetics*. pp. 93–107. Springer-Verlag, New York.

CARSON, H. L., HARDY, D. E., SPIETH, H. T. and STONE, W. S. (1970). The evolutionary biology of the Hawaiian Drosophilidae. *In*: M. K. Hecht and W. C. Steere (eds.), *Essays in evolution and genetics in honor of Theodosius Dobzhansky*, pp. 437–543: Appleton-Century-Crofts, New York.

CHANG, H. and MILLER, D. D. (1978). Courtships and mating sounds in species of the *Drosophila affinis* subgroup. *Evolution* **32**, 540–550.

CHEN, P. S. and BÜHLER, R. (1970). Paragonial substance (sex peptide) and other free ninhydrin-positive components in male and female adults of *Drosophila melanogaster*. *J. Insect Physiol.* **16**, 615–627.

CHEN, P. S., FALES, H. M., LEVENBOOK, L., SOKOLOSKI, E. A. and YEH, H. J. C. (1977). Isolation and characterization of a unique galactoside from male *Drosophila melanogaster*. *Biochemistry* **16**, 4080–4085.

CHIANG, H. C. and HODSON, A. C. (1950). An analytical study of population growth in *Drosophila melanogaster*. *Ecol. Monogr.* **20**, 175–206.

CHINNICI, J. P. (1975). Ineffective selection for sexual isolation in *Drosophila melanogaster*. *Sci. Biol. J.* **1**, 102–105.

COHET, Y. and DAVID, J. R. (1980). Geographical divergence and sexual behaviour: comparison of mating systems in French and Afro-tropical populations of *Drosophila melanogaster*. *Genetica* **54**, 161–165.

CONNOLLY, K. (1966). Locomotor activity in *Drosophila*. II. Selection for active and inactive strains. *Anim. Behav.* **14**, 444–449.

CONNOLLY, K. and COOK, R. (1973). Rejection responses by female *Drosophila melanogaster*: their ontogeny, causality, and effects upon the behaviour of the courting male. *Behaviour* **44**, 122–146.

CONNOLLY, K., BURNET, B. and SEWELL, D. (1969). Selective mating and eye pigmentation: An analysis of the visual component in the courtship behavior of *Drosophila melanogaster*. *Evolution* **23**, 548–559.

CONNOLLY, K., BURNET, B., KEARNEY, M. and EASTWOOD, L. (1974). Mating speed and courtship behaviour of inbred strains of *Drosophila melanogaster*. *Behaviour* **48**, 61–74.

COOK, R. M. (1973a). Courtship processing in *Drosophila melanogaster*. II. An adaptation to selection for receptivity to wingless males. *Anim. Behav.* **21**, 349–358.

COOK, R. M. (1973b). Physiological factors in the courtship processing of *Drosophila melanogaster*. *J. Insect. Physiol.* **19**, 397–406.

COOK, R. M. (1975). Courtship of *Drosophila melanogaster*: rejection without extrusion. *Behaviour* **52**, 155–171.

COOK, R. M. (1978). The reproductive behaviour of gynandromorphic *Drosophila melanogaster*. *Z. Naturforsch.* **32c**, 744–754.

COOK, R. M. (1979). The courtship tracking of *Drosophila melanogaster*. *Biol. Cybernetics* **34**, 91–106.

COOK, R. M. (1980). The extent of visual control in the courtship tracking of *D. melanogaster*. *Biol. Cybernetics* **37**, 41–51.

COOK, R. and COOK, A. (1975). The attractiveness to males of female *Drosophila melanogaster*: effects of mating, age and diet. *Anim. Behav.* **23**, 521–526.

COWLING, D. E. and BURNET, B. (1981). Courtship songs and genetic control of their acoustic characteristics in sibling species of the *Drosophila melanogaster* subgroup. *Anim. Behav.* **29**, 924–935.

COYNE, J. A. and GRANT, B. (1972). Disruptive selection on I-maze activity in *Drosophila melanogaster*. *Genetics* **71**, 185–188.

CROSSLEY, S. A. (1970). Mating reactions of certain mutants. *Dros. Inf. Serv.* **45**, 170.

CROSSLEY, S. A. (1974). Changes in mating behavior produced by selection for ethological isolation between ebony and vestigial mutants of *Drosophila melanogaster*. *Evolution* **28**, 631–647.

DARWIN, C. (1871). *The descent of man, and selection in relation to sex*. Burt, New York.

DAVENPORT, C. B. (1941). The early history of research with *Drosophila*. *Science* **93**, 305–306.

DIJKEN, F. R. VAN and SCHARLOO, W. (1979). Divergent selection on locomotor activity in *Drosophila melanogaster*. II. Test for reproductive isolation between selected lines. *Behav. Genet.* **9**, 555–561.

DOBZHANSKY, TH. (1937). *Genetics and the origin of species*. Columbia University Press, New York.

DOBZHANSKY, TH. (1970) *Genetics of the evolutionary process*. Columbia University Press, New York.

DOBZHANSKY, TH. and PAVAN, C. (1950). Local and seasonal variations in relative frequencies of species of *Drosophila* in Brazil. *J. Anim. Ecol.* **19**, 1–14.

DOBZHANSKY, TH. and DA CUNHA, B. (1955). Differentiation of nutritional preferences in Brazilian species of *Drosophila*. *Ecology* **36**, 34–39.

DOBZHANSKY, TH. and PAVLOVSKY, O. (1967). Repeated mating and sperm mixing in *Drosophila pseudoobscura*. *Amer. Nat.* **101**, 527–533.

DOBZHANSKY, TH., EHRMAN, L. and KASTRITSIS, P. A. (1968). Ethological isolation between sympatric and allopatric species of the obscura group of *Drosophila*. *Anim. Behav.* **16**, 79–87.

DOW, M. A. (1976). Selection for mating success of yellow mutant *Drosophila melanogaster* behavioral changes. *Behav. Biol.* **16**, 233–239.

DOW, M. A. and SCHILCHER, F. VON (1975). Aggression and mating success in *Drosophila melanogaster*. *Nature Lond.* **254**, 511–512.

DREYFUS, A. (1948). Analysis of sexual isolation between *Drosophila paranaesis* and *D. pararepleta*. *Proc. 8th Int. Cong. Genet.*, pp. 564–565.

EHRMAN, L. (1964). Genetic divergence of M. Vetukhiv's experimental populations of *Drosophila pseudoobscura*. *Genet. Res.* **5**, 150–157.

EHRMAN, L. (1965). Direct observation of sexual isolation between allopatric and between sympatric strains of different *Drosophila paulistorum* races. *Evolution* **19**, 459–464.

EHRMAN, L. (1966). Mating success and genotype frequency in *Drosophila*. *Anim. Behav.* **14**, 332–339.

EHRMAN, L. (1967). Further studies on genotype frequency and mating success in *Drosophila*. *Amer. Nat.* **101**, 415–424.

EHRMAN, L. (1969a). The sensory basis of mate selection in *Drosophila*. *Evolution* **23**, 59–64.

EHRMAN, L. (1969b). Genetic divergence in M. Vetukhiv's experimental populations of *Drosophila pseudoobscura*. 5. A further study of rudiments of sexual isolation. *Amer. Midl. Nat.* **82**, 272–276.

EHRMAN, L. (1972a). A factor influencing the rare male mating advantage in *Drosophila*. *Behav. Genet.* **2**, 69–78.

EHRMAN, L. (1972b). Rare male advantages and sexual isolation in *Drosophila immigrans*. *Behav. Genet.* **2**, 79–84.

EHRMAN, L. and PARSONS, P. A. (1976). *The Genetics of Behavior*. Sinauer Associates, Sunderland, Massachusetts.

EHRMAN, L. and PARSONS, P. A. (1980). Sexual isolation among widely distributed populations of *Drosophila immigrans*. *Behav. Genet.*, in press.

EHRMAN, L. and PETIT, C. (1968). Genotype frequency and mating success in the *willistoni* species group of *Drosophila*. *Evolution* **22**, 649–658.

EHRMAN, L. and PROBBER, J. (1978). Rare *Drosophila* males: The mysterious matter of choice. *Amer. Sci.* **66**, 216–222.

EHRMAN, L. and SPIESS, E. B. (1969). Rare-type mating advantage in *Drosophila*. *Amer. Nat.* **103**, 675–680.

EHRMAN, L., SPASSKY, B., PAVLOVSKY, O. and DOBZHANSKY, TH. (1965). Sexual selection, geotaxis, and chromosomal polymorphism in experimental populations of *Drosophila pseudoobscura*. *Evolution* **19**, 337–346.

EIBL-EIBESFELDT, I. (1970). *Ethology—The Biology of Behavior*. Holt, Rinehart and Winston, New York.

ELENS, A. A. and WATTIAUX, J. M. (1964). Direct observation of sexual isolation. *Dros. Inf. Serv.* **39**, 118–119.

ELLIS, L. B. and KESSLER, S. (1975). Differential post-eclosion housing experiences and reproduction in *Drosophila*. *Anim. Behav.* **23**, 949–952.

EOFF, M. (1973). The influence of being cultured together on hybridization between *D. melanogaster* and *D. simulans*. *Amer. Nat.* **107**, 247–255.
EOFF, M. (1975). Artificial selection in *Drosophila melanogaster* females for increased and decreased sexual isolation from *D. simulans* males. *Amer. Nat.* **109**, 225–229.
EOFF, M. (1977). Artificial selection in *Drosophila simulans* males for increased and decreased sexual isolation from *Drosophila melanogaster* females. *Amer. Nat.* **111**, 259–266.
EWING, A. W. (1964). The influence of wing area on the courtship behaviour of *Drosophila melanogaster*. *Anim. Behav.* **12**, 316–320.
EWING, A. W. (1969). The genetic basis of sound production in *Drosophila pseudoobscura* and *D. persimilis*. *Anim. Behav.* **17**, 555–560.
EWING, A. W. (1970). The evolution of courtship songs in *Drosophila*. *Rev. Comp. Anim. (Paris)* **4**, 3–8.
EWING, A. W. (1979). Complex courtship songs in the *Drosophila funebris* species groups: Escape from an evolutionary bottleneck. *Anim. Behav.* **27**, 343–349.
EWING, A. W. and BENNET-CLARK, H. C. (1968). The courtship songs of *Drosophila*. *Behaviour* **31**, 288–301.
EWING, A. W. and MANNING, A. (1963). The effect of exogenous scent on the mating of *Drosophila melanogaster* (Meigen). *Anim. Behav.* **11**, 596–598.
EWING, A. W. and MANNING, A. (1967). The evolution and genetics of insect behaviour. *Ann. Rev. Entomol.* **12**, 471–494.
FALCONER, D. S. (1960). *Introduction to quantitative genetics*. The Ronald Press Co., New York.
FISHER, R. A. (1958). *The genetical theory of natural selection*. Dover Publications, N.Y.
FOWLER, G. L. (1973). Some aspects of the reproductive biology of *Drosophila*: sperm storage and sperm utilization. *Adv. Genet.* **17**, 293–360.
FUERST, P. A., PENDLEBURG, W. W. and KIDWELL, J. F (1973). Propensity for multiple mating in *Drosophila melanogaster* females. *Evolution* **27**, 265–268.
FUKATAMI, A. and MORIWAKI, D. (1970). Selection for sexual isolation in *Drosophila melanogaster* by a modification of Koopman's method. *Jap. J. Genet.* **45**, 193–204.
GEER, B. W. and GREEN, M. M. (1962). Genotype, phenotype, and mating behavior of *Drosophila melanogaster*. *Amer. Nat.* **96**, 175–181.
GOUX, J. M. and ANXOLABEHERE, D. (1980). The measurement of sexual isolation and selection: a critique. *Heredity* **45**, 255–262.
GRANT, B. and METTLER, L. E. (1969). Disruptive and stabilizing selection on the "escape" behaviour of *Drosophila melanogaster*. *Genetics* **62**, 625–637.
GROMKO, M. H. and PYLE, D. W. (1978). Sperm competition, male fitness, and repeated mating by female *Drosophila melanogaster*. *Evolution* **32**, 588–593.
GROSSFIELD, J. (1966). The influence of light on the mating behavior of *Drosophila*. *Univ. Texas Publ.* **6615**, 147–176.
GROSSFIELD, J. (1968). The relative importance of wing utilization in light dependent courtship in *Drosophila*. *Univ. Texas Publ.* **6818**, 147–156.
GROSSFIELD, J. (1970). Species differences in light-influenced behavior in *Drosophila*. *Amer. Nat.* **104**, 307–309.
GROSSFIELD, J. (1971). Behavioral differentiation of three races of *Drosophila auraria*. *J. Hered.* **62**, 117–118.
GROSSFIELD, J. (1972). A new class of light dependent behavior in *Drosophila*. *Dros. Inf. Serv.* **48**, 72–73.
GROSSFIELD, J. and ROCKWELL, R. F. (1979). Courtship behavior of endemic Australian *Drosophila*. I. *Scaptodrosophila*: *lativittata* and *fumida* groups. *Amer. Midl. Nat.* **101**, 257–268.

HALL, J. (1977). Portions of the central nervous system controlling reproductive behavior in *Drosophila melanogaster*. *Behav. Genet.* **7**, 291–312.
HALL, J. C. (1979). Control of male reproductive behavior by the central nervous system of *Drosophila*: Dissection of a courtship pathway by genetic mosaics. *Genetics* **92**, 437–457.
HALL, J. C. and GREENSPAN, R. J. (1979). Genetic analysis of *Drosophila* neurobiology. *Ann. Rev. Genet.* **13**, 127–195.
HALL, J. C., SIEGEL, R. W., TOMPKINS, L. and KYRIACOU, C. P. (1980a). Neurogenetics of courtship in *Drosophila*. *Stadler Sympos.* **12**, 43–82.
HALL, J. C., TOMPKINS, L., KYRIACOU, C. P., SIEGEL, R. W., SCHILCHER, F. V. and GREENSPAN, R. C. (1980b). Higher behavior in *Drosophila* analyzed with mutations that disrupt the structure and function of the nervous system. *In*: 'Development and neurobiology of the nervous system' (O. Siddiqi, P. Babn, L. M. Hall and J. C. Hall eds.), Plenum, New York.
HERTWECK, H. (1931). Anatomie und Variabilitat des Nervensystems und der Sinnesorgane von *Drosophila melanogaster* (Meigen). *Z. wiss. Zool.* **139**, 559–663.
HODOSH, R. J., RINGO, J. M. and McANDREW, F. T. (1979). Density and lek displays in *Drosophila grimshawi* (Diptera: Drosophilidae). *Z. Tierpsych.* **49**, 164–172.
HOENIGSBERG, H. F. and KOREF-SANTIBAÑEZ, S. (1960). Courtship and sensory preference in inbred lines of *Drosophila melanogaster*. *Evolution* **14**, 1–7.
HOENIGSBERG, H. F., CHEJNE, A. J. and HORTOGAGJI-GERMAN, E. (1956). Preliminary report on artificial selection towards sexual isolation in *Drosophila*. *Z. Tierpsych.* **23**, 129–135.
HOTTA, Y. and BENZER, S. (1972). Mapping of behaviour in *Drosophila* mosaics. *Nature, Lond.* **240**, 527–535.
HOTTA, Y. and BENZER, S. (1976). Courtship in *Drosophila* mosaics sex-specific foci for sequential action patterns. *Proc. Natl. Acad. Sci. U.S.A.* **73**, 4154–4158.
HUTCHINSON, G. E. (1978). "An Introduction to Population Ecology." Yale University Press, New Haven.
HUXLEY, J. S. (1940). Towards the new systematics. *In*: *The New Systematics* (Julian Huxley, ed.) pp. 1–46, Oxford University Press, London.
JACOBS, M. E. (1960). Influence of light on mating of *Drosophila melanogaster*. *Ecology* **41**, 182–188.
JACOBS, M. E. (1961). The influence of light on gene frequency changes in laboratory populations of ebony and non-ebony *D. melanogaster*. *Genetics* **46**, 1089–1095.
JACOBS, M. E. (1978). The influence of beta-alanine on mating and territorialism in *Drosophila melanogaster*. *Behav. Genet.* **8**, 487–502.
JALLON, J.-M. and HOTTA, Y. (1979). Genetic and behavioral studies of female sex appeal in *Drosophila*. *Behav. Genet.* **9**, 257–276.
JALLON, J.-M., ANTONY, C. and IWATSUBO, T. (1981). Elements of chemical communication between drosophilids and their modulation. Taneguchi Sympos. Biophys. Kyoto.
JOHNSTON, J. S. and HEED, W. B. (1976). Dispersal of desert-adapted *Drosophila*: The saguaro-breeding *D. nigrospiracula*. *Amer. Nat.* **110**, 629–651.
KANESHIRO, K. Y. (1976). Ethological isolation and phylogeny in the planitibia subgroup of Hawaiian *Drosophila*. *Evolution* **30**, 740–745.
KAWANISHI, M. and WATANABE, T. K. (1981). Genes controlling courtship song and mating preference in *Drosophila melanogaster*, *Drosophila simulans*, and their hybrids. *Evolution* **35**, 1128–1133.
KELLOGG, F. F., FRIZEL, D. E. and WRIGHT, R. H. (1962). The olfactory guidance of flying insects. IV. *Drosophila*. *Can. Entomol.* **94**, 884–888.

KESSLER, S. (1966). Selection for and against ethological isolation between *Drosophila pseudoobscura* and *Drosophila persimilis*. *Evolution* 20, 634–645.

KNIGHT, G. R., ROBERTSON, A. and WADDINGTON, C. H. (1956). Selection for isolation within a species. *Evolution* 10, 14–22.

KOOPMAN, K. F. (1950). Natural selection for reproductive isolation between *Drosophila pseudoobscura* and *Drosophila persimilis*. *Evolution* 4, 135–138.

KOREF-SANTIBAÑEZ, S. (1963). Courtship and sexual isolation in five species of the *mesophragmatica* group of the genus *Drosophila*. *Evolution* 17, 99–106.

KOREF, SANTIBAÑEZ, S. (1972a). Courtship behavior in the semispecies of the superspecies *Drosophila paulistorum*. *Evolution* 26, 108–115.

KOREF-SANTIBAÑEZ, S. (1972b). Courtship interaction in the semispecies of *Drosophila paulistorum*. *Evolution* 26, 326–333.

KOREF-SANTIBAÑEZ, S. and DEL SOLAR, E. (1961). Courtship and sexual isolation in *Drosophila pavani* Brnic and *Drosophila gaucha* Jaeger and Salzano. *Evolution* 15, 401–406.

KOREF-SANTIBAÑEZ, S. and WADDINGTON, C. H. (1958). The origin of isolation between different lines within a species. *Evolution* 12, 485–493.

KVELLAND, I. (1965). Some observations on the mating activity and fertility of *Drosophila melanogaster* males. *Hereditas* 53, 281–305.

KYRIACOU, C. P. and HALL, J. C. (1980). Circadian rhythm mutations in *Drosophila melanogaster* affect short-term fluctuations in the male's courtship song. *Proc. Natl. Acad. Sci. USA* 77, 6729–6733.

LABEYRIE, V. (1978). The significance of the environment in the control of insect fecundity. *Ann. Rev. Entomol.* 23, 69–89.

LANDE, R. (1980). Sexual dimorphism, sexual selection, and adaptation in polygenic characters. *Evolution* 34, 292–305.

LEFEVRE, G. and JONSSON, U. B. (1962). Sperm transfer, storage, displacement, and utilization in *Drosophila melanogaster*. *Genetics* 47, 1719–1736.

LEMOLI, F. and MAINARDI, M. (1972). Effect of recent experiences with reproductive isolation between *Drosophila melanogaster* and *Drosophila simulans*. *1st Lombardo Accad. Sci. Lett. Rend. Sci. Biol. Med. B.* 106, 29–35.

LEVENE, H. (1949). A new measure of sexual isolation. *Evolution* 3, 315–321.

LEVENE, H. and DOBZHANSKY, TH. (1945). Experiments on sexual isolation in *Drosophila*. V. The effect of varying proportions of *Drosophila pseudoobscura* and *Drosophila persimilis* on the frequency of insemination in mixed populations. *Proc. Natl. Acad. Sci. U.S.A.* 31, 274–281.

LEWONTIN, R. (1974). *The genetic basis of evolutionary change*. Columbia Univ. Press, New York.

LONG, C. E., MARKOW, T. A. and YAEGER, P. (1980). Relative male age, fertility, and competitive mating success in *Drosophila melanogaster*. *Behav. Genet.* 10, 163–170.

LUTZ, F. E. (1914). Biological notes concerning *Drosophila ampelophila*. *J. N.Y. Entomol. Soc.* 22, 134–138.

MALOGOLOWKIN-COHEN, CH., SIMMONS, A. S. and LEVENE, H. (1965). A study of sexual isolation between strains of *Drosophila paulistorum*. *Evolution* 19, 95–103.

MANNING, A. (1959a). The sexual behaviour of two sibling *Drosophila* species. *Behaviour* 15, 123–145.

MANNING, A. (1959b). The sexual isolation between *D. melanogaster* and *D. simulans*. *Anim. Behav.* 7, 60–65.

MANNING, A. (1962a). The control of sexual receptivity in female *Drosophila*. *Anim. Behav.* 10, 384.

MANNING, A. (1962b). A sperm factor affecting the receptivity of *Drosophila melanogaster* females. *Nature, Lond.* **194**, 252–253.

MANNING, A. (1965). *Drosophila* and the evolution of behaviour. *In*: J. D. Carthy and C. L. Duddington (eds.), *Viewpoints in biology*, Vol. 4, 125–169, Butterworths, London.

MANNING, A. (1966). Corpus allatum and sexual receptivity in female *Drosophila melanogaster*. *Nature, Lond.* **211**, 1321–1322.

MANNING, A. (1967a). Antennae and sexual receptivity in *Drosophila melanogaster*. *Science* **158**, 136–137.

MANNING, A. (1967b). The control of sexual receptivity in female *Drosophila*. *Anim. Behav.* **15**, 239–250.

MANNING, A. (1967c). Genes and the evolution of insect behavior. *In*: J. Hirsch (ed.), *Behavior-genetic analysis*. McGraw Hill, N.Y.

MARKOW, T. (1981). Mating preferences are not predictive of the direction of evolution in experimental populations of *Drosophila*. *Science* **213**, 1405–1407.

MARKOW, T. and HANSON, S. (1981). Multivariate analysis of *Drosophila* courtship. *Proc. Natl. Acad. Sci. USA* **78**, 430–434.

MARKOW, T. A., QUAID, M. and KERR, S. (1978). Male mating experience and competitive courtship success in *Drosophila melanogaster*. *Nature, Lond.* **276**, 821–822.

MATHER, K. and HARRISON, B. J. (1949). The manifest effect of selection. *Heredity*, **3**, 131–162.

MAYNARD-SMITH, J. (1956). Fertility, mating behaviour and sexual selection in *Drosophila pseudoobscura*. *J. Genet.* **54**, 261–279.

MAYR, E. (1942). *Systematics and the origin of species*. Columbia Univ. Press, New York.

MAYR, E. (1946a). Experiments on sexual isolation in *Drosophila*. VI. Isolation between *Drosophila pseudoobscura* and *Drosophila persimilis* and their hybrids. *Proc. Natl. Acad. Sci. U.S.A.* **32**, 37–59.

MAYR, E. (1946b). Experiments on sexual isolation in *Drosophila*. VII. The nature of isolating mechanisms between *Drosophila pseudoobscura* and *Drosophila persimilis*. *Proc. Natl. Acad. Sci. U.S.A.* **32**, 128–137.

MAYR, E. (1950). The role of the antennae in the mating behavior of female *Drosophila*. *Evolution* **4**, 149–154.

MAYR, E. (1963). *Animal speciation and evolution*. The Belknap Press, Cambridge, Mass.

MAYR, E. and DOBZHANSKY, TH. (1945). Experiments on sexual isolation in *Drosophila*. IV. Modification of the degree of isolation between *Drosophila pseudoobscura* and *Drosophila persimilis* and sexual preference in *Drosophila prosaltans*. *Proc. Natl. Acad. Sci. U.S.A.* **31**, 75–82.

MERLE, J. (1969). Fonctionment ovarien et receptivite sexuelle de *Drosophila melanogaster* apres implantation de fragments de l'appareil genitale male. *J. Insect. Physiol.* **14**, 1159–1168.

MERRELL, D. J (1949). Selective mating in *Drosophila melanogaster*. *Genetics* **34**, 370–389.

MERRELL, D. J. (1950). Measurement of sexual isolation and selective mating. *Evolution* **4**, 326–331.

MERRELL, D. J. (1954). Sexual isolation between *Drosophila persimilis* and *Drosophila pseudoobscura*. *Amer. Nat.* **88**, 93–99.

MERRELL, D. J. (1960). On mating preferences in *Drosophila*. *Evolution* **14**, 525–526.

MILANI, R. (1950). Release of courtship display in *subobscura* males stimulated with dummies. *Dros. Inf. Serv.* **24**, 88.

MILKMAN, R. and ZEITLER, R. R. (1974). Concurrent multiple paternity in natural and laboratory populations of *Drosophila melanogaster*. *Genetics* **78**, 1191–1193.

MILLER, D. D. (1950). Mating behavior of *Drosophila affinis* and *Drosophila algonquin*. *Evolution* **4**, 123–134.
MILLER, D. D., GOLDSTEIN, R. B. and PATTY, R. A. (1975). Semispecies of *Drosophila athabasca* distinguishable by male courtship sounds. *Evolution* **29**, 531–544.
MULLER, H. J. (1942). Isolating mechanisms, evolution, and temperature. *Biol. Symp.* **6**, 71–125.
NARDA, R. D. (1966). Analysis of the stimuli involved in courtship and mating in *D. malerkotliana* (Sophophora, Drosophila). *Anim. Behav.* **14**, 378–383.
NISSANI, M. (1977). Gynandromorph analysis of some aspects of sexual behavior of *Drosophila melanogaster*. *Anim. Behav.* **25**, 555–566.
O'DONALD, P. (1962). The theory of sexual selection. *Heredity* **17**, 541–552.
O'DONALD, P. (1967). A general model of sexual selection and natural selection. *Heredity* **22**, 499–518.
O'DONALD, P. (1973). Models of sexual and natural selection in polygamous species. *Heredity* **31**, 145–156.
O'DONALD, P. (1980). Genetic models of sexual selection. Cambridge University Press, Cambridge.
O'HARA, E., PRUZAN, A. and EHRMAN, L. (1976). Ethological isolation and mating experience in *Drosophila paulistorum*. *Proc. Natl. Acad. Sci. U.S.A.* **73**, 975–976.
OHTA, A. T. (1978). Ethological isolation and phylogeny in the *grimshawi* species complex of Hawaiian *Drosophila*. *Evolution* **32**, 485–492.
OHTA, A. T. (1980). Coadaptive gene complexes in incipient species of Hawaiian *Drosophila*. *Amer. Nat.* **115**, 121–131.
PARSONS, P. A. (1965). Assortative mating for a metrical characteristic in *Drosophila*. *Heredity* **20**, 161–167.
PARSONS, P. A. (1972). Variations between strains of *Drosophila melanogaster* and *D. simulans* in giving offspring in interspecific crosses. *Can. J. Genet. Cytol.* **14**, 81–93.
PATTERSON, J. T. (1943). The Drosophilidae of the Southwest. *Univ. Texas Publ.* **4313**, 1–327.
PATTERSON, J. T. (1947). The insemination reaction and its bearing on the problem of speciation in the mulleri subgroup. *Univ. Texas Publ.* **4720**, 41–77.
PATTERSON, J. T. and STONE, W. S. (1949). The relationship of *novamexicana* to the other members of the *virilis* group. *Univ. Texas. Publ.* **4920**, 7–17.
PATTERSON, J. T. and STONE, W. S. (1952). *Evolution in the genus Drosophila*. Macmillan, N.Y.
PATTERSON, J. T., MCDONALD, L. W. and STONE, W. S. (1947). Sexual isolation between members of the *virilis* group. *Univ. Texas Publ.* **4720**, 7–31.
PETIT, C. (1958). Le déterminisme génétique et psychophysiologique de la competition sexuelle chez *Drosophila melanogaster*. *Bull. Biol. Fr. et Belg.* **92**, 248–329.
PETIT, C. (1959). Les factors génétique de la competition sexuelle entre une forme mutante et son allelomorph sauvage chez *Drosophila melanogaster*. *Ann. Genet.* **35**, 83–87.
PETIT, C. and EHRMAN, L. (1969). Sexual selection in *Drosophila*. *In*: Th. Dobzhansky, M. K. Hecht and W. C. Steere (eds.), *Evolutionary Biology* **2**, 157–191. New York: Appleton-Century-Crofts.
PETIT, C., KITAGAWA, O. and TAKAMURA, T. (1976). Mating systems between Japanese and French geographic strains of *Drosophila melanogaster*. *Jap. J. Genet.* **51**, 99–108.
PHILIP, U., RENDEL, J. M., SPURWAY, H. and HALDENE. J. B. S. (1944). Genetics and karyology of *Drosophila subobscura*. *Nature, Lond.* **154**, 260–262.
PIPKIN, S. B. (1965). The influence of adult and larval food habits on population size of neotropical ground-feeding *Drosophila*. *Amer. Midl. Nat.* **74**, 1–27.

PLUNKA, S. M. and POTTER, J. H. (1977). The effect of mixed culturing on reproductive isolation between *Drosophila pallidosa* and *Drosophila ananassae*. *Amer. Nat.* **111**, 598–603.

POT, W., VAN DELDEN, W. and KRUIJT, J. P. (1980). Genotypic differences in mating success and the maintenance of the alcohol dehydrogenase polymorphism in *Drosophila melanogaster*: No evidence for overdominance or rare genotype mating advantage. *Behav. Genet.* **10**, 43–58.

POWELL, J. R. (1978). The founder-flush speciation theory: An experimental approach. *Evolution* **32**, 465–474.

POWELL, J. R. and MORTON, L. (1979). Inbreeding and mating patterns in *Drosophila pseudoobscura*. *Behav. Genet.* **9**, 425–431.

PRAKASH, S. (1967). Association between mating speed and fertility in *Drosophila robusta*. *Genetics* **57**, 655–663.

PRAKASH, S. (1972). Origin of reproductive isolation in the absence of apparent genic differentiation in a geographic isolate of *Drosophila pseudoobscura*. *Genetics* **72**, 143–155.

PROUT, T. and BUNDGAARD, J. (1977). The population genetics of sperm displacement. *Genetics* **85**, 95–124.

PRUZAN, A. (1973). Age and experience and rare male advantages in *Drosophila pseudoobscura*. *Behav. Genet.* **3**, 412–413.

PRUZAN, A. (1976). Effects of age, rearing, and mating experiences on frequency dependent sexual selection in *Drosophila pseudoobscura*. *Evolution* **30**, 130–145.

PRUZAN, A., APPLEWHITE, P. B. and BUCCI, M. J. (1977). Protein synthesis inhibition alters *Drosophila* mating behavior. *Pharmacol. Biochem. Behav.* **6**, 355–358.

PRUZAN, A. and EHRMAN, L. (1974). Age, experience, and rare-male mating advantages in *Drosophila pseudoobscura*. *Behav. Genet.* **4**, 159–164.

PRUZAN-HOTCHKISS, A., DEJIANNE, D. and FARO, S. H. (1981). Sperm utilization in once- and twice-mated *Drosophila pseudoobscura* females. *Amer. Nat.* **118**, 37–45.

PYLE, D. W. and GROMKO, M. H. (1978). Repeated mating by female *Drosophila melanogaster*: The adaptive importance. *Experientia* **34**, 449–450.

PYLE, D. W. and GROMKO, M. H. (1981). Genetic basis for repeated mating in *Drosophila melanogaster*. *Amer. Nat.* **117**, 133–146.

QUINN, W. G., SZIBER, P. P. and BOOKER, R. (1979). The *Drosophila* memory mutant *amnesiac*. *Nature, Lond.* **277**, 212–214.

REED, S. C. and REED, W. E. (1950). Natural selection in laboratory populations of *Drosophila*. Competition between a white-eye gene and its wild type allele. *Evolution* **4**, 34–42.

RENDEL, J. M. (1945). Genetics and cytology of *Drosophila subobscura*. II. Normal and selective matings in *Drosophila subobscura*. *J. Genet.* **46**, 287–302.

RENDEL, J. M. (1951). Mating of ebony vestigial and wild type *Drosophila melanogaster* in light and dark. *Evolution* **5**, 226–230.

RICHMOND, R. C. and EHRMAN, L. (1974). The incidence of repeated mating in the superspecies *Drosophila paulistorum*. *Experientia* **30**, 489–490.

RICHMOND, R. C., GILBERT, D. G., SHESHAN, K. B., GROMKO, M. H. and BUTTERWORTH, F. M. (1980). Esterase 6 and reproduction in *Drosophila melanogaster*. *Science* **207**, 1483–1485.

RINGO, J. M. (1974). Behavioral characters distinguishing two species of Hawaiian *Drosophila*, *D. grimshawi* and *D. pullipes* (Diptera: Drosophilidae). *Ann. Entomol. Soc. Amer.* **67**, 823.

RINGO, J. M. (1976). A communal display in Hawaiian *Drosophila* (Diptera: Drosophilidae). *Ann. Entomol. Soc. Amer.* **69**, 209–214.

RINGO, J. M. (1977a). The influence of visual and olfactory stimuli on jousting behaviour in *Drosophila grimshawi* (Diptera: Drosophilidae). *Anim. Behav.* **25**, 275–280.
RINGO, J. M. (1977b). Why 300 species of Hawaiian *Drosophila*? The sexual selection hypothesis. *Evolution* **31**, 694–696.
RINGO, J. M. (1978). The development of stereotyped displays in *Drosophila*. *In*: G. Burghardt and M. Bekoff (eds.), *The development of behavior: Comparative and evolutionary aspects*. Garland, New York.
RINGO, J. M. and HODOSH, R. J. (1978). A multivariate analysis of behavioral divergence among closely related species of endemic Hawaiian *Drosophila*. *Evolution* **32**, 389–397.
RINGO, J. M. and PRATT, N. P. (1978). A juvenile hormone analog induces precocial sexual behavior in *Drosophila grimshawi*. *Ann. Entomol. Soc. Amer.* **71**, 264–266.
RINGO, J. M. and WOOD, D. (1980). Frequencies of mating among geographic populations of *Drosophila simulans*. *Behav. Genet.* **10**, 492.
SCHAFFER, H. E. (1968). A measure of discrimination in mating. *Evolution*, **22**, 125–129.
SCHILCHER, F. VON. (1976a). The role of auditory stimuli in the courtship of *Drosophila melanogaster*. *Anim. Behav.* **24**, 18–26.
SCHILCHER, F. VON. (1976b). The function of pulse song and sine song in the courtship of *Drosophila melanogaster*. *Anim. Behav.* **24**, 622–625.
SCHILCHER, F. VON and DOW, M. (1977). Courtship behavior in *Drosophila* sexual isolation or sexual selection. *Z. Tierpsych.* **43**, 304–310.
SCHILCHER, F. VON and HALL, J. C. (1979). Neural topography of courtship song in sex mosaics of *Drosophila melanogaster*. *J. Comp. Physiol.* A **129**, 85–95.
SCHILCHER, F. VON and MANNING, A. (1975). Courtship song and mating speed in hybrids between *Drosophila melanogaster* and *Drosophila simulans*. *Behav. Genet.* **5**, 395–404.
SCHNEIDERMAN, H. A. (1972). Insect hormones and insect control. *In*: J. J. Menn and M. Beroza (eds.), *Insect juvenile hormones*. Academic Press, New York and London.
SCIANDRA, R. J. and BENNETT, J. (1976). Behavior and single gene substitution in *Drosophila melanogaster*. I. Mating and courtship differences with *white*, *cinnabar*, and *brown* loci. *Behav. Genet.* **6**, 205–218.
SEARS, J. W. (1947). Relationships within the *quinaria* species group of *Drosophila*. *Univ. Texas Publ.* **4720**, 137–156.
SHOREY, H. H. (1962). Nature of the sound produced by *Drosophila melanogaster* during courtship. *Science* **137**, 677–678.
SHOREY, H. H. and BARTELL, R. J. (1970). Role of a volatile sex pheromone in stimulating male courtship behaviour in *Drosophila melanogaster*. *Anim. Behav.* **18**, 159–164.
SIEGEL, R. W. and HALL, J. C. (1979). Conditioned responses in courtship behavior of normal and mutant *Drosophila*. *Proc. Natl. Acad., Sci. USA* **76**, 3430–3434.
SKRZIPEK, VON K. H., KRÖNER, B. and HAGER, H. (1979). Aggression bei *Drosophila melanogaster* — Laboruntersuchungen. *Z. Tierpsych.* **49**, 87–103.
SMITH, J. M. (1956). Fertility, mating behaviour and sexual selection in *Drosophila subobscura*. *J. Genet.* **54**, 261–279.
SNEDECOR, G. W. and COCHRAN, W. G. (1967). *Statistical methods*. Iowa State Univ. Press, Ames, Iowa.
SOKAL, R. R. and ROHLF, F. J. (1981). *Biometry*, 2nd edition. Freeman, San Francisco.
SOLAR, E. DEL. (1966). Sexual isolation caused by selection for and against positive and negative phototaxis and geotaxis in *Drosophila pseudoobscura*. *Proc. Natl. Acad. Sci. U.S.A.*, **56**, 484–487.
SPIESS, E. B. (1968a). Low frequency advantage in mating of *Drosophila pseudoobscura* karyotypes. *Amer. Nat.* **102**, 363–379.

SPIESS, E. B. (1968b). Courtship and mating time in *Drosophila pseudoobscura. Anim. Behav.* 16, 470–479.
SPIESS, E. B. (1970). Mating propensity and its genetic basis in *Drosophila. In*: M. K. Hecht and W. C. Steere (eds.), *Essays in evolution and genetics in honor of Theodosius Dobzhansky*, North Holland Publishing Co., Amsterdam.
SPIESS, E. B. (1982). Minority mating advantage of certain eye color mutants of *Drosophila melanogaster.* III. Female discrimination and genetic background. *Genetics* 12, 209–221.
SPIESS, E. B. and CARSON, H. (1981). Sexual selection in *Drosophila silvestris* of Hawaii. *Proc. Natl. Acad. Sci. USA* 78, 3088–3092.
SPIESS, E. B. and DAPPLES, C. C. (1980). A model of fly mating intensity, not behavior. *Amer. Nat.* 118, 307–315.
SPIESS, E. B. and KRUCKEBERG, J. F. (1980). Minority advantage of certain eye color mutants of *Drosophila melanogaster.* II. A behavioral basis. *Amer. Nat.* 115, 307–327.
SPIESS, E. B. and SPIESS, L. D. (1969). Mating propensity, chromosomal polymorphisms, and dependent conditions in *Drosophila persimilis,* part 2: Factors between larvae and between adults. *Evolution* 23, 225–226.
SPIESS, E. B. and SCHWER, W. A. (1978). Minority mating advantage of certain eye color mutants of *Drosophila melanogaster.* I. Multiple choice and single female tests. *Behav. Genet.* 8, 155–168.
SPIESS, L. D. and SPIESS, E. B. (1969). Minority advantage in interpopulational matings of *Drosophila persimilis. Amer. Nat.* 103, 155–173.
SPIETH, H. T. (1947). Sexual behavior and isolation in *Drosophila.* I. The mating behavior of the *willistoni* group. *Evolution* 1, 17–31.
SPIETH, H. T. (1949). Sexual behavior and isolation in *Drosophila.* II. The interspecific mating behavior of species of the *willistoni* group. *Evolution* 3, 67–81.
SPIETH, H. T. (1951). Mating behaviour and sexual isolation in the *Drosophila virilis* species group. *Behaviour* 3, 105–145.
SPIETH, H. T. (1952). Mating behaviour within the genus *Drosophila* (Diptera). *Bull. Amer. Mus. Natur. Hist.* 99, 395–474.
SPIETH, H. T. (1957). *Drosophila* of the Itasca Park Minnesota region. *N.Y. Entomol. Soc.* 55, 89–96.
SPIETH, H. T. (1958). Behavior and isolating mechanisms. *In*: A. Roe and G. C. Simpson (eds.), *Behavior and evolution,* 363–389, Yale Univ. Press, New Haven, Connecticut.
SPIETH, H. T. (1966a). *Drosophila* mating behavior: The behavior of *ananassae* and some *ananassae*-like flies of the Pacific. *Univ. Texas Publ.* 6615, 133–145.
SPIETH, H. T. (1966b). Courtship behavior of endemic Hawaiian *Drosophila. Univ. Texas Publ.* 6615, 245–313.
SPIETH, H. T. (1966c). Drosophilid mating behavior: the behavior of decapitated females. *Anim. Behav.* 14, 226–235.
SPIETH, H. T. (1968). The evolutionary implications of sexual behavior in *Drosophila. Evol. Biol.* 2, 157–191.
SPIETH, H. T. (1969). Courtship and mating behavior of the *Drosophila nasuta* subgroup of species. *Univ. Texas Publ.* 6918, 255–270.
SPIETH, H. T. (1970). *In: Essays in evolution and genetics in honor of Theodosius Dobzhansky* (M. K. Hecht and W. C. Steere, eds.). Appleton-Century-Crofts, N.Y., pp. 438–449, 469–491.
SPIETH, H. T. (1974a). Courtship behaviour in *Drosophila. Ann. Rev. Entomol.* 19, 385–405.
SPIETH, H. T. (1974b). Mating behavior and evolution of the Hawaiian *Drosophila. In*: M. J. D. White (ed.), *Genetic mechanisms of speciation in insects.* Australia and New Zealand Book Co., Sydney.

SPIETH, H. T. (1978). Courtship patterns and evolution of the *Drosophila adiastola* and *plantitibia* species subgroups. *Evolution* **32**, 435–451.
SPIETH, H. T. and HSU, T. C. (1950). The influence of light on the mating behavior of seven species of the *Drosophila melanogaster* species group. *Evolution* **4**, 316–325.
STALKER, H. D. (1942). Sexual isolation in the species complex *Drosophila virilis*. *Genetics* **27**, 238–257.
STREISINGER, G. (1948). Experiments on sexual isolation in *Drosophila*. IX. Behavior of males with etherized females. *Evolution* **2**, 187–188.
STRÖMNAES, O. and KVELLAND, I. (1962). Sexual activity of *Drosophila melanogaster* males. *Hereditas* **48**, 442–470.
STURTEVANT, A. H. (1915). Experiments on sex recognition and the problem of sexual selection in *Drosophila*. *J. Anim. Behav.* **5**, 351–366.
STURTEVANT, A. H. (1920). Genetic studies on *Drosophila simulans*. I. Introduction. Hybrids with *D. melanogaster*. *Genetics* **5**, 488–500.
STURTEVANT, A. H. (1921). The North American species of *Drosophila*. *Carnegie Inst. Washington Publ.* **301**, 1–150.
STURTEVANT, A. H. (1942). The classification of the genus *Drosophila* with description of nine new species. *Univ. Texas Publ.* **4213**, 5–66.
SUSTARE, D. (1978). Systems diagrams. *In*: P. W. Colgan (ed.), *Quantitative ethology*. John Wiley and Sons, New York.
TAN, C. C. (1946). Genetics of sexual isolation between *Drosophila pseudoobscura* and *Drosophila persimilis*. *Genetics* **31**, 558–573.
TAYLOR, G. E. (1975). Differences in mating propensity: some models for examining the genetic consequences. *Behav. Genet.* **V**, 381–393.
THODAY, J. M. (1964). Effects of selection for genetic diversity. *Proc. 11th Int. Congr. Genet.* **3**, 533–540.
THODAY, J. M. and GIBSON, J. B. (1962). Isolation by disruptive selection. *Nature, Lond.* **193**, 1164–1166.
THODAY, J. M. and GIBSON, J. B. (1970). The probability of isolation by selection. *Amer. Nat.* **104**, 219–230.
TRIVERS, R. L. (1972). Parental investment and sexual selection. *In*: B. Campbell (ed.), *Sexual selection and the descent of man, 1871–1971*. Aldine Publishing Co., Chicago.
WALDRON, I. (1964). Courtship sound production in two sympatric sibling *Drosophila* species. *Science* **144**, 191–193.
WALLACE, B. (1954). Genetic divergence of isolated populations of *Drosophila melanogaster*. *Proc. IX Int. Congr. Genet., Caryologia* **6** (Suppl.), 761–764.
WALLACE, B. and DOBZHANSKY, TH. (1946). Experiments on sexual isolation in Drosophila. VIII. Influence of light on the mating behavior of *D. subobscura*, *D. persimilis*, and *D. pseudoobscura*. *Proc. Natl. Acad. Sci. U.S.A.* **32**, 226–234.
WASSERMAN, M. and KOEPFER, H. R. (1980). Does asymmetrical mating preference show the direction of evolution? *Evolution* **34**, 1116–1124.
WATANABE, T. K. and KAWANISHI, M. (1979). Mating preference and the direction of evolution in *Drosophila*. *Science* **205**, 906–907.
WHEELER, M. R. (1947). The insemination reaction in intraspecific mating of *Drosophila*. *Univ. Texas Publ.* **4720**, 78–115.
WHEELER, M. R. (1959). A nomenclatural study of the genus *Drosophila*. *Univ. Texas Publ.* **5914**, 181–205.
WILSON, E. O. (1975). *Sociobiology the new synthesis*. The Belknap Press, Cambridge, Massachusetts.

WOOD, D. F. and RINGO, J. M. (1980). Male mating discrimination in *Drosophila melanogaster*, *D. simulans*, and their hybrids. *Evolution* **34**, 320–329.

WOOD, D., RINGO, J. M. and JOHNSON, L. L. (1980). Analysis of courtship sequences of the hybrids between *Drosophila melanogaster* and *Drosophila simulans*. *Behav. Genet.* **10**, 459–466.

WOOLF, C. M. (1968). *Principles of Biometry*. D. Van Nostrand Co., Inc., N. J.

YULE, G. U. (1912). On the methods of measuring association between two attributes. *J. Roy. Statis. Soc.* **75**, 579–642.

ZOUROS, E. (1981). The chromosomal basis of sexual isolation in two sibling species of *Drosophila*: *D. arizonensis* and *D. mojavensis*. *Genetics* **97**, 703–718.

ZOUROS, E. and D'ENTREMENT, C. J. (1980). Sexual isolation among populations of *Drosophila mohavensis*: Response to pressure from a related species. *Evolution* **34**, 421–430.

24. Interspecific Competition

J. S. F. BARKER

*Department of Animal Science,
University of New England,
Armidale, N.S.W., Australia*

I. Introduction	285
A. The Concept of Competition	286
B. The Nature of Competition	286
II. Competitive Exclusion or Coexistence	287
A. Principle of Competitive Exclusion	287
B. Possible Outcome of Competition	289
C. Exclusion and Coexistence in Laboratory Populations of *Drosophila*	291
III. Competition in Natural Populations of *Drosophila*	296
A. Detection and Importance of Competition	296
B. Indirect Evidence of Competition	298
C. Direct Evidence of Competition	300
D. Microhabitat Divergence and Competition	303
IV. *Drosophila* as a Laboratory Population Model	305
A. Ecology of Competition	305
B. Competitive Ability as a Measure of Population Fitness	320
C. Changes in Competitive Ability in Two-species Populations	325
D. Competition Involving More than Two Species	328
V. Implications in Evolution, Ecology and Biological Control	330
Acknowledgements	333
References	333

I. Introduction

It is hardly to be questioned, therefore, that the problem of the growth of a mixed population of two species is worthy of a very careful study so that it could give a sound basis for the development of complicated genetic theories of natural selection. (Gause and Witt, 1935).

In the period since this observation was made, many field, laboratory and theoretical studies have been done, but the integration of population genetics and ecology into a coherent evolutionary biology of populations remains to be achieved. The ecological problem is understanding the factors determining the numbers of animals or plants, while the genetical problem is understanding the factors determining the kinds. While the disciplines have been largely disjunct (Sammeta and Levins, 1970), the problems are interdependent in that changes in numbers may lead to changes in kinds, and vice versa (Birch, 1960).

Studies of the effects of association of two or more species, which may involve competition among them, are but part of the total effort devoted to analysis of these interdependent problems, although it should be noted that many workers, from Darwin (1859) on, have attributed to competition between closely related species a major role in natural selection.

A. The Concept of Competition

The term "competition" has been used in biology to refer to a variety of phenomena and there has been extensive discussion and review of its meaning and significance (e.g., Birch, 1957; Milne, 1961; Donald, 1963). Some workers (e.g., Ehrlich and Holm, 1962) even consider that it has lost its usefulness as a scientific term. However, Bakker (1961), Ayala (1970a) and Barker and Podger (1970a) are agreed that the term is useful when applied in its original strict meaning, and have used it in the sense of the first meaning of Birch (1957): "Competition occurs when a number of animals (of the same or of different species) utilize common resources the supply of which is short; or if the resources are not in short supply, competition occurs when the animals seeking that resource nevertheless harm one or other in the process."

The "short supply" in this definition must be seen in terms of availability of the resource to the animals, and may be an absolute shortage—when there is simply not enough of the resource available for all of the animals present, or a relative shortage—when there is ample resource available for all, but animals seeking the resource choose to fight over single resource items.

Clearly there is more than one form of competition implicit in this definition, so that it is necessary to ask: What is the nature of competition and what types of competition may be recognized? And also: What is a common resource that can be considered an object of competition? How can the action of competition be detected and evaluated, and what constitutes adequate proof of competition?

B. The Nature of Competition

Park (1954) distinguished *exploitation* and *interference* as components of competition. While both may be operating in some cases of competition, they are distinct processes that may occur in isolation from each other, and hence are better described as categories rather than components. The meanings of exploitation and interference are inherent in the two parts of Birch's definition of competition. Thus exploitation refers to the use of a limited resource by competing individuals, while interference refers to any activity which directly or indirectly reduces a competitor's access to a resource. Where interspecific competition involves interference, the resource will be used less efficiently than when only one species is present.

A different distinction of aspects of competition, *viz. contest* and *scramble*, was made by Nicholson (1954), depending on the way of utilization of the resource. If animals contest for a resource, each one is either fully successful or unsuccessful, *i.e.* each resource unit is completely utilized by one of the competitors. If animals scramble for a resource, each one tries to acquire as much as possible up to some necessary minimum. Thus the resource is shared in various proportions by the competitors. Some may not acquire sufficient for survival, so that the portion of the resource they have used is wasted.

Various interpretations of these aspects of competition are to be found in the literature. Thus Miller (1967) equates exploitation and scramble, while at the other extreme, de Jong (1976) considers that "both scramble and contest competition can be by the mechanisms of exploitation and interference". However, as Birch (1957) points out, contest invariably involves interference of some sort. Therefore, taking account of the available levels of a resource, the possible mechanisms of the competition process are shown in Fig. 1.

II. Competitive Exclusion or Coexistence

A. Principle of Competitive Exclusion

The idea that two species competing for the same limiting resources cannot coexist indefinitely, generally referred to as the Principle of Competitive Exclusion (Hardin, 1960; Hutchinson, 1965; Ayala, 1970a), has generated vast controversy and a great deal of confusion. Some workers (e.g. Hardin, 1960; Hutchinson, 1965; DeBach, 1966) consider it to be of fundamental importance in ecology and evolution; others think that it is either invalid or trivial, depending on how it is formulated (e.g., Andrewartha and Birch, 1954; Cole, 1960), or misleading (Darlington, 1972). Some of the confusion has arisen because it has been defined in different ways—it means different things to different people. On the other hand, some has been due to the principle being regarded as a theory, on which predictions or falsifiable inferences may be based. However, as emphasized by Peters (1976), the

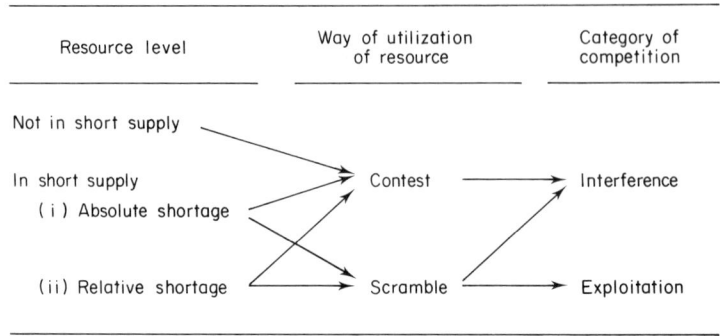

FIG. 1. Possible relationships between resource level, its way of utilization and category of competition.

principle is not subject to empirical falsification and is incapable of prediction; it is a tautology.

All aspects of the controversy surrounding this idea cannot be reviewed here, but Hutchinson (1975) has considered the history of the idea, while DeBach (1966), Miller (1967), Ayala (1970a) and Darlington (1972) provide more detailed review and comment, and aspects of the theory have been discussed by MacArthur (1972).

The principle of competitive exclusion is sometimes referred to as Gause's principle, as his work (1934) certainly provided much of the stimulus. The principle is, as acknowledged by Gause, based on a model of interspecific competition proposed independently by Lotka (1925) and Volterra (1926). The model is a simple one which assumes linear competitive interactions, both intra- and interspecific, and that the species are competing for a single limiting factor or resource. Thus the instantaneous rates of growth of the two species are given by the differential equations:

$$\frac{dN_1}{dt} = r_1 N_1 \frac{K_1 - N_1 - \alpha N_2}{K_1}$$

$$\frac{dN_2}{dt} = r_2 N_2 \frac{K_2 - N_2 - \beta N_1}{K_2}$$

where N_1 and N_2 are the numbers of the two species, r_1 and r_2 their rates of increase when there are no limiting resources, and K_1 and K_2 the maximum numbers of each species growing alone when the limiting resource is saturated. β/K_2 is the inhibitory effect of species 1 on the growth of species 2, and α/K_1 is the reciprocal effect of species 2 on species 1. This model is variously referred to in the literature as the Lotka–Volterra model (or equations), as the Gause equations or as the Gause and Witt equations.

While Gause's experiments did not prove the principle of competitive exclusion (see Peters, 1976), as some have claimed, he did demonstrate that when two species of protozoans were placed in competition for food in a laboratory environment, one species always became extinct. Numerous subsequent laboratory studies, particularly using granivorous insects and *Drosophila*, have shown that when two species are placed together in a closed environment, then in from a few to many generations, one species usually is eliminated. As Miller (1967) points out, these experiments confirm that competitive exclusion can be demonstrated, but he also claims that there has been a tendency "to ignore the more interesting fact that co-existence for many generations has been repeatedly observed in highly simplified laboratory environments". To a large extent, this simply reverses the problem. How many generations are sufficient for it to be said that two species are coexisting? Can two species be said to be coexisting while the frequency of one is steadily decreasing?

B. Possible Outcome of Competition

In natural populations, any one species is associated with many other species, some of which may compete with it for limited resources. The problem then is to reconcile this natural association of species with the principle of competitive exclusion and with laboratory studies showing that competitive exclusion can be demonstrated. What is needed is an explanation of the fact that competitors do coexist.

Much of the discussion of the competitive exclusion principle has been unrealistic in ignoring the genetic composition of the competing species, and ignoring the possibility of genetic change that could be a direct result of the competition, and that could change the outcome of competition. Studies in *Drosophila* analysing such changes are discussed in Section IV, C. However, given that evolutionary changes may occur, the possible outcomes when two species become associated include (Barker, 1973a):

1. Elimination of one species (competitive exclusion)

(a) Elimination of one species without any evolutionary changes in either species.

(b) Elimination of one species with *natural selection for increased competitive ability* in the successful species during the competitive process. Increased competitive ability may result from changes which allow more effective *exploitation* of the environment through an increased utilization of a limiting or of an alternate resource, or from changes which lead to increased *interference*.

(c) Elimination of one species with the rate of change in frequency of this species increasing during the competitive process due to genetic change causing decreased competitive ability in the unsuccessful species. This possibility would seem unlikely, but is included for completeness.

2. *Coexistence due to environmental or extrinsic factors*

(a) Coexistence because of *rapidly and randomly changing environmental conditions* (Hutchinson, 1961). Such coexistence might be expected to be unstable, and if the species do compete and differ in competitive ability, one will eventually be eliminated (provided evolutionary changes of types 3b(i) or 3b(ii) did not occur).

(b) Coexistence because *competitive equilibrium is prevented by periodic population reductions and environmental fluctuations* (Huston, 1979).

(c) Coexistence because of *cyclical changes in the environment* favouring first one species, then the other (Merrell, 1951).

(d) Coexistence because of *stabilization by predation* (Paine, 1966; May, 1973; Connell, 1975).

(e) Coexistence because of differential *frequency-dependent predation* (apostatic selection—Clarke, 1962).

3. *Coexistence due to intrinsic factors*

(a) With no evolutionary change in either species during association:
   (i) Coexistence because the species *do not compete* for the same limiting resources.
   (ii) Coexistence because of *frequency-dependent competitive fitnesses* (de Wit, 1960; Antonovics and Ford, 1972).
   (iii) Coexistence because of *density-dependent competitive fitnesses* (Bulmer, 1974).
   (iv) Coexistence because of the *reciprocal competitive abilities* of the two species at different life cycle stages, one species being better at one stage, but worse at another (Ayala, 1970a). While theoretically possible and included for completeness, the conditions for a "stable" coexistence are very restrictive—competitive abilities at the two stages must, on average over generations, be exact reciprocals, and even then continued coexistence would depend on the magnitude of the absolute competitive ability relative to its variance over generations.

(b) With evolutionary change in either or both species during association:
  (i) Coexistence because the otherwise inferior competitive ability of one species is balanced by an advantageous effect due to the life processes of the other species—*facilitation*, or if each species beneficially affects the other, *mutual facilitation* (Bos et al., 1977). If facilitation were present initially, this mechanism would be classified under 3(a).
  (ii) Coexistence because of evolutionary changes in either or both species leading to decreased niche overlap—*ecological divergence*, subsumed under *character displacement* (Brown and Wilson, 1956).
  (iii) Coexistence because of *genetic feedback* (Pimentel, 1968)—selection for increased competitive ability in the rarer species. This would lead to fluctuations in numerical dominance rather than a "stable" coexistence, although the magnitude of the oscillations might be expected to decline. In a laboratory system, such changes could not continue indefinitely, and if the only changes were for increased competitive ability, the long-term outcome must be competitive exclusion when one species can no longer match the then superior competitive ability of the other by further change.

C. EXCLUSION AND COEXISTENCE IN LABORATORY POPULATIONS OF *DROSOPHILA*

Laboratory studies of interspecific competition in continuous populations of *Drosophila* were initiated by L'Héritier and Teissier (1935), using the population cage technique that they had developed (1933) for studies of intraspecific selection among genotypes. The species used were *D. melanogaster* and *D. funebris*, and they observed that the populations reached an equilibrium with a high proportion of *D. melanogaster*.

Experiments of Zimmering (1948) used "interspecific competition as a means of determining the relative fitness of different species or strains of the same species". Two wild-type strains of *D. pseudoobscura* and a wild-type and three mutant strains of *D. melanogaster* were used. In all cages, one or the other species was excluded rapidly, the maximum time being 110 days (i.e. only about five generations, Barker, 1962). Wild-type and strains of *D. melanogaster* homozygous for one or for three mutants excluded *D. pseudoobscura*, but the strain of *D. melanogaster* homozygous for six mutants was excluded by *D. pseudoobscura*; that is, differences among strains in competitive ability were demonstrated. This aspect of interspecific competition is taken up in detail in Section IV, B.

The apparent coexistence of *D. melanogaster* and *D. funebris* observed by L'Héritier and Teissier (1935) was studied further by Merrell (1951), using the population bottle technique of Reed and Reed (1948). When food was added at regular intervals, and population counts made at about the same relative time in the cycle of food addition, there was an apparent fairly stable equilibrium with *D. melanogaster* predominating. Merrell showed that *D. melanogaster* was favoured by the addition of fresh food, while the proportion of *D. funebris* increased with the age of the food, so that coexistence was due to fluctuations in the environment which alternately favoured the two species. Long-term coexistence ("more than 100 generations") of wild-type *D. simulans* and a strain of *D. melanogaster* homozygous for the mutant *white*, with an equilibrium frequency of the latter of about 3% was reported by Goldstein (1953), although no analysis of the nature of the equilibrium was made.

A number of studies of interspecific competition in *Drosophila*, like those of Zimmering (1948) and Goldstein (1953), have used mutant strains of one or both species, and it is rather curious that such studies are largely ignored in the ecological literature. Apparently mutant strains are considered to be aberrant and irrelevant to the real world. Yet many of the "wild-type" strains used in laboratory studies have been ones descended from single wild-caught females (isofemale strains) or inbred laboratory strains. As such, they are no more representative of the species than are mutant strains. Further, there has been a tendency to ignore genetic variation within the species, and to generalize on competition between species A and species B from the results of a particular experiment, when the results of course apply only to the particular strains of each species used in that experiment. Comparative studies using a variety of strains are essential before general conclusions about the nature of the competitive process can be drawn, and then the emphasis could well be on the magnitude of genetic variation in competitive ability.

To return from this digression to the question of exclusion or coexistence, Merrell (1951) concluded that the coexistence between *D. melanogaster* and *D. funebris* did not constitute an exception to the principle of competitive exclusion, because the two species were alternately favoured by the fluctuating environment. On the other hand, Ayala (1969a, 1970a, 1971a, 1972) has interpreted observed apparent coexistence of *D. serrata* with *D. pseudoobscura* and of *D. willistoni* with *D. pseudoobscura* as invalidating the principle of competitive exclusion. All populations were maintained in bottles using the serial transfer technique (Buzzati-Traverso, 1955; Ayala, 1965). In the initial experiment (Ayala, 1969a), stability of the apparent coexistence was disturbed by a change in the temperature at which the populations were maintained (from 23·5°C to 23°C), and as the populations were studied for only 40 weeks, I for one (Barker, 1973a) was not convinced that a stable equilibrium coexistence had been demonstrated. In later evidence (Ayala, 1972) for populations of *D. pseudoobscura* and *D. willistoni*, a stable coexistence clearly was maintained (Fig. 2).

The results of these experiments, or rather Ayala's interpretation of them, led to considerable discussion, and to model-building and development of the theory of interspecific competition in *Drosophila*. Gause (1970) was the first to criticize Ayala's interpretation, claiming that the experimental environment contained two niches—a solid phase where larvae of *D. serrata* were superior, and an aerial phase where adults of *D. pseudoobscura* were superior—and that coexistence therefore would be predicted. In reply, Ayala (1970b) disagreed, arguing that the two species compete for limiting resources in both phases. Borowsky (1971) analysed the magnitude and pattern of temporal fluctuations in adult numbers in Ayala's populations, and concluded that a stable equilibrium was not attained and that the decrease in population numbers in the mixed species populations, as compared with that expected from single-species populations, was due to short-term fluctuations in temperature. This latter was shown to be unlikely by Ayala (1971b). However, he did not comment on the question of whether the equilibrium was stable or not, reasserting that the species did coexist and that the results were not consistent with the necessary conditions for a stable equilibrium in the Gause and Witt (1935) equations.

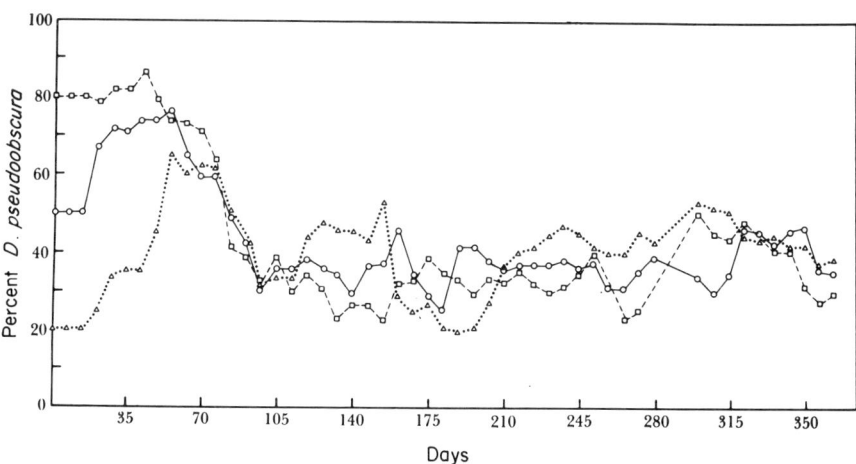

FIG. 2. Frequency of *D. pseudoobscura* in competition with *D. willistoni* in three populations started with different frequencies (From Ayala, 1972).

However, as pointed out by Antonovics and Ford (1972), the observed coexistence reflects more on the applicability of these equations, which refer to competition for a *single* limiting resource. Rather, they emphasized the possibility of frequency-dependent relationships between the species, which could account for the results observed. The question of the applicability of the equations also was raised by Gilpin and Justice (1972) and Coleman and Gomatam (1972). The former developed a phenomenological non-linear model based on Ayala's data; the latter introduced a new non-linear model whose equations could be solved analytically. Both showed a stable equilibrium to be possible, and both noted that the necessary and sufficient condition for the stability of a two-species competitive equilibrium was that the product of the coefficients of intraspecific competition be greater than the product of the coefficients of interspecific competition.

Ten new models were introduced and tested against experimental data by Ayala *et al.* (1973). Emphasis was no longer placed on any alleged lack of validity of the principle of competitive exclusion, but more realistically on the realization that "the widely used Lotka–Volterra model of competition between species cannot account for the process of competition as studied in experimental systems with *Drosophila* species". Experiments to test the models were of two types: (1) continuous one or two-species populations, which allow estimation of the carrying capacity of each species and the numbers of the two species at the point of stable equilibrium, and (2) single-generation experiments to measure changes in numbers in populations started at many different initial densities and frequencies of the two species. The results from one such set of experiments, using one strain of each of the species *D. pseudoobscura* and *D. willistoni*, are shown in Fig. 3. Such data provide for very adequate testing of the models, but it must be noted that analysis is restricted to systems that do reach a stable equilibrium. Two of the models, each with four parameters (one more than the Lotka–Volterra model) gave a good fit to the data summarized in Fig. 3.

In relation to the earlier comment on the use of mutant and inbred strains in laboratory studies of interspecific competition, much of the discussion of Ayala's work has referred to competition between the species *D. pseudoobscura* and *D. willistoni*. Yet only one strain of each species was studied, and the strain of *D. willistoni* (M11) was homozygous for the second chromosome, and more than 90% of the rest of its genome came from laboratory stocks that were probably highly inbred (Mourão and Ayala, 1971).

Of the two best models identified by Ayala *et al.* (1973), one was rejected later by Gilpin and Ayala (1973) on the basis of lack of biological reality. For the remaining model, they conclude by saying that if it "proves equally

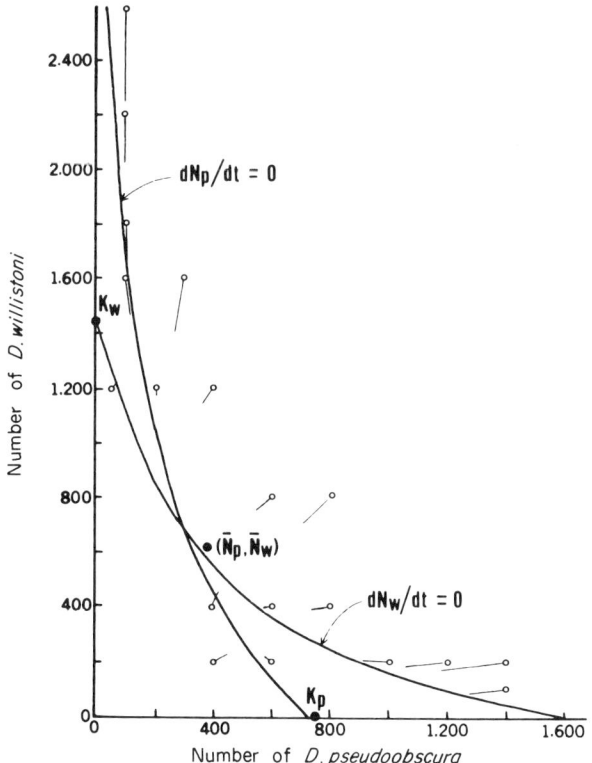

FIG. 3. The phase plane description of the dynamics of a *D. pseudoobscura* and *D. willistoni* competitive system. The *open circles* represent initial densities of the two species in Type (2) experiments. The *directed line segments* (or *vectors*) show the changes in the system after one week (*Vectors* reduced to one third their true length for clarity). The *solid lines* are the zero isoclines for the two species. They separate the phase plane region of positive growth from that of negative growth and have been drawn by visual inspection of the data. The *solid circles* represent the carrying capacities (K_p and KINw) and equilibrium point ($\bar{N}_p$, $\bar{N}_w$) of the system as determined by Type (1) experiments (from Gilpin and Ayala, 1973).

useful with other organisms, ... (it) may be advanced as a globally accurate model of intraspecific and interspecific competition."

Clearly Ayala *et al.* (1973) and Gilpin and Ayala (1973) have extended our understanding of interspecific competition in laboratory populations of *Drosophila* by the combination of experimentation and modelling. However, there are two restrictions which must be noted and which emphasize the need for further study. Firstly, their model of choice is based on experimentation with only one strain of each of two species, and one of these highly inbred. It would seem rather presumptive to suggest "a globally accurate model" on such a limited sample. Even for competition in *Drosophila*, its generality remains to be tested for other strains of the same two species, and for other species. Further, the model is a linear model of interspecific competition, although the intraspecific competition is non-linear. Thus it may fail to be a general model as work with species other than *Drosophila* (Neill, 1974; Smith-Gill and Gill, 1978) has shown that non-linear interactions between species may be very important in many competitive relationships, while for cactophilic species of *Drosophila*, Mangan (1978) (see Section III, C) found non-linear interspecific interactions for particular fitness components that would contribute to (or be components of) competitive ability. Secondly, while their model may fit their data, it does not provide direct information on the nature of the competitive process. In other analyses of what is essentially the same data, Ayala (1971a) reported that the outcome of larval competition was strongly frequency-dependent and that, at the equilibrium, the advantage of *D. pseudoobscura* as adults was exactly compensated by their disadvantage at the larval stage.

A continuing feedback between experimentation and model-building is essential, and the former must include studies of the biological mechanisms of the competitive process and the dynamics of competition, as well as competitive equilibria.

III. Competition in Natural Populations of *Drosophila*

A. Detection and Importance of Competition

In discussing the concept of competition (Section I, A) the problem was raised as to how the action of competition can be detected and evaluated, and what constitutes adequate proof of competition. Directly identifying the process of competition in natural populations generally will be difficult, because, as noted by Simberloff (1978), "competition, particularly exploitation rather than interference competition, is a low-visibility phenomenon." Thus presumed effects of past competition usually are inferred, rather than present competitive interactions being observed. Such inferences may be based on patterns of geographical or habitat distribution and of niche relationships (MacArthur, 1972). Thus mutually exclusive ranges of species of apparently similar ecology may be taken as evidence for the operation of interspecific competition. This evidence would be much stronger if it could be demonstrated that a species would occupy a broader range or habitat in the absence of the presumed competitor (Connell, 1975).

For *Drosophila* species, controlled field experiments involving removal of some presumed competitor species would be difficult, to say the least, although addition of a presumed competitor may be easier. Further, mutual exclusion is likely to be at the microhabitat level rather than the geographical range of the species, so that evidence for extension in a broader niche would have to be sought at that level. The uncertainties in inferring competitive effects from species distribution patterns or niche relationships have been discussed by Miller (1967), who concluded that "species which compete primarily through exploitation are more often controlled by physical factors which affect their distribution and abundance and are less likely to be spatially exclusive, except at the level of microhabitat distribution. They are also more likely to show complete differentiation in some critical element of their fundamental niches." Of course, while physical factors may control the geographical distribution of species such as *Drosophila*, and their abundance within this distribution, the critical point is that associations will be at the microhabitat level, and there may be competitive exclusion at this level. In addition, it should be recognized that any differentiation in elements of the fundamental niches may itself have evolved because of past competitive associations (Section II, B).

Nevertheless, conventional wisdom of invertebrate zoologists seems to be that interspecific competition is not of major importance in natural populations. For *Drosophila*, this is supported by Sokoloff (1964, 1966), who concluded that *D. pseudoobscura* and *D. persimilis* were subject to minimum intra- and interspecies competition under natural conditions. Sokoloff suggested that larval competition in the slime flux breeding site of these species may be precluded by predators and associated species that also utilize the slime flux, preventing excessive oviposition by *Drosophila*. Comparable studies of other species have not been made, so the generality of Sokoloff's conclusion remains to be determined. Nevertheless, it should be noted that there are some *Drosophila* species that are not subject to interspecific competition, at least with other species of *Drosophila*, viz. those species that are uniquely adapted to a specific niche, such as *D. pachea* to senita cactus (*Lophocereus schottii*), *D. flavopilosa* to the flowers of *Cestrum parqui*, *D. carcinophila* to the external nephric groove of the land crab *Gecarcinus ruricola* (Carson, 1971). Clearly these are special cases.

My own impression, based on extensive collecting of wild adults, and collection in the laboratory of adults emerging from natural substrates, is that the cosmopolitan species *D. melanogaster* and *D. simulans* and the cactophilic species *D. buzzatii* and *D. aldrichi* are subject to intraspecific competition, and given that both species of each of these pairs often emerge from the same substrate, interspecific competition may be inferred.

B. Indirect Evidence of Competition

As noted previously, inferences of competitive interactions, based on species distributions, species diversity and niche relationships, have been made far more often than they have been demonstrated.

The essential argument is that interspecific competition reduces fitness of at least one of the species, resulting in competitive exclusion (extinction) or competitive displacement (change in resource utilization). The simple expectations then are that competitive exclusion at the geographic level will lead to disjunct species distributions, and at the community level to decreased species diversity. On the other hand, competitive displacement, operating only at the community level, will lead to increased species diversity, decreased niche overlap, and may involve character displacement for either morphological, physiological, ecological or behavioural characters. Undoubtedly, competition has been an important force in the evolution of natural communities, but while all of the above expectations may be reasonable, the problem remains that all may be explained by mechanisms other than interspecific competition.

On the basis of distribution and abundance data for *D. athabasca* and *D. affinis* on the mainland and on large and small islands in the vicinity of Deer Isle, Maine, U.S.A., Jaenike (1978) argued that these two species compete in nature. The evidence is circumstantial and the argument depends largely on the observation that the frequency of *D. athabasca* decreases as island size decreases. As nothing significant is known of the natural breeding sites of these two species, the observed distributions and abundances could be quite independent of interspecific competition and due to ecological variation that is correlated with island size. The conclusion that competition plays any role at all is completely inappropriate without initial investigation to show that the substrates of the two species do indeed overlap.

A theoretical approach to the evaluation of competition between species in a community was introduced by Levins (1968). He suggested that habitat separation may provide a measure of niche overlap, related various measures of overlap to the competition coefficients (α's) of the Lotka–Volterra equations, and defined the community matrix as the matrix of pairwise competition coefficients. If the niche overlap of two species, for whatever the niche variables considered, is not zero, competition is inferred. Richardson and Smouse (1975) used this theory in a study of three species of Hawaiian *Drosophila* (*D. imparisetae*, and two sibling species *D. mimica* and *D. kambysellisi*). The first two species utilize *Sapindus* fruit for a larval substrate, while the last uses *Pisonia* leaves. Results (Table I) were interpreted to indicate competition among these species, with least overlap ($\equiv$ least competition) between the two sibling species. In addition, they

TABLE I. Estimated community matrix for three species of Hawaiian *Drosophila* (from Richardson and Smouse, 1975)

	D. mimica		D. kambysellisi		D. imparisetae	
	Males	Females	Males	Females	Males	Females
D. mimica						
Males	1·0000	0·6616	0·4601	0·3605	0·4901	0·4975
Females	0·6616	1·0000	0·4958	0·4302	0·5608	0·6106
D. kambysellisi						
Males	0·4601	0·4958	1·0000	0·6964	0·4987	0·4969
Females	0·3605	0·4302	0·6964	1·0000	0·4825	0·4674
D. imparisetae						
Males	0·4901	0·5608	0·4987	0·4825	1·0000	0·7226
Females	0·4975	0·6106	0·4969	0·4674	0·7226	1·0000

suggested that *D. imparisetae* may have acted as an ecological wedge that contributed to habitat displacement and reproductive isolation of the two sibling species; that is, past interspecific competition contributed to sympatric speciation.

Laboratory studies of interspecific competition in *Drosophila* (Section II, C) have shown the inadequacy of the Lotka–Volterra model and the need to include non-linear parameters to account for higher-order interactions, while the community matrix and related concepts have been criticized by Heck (1976) as unrealistic and potentially misleading. Richardson and Smouse (1975) recognized the simplifying assumptions in their use of the community matrix, but said "the patterns are clear enough" and "we suspect the basic conclusions are qualitatively correct, even if numerically approximate." Nevertheless, without more detailed ecological analysis of the nature of any competition between these species, their conclusions must be viewed with caution. Some ecological evidence is available showing that, although both *D. imparisetae* and *D. mimica* use *Sapindus* fruit as larval substrate, they are at least partially non-overlapping. The former prefers wounded decaying parts of the fruit for oviposition, while the latter prefers undamaged, green tissue, and analysis of larval mouth hooks suggests that the larvae remain largely separate in these different tissues (Mangan, personal communication).

Competition between *D. melanogaster* and *D. simulans* may be inferred from changes in their relative proportions in natural populations. Thus Tantawy *et al.* (1970) found for one locality in Egypt that the relative frequencies of these two species changed over a ten year period from an average of about 70% *D. melanogaster* to one of more than 90% *D. simulans*. A similar change in the relative proportions of these two species, but over a much shorter time span, was reported by Hoenigsberg (1968).

Direct evidence for competition between these two species may come from a natural experiment in Japan, where *D. simulans* had not been recorded prior to 1972 (Watanabe and Kawanishi, 1976). Subsequent collections (Kawanishi and Watanabe, 1977) have shown site and seasonal differences in the relative frequency of the two species, and future collections to record any increase in the distribution of *D. simulans* and any consequent decrease in the *D. melanogaster* population will be most interesting.

C. Direct Evidence of Competition

Carson (1971) proposed that the major specificity of the ecology of *Drosophila* relates to the niche in which the female of the species deposits her eggs. Unfortunately, the species that are best known genetically and that have been used extensively in studies of the genetics of natural populations, such as *D. melanogaster*, *D. simulans*, *D. pseudoobscura* and *D. willistoni*, are domestic or are generalists as regards oviposition site, or both. While the genetic basis of generalism may include microhabitat adaptations allowing competitive coexistence, natural populations of these species are less amenable to studies of competitive interactions than are species adapted to specific habitats.

However, Brncic (1962) estimated that at least 90% of *Drosophila* species are "delicately adjusted to specialized environments", and Carson (1971) emphasized that a considerable amount of basic information on *Drosophila* ecology is available. Thus, where two or more species exist in the one community, and are known to utilize the same or similar and relatively specific breeding sites, detailed analyses are possible, and have been initiated for the Sonoran Desert cactophilic *Drosophila* and some of the Hawaiian species.

Extensive study of the Sonoran Desert *Drosophila* has been done by W. B. Heed and his colleagues, their aim being "to uncover the forces maintaining each species in its habitat and to examine the genetic elements that presumably respond to and guide them" (Heed, 1979). The evidence for interspecific competition as one force maintaining each species in its habitat will be discussed here; the broader aspects of these studies are considered elsewhere in this series. Four species of *Drosophila* endemic to the Sonoran Desert breed in fermenting stems (rot pockets) of the giant cacti (tribe *Pachycereae*) and show a high degree of host plant specificity with little species overlap (Table II). *D. pachea* is restricted to senita because this

TABLE II. The major host plants of the four endemic species of *Drosophila* of the Sonoran Desert (after Heed, 1979)

Host plant	Abundance	*Drosophila* species overlap	Resident species
Lophocereus schottii (senita)	+++	No	*D. pachea*
Machaerocereus gummosus (agria)	+[a]	No	*D. mojavensis*
Stenocereus thurberi (organ pipe)	+++	No	*D. mojavensis*
Carnegiea gigantea (saguaro)	+++	Yes	*D. nigrospiracula*
Pachycereus pringlei (cardón)	+[a]	Yes	*D. nigrospiracula*
Saguaro soil	+++	No	*D. mettleri*
Cardón soil	+[a]	No	*D. mettleri*

[a] Host plant rare, but regularly used when present.

cactus contains a unique sterol necessary for the development and fertility of *D. pachea* (Heed and Kircher, 1965; Heed and Russell, 1968). Further, none of eight other *Drosophila* species which inhabit the Sonoran Desert can utilize senita because of toxic alkaloids present in it (Kircher *et al.*, 1967). For the other three species, observed host plant preferences might be due to nutritional deficiency of a host plant for non-resident species, or to competitive exclusion by the resident species in each host plant. Using artificial rots of cactus in the laboratory, Fellows and Heed (1972) showed both factors to be involved and, in particular, that *D. nigrospiracula* was capable of competitively excluding *D. mojavensis* from the saguaro niche. Collection records and field tests showed *D. nigrospiracula* to be fairly tightly bound to its common host saguaro, and it showed active discrimination between host and non-host plants. Thus *D. nigrospiracula* females would be unlikely to oviposit on organ pipe (the common host plant of *D. mojavensis*). However, should they do so, this would not impose competitive effects on *D. mojavensis*, as Fellows and Heed (1972) found that *D. nigrospiracula* suffered mass first-instar larval mortality on organ pipe.

In a series of elegant experiments, Mangan (1978) has extended these studies for competition of *D. nigrospiracula* with either *D. mojavensis* or *D. mettleri*. These experiments were done in the laboratory but using substrates prepared to mimic as closely as possible natural cactus necroses, and eggs used to initiate the experiments were laid by flies not more than one generation removed from the natural population. This experimental strategy, together with prior knowledge of the ecology of the species, allowed direct interpretation of competitive interactions. Further, density and competitive effects were assayed in terms of pre-adult survival, body size (as an index of potential fecundity) and developmental time. Only saguaro (the natural substrate of *D. nigrospiracula*) was used, and the results are summarized in Table III for both intra- and interspecific effects on fitness components. Results are given for each of the three species (as dependent species) for intraspecific competition and for the effects of interspecific competition imposed by the other species (independent species). Except as indicated in Table III, the specification of body size, development time or pre-adult survival in a cell means that that component decreased with increasing density (i.e. a significant competitive effect). On old cactus, all species showed significant effects of increasing density (intraspecific competition) on all three fitness components. *D. mojavensis* was clearly the weakest competitor, but both it and *D. mettleri* imposed some competitive stress on *D. nigrospiracula*. On fresh cactus, there were fewer significant interactions and, in particular, less effect on survival presumably because of the higher nutrient status, and the two potential saguaro rot competitors reacted to high *D. nigrospiracula* densities by changes in the fitness components for which each is most different from *D. nigrospiracula*. Clearly the interspecific interactions are dependent on the nature of the substrate. In both substrates, however, many of the interactions were

TABLE III. Significant interactions between *D. nigrospiracula* and its potential competitors, *D. mojavensis* and *D. mettleri*, for three components of fitness (after Mangan, 1978)

Dependent species	Independent species		
	D. nigrospiracula	*D. mettleri*	*D. mojavensis*
(i) Old cactus			
D. nigrospiracula	size/dev./sur.[a]	sur.	size/dev.
D. mettleri	none	size/dev./sur.	—
D. mojavensis	size/dev./sur.	—	size/dev./sur.
(ii) Fresh cactus			
D. nigrospiracula	size/dev.	size	dev.
D. mettleri	size[b]/dev.	size/dev.	—
D. mojavensis	size/sur.	—	size/sur.[b]

[a] Dev. = development time, sur. = pre-adult survival. [b] Increased with increasing density.

non-linear, and these were interpreted as being primarily due to heterogeneous resource availability. The nature of possible interspecific competition between the saguaro rot resident species *D. nigrospiracula* and two of its potential competitors is clearly indicated in these experiments, and Mangan (1978) concludes that "competition is probably important in maintaining the niche separation, but adaptation to the general natures of these substrates, including distribution, size, and nutritional value, are more likely the forces initiating habitat selection."

In a comparative study, Mangan (1978) also considered the role of interspecific competition in the evolution of a group of eight species of Hawaiian *Drosophila*. These species all utilize decaying *Cheirodendron* leaves as larval substrate, so that the problem was to assess the nature of possible competitive interactions among species apparently coexisting in the same microhabitat. Measurements of body size, ovariole number and egg size were made on field collected adults, while for collected leaves, numbers of eggs, puparia and emerging adults were counted, body size of adults measured and development time and percentage leaf consumption estimated. Field adult collections indicated that the species show some temporal and spatial segregation, while data from the leaf collections gave little evidence for interspecific competition affecting adult size or developmental time, even though there was clear evidence for nutrient depletion of larval substrates. These results indicating lack of interaction led to morphological analyses of larval mouthparts to investigate feeding differences and resource partitioning. Within each of the species groups, closely related species pairs show divergence in mouth hook morphology. Habitat selection, both for the macrohabitat (temporal and spatial variation) and the microhabitat (size, thickness, stage of decay of leaf), is the main factor now allowing coexistence, but the role of interspecific competitive stress as a selective force in the evolution of these species seems certain.

D. Microhabitat Divergence and Competition

The ecological or microhabitat separation of sympatric species is generally assumed to be due to natural selection acting to minimize competition. While critical evidence is only just becoming available—as in the Sonoran Desert and Hawaiian *Drosophila* studies above, *Drosophila* workers have sought evidence for niche separation in adult food preferences to determine the extent to which competition in fact was minimized. Thus Dobzhansky and Pavan (1950) observed that populations of *Drosophila* in Brazil occur on a large variety of fruits, but that most species preferred some fruits to others, and different species often differed in their preferences. These preferences of course are not absolute, so that adults could be in competition for food, and if females of two or more species oviposit on the same substrate, their larval progeny may compete. On the other hand, if the species were actually utilizing different microflora in the substrate, they would not be competing for food at all.

Yeast–*Drosophila* relationships are discussed in detail by Begon in Chapter 17, but some aspects relevant to interspecific competition are discussed here. There is ample evidence that different yeasts do show differential attractiveness to sympatric *Drosophila* species in their natural habitats, and that the spectrum of yeast species isolated from adults is different for different *Drosophila* species (da Cunha *et al.*, 1951, 1957; Dobzhansky and da Cunha, 1955; Dobzhansky *et al.*, 1956; Barker, 1977; Barker *et al.*, 1981). Again, these preferences are not absolute, but even if two sympatric species preferred the same yeast, there need be no competition for that resource if it were not in short supply. However, given the ephemeral nature of many *Drosophila* feeding sites, it would seem reasonable to expect that, at least sometimes, adult food resources will be in short supply. Apart from competition for food, adult females may compete for oviposition sites. For neotropical *Drosophila* in Panama, Pipkin (1965) observed that the number of species netted from a given fruit greatly exceeded the number of species bred from a limited volume of the fruit, and inferred interspecific interference during oviposition. Such interference is possible and further information would be valuable, but the feeding and breeding sites of some *Drosophila* species are known to be different (Carson *et al.*, 1956), and females may utilize a wider range of substrates for feeding than for breeding.

While there is at least partial niche separation for food among adults of sympatric species, larval niche separation and/or competition is probably more important in determining community structure. As emphasized previously, larval niche separation may be essentially complete because of adult female oviposition site preferences, yet many breeding sites have been shown to support two or more *Drosophila* species (e.g. Carson, 1951; Heed, 1957; Pipkin, 1965; Buruga and Olembo, 1971; Lachaise, 1974; Lachaise and Tsacas, 1974; Montgomery, 1975; Kimura, 1976). In such cases, there may be temporal or spatial separation of larvae, but Pipkin (1965) found species of the same species group, and even sibling species, showing synchronous or closely overlapping development in the same substrate. Differences in microflora utilization could be important in such cases in reducing interspecific competition, particularly if there were differences among microflora species in their nutritive value to different *Drosophila* species, and provided the *Drosophila* larvae could discriminate among microflora species during feeding. The possible importance of differential larval utilization of different yeasts was indicated by Wagner (1944, 1949) who showed, for five species of the *mulleri* group, differences in ability to complete larval development on a set of eight yeast species isolated from fruits of *Opuntia lindheimeri*—the natural breeding site of all the *Drosophila* species except *D. buzzatii*. *D. mulleri*, the predominant species in the area where the yeasts were collected, completed development on all eight yeasts, while *D. aldrichi* developed normally on only five of them and *D. mojavensis* on six. *D. arizonensis* and *D. buzzatii* were tested on six of the yeasts; the former developed normally on all six, but the latter on only four of them. Similar differences were found by Dudgeon (1954) for three species of the *virilis* group.

Lindsay (1958) and Cooper (1960) found that larvae of *D. pseudoobscura* and *D. persimilis* showed yeast preferences in laboratory experiments, that is, *Drosophila* larvae can discriminate among yeast species. The extent to which they exhibit preferences in a natural substrate and the degree of differential utilization of yeast species by larvae of different *Drosophila* species feeding in the same substrate remain unknown. Detailed microhabitat studies are necessary to define niche separation and/or potential interspecific competition at this level.

IV. *Drosophila* as a Laboratory Population Model

Two genera of insects, *viz. Drosophila* and *Tribolium* have been extensively used in laboratory studies of interspecific competition. The *Tribolium* studies, largely done by Thomas Park and his colleagues, have been consistently referred to by Park under the rubric "The *Tribolium* model" (Neyman *et al.*, 1956; Park, 1962; Mertz, 1972). In other words, these laboratory population studies (including interspecific competition) have been seen not as simplified models of the real world, but as quantitative, biological models. Further, in one important aspect, the model does simulate the real world because the trophic environment in the laboratory—finely milled flour—is also the natural environment.

In contrast, studies of competition between *Drosophila* species have not had a consistent underlying theme, and their relevance to the real world is perhaps more difficult to assess. Nevertheless, they are complementary to the *Tribolium* studies, and together with studies of intraspecific competition, they are particularly so in allowing more detailed analysis of the genetic basis of competitive ability.

A. Ecology of Competition

1. Competition for what?

Laboratory studies using *Drosophila*, both intra- and interspecific, have largely assumed that in the simplified environment of laboratory culture, any competition would be expressed during the larval stage through direct exploitation of food resources. Bakker (1961) suggested that competitive advantages at this stage could be due to (1) a high rate of feeding, (2) a short duration of the moulting periods, (3) a low food requirement, (4) a high initial larval weight, and (5) resistance to possible disoperative (i.e. disadvantageous) effects resulting from crowding. For two strains of *D. melanogaster*, he found the outcome of competition for food to be largely dependent on the rate of feeding.

As well as competition for food, however, other kinds of interaction have been described in which products of larval metabolism released into the food medium differentially affect larval survival and development. By growing larvae of one strain in the presence of accumulated metabolic products of other strains (i.e. on "conditioned" medium) and measuring survival to the adult stage, Weisbrot (1966) found that different biotic residues produce differential survival rates. The survival of *D. pseudoobscura* was decreased on *D. melanogaster* conditioned medium (interference), while media conditioned by *D. pseudoobscura* increased the survival of *D. melanogaster* (facilitation). Bakker (1969) argued that differences in food requirement and acquisition could explain the outcome of these experiments, and that it was unnecessary to invoke interference or facilitation. However, in some cases at least, these would seem to be real phenomena. For example, Budnik and Brncic (1975a,b) used unconditioned, homotypically conditioned and heterotypically conditioned media, but added fresh yeast every two days to minimize competition for food. In these experiments, significant decreases in pre-adult viability on conditioned medium were found for a number of species combinations which, because of the excess food, were attributed to interspecific interference due to toxicity of accumulated waste products. Later, Budnik (1977) suggested that the competitive situation may be even more complex, with interference due to larval biotic residues sometimes compensated by a mechanical facilitation due to the tunnelling of the medium by larvae. However, interference effects due to medium conditioning have not always been found, e.g. Dolan and Robertson (1975) for strains of *D. melanogaster*, and as they emphasize, the effect may be dependent on the medium used, the population density, and the length of the conditioning period.

Genetic variation within species in ability to impose interference effects on another species has not yet been investigated, nor is there any data on the isolation and identification of the relevant metabolites excreted into the medium. Clearly further investigation is necessary in order to understand the basis of this phenomenon, which as Budnik and Brncic (1975b) point out, could be of importance in the regulation of population size and the relative frequency of species in nature.

Facilitation between species at the larval stage has been clearly demonstrated by Bos *et al.* (1977). *D. melanogaster* and *D. simulans* differ in their ability to utilize *erg* mutants of *Saccharomyces cerevisiae* which have an abnormal pattern of sterol biosynthesis (Bos *et al.*, 1976). In pure cultures, *D. melanogaster* do not complete development to the adult stage, and *D. simulans* have only a low pre-adult viability. In mixed cultures, the pre-adult viability of each species is significantly increased by the presence of the other, i.e. mutual facilitation.

While most attention has been devoted to competition at the larval stage, the outcome of competition also could be affected by interactions at other life cycle stages. Ayala (1967, 1968) has found for intraspecific competition in a number of species of *Drosophila*, that in the high adult density populations of the serial transfer technique, adult population numbers are limited by space. Differential effects of this type for different species would affect the outcome of their interspecific competition. The possibility of interspecific interference between females during oviposition on natural substrates has been noted previously (Section III, D). No evidence for such interference has been produced for laboratory populations, but del Solar and Palomino (1966) and del Solar (1968) have shown that females of *D. pseudoobscura* and *D. melanogaster* have a "gregarious" tendency in oviposition. This tendency in populations that were not crowded, or the high adult densities in some laboratory population studies, could promote an interference effect during oviposition. Further, del Solar and Palomino (1966) showed that *D. melanogaster* females preferred to oviposit in medium already occupied by larvae, regardless of whether these larvae were of the same or a different species. Such eggs laid into inhabited medium are subject to the risk of destruction, through cannibalism or burial, by larvae already present (Moth and Barker, 1976). Thus for two species in competition, one showing a higher tendency to oviposit in medium already inhabited by larvae would, other things being equal, be at a competitive disadvantage. Competitive effects also may be mediated by interspecific differences in pupation site, and interaction between larvae and pupae. For *D. melanogaster* and *D. simulans*, Sameoto and Miller (1968) and Barker (1971) found that a much higher proportion of the latter species pupate on the medium surface, and that many of these are buried by larval activity and die. Thus for pupal survival as a factor affecting competitive outcome, *D. simulans* is at a disadvantage.

Even in the simplified environment of *Drosophila* laboratory culture, many factors may affect competitive outcome. Clearly this outcome cannot be predicted simply on the basis of food exploitation (as modelled by de Jong, 1976), nor is it likely that it can be predicted from the performance of each species in pure culture.

2. Niche specification and heterogeneity of resources

Within the apparently uniform environment of a laboratory *Drosophila* culture, there may be differences in the realized niches of competing species. Ayala (1972) has discussed the possibility that niche heterogeneity can lead to frequency-dependent fitnesses among competing species, even if every species is able to exploit every niche. Subsequently, DeBenedictis (1977) re-analysed some of Ayala's data (Ayala, 1971a; Ayala *et al.*, 1973) and demonstrated that the interspecific competition was frequency-dependent; the competition coefficients (α's) of the Lotka–Volterra equations changed with frequency of the competitor.

A number of examples of heterogeneity of resources and differences in realized niches between *Drosophila* species have been reported. Differences in the ability of *D. melanogaster* and *D. funebris* to utilize fresh and old medium (Merrell, 1951) have been noted previously (Section II, C). Erk and Sang (1966) found that the quantitative nutritional requirements for pre-adult development in axenic culture were different for *D. melanogaster* and *D. simulans*. Micro-ecological differences between these two species were found by Moore (1952a), who demonstrated that *D. simulans* was better able to utilize for oviposition either the centre of the food medium surface or medium having a surface crust. Similar differences in oviposition site preferences for different strains of these two species were found by Barker (1971), but clear differences in larval and pupal distributions also were demonstrated. Egg distribution on the medium surface was markedly non-random, and there were significant differences between species. Both species preferred the periphery of the medium, but *D. simulans* laid a higher proportion in the centre of the medium than did *D. melanogaster*. Subsequent larval dispersal still did not lead to a random distribution through the medium, as although both species were found predominantly in the upper half of the medium, the proportion of *D. simulans* in the lower half was about three times that for *D. melanogaster*. Further, *D. melanogaster* larvae remained predominantly around the periphery of the medium. A significant effect of interspecific association was found for the joint egg and young larvae distribution, in that *D. simulans* in mixed species populations showed an increased preference, as compared with pure cultures, for the centre of the medium where *D. melanogaster* was at low frequency.

In addition, the nature of the interaction between the species changed during the developmental period, i.e. heterogeneity of resource utilization over time. Clearly the "apparently uniform environment" is in fact far from uniform, and subtle differences in resources utilization could have a significant impact on the outcome of interspecific competition.

3. Environmental factors affecting competition

Interspecific competition studies have used one or both of two techniques. In the first, populations are maintained in cages, or in some way which allows for continuous population maintenance with either overlapping or discrete generations, and the outcome of competition is assessed in terms of changing frequencies of the two species. Competition at all life cycle stages affects the results, but their effects are confounded. In the second, competition is studied only through part of one generation, with the aim of analysing the physiological basis of competition, and the effects of competition on or expressed through individual fitness components. Such experiments may be initiated either with eggs or larvae, so that only pre-adult competition is studied, or with adults to study effects of competition on fecundity and adult viability, and may be done in conjunction with population cage experiments to assist in interpretation of the competition process. Both techniques are considered here in reviewing the effects of environmental factors on competition.

(a) Temperature

As different species may differ in their temperature tolerances, effects of temperature on competitive outcome are not surprising, and the physiological basis of any effect is of most interest, particularly in relation to adaptation in natural populations. For *D. melanogaster* and *D. simulans* in population cages (Moore, 1952a; Tantawy and Soliman, 1967), the latter species is usually excluded at 25°C, but at 15°C, the reverse may occur (Fig. 4). As 25°C is a more extreme temperature for *D. simulans* than it is for *D. melanogaster* (Hosgood and Parsons, 1966), the exclusion of *D. simulans* at this temperature might be expected, but the physiological basis of the differences in competitive ability at 15°C and 25°C is not at all clear. Moore (1952a) considered that the superiority of *D. simulans* at 15°C was due to a shorter development time than that of *D. melanogaster*, while it was slightly longer at 25°C. Tantawy and Soliman (1967) found the developmental time of *D. simulans* to be shorter at both temperatures, but with the difference more marked at 15°C. The fecundity of *D. simulans*, relative to that of *D. melanogaster*, was shown by Moore (1952a) to be less at 15°C than at 25°C, while Tantawy and Soliman (1967) inferred the reverse to be true.

Such differences between different studies may be due to differences in experimental techniques or in other environmental variables, but are most likely due to genetic differences between the strains used. They serve to re-emphasize the point made earlier (Section II, C) that results from any one experiment are strain-specific, that care therefore is needed in

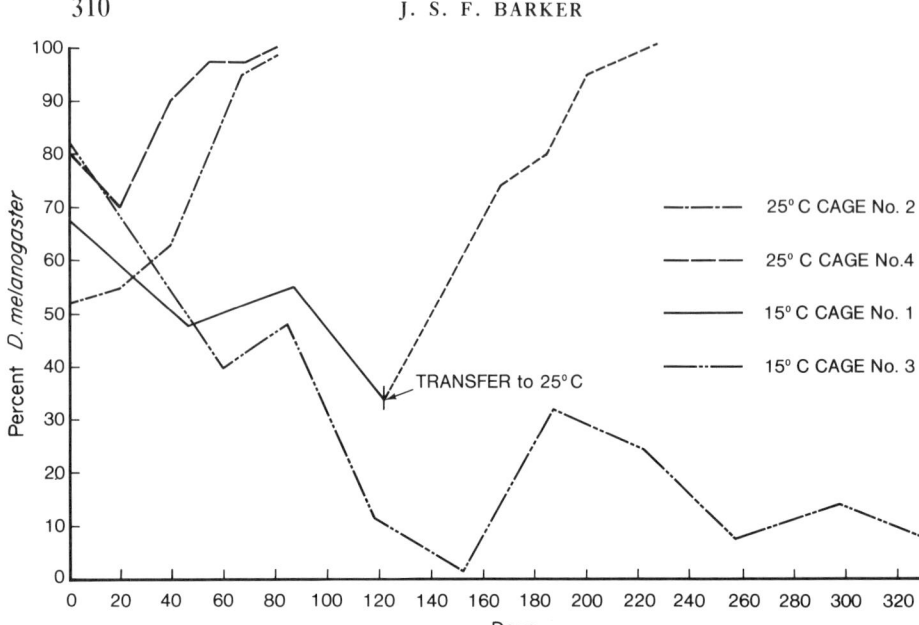

FIG. 4. Changes in the percentage of *D. melanogaster* when in competition with *D. simulans* in population cages at 15°C and 25°C (adapted from Tantawy and Soliman, 1967).

extrapolating from strain to species, and that comparative studies using a variety of strains are essential before general conclusions are drawn.

Considerable study has been devoted to competition in *D. melanogaster* and *D. simulans*, but almost exclusively at 25°C. Given the differences between experiments in the effects of temperature on some fitness components, the differences in geographical distribution and seasonal variations in numbers in natural populations that are related to temperature (Parsons, 1975), and the ease of laboratory manipulation of temperature as an environmental variable, it is surprising that more work has not been done to clarify the physiological basis of differences in competitive ability at different temperatures. Comparative study of a number of strains, recently collected from different geographical and climatic regions, clearly is necessary.

For competition between *D. serrata* and *D. pseudoobscura*, Ayala (1966) found that at 19°C, the former species was excluded, while at 25°C (Ayala, 1969b), *D. pseudoobscura* was excluded. While only one strain of each species was used in the 19°C experiment, and it is not stated whether these same strains were included in the 25°C experiment, these results led to Ayala's study of competition between these species at the intermediate temperature of 23.5°C, and the demonstration of coexistence at that temperature (Ayala,

1969a), as discussed in Section II, C. Similar reversals of competitive success at different temperatures have been reported for other species. Ayala (1966) also showed, again for one strain each of *D. nebulosa* and *D. serrata*, that at 25°C the former species rapidly became dominant and maintained a frequency greater than 95% from the fifth week to termination of the population after 26 weeks. At 19°C, however, both species remained at intermediate frequencies to week 26, with *D. nebulosa* predominating slightly.

Competition between *D. willistoni* and *D. pseudoobscura* was studied at three temperatures by Mourão and Ayala (1971) using four strains of the former species and one of the latter. Differences among *D. willistoni* strains were found at each temperature, but at 25°C and 22°C, *D. willistoni* were generally stronger competitors, while at 20°C, *D. pseudoobscura* generally had a competitive advantage. An additional complexity in evaluating temperature effects, and the necessity of considering intraspecific genetic variation, is illustrated by the finding of a significant strain × temperature interaction.

(b) Population density
In population cage and population bottle experiments using *D. melanogaster* and *D. simulans*, Barker (1963a) found that although the latter species was always eliminated, the competitive ability of *D. melanogaster* was generally higher for larger initial population numbers. That is, at larger initial population size, *D. melanogaster* gained an additional advantage in the first generation of the competition.

Most studies of the effects of population density, however, have been ones where larval or adult numbers were varied and effects on individual components of fitness investigated. Effects of variation in larval density on pre-adult viability, developmental time and body size of emerging adults have been studied extensively, for different genotypes within a species (reviewed elsewhere in this volume) and for different species in both pure and mixed populations. While it might be argued that such studies are largely irrelevant to understanding the dynamics of natural populations (Section III, A), they are essential for the analysis of the dynamics of laboratory populations. In general, the patterns of change in fitness components with increasing larval density have been similar in all studies (reviewed by Andersen, 1960; Klomp, 1964). Pre-adult viability either remains constant or increases slightly from low to some intermediate density, but then decreases to zero at sufficiently high density. Mean developmental times increase to a maximum as zero viability is approached. Mean size or body weight decreases to a minimum, but some small increase may occur at extremely high density.

In analysing interspecific competition, both species may show these general changes, but the patterns may differ for the two species, so that competitive ability varies with density. Further, much of the published data refers to the effects of density in single-strain cultures only. Two or more strains may be compared in this way, but such comparisons cannot necessarily be expected to predict the results of interstrain competition. Also, it is important to recognize that larval competition is only one phase of the association of two species, and that therefore results of such studies cannot necessarily predict competitive effects in continuous populations.

In the first extensive study of larval density effects in interspecific competition, Sokoloff (1955) chose to use the three closely related species *D. pseudoobscura*, *D. persimilis* and *D. miranda*. At low densities in both pure and mixed species cultures, *D. pseudoobscura* and *D. persimilis* were similar in pre-adult survival and developmental rate. At high densities in pure cultures, *D. persimilis* showed higher pre-adult survival and faster development, and this superiority was increased in mixed species cultures. Both these species were superior to *D. miranda* in pure cultures at low and at high densities, while in mixed species cultures, *D. miranda* showed even slower development and lower survival at low densities, and zero survival at high densities. Thus in all cases, the order of these species in competitive ability remained the same in pure and in mixed species cultures, although differences were greater in the latter, and, for all species, competitive ability changed with changes in larval density.

Studies of competition between the sibling species *D. melanogaster* and *D. simulans* at various larval densities were initiated by Miller (1964a), using the same strains as those of Moore (1952a), and results were very similar to those of Sokoloff (1955) for the sibling species *D. pseudoobscura* and *D. persimilis*. In single-species cultures of up to 120 larvae per vial, the two species were remarkably similar and showed comparable changes in body weight, developmental time and viability as density increased. At higher densities, *D. melanogaster* continued to extend its larval period in response to increased density and had significantly higher pre-adult survival than *D. simulans*. The outcome of interspecific larval competition was predictable from these results, assuming that the interactions between members of the two species were no different from the interactions within species (Miller, 1964a,b).

For other strains of *D. melanogaster* and *D. simulans*, however, Barker and Podger (1970a) found that results observed in mixed cultures generally were not predictable from results in single-species cultures, particularly at high larval densities. The general effects of density on both species were similar, but there were significant differences between the species that would cause differences in competitive outcome at different densities. The lack of predictability from single-species to mixed species cultures in this experiment, as compared with those of Miller (1964a,b), probably was due to the *D. melanogaster* strain having significantly higher competitive ability than the *D. simulans* strain (Barker, 1963a,b), while the strains used by Miller were of similar average competitive ability (Miller, 1964b). Further study of a range of stains of both species is necessary to test the generality of this conclusion. In relation to the interpretation of results in continuous population experiments, it should be noted that Barker and Podger (1970a) replicated their basic design in time (three sets, separated at initiation by intervals of one week), and that this variable had highly significant effects on all fitness components. Therefore they emphasized the need to include the time variable in analytical experiments designed to elucidate the ecological determinants of interspecific competitive outcome, and concluded that the details of the process and the outcome of interspecific competition in individual continuous populations were not likely to be simply predictable in terms of any generalized "coefficients of competition".

Further, Barker and Podger (1970b) studied the effects of variation in larval density on fecundity, hatchability and adult viability of emerging adults. There were significant differences between the species for all components, but only fecundity showed significant effects of larval density—probably mediated through adult body weight. Again results in mixed cultures were not predictable from those in single species cultures; in particular, *D. melanogaster* females raised in mixed cultures were less fecund than those from pure cultures, while *D. simulans* showed the reverse effect.

Effects of adult density on progeny productivity of *Drosophila* were studied by Pearl and Parker (1922), but in analytical studies of competition, little attention has been devoted to variation in adult density and its effects on adult fitness components of fertility, hatchability, fecundity and viability. Studies of the effects of larval density in *D. melanogaster* and *D. simulans* by Barker and Podger (1970a,b) were the first part of a comparative analytical study of the process of interspecific competition in laboratory *Drosophila* populations, and subsequent studies have considered effects of adult density. For single-species cultures only, Barker (1973b) found for both species that as adult density increased, progeny number per culture increased to a maximum and then decreased, but the average number of progeny per female decreased rapidly from the lowest density. The cause of this decreased progeny number per female, however, was argued to be different for the two species. For *D. simulans*, it was apparently due to decreased fecundity per female while, for *D. melanogaster*, it was mainly due to reduced immature stage survival in the progeny as a result of increased larval crowding. As the decreased average fecundity per female could result either from a decrease in the average number of eggs per female or from a

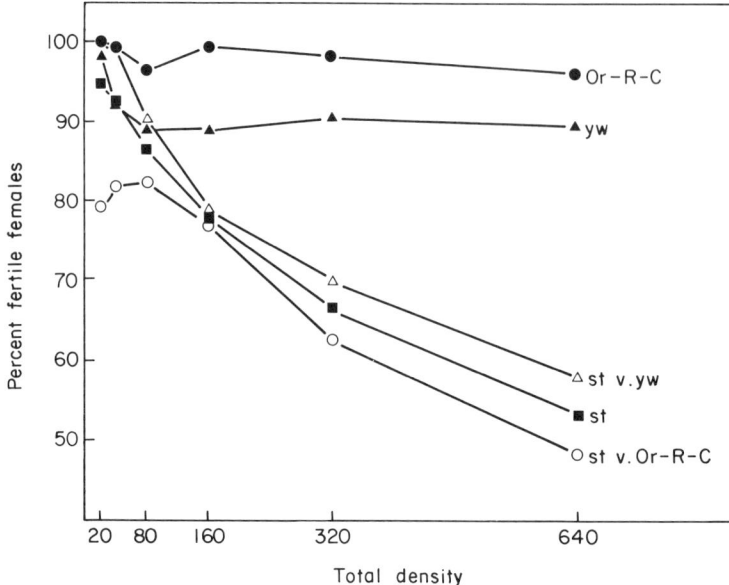

FIG. 5. Least squares means of the per cent fertile females at varying densities for *D. melanogaster* (*Or-R-C* and *y w*) and *D. simulans* (*st*) averaged over pure and mixed-species treatments, and for *D. simulans st* when in mixed-species cultures with *Or-R-C* (*st* v. *Or-R-C*) and *y w* (*st* v. *y w*) (from Moth, 1974).

proportion of the females laying no eggs, Moth (1974) studied fertility in the same two strains* and an additional strain of *D. melanogaster* marked with the mutants yellow (*y*) and white (*w*), at various adult densities in single and mixed species populations. Increasing adult density significantly reduced the fertility of *D. simulans* females, but not that of either of the *D. melanogaster* strains (Fig. 5). Thus at least part of the reduction in average female fecundity of *D. simulans* (Barker, 1973b) resulted from some females remaining unmated and contributing no progeny. In mixed species populations, there were additional effects not predictable from single-species results, in that one strain of *D. melanogaster* (the high fitness wild-type Oregon-R-C) significantly reduced *D. simulans* fertility because of physical interference, particularly at low densities.

Other experiments in our laboratory have studied the effects of variation in adult density, for these same three strains, on egg hatchability (Moth, 1977), fecundity (Moth and Barker, 1981) and adult viability (Moth and Barker, 1977). Increased adult density significantly decreased egg hatcha-

* The *D. simulans* strain is referred to by Moth (1974) as scarlet (*st*) but it is the same strain as that called vermilion by Barker (1973b) and in all our previous studies (see Barker, 1973a).

bility for *D. simulans st* and *D. melanogaster Or-R-C*, but not for *D. melanogaster y w*; significantly decreased fecundity for all strains, similarly for *st* and *y w*, but more severely for *Or-R-C*; and significantly decreased adult viability, similarly for the two *D. melanogaster* strains, but to a greater extent for *D. simulans st*. Previous studies in single-species populations have considered only effects on fecundity, which has generally been shown to decrease as density increased. Sameoto and Miller (1966) for one strain of *D. melanogaster*, observed no effect of density although their results suggest that at higher densities than those tested, fecundity may have been decreased.

Comparison of our results in single and in mixed species populations for fecundity and viability is not straightforward as a number of different proportions of the two species were used in a series of experiments. Nevertheless, there were differences which themselves were different for different components and for the different strains. The viability and fecundity of *D. simulans st* were significantly higher in mixed populations with *D. melanogaster y w*, and its fecundity was probably higher in mixed populations with *Or-R-C*, as was found by Barker and Podger (1970b). All effects on fecundity were explainable in terms of differing disruption at the oviposition site, due to activity differences among the three strains. The two *D. melanogaster* strains showed no changes in fecundity and viability (except possibly for *y w* viability) in mixed populations with *D. simulans st*, as compared with single-species populations. *D. melanogaster y w* was the only strain to show differences for hatchability, this being higher in single-species populations. Given the significant effects of time (micro-environmental variation) in all these experiments as previously noted by Barker and Podger (1970a), prediction of competitive outcome for these strains clearly would not be simple.

These experiments were done to provide a detailed analysis of effects of variation in density on individual fitness components, and to compare responses in single and mixed species populations for particular strains of *D. melanogaster* and *D. simulans*. A completely different approach was taken by Wallace (1974), who used eight species of *Drosophila* in single and mixed species cultures initiated with varying numbers of adults, with competition effects assessed in terms of progeny numbers. Effects mediated through all components of the life cycle therefore were confounded, but the aim was basically to explore the nature of intra- and interspecific competition as expressed in a variety of species. As might be expected, the outcome of mixed species competition depended mainly on the two species concerned, and the competitive abilities of the species differed considerably, while intra- and interspecific competition differed for a number of species pairs. Replacement series analyses (de Wit, 1960) showed that members of individual species pairs either enhanced or interfered with each other's development, or did neither. Budnik and Brncic (1974) suggested that interference may be more common between species that are never sympatric than between species that may be naturally sympatric. Classification of the species pairs studied by Wallace as sympatric or not is not simple, as a number are cosmopolitans. However, assuming that cosmopolitan species are usually found in association with human habitation, and the remainder in natural habitats, some pairs can be classified as "probably not sympatric", in the sense of not utilizing the same breeding sites (Table IV).

In Wallace's data, four species pairs showed evidence of interference, two of which are not sympatric (*D. melanogaster, D. willistoni* and *D. simulans, D. pseudoobscura*) while two are or may be sympatric (*D. melanogaster, D. simulans* and *D. melanogaster, D. virilis*). For comparison, among the seven species pairs that showed enhancement or facilitation, four pairs are not sympatric and three are, while among the seven species pairs that apparently had no effect on each other, four are not sympatric and three are. A number of the species pairs involve two cosmopolitan species, which may not be appropriate in this comparison, but if these are excluded, all remaining pairs are not sympatric. These data do not support the suggestion of Budnik and Brncic (1974). Again, however, it should be noted that Wallace's study used just one strain of each species and that the relative competitive abilities observed for some pairs of these strains did differ from those for other strains of the same species in other studies, so that further study using a number of strains of each species would be necessary to test the generality of the suggestion of Budnik and Brncic (1974). However, it is interesting that Wallace (1974) found evidence for partly non-overlapping niches in the

TABLE IV. Classification of the species pairs studied by Wallace (1974) as probably sympatric or probably not sympatric in natural populations

Probably sympatric		Probably not sympatric	
D. melanogaster—*D. simulans*		*D. pseudoobscura*—*D. melanogaster*	
D. melanogaster	*D. virilis*	*D. pseudoobscura*	*D. simulans*
D. melanogaster	*D. funebris*	*D. pseudoobscura*	*D. virilis*
D. melanogaster	*D. ananassae*	*D. pseudoobscura*	*D. funebris*
D. simulans	—*D. virilis*	*D. pseudoobscura*	*D. ananassae*
D. simulans	*D. funebris*	*D. pseudoobscura*	*D. nebulosa*
D. simulans	*D. ananassae*	*D. pseudoobscura*	*D. willistoni*
D. virilis	—*D. funebris*	*D. willistoni*	—*D. melanogaster*
		D. willistoni	*D. funebris*
		D. willistoni	*D. ananassae*

culture vial environment only among species pairs showing interference or no effect on each other.

(c) Species frequency

Ayala (1971a) reported that competition between *D. pseudoobscura* and *D. willistoni* was frequency-dependent, and Ayala (1972) stated that the necessary and sufficient conditions for stable coexistence of two competing species are that (i) the competitive fitnesses of two species are inversely related to their relative frequencies, and (ii) there is a frequency at which the two species have identical fitnesses. This definition of frequency-dependent competition has been criticized by DeBenedictis (1977) as trivial, in making any stable multispecies competitive system frequency-dependent and in providing no real alternatives, and also he showed that the de Wit replacement series analysis used by Ayala (1971a) is incapable of detecting frequency-dependent competition. DeBenedictis (1977) proposed a more general definition: competition is frequency dependent whenever per individual competitive effects vary with species frequency; and he re-analysed Ayala's data to estimate the competition coefficients for each species at various frequencies. In general, the competition was found to be frequency-dependent.

In his analyses of competition using eight species in small laboratory cultures, Wallace (1974) varied species frequency as well as density, and also used the de Wit replacement series and ratio diagrams in part of his analysis. DeBenedictis (1977) pointed out that interpretation of ratio diagram results requires caution, but this analysis did suggest again that *D. pseudoobscura* and *D. willistoni* would establish a stable equilibrium if competing in laboratory populations, and that *D. pseudoobscura–D. virilis* and *D. funebris–D. virilis* would also do so. However, Wallace (1974) did not claim that these three species pairs showed frequency-dependent competition. Rather, he argued that where competition between individuals of two different species is different from that between individuals of the same species, then there must be more than one resource available, and the ease with which the two species use these resources differs. Frequency-dependent selection and stable equilibria in mixed populations then may result. However, not all of the three species pairs, where stable equilibria were suggested by the ratio diagram analysis, showed evidence of partially non-overlapping resource utilization in Wallace's analyses, but other species pairs did. Thus the results are not unequivocal, but the experiments were primarily exploratory. In view of Wallace's (1974) conclusion that frequency-dependent selection is precisely the sort of evidence needed to demonstrate that two species are not competing identically for identical resources, and of DeBenedictis' definition and reanalysis of Ayala's data, the concept of frequency-dependent competition is open to intensive investigation.

In an earlier analysis of the effect of species frequency in competition, Narise (1965) claimed that for one strain each of *D. melanogaster* and *D. simulans*, competition was frequency-dependent with competitive ability increasing as species frequency increased. However, Putwain *et al.* (1967) showed this conclusion to be incorrect, while DeBenedictis (1977) found the data to be unsuitable to test for frequency-dependence.

In the analytical studies of competition between *D. melanogaster* and *D. simulans* done in the author's laboratory, effects of species frequency on individual fitness components were investigated in the same experiments where density was varied (Section (b) above). In experiments with varying larval density, significant effects of species frequency were found for pre-adult viability at high densities, for developmental time at low to intermediate densities, and for adult body weight at intermediate densities (Barker and Podger, 1970a). For all three fitness components, the performance of *D. simulans* improved as its frequency decreased, but responses for *D. melanogaster* were more complex. Thus its pre-adult viability increased as its frequency decreased, but developmental time increased as frequency decreased at high densities, and showed a maximum at intermediate frequencies at low to intermediate densities, while body weight decreased as frequency decreased. As developmental time and adult body size are more sensitive than viability to increase in larval density, the competitive effect of *D. simulans* on itself is greater than that of *D. melanogaster* on *D. simulans*, while the reverse is true for *D. melanogaster*. No effects of species frequency were found for either *D. melanogaster* strain in the adult density experiments, while effects for *D. simulans* depended on the competing strain. Viability and fecundity both increased as *D. simulans* frequency decreased when competing with $y \, w$, but did not change when with *Or-R-C*.

Thus the available evidence indicates that competitive effects may change as the relative proportions of two species change. However, such frequency effects may be mediated through different fitness components for different strains within a species, while any effects exhibited by a particular strain of one species may differ according to the strain of the species with which it is competing.

(d) Medium and other variables
Medium age effects on competition between *D. melanogaster* and *D. funebris* (Merrell, 1951) have been referred to previously (Section II, C). Competition between *D. melanogaster* and *D. simulans* in population cages was shown by Moore (1952a) to be affected by a number of medium-related variables:

the former species was better able to utilize old cage cups from which flies had already emerged and it continued oviposition for longer into cups containing eggs and young larvae, but it oviposited less readily onto medium with a dry surface crust than did *D. simulans*. Medium quality and amount of live yeast seeded onto the medium significantly affected the competition between *D. melanogaster* and *D. simulans* in population bottles (Claringbold and Barker, 1961; Barker, 1963a), while, in population cages, each of three strains of *D. pseudoobscura* competing with *D. nebulosa* showed higher competitive ability on a dead yeast fortified medium than on medium without dead yeast (Barker, 1965).

The importance of yeasts in the nutrition of *Drosophila*, and the differential attractiveness of different yeasts to sympatric *Drosophila* species have been discussed in Section III, D. Thus one might expect the relative competitive abilities of *Drosophila* species in laboratory populations to vary depending on the yeast species used in the culture, and that such studies could have considerable significance in analysing the realized niches of sympatric *Drosophila* species in natural populations. Yet few studies have been done. For larval competition between *D. melanogaster* and *D. simulans*, El-Helw and Ali (1970) showed that larval viability and adult body weight of both species in pure and mixed cultures were lower in cultures supplemented with *Schizosaccharomyces pombe*, as compared with *Saccharomyces cerevisiae*. In mixed cultures, the outcome of competition (i.e. proportion of each species in the emerging adults) was significantly affected by yeast species, with *D. simulans* showing higher competitive ability on *S. pombe*. In a similar experiment using naturally occurring yeast species (*Candida pulcherima* and *Nadsonia elongata*) collected in the same general area as the *Drosophila* strains, El-Helw *et al.* (1972) found differences between *Drosophila* species and yeasts for larval viability and adult body weight, but the competitive abilities of the *Drosophila* species were approximately the same in cultures with either yeast species.

Rizki and Davis (1953) showed that light is an environmental factor affecting the degree of interspecific competition in *Drosophila*. Larvae of *D. melanogaster* tended to pupate away from the light source, while those of *D. willistoni* showed the reverse tendency. In mixed species populations, these tendencies were intensified, so that interspecific interactions between these species would be reduced, at least in the late larval stage. Effects of a daily 12 hour light: 12 hour dark cycle on fecundity were studied by Moth and Barker (1981) in the experiment with varying adult density and species frequency discussed previously. All three strains laid more than 50% of their eggs during the light period and interspecific competition effects were found in that *D. simulans st* laid a higher proportion in the light period when competing against *D. melanogaster Or-R-C* than when against

D. melanogaster y w. Caution is therefore indicated in designing and interpreting competition experiments initiated with adults, as one species may be at a greater relative advantage in cultures set up near the beginning of a light phase, as compared with ones set up near the end.

The final environmental variable to be noted is that of ionizing radiation, shown by Blaylock (1969) to influence the outcome of competition between *D. melanogaster* and *D. simulans*.

B. COMPETITIVE ABILITY AS A MEASURE OF POPULATION FITNESS

The Darwinian fitness of a genotype involves an intraspecific comparison and can be defined explicitly as "the average contribution which the carriers of a genotype, or of a class of genotypes, make to the gene pool of the following generation relative to the contributions of other genotypes" (Dobzhansky, 1968, p. 17). In contrast, the fitness of a population, while also a concept of importance to evolutionary biologists, is less amenable to explicit definition, and has been the subject of extensive discussion (e.g., Lewontin, 1957; Ayala, 1969c).

Partly by analogy with the fitness of a genotype, one possible approach is to define population fitness in terms of the outcome of competition with a non-interbreeding population, i.e. interspecific competitive ability.

1. Definitions and methods of estimation

Claringbold and Barker (1961) defined a model for estimating interspecific competitive ability as a measure of population fitness of *D. melanogaster* populations when in competition with a standard strain of *D. simulans*. In experimental populations, relative frequencies of the two species are determined by census of living adults at intervals of the mean generation length. Where $\bar{p}_{Mi}$ is the proportion of population M (*D. melanogaster*), and $\bar{p}_{Si}$ the proportion of population S (*D. simulans*), in generation i, then:

and
$$\bar{p}_{M(i+1)}/\bar{p}_{Mi} = n_M$$

where
$$\bar{p}_{S(i+1)}/\bar{p}_{Si} = n_S$$

$$\bar{p}_{Mi} + \bar{p}_{Si} = 1.$$

Let
$$r_i = \bar{p}_{Mi}/\bar{p}_{Si}$$

Then the relative fitness of population M to population S is given by:
$$W_{MS} = n_M/n_S,$$
$$= \bar{p}_{Si}\, \bar{p}_{M(i+1)}/\bar{p}_{Mi}\, \bar{p}_{S(i+1)}$$

$$= r_{(i+1)}/r_i$$
$$= (r_i/r_0)^{1/i}$$

The range of variation of relative fitness so defined is:

$$\infty > W_{MS} > 0.$$

An alternative scale of measurement of relative fitness is given by:

$$f_{MS} = \log W_{MS}$$

so that equal fitness of the two species is defined by $f_{MS} = 0$, with the range of variation symmetrically disposed about this point. f_{MS} is the difference between the Malthusian parameters of the two competing species, and a maximum likelihood procedure for the estimation of this relative fitness, and the standard error of the estimate was given.

Ayala (1969b) used the same basic approach but, assuming the rate of change in numbers for each species to be constant for any fixed interval of time, he defined his model in terms of time intervals not restricted to the mean generation length. Then if s_i is the number of one species, and p_i the number of the second species at time i, the regression coefficient of $\log [s_i/p_i]$ on time is an estimate of m_{sp} ($\equiv f_{MS}$ of Claringbold and Barker, 1961), i.e. the relative fitness of species p to species s. Further, the relative fitness of one population of species p (p_j) to another population of the same species (p_i) is given as:

$$m_{p_i p_j} = m_{sp_j} - m_{sp_i}$$

which can be estimated as the difference between two coefficients of regression.

2. Experimental results

In a series of experiments, Barker (1963a,b, 1967a) evaluated the model of Claringbold and Barker (1961), with some experiments using population cages, but most done with population bottles. Effects of variation in initial population numbers, initial frequencies of the two species, medium quality, amount of live yeast and of replication in time were tested, and the model was applied to a variety of strains of *D. melanogaster* differing widely in fitness. In these experiments, the fitness estimate was significantly affected by a number of the experimental variates, but was additive over those examined. In many of the populations, however, considerable heterogeneity was observed, i.e. generation to generation fluctuation in the species frequencies. This could arise if the model did not adequately describe the interaction between the species, or if the relative fitnesses varied in time. As there was no evidence of systematic deviations of observed proportions from those expected, nor evidence for any factor that would give rise to non-randomness in the interspecific competition, the model gave an adequate description of the interaction between the two species. Highest heterogeneity was observed for *D. melanogaster* strains of intermediate or low fitness (Barker, 1963b), and many of these populations showed an increase in *D. melanogaster* frequency for two or three generations, followed by a decrease in frequency in most populations and indeterminacy in the final outcome. This heterogeneity and indeterminacy in final outcome is illustrated in Fig. 6, which shows the results for six replicate populations of one strain of *D. melanogaster*. Further evaluation (Barker, 1967a) led to the suggestion that variation in larval density could be the most important factor contributing to generation-to-generation heterogeneity, and then to the analytical experiments on effects of variation in larval and adult density and species frequency on individual components of fitness (Barker and Podger, 1970a,b; Moth, 1974, 1977; Moth and Barker, 1976, 1977, 1981).

The estimation procedure of Ayala (1969b) has not been evaluated for effects of variation in environmental variables, except for temperature (Mourão and Ayala, 1971). The method has the apparent advantage that relative fitness is estimated simply by a regression coefficient, but as Ayala (1970c) points out the consecutive observations of each population are not statistically independent. In fact, in the experiments where the method has been applied, populations were maintained using the serial transfer technique, and counts of the numbers of each species were made only one week apart. Generation to generation heterogeneity within populations (specifically analysed in the model of Claringbold and Barker, 1961) also is obvious in the results presented for some populations. For both these reasons, the standard errors of the estimates of relative fitness will be underestimated. Nevertheless, significant differences in competitive ability

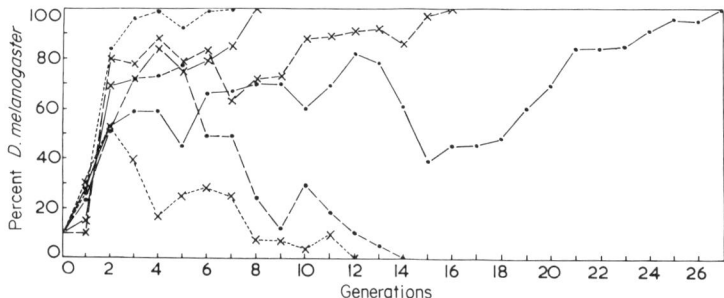

FIG. 6. Changes in percentage *D. melanogaster* for six replicate populations of one strain when in competition with the standard strain of *D. simulans* (from Barker, 1963b).

have been shown for genetically different strains of *D. pseudoobscura* (Ayala, 1969b), for strains of *D. serrata* of different geographic origin (Ayala, 1970c), and for genetically different strains of *D. willistoni*, but with a significant interaction between strain and temperature (Mourão and Ayala, 1971).

3. Comparisons with other measures of fitness

In these models, interspecific competitive ability has been taken as a measure of population fitness. It is recognized that an estimate is valid only for the specific conditions under which it was made, and is not a measure of population fitness defined in any general way (Barker, 1967b; Ayala, 1970c). However, an important part of the environment of a population is the presence of other species that are potential competitors for available resources of living space and food, so that interspecific competitive ability is at least a significant component of population fitness.

Population fitness also has been measured in terms of probability of survival, average population size, biomass produced per unit time and the innate capacity for increase. These measures are all made in single species populations, so that it is of interest to compare them with estimates based on interspecific competitive ability. For one strain each of *D. melanogaster* and *D. simulans* the former had significantly higher competitive ability (Barker, 1963a,b). In single species populations, however, the probability of population survival was lower for *D. melanogaster*, yet this was probably a direct result of its higher productivity (Barker, 1967b).

Various measures of fitness have been used for chromosomally polymorphic and monomorphic populations of *D. pseudoobscura*: biomass productivity and population numbers in cages (Beardmore *et al.*, 1960) and in bottles (Dobzhansky and Pavlovsky, 1961), the innate capacity for increase (Dobzhansky *et al.*, 1964), and interspecific competitive ability (Strickberger, 1963; Ayala, 1969b). The polymorphic populations (AR/CH) generally were fitter than the two monomorphic (AR and CH), but the relative fitnesses of the latter were not consistent over all measures.

For geographic strains of *D. serrata*, Ayala (1965, 1970c) measured fitness in single species populations and in interspecific competition with three other species (Table V). There was generally good agreement between competitive ability and average population size, while productivity gave somewhat different results. However, no relationship between competitive ability and population size was found for four strains of *D. willistoni* by Mourão and Ayala (1971), although in this case there was a significant positive correlation between interspecific competition ability and the Darwinian fitness of the strains (measured by their performance in

TABLE V. Summarized results for the comparative fitnesses of three strains of *D. serrata* (Sydney—S, Cooktown—C and Popondetta—P) as measured by competitive ability, and productivity and average population size in single species populations (Ayala, 1965, 1970c)

	Temperature	
Method of estimation	25°C	19°C
Competitive ability—		
against *D. pseudoobscura*	C>P=S	
against *D. melanogaster*	C=P=S	
against *D. nebulosa*		C>S>P
Productivity—		
Individuals per food unit	C=S>P	C=S>P
Biomass per food unit	C=S>P	S=C>P
Average population size	C>P=S	C>S>P

competition with genetically different flies of the same species, Mourão *et al.*, 1972).

While there is some general consistency in the results of these experiments, the results serve to emphasize that any one measure of population fitness is specific. Different measures may be used for particular purposes, but estimates of more than one of them, preferably under a variety of environmental regimes, would be necessary for a full evaluation of the relative fitness of different populations.

4. Fitness components and prediction of competitive ability

The complexity of the process of interspecific competition was pointed out by Zimmering (1948), and it is clear that no simple relationship should be expected between any one component of fitness and competitive ability. Nevertheless, analyses of such relationships are necessary to determine whether particular components are of major importance, and to understand the nature of the process of interspecific competition. Ideally, of course, the components should be estimated in mixed species populations under environmental conditions the same as used in measuring competitive ability, and for an appropriate range of population densities and species frequencies. Although not achieving this ideal, the analytical experiments in the author's laboratory were designed from this viewpoint. The aim has been to provide comprehensive data for a small number of strains of *D.*

melanogaster in competition with one strain of *D. simulans*, and then to use the data in a computer simulation model. If the model using these data provides a satisfactory simulation of competitive outcome for these strains, then it can be used in a parameter sensitivity analysis to further investigate the interspecific competition process and to provide feedback to laboratory studies.

Some comparisons of fitness components measured in single-species populations and competitive ability have been made. Barker (1963b) found no relationship between hatchability or developmental time and competitive ability for eleven strains of *D. melanogaster*. Selection for high productivity in *D. melanogaster* and *D. simulans* was done by Tantawy *et al.* (1976), and unselected and selected populations were measured for some fitness components and tested in one generation interspecific competition. Results as presented are not readily interpreted, but it was argued that productivity is not a major component determining competitive outcome. For *D. funebris* and *D. virilis*, the former showed significantly higher lifetime egg production per female, egg production per female per day, hatchability and longevity in pure cultures, yet was rapidly eliminated by *D. virilis* in population cages (Tantawy and El-Wakil, 1970). Van Delden (1968) measured a variety of fitness components and other measures of fitness, including one-generation competition with *D. simulans*. Specific comparisons of competitive ability with the other measures for the nine populations studied were not made, but their ranking for the different measures varied widely. Van Delden (1968) argued that estimates of population fitness based on only one measure have a restricted value, and advocated a more general fitness measure based on relative performance in a number of environments and on a combination of criteria.

C. Changes in Competitive Ability in Two-species Populations

Lewontin (1968) has stated "A proper understanding of the interactions between species in a community, classically the province of ecology, is impossible without a consideration of the evolution and variation of the species themselves. Species are not static entities with fixed relations to the environment, but plastic elements, changing their genetic constitution under the influence of the physical factors of the environment and of the interactions with other species." Effects of competitive interactions with other species have been studied in laboratory *Drosophila* populations, and the possible outcomes when two species become associated, including genetic changes in either or both species, have been outlined in Section II, B. Lawlor and Maynard Smith (1976) have shown theoretically that if two species compete for two renewable resources, coevolution through natural selection is expected to lead to their divergence and to reduce the degree of competition between them.

Competition between *D. melanogaster* and *D. simulans* in population cages was studied by Moore (1952b). The former species was superior and eliminated *D. simulans* in 19 of 20 populations within about 100 days. In the exceptional cage, however, *D. melanogaster* increased in frequency to day 73, decreased to day 218, and then increased to 100% by day 375. Moore showed that the decrease in frequency from days 73 to 218 was due to improved competitive ability of *D. simulans*, and by successively selecting *D. simulans* populations that had competed with *D. melanogaster* for up to 500 days, demonstrated further increases in competitive ability. In six experimental populations (Ayala, 1966), where one species was always *D. serrata*, the other being *D. pseudoobscura*, *D. nebulosa* or *D. melanogaster*, the frequency of *D. serrata* initially decreased, but it later increased in three populations. While it was not demonstrated that genetic changes had occurred which improved the interspecific competitive ability of *D. serrata* in these populations, Ayala argued that this was the most likely explanation, and initiated further experiments with *D. serrata* and *D. nebulosa* (Ayala, 1969d). Only two populations were used, and remarkably—in view of later experiments discussed below—a reversal of dominance was observed in one population where *D. serrata* frequency decreased to about 20% by week 17, and then increased to about 90% by week 31 (Fig. 7). Subsequent tests

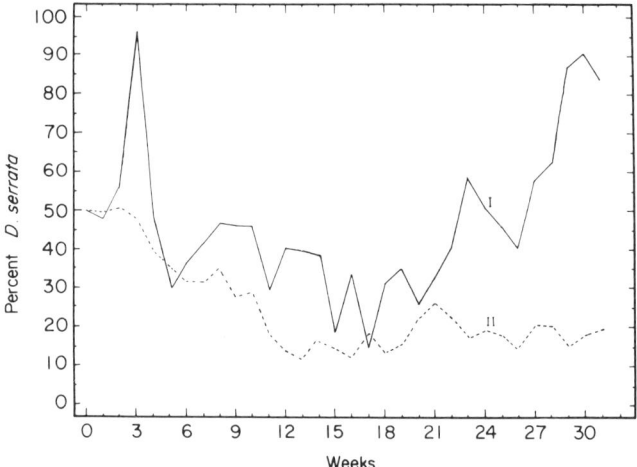

FIG. 7. Changes in percentage *D. serrata* when in competition with *D. nebulosa* in two experimental populations (from Ayala, 1969d).

showed that the interspecific competitive ability of *D. serrata* had increased in this population. Competitive ability of *D. nebulosa* also increased in both populations, perhaps due simply to adaptation to the experimental environment.

Another example of possible evolution of competitive ability was given by Gibo (1972), again during competition between *D. melanogaster* and *D. simulans*. Populations initiated with small numbers showed variability in competitive outcome, but when these populations were mixed after a number of generations, the percentage of *D. melanogaster* decreased more slowly than in populations initiated with large numbers. This result was interpreted as indicating that the competitive interactions of the two species had undergone a rapid and profound change. However, Barker (1963a) found for competing populations of these two species that initial population number had significant effects on competitive ability, so that a "rapid change" in competitive ability is not necessarily indicated by Gibo's results. Further, the apparent change in the competitive ability of *D. melanogaster* in the populations derived from the mix of the small initial number populations could have been due to adaptation to the experimental population cage conditions. Extraction of mixed populations from the large initial number populations would have been necessary to eliminate this possibility.

In experiments using discrete generations, and designed specifically to evaluate the effects of species association, neither consistent nor rapid changes in competitive ability have been found (Futuyma, 1970; Van Delden, 1970; Barker, 1973a). In the experiments of Futuyma (1970), populations of *D. melanogaster* were exposed to competition with *D. simulans* for 10 generations, the latter species being taken from control single-species stocks in each generation. Evidence for change in *D. melanogaster* was obtained in only 10 of 28 comparisons of control populations with those exposed to interspecific competition, and for these 10, a variety of responses was observed. Using different strains of the same two species, Van Delden (1970) found increased competitive ability of *D. melanogaster* in each of two populations, but only after 65 generations of association. Barker (1973a) used a similar design to the above two studies and different strains again of *D. melanogaster* and *D. simulans*. Four replicate populations were initiated, and no changes in competitive ability were detected in three of these populations prior to their loss at generations 20, 29 and 30. In the remaining population, *D. simulans* apparently had increased in competitive ability by generation 42, and subsequent tests showed that it had evolved some facilitation with or ecological divergence from *D. melanogaster*. However, the three *D. simulans* and one *D. melanogaster* single-species control populations studied in these tests also showed apparent changes in competitive ability, while Hedrick (1972) reported evolution of increased competitive ability in a sample of the same *D. melanogaster* strain, but again during maintenance as a stock culture. Given such changes in control populations, understanding of evolutionary changes in mixed-species populations clearly depends on detailed analyses to determine the nature of any changes that do occur. Both Futuyma (1970) and Barker (1973a) emphasized the variability in changes observed, while the main generalities to emerge from these studies are that response to natural selection for characters relating to interspecific competition and resource utilization is slow, and the genetic variance for such characters is highly non-additive. Clearly the simple expectations discussed by Ayala (1969d), Futuyma (1970) and Barker (1973a) and the theory of Lawlor and Maynard Smith (1976) only touch on the problem, and further experimentation is necessary. Such experimentation should specifically consider effects of population density and species frequency, and tests for evolutionary changes should be designed to distinguish increased competitive ability due to increased exploitation from that due to increased interference and both of these from ecological divergence (as a result of selection for avoidance of competition).

D. Competition Involving More Than Two Species

For competition between only two species, the evidence already reviewed indicates that the outcome of competition in mixed-species populations generally will not be simply predictable from single-species population performance. Further, the nature of interactions between a particular pair of species that affect competitive outcome may be specific to that pair, so that information on the competitive outcome for certain pairs of species may not allow prediction for other combinations of these same species. For example, if species A excludes species B, and species B excludes species C, the obvious simple prediction would be that species A would exclude species C. However, depending on the nature of the interspecific interactions, it could be that species C would exclude species A. The predictability of such pairwise interactions has been studied by Goodman (1979) for 19 strains of *Drosophila* involving 15 species. Surprisingly, in view of the complexity of the competition process emphasized in this review, these species showed a distinct competitive hierarchy. While it is clear that even in the simplified environment of *Drosophila* laboratory culture there are many opportunities for resource partitioning and species-specific differentiation of limiting factors and modes of competition (Section IV, A), these results suggest that the strains tested have basically the same mode of competition and resource exploitation, presumably evolved through many generations of isolation in

24. INTERSPECIFIC COMPETITION

laboratory culture. Thus testing of strains newly collected from wild populations, or strains maintained in multi-species laboratory populations should give a less hierarchical system of relationships.

Clearly there are many aspects of two-species systems that require further theoretical development and experimental study, but extension to multi-species systems also is necessary. Such systems more closely approximate natural populations, and competitive outcomes in them need not follow from considerations of the dynamics of the included two-species systems.

An example of stable coexistence of three competing species (Fig. 8) was given by Ayala (1972), but the included two-species systems were not studied. Richmond *et al.* (1975) outlined some theoretical expectations for three-species systems, and studied all possible one-, two- and three-species combinations of *D. nebulosa*, *D. willistoni* and *D. pseudoobscura*, with two replicates of each of the two-species systems and three of the three-species at each of two temperatures. In the two-species populations at 19°C, *D. pseudoobscura* eliminated *D. nebulosa*, and came to equilibrium with *D.*

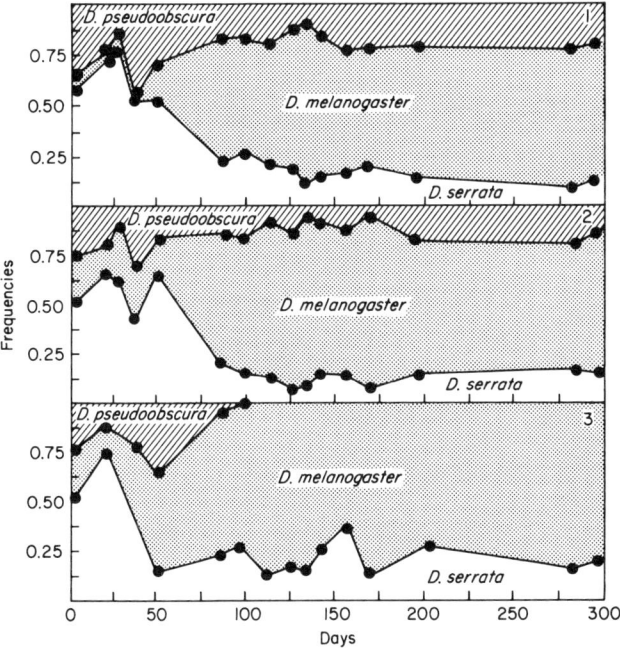

FIG. 8. Frequencies of three species of *Drosophila* in three separate experiments. At any given time, the distance from the lower coordinate to the first curve gives the frequency of *D. serrata*; the distance between the two curves gives the frequency of *D. melanogaster*; the remainder is the frequency of *D. pseudoobscura* (from Ayala, 1972).

willistoni in one population and eliminated it in the other. *D. nebulosa* and *D. willistoni* coexisted. In three-species populations, *D. pseudoobscura* eliminated both of the other species. At 22°C, *D. pseudoobscura* eliminated *D. nebulosa* and came to equilibrium with *D. willistoni*. *D. willistoni* came to equilibrium with *D. nebulosa* in one population and eliminated it in the other. In the three-species populations, *D. nebulosa* was eliminated, and *D. pseudoobscura* and *D. willistoni* were still present at the termination of the experiment, although not in a stable equilibrium. At both temperatures then, results of the two-species populations gave qualitatively adequate predictions of the three-species outcome.

Although not directly a study of interspecific competition in multi-species systems, Wallace's (1976) ingenious experimental simulation of island biogeography necessarily included competition as a component contributing to species extinction on individual islands. The average number of species per island at equilibrium clearly depended on the presence or absence of *D. melanogaster* in the pool of species that could migrate to the islands, being higher in the absence of *D. melanogaster*. In addition, a multitude of other interspecific interactions were altered by the removal of *D. melanogaster* from the species pool.

Detailed studies of two-species systems, particularly of the mechanisms of the competitive process, should continue to increase our understanding of competition. It is clear that these systems are complex enough, so that one might hesitate to consider investigation of multi-species systems. But natural communities are multi-species associations, and the ultimate aim of studies of interspecific competition in *Drosophila* must be to contribute to the understanding of the dynamics of communities.

V. Implications in Evolution, Ecology and Biological Control

As pointed out in Section I, many workers have attributed to competition between closely related species a major role in natural selection. Recently diverged species are expected to be similar in resource requirements, i.e. partially overlapping niches, and where such species come into contact, interspecific competition is predicted to contribute to niche separation, which may involve evolutionary changes in morphological, behavioural, ecological or physiological traits. Although many pairs of sympatric, sibling species are known in *Drosophila*, very few studies of niche separation have been done. Kaneshiro *et al.* (1973) showed that two species of Hawaiian *Drosophila* (*D. heedi* and *D. silvarentis*), which are closely related, are chromosomally homosequential, are morphologically very similar and are abundant in a zone of sympatry on the island of Hawaii, have almost completely separate oviposition and larval sites. Both species are dependent on sap flows from a single species of tree, *Myoporum sandwicense* A. Gray (D.C.). However, *D. silvarentis* breeds on exposed parts of trunks and branches wet by flux emerging from the tree, while *D. heedi* oviposits on soil or tree litter on the ground wet by flux dripping from above. As Kaneshiro *et al.* (1973) point out, a shift in oviposition site can be one of the most crucial phases of adaptive separation of two closely related *Drosophila* species, but the important question is how such niche separation is initially accomplished and how it is maintained thereafter. They do not postulate interspecific competition, but state that it is not known whether initial allopatry is required before such niche separation can occur.

Another interesting example of *Drosophila* niche separation is given by Lachaise (1977). African species of *Lissocephala*, the most primitive genus of the Drosophilidae, are strictly specialized on a number of species of *Ficus*, breeding inside immature figs. Lachaise (1977) suggests that selection for avoidance of competition from *Zaprionus* and *Drosophila* species which breed in the ripening exocarp of the fig has led to the niche separation in larval feeding site. While there is considerable diversity of both *Lissocephala* and *Ficus* species, and coevolution is conjectured, more than one species of *Lissocephala* may utilize a single species of *Ficus*. The question remains as to the extent of competition or niche separation between these *Lissocephala* species.

The concept that interspecific competition will contribute to niche separation and differential adaptation is attractive, but while the studies discussed here demonstrate niche separation, they provide no evidence as to how it arose. Detailed field and laboratory studies (such as those of Mangan, 1978) discussed in Section III, C point the way to appropriate analysis of the importance of interspecific competition. Laboratory studies specifically designed to investigate evolutionary changes towards increased competitive ability or avoidance of competition (Section IV, C) have given equivocal results, and further work in this area is clearly indicated. However, it should be noted that these studies have all used *D. melanogaster* and *D. simulans*, and it could be that it would be more appropriate to use pairs of species that are closely related, but which are not naturally sympatric. In natural populations, comparisons of ecological relationships and morphological, behavioural and physiological traits in sympatric and allopatric populations could provide critical information. In Australia, two cactophilic *Drosophila* species of the *mulleri* subgroup (*D. buzzatii* and *D. aldrichi*) have successfully colonized the *Opuntia* cactus niche, most likely being introduced during the *Opuntia* biological control program in the 1920's (Barker and Mulley, 1976; Mulley and Barker, 1977). *D. buzzatii* is widespread throughout the world (Carson and Wasserman, 1965), but *D. aldrichi* otherwise occurs only in the southern U.S.A. and Mexico, so that these species are sympatric only in Australia, and became so only some 50–60 years ago. Both species are apparently specific to the cactus niche, with the *D. aldrichi* distribution totally included within that of *D. buzzatii*. In those localities where both species exist, adults of both have been collected from single cactus rots, both in the natural populations and as emergences from rotting cladodes brought back to the laboratory. A "natural experiment" is thus in progress; whether evolutionary changes in either species in the sympatric populations yet can be detected remains to be determined.

Since mathematical models of competition were first developed by Volterra (1926), much of the ecological literature has dealt with the manner in which interspecific competition limits the number of species that can coexist in a given community. Thus Gause (1936) stated that a community "is already organized in the sense that its membership is a limited one". This limitation in the number of species "is apparently connected with the limited number of 'ecological niches' which can be utilized by different species without expelling one another", and thus the structure of a community was seen to be the outcome of a long process of competition and selection. These ecological implications overlap broadly with the evolutionary ones already discussed, and are implicit in much of the discussion in this review. In addition, while it has been stated before, it is worth re-emphasizing that caution is necessary in interpreting and extrapolating from laboratory studies of interspecific competition. One does not study competition between species, but between specific strains of each species. Observed results could be quite unique to the particular strains used; genetic variation within species cannot be ignored.

The control of insect pests is discussed in Chapter 26, but one aspect of interspecific competition in relation to pest control should be noted here, particularly as it stems from the one multi-species study that has been done (Wallace, 1976), and so further emphasizes the need for work in the more complex multi-species systems. As Wallace (1976) points out, his result that a species which is easily eliminated by a second may thrive in its absence, raises a paradoxical problem. Wallace goes on: "If among many potential insect pests one predominates, the economics of control dictate that it be controlled first because it causes the damage. Nevertheless, the ecology of control dictates that the lesser pests be controlled (eradicated, if possible) first because (1) in the absence of the major pest, a lesser one will expand to take its place and (2) the eradication (by the release of sterile males, at any rate) of a small population is much easier than that of a large, thriving population. The paradox arises, however, when seeking support for an eradication program whose early efforts would be spent not on 'real' pests but, rather, on the elimination of insect species none of which is causing appreciable harm."

That an esoteric laboratory *Drosophila* study of island biogeography can raise a practical problem of such magnitude would seem an appropriate note on which to conclude this review.

ACKNOWLEDGEMENTS

I am indebted to Dr. M. R. Wheeler and Dr. I. R. Bock for comment on the classification of species pairs as probably sympatric or probably not sympatric in Section IV, A 3(b), and to Dr. F. W. Nicholas for comments on an early draft.

References

ANDERSEN, F. S. (1960). Competition in populations consisting of one age group. *Biometrics* **16**, 19–27.

ANDREWARTHA, H. G. AND BIRCH, L. C. (1954). "The Distribution and Abundance of Animals". Univ. Chicago Press, Chicago, Illinois.

ANTONOVICS, J. and FORD, H. (1972). Criteria for the validation or the invalidation of the competitive exclusion principle. *Nature, Lond.* **237**, 406–408.

AYALA, F. J. (1965). Relative fitness of populations of *Drosophila serrata* and *Drosophila birchii*. *Genetics* **51**, 527–544.

AYALA, F. J. (1966). Reversal of dominance in competing species of *Drosophila*. *Amer. Nat.* **100**, 81–83.

AYALA, F. J. (1967). Dynamics of populations. II. Factors controlling population growth and population size in *Drosophila pseudoobscura* and in *Drosophila melanogaster*. *Ecology* **48**, 67–75.

AYALA, F. J. (1968). Environmental factors limiting the productivity and size of experimental populations of *Drosophila serrata* and *D. birchii*. *Ecology* **49**, 562–565.

AYALA, F.J. (1969a). Experimental invalidation of the principle of competitive exclusion. *Nature, Lond.* **224**, 1076–1079.

AYALA, F. J. (1969b). Genetic polymorphism and interspecific competitive ability in *Drosophila*. *Genet. Res.* **14**, 95–102.

AYALA, F. J. (1969c). An evolutionary dilemma: Fitness of genotypes versus fitness of populations. *Can. J. Genet. Cytol.* **11**, 439–456.

AYALA, F. J. (1969d). Evolution of fitness. IV. Genetic evolution of interspecific competitive ability in *Drosophila*. *Genetics* **61**, 737–747.

AYALA, F. J. (1970a). Competition, coexistence, and evolution. *In* "Essays in Evolution and Genetics in Honor of Theodosius Dobzhansky" (M. K. Hecht and W. C. Steere, eds), pp. 121–158. Appleton-Century-Crofts, New York.

AYALA, F. J. (1970b). Invalidation of principle of competitive exclusion defended. *Nature, Lond.* **227**, 89–90.

AYALA, F. J. (1970c). Population fitness of geographic strains of *Drosophila serrata* as measured by interspecific competition. *Evolution* **24**, 483–494.

AYALA, F. J. (1971a). Competition between species: Frequency dependence. *Science* **171**, 820–824.

AYALA, F. J. (1971b). Environmental fluctuations and population size. *Nature, Lond.* **231**, 112–114.

AYALA, F. J. (1972). Competition between species. *Amer. Sci.* **60**, 348–357.

AYALA, F. J., GILPIN, M. E. and EHRENFELD, J. G. (1973). Competition between species: theoretical models and experimental tests. *Theor. Populat. Biol.* **4**, 331–356.

BAKKER, K. (1961). An analysis of factors which determine success in competition for food among larvae of *Drosophila melanogaster*. *Archs néerl. Zool.* **14**, 200–281.

BAKKER, K. (1969). Selection for rate of growth and its influence on competitive ability of larvae of *Drosophila melanogaster*. *Neth. J. Zool.* **19**, 541–595.

BARKER, J. S. F. (1962). The estimation of generation interval in experimental populations of *Drosophila*. *Genet. Res.* **3**, 388–404.

BARKER, J. S. F. (1963a). The estimation of relative fitness of *Drosophila* populations. II. Experimental evaluation of factors affecting fitness. *Evolution* **17**, 56–71.

BARKER J. S. F. (1963b). The estimation of relative fitness of *Drosophila* populations. III. The fitness of certains strains of *Drosophila melanogaster*. *Evolution* **17**, 138–146.

BARKER, J. S. F. (1965). The estimation of relative fitness of *Drosophila* populations. IV. Experiments on *D. pseudoobscura*. *Genetics* **51**, 747–759.

BARKER, J. S. F. (1967a). The estimation of relative fitness of *Drosophila* populations. V. Generation interval and heterogeneity in competition. *Evolution* **21**, 299–309.

BARKER, J. S. F. (1967b). The fitness of single-species populations of *Drosophila*. *Evolution* **21**, 606–619.

BARKER, J. S. F. (1971). Ecological differences and competitive interactions between *Drosophila melanogaster* and *Drosophila simulans* in small laboratory populations. *Oecologia* **8**, 139–156.

BARKER, J. S. F. (1973a). Natural selection for coexistence or competitive ability in laboratory populations of *Drosophila*. *Egypt. J. Genet. Cytol.* **2**, 288–315.

BARKER, J. S. F. (1973b). Adult population density, fecundity and productivity of *Drosophila melanogaster* and *Drosophila simulans*. *Oecologia* **11**, 83–92.

BARKER, J. S. F. (1977). Cactus breeding *Drosophila*—A system for the measurement of natural selection. *In* "Measuring Selection in Natural Populations" (F. B. Christiansen and T. Fenchel, eds) Lecture Notes in Biomathematics, Vol. 19, pp. 403–430. Springer-Verlag, Berlin.

BARKER, J. S. F. and MULLEY, J. C. (1976). Isozyme variation in natural populations of *Drosophila buzzatii*. *Evolution* **30**, 213–233.

BARKER, J. S. F. and PODGER, R. N. (1970a). Interspecific competition between *Drosophila melanogaster* and *Drosophila simulans*: Effects of larval density on viability, developmental period and adult body weight. *Ecology* **51**, 170–189.

BARKER, J. S. F. and PODGER, R. N. (1970b). Interspecific competition between *Drosophila melanogaster* and *Drosophila simulans*: Effects of larval density and short-term adult starvation on fecundity, egg hatchability and adult viability. *Ecology* **51**, 855–864.

BARKER, J. S. F., TOLL, G. L., EAST, P. D. and WIDDERS, P. R. (1981). Atttaction of *Drosophila buzzatii* and *D. aldrichi* to species of yeasts isolated from their natural environment. II. Field experiments. *Aust. J. Biol. Sci.* **34**, 613–624.

BEARDMORE, J. A., DOBZHANSKY, TH. and PAVLOVSKY, O. A. (1960). An attempt to compare the fitness of polymorphic and monomorphic experimental populations of *Drosophila pseudoobscura*. *Heredity* **14**, 19–33.

BIRCH, L. C. (1957). The meanings of competition. *Amer. Nat.* **91**, 5–18.

BIRCH, L. C. (1960). The genetic factor in population ecology. *Amer. Nat.* **94**, 5–24.

BLAYLOCK, B. G. (1969). Effects of ionizing radiation on interspecific competition. *In*: Symposium on Radioecology. Proceedings of the 2nd National Symposium, Ann Arbor, Michigan, May 15–17, 1967. pp. 61–67.

BOROWSKY, R. (1971). Principle of competitive exclusion and *Drosophila*. *Nature, Lond.* **230**, 409–410.

Bos, M., Burnet, B., Farrow, R. and Woods, R. A. (1976). Development of *Drosophila* on sterol mutants of the yeast *Saccharomyces cerevisiae*. *Genet. Res.* **28**, 163–176.

Bos, M., Burnet, B., Farrow, R. and Woods, R. A. (1977). Mutual facilitation between larvae of the sibling species. *Drosophila melanogaster* and *D. simulans*. *Evolution* **31**, 824–828.

Brncic, D. (1962). Chromosomal structure of populations of *Drosophila flavopilosa* studied in larvae collected in their natural breeding sites. *Chromosoma* **13**, 183–195.

Brown, W. L. and Wilson, E. O. (1956). Character displacement. *Syst. Zool.* **5**, 49–64.

Budnik, M. (1977). The inhibition of *Drosophila pavani* preadult viability by different concentrations of larval biotic residues. *Cienc. Cult. (Sao Paulo)* **29**, 675–676.

Budnik, M. and Brncic, D. (1974). Preadult competition between *Drosophila pavani* and *Drosophila melanogaster, Drosophila simulans* and *Drosophila willistoni*. *Ecology* **55**, 657–661.

Budnik, M. and Brncic, D. (1975a). Response of *Drosophila pavani, Drosophila gaucha* and their hybrids to larval biotic residues. *Experientia* **31**, 781–782.

Budnik, M. and Brncic, D. (1975b). Effects of larval biotic residues on viability in four species of *Drosophila*. *Evolution* **29**, 777–780.

Bulmer, M. G. (1974). Density-dependent selection and character displacement. *Amer. Nat.* **108**, 45–58.

Buruga, J. H. and Olembo, R. J. (1971). Plant food preferences of some sympatric Drosophilids of tropical Africa. *Biotropica* **3**, 151–158.

Buzzati-Traverso, A. A. (1955). Evolutionary changes in components of fitness and other polygenic traits in *Drosophila melanogaster* populations. *Heredity* **9**, 153–186.

Carson, H. L. (1951). Breeding sites of *Drosophila pseudoobscura* and *Drosophila persimilis* in the transition zone of the Sierra Nevada. *Evolution* **5**, 91–96.

Carson, H. L. (1971). The ecology of *Drosophila* breeding sites. Univ. of Hawaii: Harold L. Lyon Arboretum Lecture No. 2, 27 pp.

Carson, H. L., Knapp, E. P. and Phaff, H. J. (1956). Studies on the ecology of *Drosophila* in the Yosemite region of California. III. The yeast flora of the natural breeding sites of some species of Drosophila. *Ecology* **37**, 538–544.

Carson, H. L. and Wasserman, M. (1965). A widespread chromosomal polymorphism in a widespread species, *Drosophila buzzatii*. *Amer. Nat.* **99**, 111–115.

Claringbold, P. J. and Barker, J. S. F. (1961). The estimation of relative fitness of *Drosophila* populations. *J. Theo. Biol.* **1**, 190–203.

Clarke, B. (1962). Balanced polymorphism and the diversity of sympatric species. *Systematics Assoc. Publ.* **4**, 47–70.

Cole, L. C. (1960). Competitive exclusion. *Science* **132**, 348–349.

Coleman, T. P. and Gomatam, J. (1972). Application of a new model of species competition to *Drosophila*. *Nature: New Biol.* **239**, 251–253.

Connell, J. H. (1975). Some mechanisms producing structure in natural communities: a model and evidence from field experiments. *In*: "Ecology and Evolution of Communities" (M. L. Cody and J. M. Diamond, eds), pp. 460–490. The Belknap Press of Harvard University Press, Cambridge, Mass.

Cooper, D. M. (1960). Food preferences of larval and adult *Drosophila*. *Evolution* **14**, 41–55.

da Cunha, A. B., Dobzhansky, Th. and Sokoloff, A. (1951). On food preferences of sympatric species of *Drosophila*. *Evolution* **5**, 97–101.

da Cunha, A. B., El-Tabey Shehata, A. M. and de Oliveira, W. (1957). A study of the diets and nutritional preferences of tropical species of *Drosophila*. *Ecology* **38**, 98–106.

DARLINGTON, P. J., JR. (1972). Competition, competitive repulsion and coexistence. *Proc. Natl. Acad. Sci. U.S.A.* **69**, 3151–3155.

DARWIN, C. (1859). "On the Origin of Species by means of Natural Selection, or the Preservation of Favoured Races in the Struggle for Life". John Murray, London.

DE BACH, P. (1966). The competitive displacement and coexistence principles. *A. Rev. Ent.* **11**, 183–212.

DEBENEDICTIS, P. A. (1977). The meaning and measurement of frequency-dependent competition. *Ecology* **58**, 158–166.

DE JONG, G. (1976). A model of competition for food. I. Frequency-dependent viabilities. *Amer. Nat.* **110**, 1013–1027.

DEL SOLAR, E. (1968). Selection for and against gregariousness in the choice of oviposition sites by *Drosophila pseudoobscura*. *Genetics* **58**, 275–282.

DEL SOLAR, E. and PALOMINO, H. (1966). Choice of oviposition in *Drosophila melanogaster*. *Amer. Nat.* **100**, 127–133.

DE WIT, C. T. (1960). On competition. *Versl. landbouwk. Onderz. Ned.* **66**, 1–82.

DOBZHANSKY, TH. (1968). On some fundamental concepts of Darwinian biology. *In*: "Evolutionary Biology," Vol. 2. (Th. Dobzhansky, M. K. Hecht and W. C. Steere, eds), pp. 1–34. Appleton-Century-Crofts, New York.

DOBZHANSKY, TH., COOPER, D. M., PHAFF, H. J., KNAPP, E. P. and CARSON, H. L. (1956). Studies on the ecology of *Drosophila* in the Yosemite region of California. IV. Differential attraction of species of *Drosophila* to different species of yeasts. *Ecology* **37**, 544–550.

DOBZHANSKY, TH. and DA CUNHA, A. B. (1955). Differentiation of nutritional preferences in Brazilian species of *Drosophila*. *Ecology* **36**, 34–39.

DOBZHANSKY, TH., LEWONTIN, R. C. and PAVLOVSKY, O. (1964). The capacity for increase in chromosomally polymorphic and monomorphic populations of *Drosophila pseudoobscura*. *Heredity* **19**, 597–614.

DOBZHANSKY, TH. and PAVAN, C. (1950). Local and seasonal variations in relative frequencies of species of *Drosophila* in Brazil. *J. Anim. Ecol.* **19**, 1–14.

DOBZHANSKY, TH. and PAVLOVSKY, O. (1961). A further study of fitness of chromosomally polymorphic and monomorphic populations of *Drosophila pseudoobscura*. *Heredity* **16**, 169–179.

DOLAN, R. and ROBERTSON, A. (1975). The effect of conditioning the medium in *Drosophila*, in relation to frequency-dependent selection. *Heredity* **35**, 311–316.

DONALD, C. M. (1963). Competition among crop and pasture plants. *Adv. Agron.* **15**, 1–118.

DUDGEON, E. (1954). Species differences in the utilization of wild yeasts by *Drosophila*. *Univ. Texas Publs.* **5422**, 65–97.

EHRLICH, P. R. and HOLM, R. W. (1962). Patterns and populations. *Science* **137**, 652–657.

EL-HELW, M. R. and ALI, A. M. M. (1970). Competition between *Drosophila melanogaster* and *D. simulans* on media supplemented with *Saccharomyces* and *Schizosaccharomyces*. *Evolution* **24**, 531–537.

EL-HELW, M. R., ALI, A. M. M. and MOAWAD, H. (1972). Fitness of *Drosophila melanogaster* and *D. simulans* in relation to natural genera of yeasts. *Egypt. J. Genet. Cytol.* **1**, 196–202.

ERK, F. C. and SANG, J. H. (1966). The comparative nutritional requirements of two sibling species *Drosophila simulans* and *D. melanogaster*. *J. Insect. Physiol.* **12**, 43–51.

FELLOWS, D. P. and HEED, W. B. (1972). Factors affecting host plant selection in desert-adapted cactiphilic *Drosophila*. *Ecology* **53**, 850–858.

FUTUYMA, D. J. (1970). Variation in genetic response to interspecific competition in laboratory populations of *Drosophila*. *Amer. Nat.* **104**, 239–252.
GAUSE, G. F. (1934). "The Struggle for Existence". The Williams and Wilkins Co., Baltimore.
GAUSE, G. F. (1936). The principles of biocoenology. *Q. Rev. Biol.* **11**, 320–336.
GAUSE, G. F. (1970). Criticism of invalidation of principle of competitive exclusion. *Nature, Lond.* **227**, 89.
GAUSE, G. F. and WITT, A. A. (1935). Behavior of mixed populations and the problem of natural selection. *Amer. Nat.* **69**, 596–609.
GIBO, D. L. (1972). A stabilizing interaction between the founder effect and interdeme mixing in competing populations of *Drosophila melanogaster* and *Drosophila simulans*. *Can. J. Zool.* **50**, 325–331.
GILPIN, M. E. and AYALA, F. J. (1973). Global models of growth and competition. *Proc. Natl. Acad. Sci. U.S.A.* **70**, 3590–3593.
GILPIN, M.E. and JUSTICE, K. E. (1972). Reinterpretation of the invalidation of the principle of competitive exclusion. *Nature, Lond.* **236**, 273–274, 299–301.
GOLDSTEIN, L. (1953). Recherches sur les populations mixtes expérimentales de *Drosophila melanogaster* et *Drosophila simulans*. *Int. Cong. Genet.* Bellagio (Como) Part II, 668–670.
GOODMAN, D. (1979). Competitive hierarchies in laboratory *Drosophila*. *Evolution* **33**, 207–219.
HARDIN, G. (1960). The competitive exclusion principle. *Science* **131**, 1292–1298.
HECK, K. L. (1976). Some critical considerations of the theory of species packing. *Evol. Theory* **1**, 247–258.
HEDRICK, P. W. (1972). Factors responsible for a change in interspecific competitive ability in *Drosophila*. *Evolution* **26**, 513–522.
HEED, W. B. (1957). Ecological and distributional notes on the Drosophilidae (Diptera) of El Salvador. *Univ. Texas Publs.* **5721**, 62–78.
HEED, W. B. (1979). Ecology and genetics of Sonoran Desert *Drosophila*. *In*: "Ecological Genetics: The Interface" (P. F. Brussard, ed), pp. 109–126. Springer-Verlag, Heidelberg.
HEED, W. B. and KIRCHER, H. W. (1965). Unique sterol in the ecology and nutrition of *Drosophila pachea*. *Science* **149**, 758–761.
HEED, W. B. and RUSSELL, J. S. (1968). Inability of *D. pachea* to breed in cereus cacti other than Senita. *Dros. Inf. Serv.* **43**, 94–96.
HOENIGSBERG, H. F. (1968). An ecological situation which produced a change in the proportion of *Drosophila melanogaster* to *Drosophila simulans*. *Amer. Nat.* **102**, 389–390.
HOSGOOD, S. M. W. and PARSONS, P. A. (1966). Differences between *D. simulans* and *D. melanogaster* in tolerances to laboratory temperatures. *Dros. Inf. Serv.* **41**, 176.
HUSTON, M. (1979). A general hypothesis of species diversity. *Amer. Nat.* **113**, 81–101.
HUTCHINSON, G. E. (1961). The paradox of the plankton. *Amer. Nat.* **95**, 137–145.
HUTCHINSON, G. E. (1965). "The Ecological Theatre and the Evolutionary Play". Yale Univ. Press, New Haven, Conn.
HUTCHINSON, G. E. (1975). Variations on a theme by Robert MacArthur. *In*: "Ecology and Evolution of Communities" (M. L. Cody and J. M. Diamond, eds), pp. 492–521. The Belknap Press of Harvard University Press, Cambridge, Mass.
JAENIKE, J. (1978). Ecological genetics in *Drosophila athabasca*: its effect on local abundance. *Amer. Nat.* **112**, 287–299.
KANESHIRO, K. Y., CARSON, H. L., CLAYTON, F. E. and HEED, W. B. (1973). Niche separation in a pair of homosequential *Drosophila* species from the island of Hawaii. *Amer. Nat.* **107**, 766–774.
KAWANISHI, M. and WATANABE, T. K. (1977). Ecological factors controlling the

coexistence of the sibling species *Drosophila simulans* and *D. melanogaster*. *Jap. J. Ecol.* **27**, 279–283.

KIMURA, M. T. (1976). *Drosophila* survey of Hokkaido, XXX. Microdistribution and seasonal fluctuations of drosophilid flies dwelling among the undergrowth plants. *J. Fac. Sci. Hokkaido Univ. Ser. VI, Zool.* **20**, 192–202.

KIRCHER, H. W., HEED, W. B., RUSSELL, J. S. and GROVE, J. (1967). Senita cactus alkaloids: their significance to Sonoran Desert *Drosophila* ecology. *J. Insect. Physiol.* **13**, 1869–1874.

KLOMP, H. (1964). Intraspecific competition and the regulation of insect numbers. *A. Rev. Ent.* **9**, 17–40.

LACHAISE, D. (1974). Les Drosophilidae des savanes préforestières de la région tropicale de Lamto (Cote-d'Ivoire). I.—Isolement écologique des espèces affines et sympatriques; rythmes d'activité saisonnière et circadienne; rôle des feux de brousse. *Ann. Univ. Abidjan, Ser. E, Ecol.* **7**, 7–152.

LACHAISE, D. (1977). Niche separation of African *Lissocephala* within the *Ficus* Drosophilid community. *Oecologia* **31**, 201–214.

LACHAISE, D. and TSACAS, L. (1974). Les Drosophilidae des savanes préforestières de la region tropicale de Lamto (Cote-d'Ivoire) II. Le peuplement des fruits de *Pandanus candelabrum* (Pandanacées). *Ann. Univ. Abidjan, Ser. E. Ecol.* **7**, 153–192.

LAWLOR, L. R. and MAYNARD SMITH, J. (1976). The coevolution and stability of competing species. *Amer. Nat.* **110**, 79–99.

LEVINS, R. (1968). "Evolution in Changing Environments", Princeton Univ. Press, Princeton, New Jersey.

LEWONTIN, R. C. (1957). The adaptations of populations to varying environments. *Cold Spring Harb. Symp. quant. Biol.* **22**, 395–408.

LEWONTIN, R. C. (1968). Introduction. *In*: "Population Biology and Evolution" (R. C. Lewontin, ed), pp. 1–4. Syracuse Univ. Press, Syracuse, New York.

L'HÉRITIER, PH. and TEISSIER, G. (1933). Etude d'une population de Drosophiles en equilibre. *C. r. hebd. Séanc. Acad. Sci.*, Paris **197**, 1765–1767.

L'HÉRITIER, PH. and TEISSIER, G. (1935). Recherches sur la concurrence vitale. Etude de populations mixtes de *Drosophila melanogaster* et *Drosophila funebris*. *C. r. Seanc. Soc. Biol.* **118**, 1396–1398.

LINDSAY, S. L. (1958). Food preferences of *Drosophila* larvae. *Amer. Nat.* **92**, 279–285.

LOTKA, A. J. (1925). "Elements of Physical Biology". The Williams and Wilkins Co., Baltimore.

MACARTHUR, R. H. (1972). "Geographical Ecology. Patterns in the Distribution of Species". Harper and Row, New York.

MANGAN, R. L. (1978). Competitive interactions among host plant specific *Drosophila* species. Ph.D. Dissertation, University of Arizona, Tucson.

MAY, R. M. (1973). "Stability and Complexity in Model Ecosystems". Princeton Univ. Press, Princeton.

MERRELL, D. J. (1951). Interspecific competition between *Drosophila funebris* and *Drosophila melanogaster*. *Amer. Nat.* **85**, 159–169.

MERTZ, D. B. (1972). The *Tribolium* model and the mathematics of population growth. *Ann. Rev. Ecol. Syst.* **3**, 51–78.

MILLER, R. S. (1964a). Larval competition in *Drosophila melanogaster* and *D. simulans*. *Ecology* **45**, 132–148.

MILLER, R. S. (1964b). Interspecies competition in laboratory populations of *Drosophila melanogaster* and *Drosophila simulans*. *Amer. Nat.* **98**, 221–238.

MILLER, R. S. (1967). Pattern and process in competition. *Adv. Ecol. Res.* **4**, 1–74.

MILNE, A. (1961). Defintion of competition among animals. *In*: "Mechanisms in Biological Competition" (F. L. Milthorpe, ed). Symposium XV, Soc. Exp. Biol., pp. 40–61. Cambridge Univ. Press, Cambridge.

MONTGOMERY, S. L. (1975). Comparative breeding site ecology and the adaptive radiation of picture-winged *Drosophila* (Diptera: Drosophilidae) in Hawaii. *Proc. Hawaii. ent. Soc.* **22**, 65–103.

MOORE, J. A. (1952a). Competition between *Drosophila melanogaster* and *Drosophila simulans*. I. Population cage experiments. *Evolution* **6**, 407–420.

MOORE, J. A. (1952b). Competition between *Drosophila melanogaster* and *Drosophila simulans*. II. The improvement of competitive ability through selection. *Proc. Natl. Acad. Sci. U.S.A.* **38**, 813–817.

MOTH, J. J. (1974). Density, frequency and interspecific competition: Fertility of *Drosophila simulans* and *Drosophila melanogaster*. *Oecologia* **14**, 237–246.

MOTH, J. J. (1977). Interspecific competition between *Drosophila melanogaster* and *D. simulans*: Effects of adult density, species frequency and dietary ^{32}P on egg hatchability. *Aust. J. Zool.* **25**, 699–709.

MOTH, J. J. and BARKER, J. S. F. (1976). Interspecific competition between *Drosophila melanogaster* and *Drosophila simulans*. Reduction in fecundity and destruction of eggs when the medium is inhabited by larvae. *Oecologia* **23**, 151–164.

MOTH, J. J. and BARKER, J. S. F. (1977). Interspecific competition between *Drosophila melanogaster* and *Drosophila simulans*: Effects of adult density on adult viability. *Genetica* **47**, 203–218.

MOTH, J. J. and BARKER, J. S. F. (1981). Interspecific competition between *Drosophila melanogaster* and *Drosophila simulans*: Effects of adult density, species frequency, light, and dietary phosphorus-32 on fecundity. *Physiol. Zool.* **54**, 28–43.

MOURÃO, C. A. and AYALA, F. J. (1971). Competitive fitness in experimental populations of *Drosophila willistoni*. *Genetica* **42**, 65–78.

MOURÃO, C. A., AYALA, F. J. and ANDERSON, W. W. (1972). Darwinian fitness and adaptedness in experimental populations of *Drosophila willistoni*. *Genetica* **43**, 552–574.

MULLEY, J. C. and BARKER, J. S. F. (1977). The occurrence and distribution of *Drosophila aldrichi* in Australia. *Dros. Inf. Serv.* **52**, 151–152.

NARISE, T. (1965). The effect of relative frequency of species in competition. *Evolution* **19**, 350–354.

NEILL, W. E. (1974). The community matrix and interdependence of the competition coefficients. *Amer. Nat.* **108**, 399–408.

NEYMAN, J., PARK, T. and SCOTT, E. L. (1956). Struggle for existence. The *Tribolium* model: biological and statistical aspects. *Proc. Third Berkeley Symp. math. Statist. Probab.* **4**, 41–79.

NICHOLSON, A. J. (1954). An outline of the dynamics of animal populations. *Aust. J. Zool.* **2**, 9–65.

PAINE, R. T. (1966). Food web complexity and species diversity. *Amer. Nat.* **100**, 67–75.

PARK, T. (1954). Experimental studies of interspecies competition. II. Temperature, humidity, and competition in two species of *Tribolium*. *Physiol. Zool.* **27**, 177–238.

PARK, T. (1962). Beetles, competition and populations. *Science* **138**, 1369–1375.

PARSONS, P. A. (1975). The comparative evolutionary biology of the sibling species, *Drosophila melanogaster* and *D. simulans*. *Q. Rev. Biol.* **50**, 151–169.

PEARL, R. and PARKER, S. L. (1922). On the influence of density of population upon the rate of reproduction in *Drosophila*. *Proc. Natl. Acad. Sci. U.S.A.* **8**, 212–219.

PETERS, R. H. (1976). Tautology in evolution and ecology. *Amer. Nat.* **110**, 1–12.

PIMENTEL, D. (1968). Population regulation and genetic feedback. *Science* **159**, 1432–1437.

PIPKIN, S. B. (1965). The influence of adult and larval food habits on population size of neotropical ground-feeding *Drosophila*. *Amer. Midl. Nat.* **74**, 1–27.

PUTWAIN, P. D., ANTONOVICS, J. and MACHIN, D. (1967). The effect of relative frequency of species in competition: A reappraisal. *Evolution* **21**, 638–641.

REED, S. C. and REED, E. W. (1948). Natural selection in laboratory populations of Drosophila. *Evolution* **2**, 176–186.

RICHARDSON, R. H. and SMOUSE, P. E. (1975). Ecological specialization of Hawaiian *Drosophila* II. The community matrix, ecological complementation, and phyletic species packing. *Oecologia* **22**, 1–13.

RICHMOND, R. C., GILPIN, M. E., SALAS, S. P. and AYALA, F. J. (1975). A search for emergent competitive phenomena: The dynamics of multispecies *Drosophila* systems. *Ecology* **56**, 709–714.

RIZKI, M. T. M. and DAVIS, C. G. (1953). Light as an ecological determinant of interspecific competition between *Drosophila willistoni* and *Drosophila melanogaster*. *Amer. Nat.* **87**, 389–392.

SAMEOTO, D. D. and MILLER, R. S. (1966). Factors controlling the productivity of *Drosophila melanogaster* and *D. simulans*. *Ecology* **47**, 695–704.

SAMEOTO, D. D. and MILLER, R. S. (1968). Selection of pupation site by *Drosophila melanogaster* and *D. simulans*. *Ecology* **49**, 177–180.

SAMMETA, K. P. V. and LEVINS, R. (1970). Genetics and ecology. *Ann. Rev. Genet.* **4**, 469–488.

SIMBERLOFF, D. (1978). Using island biogeographic distributions to determine if colonization is stochastic. *Amer. Nat.* **112**, 713–726.

SMITH-GILL, S. J. and GILL, D. E. (1978). Curvilinearities in the competition equations: An experiment with Ranid tadpoles. *Amer. Nat.* **112**, 557–570.

SOKOLOFF, A. (1955). Competition between sibling species of the *Pseudoobscura* subgroup of *Drosophila*. *Ecol. Monogr.* **25**, 387–409.

SOKOLOFF, A. (1964). Studies on the ecology of *Drosophila* in the Yosemite region of California. V. A preliminary survey of species associated with *D. pseudoobscura* and *D. persimilis* at slime fluxes and banana traps. *Pan-Pacif. Ent.* **40**, 203–218.

SOKOLOFF, A. (1966). Morphological variation in natural and experimental populations of *Drosophila pseudoobscura* and *Drosophila persimilis*. *Evolution* **20**, 49–71.

STRICKBERGER, M. W. (1963). Evolution of fitness in experimental populations of *Drosophila pseudoobscura*. *Evolution* **17**, 40–55.

TANTAWY, A. O. and EL-WAKIL, H. M. (1970). Studies on natural populations of *Drosophila* XI. Fitness components and competition between *Drosophila funebris* and *D. virilis*. *Evolution* **24**, 528–530.

TANTAWY, A. O., MOURAD, A. M., DAWOOD, M. M. and EL-DABBAGH, H. (1976). Studies on natural populations of *Drosophila*. XVIII. Response to selection for high productivity in *Drosophila melanogaster* and *D. simulans* and their interspecific competition. *Egypt. J. Genet. Cytol.* **5**, 1–14.

TANTAWY, A. O., MOURAD, A. M. and MASRI, A. M. (1970). Studies on natural populations of *Drosophila*. VIII. A note on the directional changes over a long period of time in the structure of *Drosophila* near Alexandria, Egypt. *Amer. Nat.* **104**, 105–109.

TANTAWY, A. O. and SOLIMAN, M. H. (1967). Studies on natural populations of *Drosophila* VI. Competition between *Drosophila melanogaster and Drosophila simulans*. *Evolution* **21**, 34–40.

VAN DELDEN, W. (1968). Fitness of experimental populations of *Drosophila melanogaster*. Doctoral Thesis, Rijksuniversiteit te Groningen.

VAN DELDEN, W. (1970). Selection for competitive ability. *Dros. Inf. Serv.* **45**, 169.

VOLTERRA, V. (1926). Variazioni e fluttuazioni del numero d'individui in specie animali conviventi. *Memorie della R. Accademia Nazionale dei Lincei*, Ser. 6, **2**, 31–113.

WAGNER, R. P. (1944). The nutrition of *Drosophila mulleri* and *D. aldrichi*. Growth of the larvae on a cactus extract and the microorganisms found in cactus. *Univ. Texas Publs.* **4445**, 104–128.

WAGNER, R. P. (1949). Nutritional differences in the *mulleri* group. *Univ. Texas Publs.* **4920**, 39–41.

WALLACE, B. (1974). Studies on intra- and inter-specific competition in *Drosophila*. *Ecology* **55**, 227–244.

WALLACE, B. (1976). The biogeography of laboratory islands. *Evolution* **29**, 622–635.

WATANABE, T. K. and KAWANISHI, M. (1976). Colonization of *Drosophila simulans* in Japan. *Proc. Jap. Acad.* **52**, 191–194.

WEISBROT, D. R. (1966). Genotypic interactions among competing strains and species of *Drosophila*. *Genetics* **53**, 427–435.

ZIMMERING, S. (1948). Competition between *Drosophila pseudoobscura* and *Drosophila melanogaster* in population cages. *Amer. Nat.* **82**, 326–330.

25. Natural and Experimental Parthenogenesis

ALAN R. TEMPLETON

*Department of Biology,
Washington University, St. Louis, Missouri, U.S.A.*

I. Introduction	343
II. The Incidence of Parthenogenesis in the genus *Drosophila*	345
III. Mechanisms of Parthenogenesis and the Genetic Consequences	348
A. General Considerations	348
B. Central Fusion	349
C. Terminal Fusion	352
D. Gamete (Pronuclear) Duplication	353
E. Mixtures of Mechanisms	354
F. Non-Disjunction and Production of Males	356
G. Sexual Reproduction and Parthenogenesis	360
IV. The Genetic Basis of Parthenogenesis	360
A. Evidence from Selection Experiments	360
B. Evidence from Crosses Between Unisexual and Bisexual Strains	363
C. Evidence from Contrasts of Isogenic Lines	363
D. Evidence from Screening Experiments	363
E. Components of the Genetic Basis of Parthenogenesis	364
V. Behavioral Genetics	369
A. Sexual Behavior and Isolation	369
B. Other Behavioral Traits	372
VI. Population Genetics	373
A. Population Genetic Theory for Automictic Thelytoky	373
B. Clonal Selection in Experimental Populations	376
C. The Unit of Selection in Experimental Populations	381
D. Total Homozygosity and Isozyme Variability	385
VII. Speciation	386
A. The Origin of Parthenogenetic Species and the Evolution of Sex	386
B. Parthenogenesis as an Experimental Model of Speciation in Sexual Populations	393
VIII. Prospects for Parthenogenetic *Drosophila*	395
References	395

I. Introduction

Parthenogenesis is the development of an unfertilized egg into a new individual. The phrase "unfertilized egg" distinguishes parthenogenesis from various forms of somatic or vegetative reproduction and from asexual reproduction in organisms that did not have sexual ancestors. However, the meaning of "individual" has ranged from a small bit of embryonic tissue to a normal adult. In this chapter, I generally discuss the parthenogenetic production of adult flies, although some sections deal with early embryonic development that is aborted before adulthood.

Parthenogenesis is a conglomeration of several distinct types of reproduction. The following is a brief classification using the terms as they are most often (but not universally) used in the insect literature.

First, the type of parthenogenesis may be classified according to the sex of the parthenogenetic offspring: thelytokous (female producing), arrhenotokous (male producing) and amphitokous (also deuterotokous and amphoterotokous; both sexes are produced). Almost all parthenogenesis in the genus *Drosophila* is thelytokous, although some amphitoky exists. A second categorization is based on the ploidy level of the parthenogenetic individuals. Parthenogenetic *Drosophila* are usually diploid, but some polyploids occur. Third, parthenogenesis may be typed according to its role in the life history of the species. The major categories follow, but note that they are not mutually exclusive. In obligate parthenogenesis, all reproduction is parthenogenetic. By necessity, this type cannot be arrhenotokous and is usually thelytokous. Facultative parthenogenesis occurs when females may either reproduce sexually or parthenogenetically. Accidental or rare facultative parthenogenesis occurring in an otherwise sexually reproducing species is known as tychoparthenogenesis. Under heterogenous or cyclical parthenogenesis, bisexual and parthenogenetic generations alternate or go through some sort of cycle. Paedogenesis (parthenogenetic progenesis), in which eggs appear in a larva and produce more larvae parthenogenetically, is frequently coupled with cyclical parthenogenesis. A final life history type is gynogenesis in which parthenogenetic development requires the penetration of the egg by a sperm, but the sperm makes no genetic contribution. Within the Drosophilidae, all parthenogenesis is either obligate or tychoparthenogenetic, although a paedogenetic drosophilid may not be recognized.

The fourth, and last, classification of parthenogenesis is based on the cytological or genetic mechanism involved. There are two basic types: ameiotic or apomictic, in which meiosis is suppressed; and meiotic, in which meiosis is retained. Meiotic parthenogenesis is further subdivided into generative parthenogenesis in which the haploid egg nucleus develops (almost always into a haploid male) and automixis in which the ploidy level is restored by post-meiotic nuclear fusions or by a suppression of meiosis I or II, but not both. Sometimes more than two nuclei fuse, resulting in polyploidy. All parthenogenesis in *Drosophila* is automictic, and involves fusion of nuclei after meiosis has been completed.

The apomictic/meiotic classification is plagued by the ambiguity of just what constitutes a suppression of meiosis. For example, in some parthenogenetic insects there is a pre-meiotic doubling of the chromosomes followed by two meiotic divisions that reduce the duplicated number to the diploid level. Moreover, in a few cases synapsis is not restricted to sister chromatids. Is this an apomictic or meiotic parthenogenesis? Similarly, Nur (1979) described a thelytokous coccid in which the chromosomes do not pair during meiotic prophase, but they do undergo the two meiotic divisions, including reduction, followed by fusion of the meiotic pronuclei that divided at meiosis I. Is this automictic or apomictic? To avoid these difficulties, Nur proposed the terms gonoid and agonoid parthenogenesis. Under gonoid parthenogenesis, the number of chromosomal elements visible in the first (or only) metaphase of oogenesis is the same as the number of chromosomes present in the oogonia (germ line). Under agonial parthenogenesis, these numbers differ. By this definition, parthenogenesis is agonoid in *Drosophila*. Nur feels this is an important distinction because gonoid thelytoky would promote heterozygosity whereas agonoid thelytoky would promote homozygosity. Unfortunately, this prediction is not true for *Drosophila*. Consequently, from an evolutionary perspective it is better simply to classify the modes into those that promote heterozygosity versus those that promote homozygosity, both of which are found in *Drosophila*. Another evolutionarily important distinction is between those modes that allow recombination between homologous chromosomes and those that do not. Once again, both types exist in parthenogenetic *Drosophila*.

II. The Incidence of Parthenogenesis in the Genus *Drosophila*

The only naturally occurring, obligate thelytokous *Drosophila* is *D. mangabeirai* from the *willistoni* group of the subgenus *Sophophora* (Carson et al., 1957). Collections from natural populations in the Caribbean, Central America, and South America are almost exclusively female, but three males have been found in a collection of 116 flies (Malogolowkin, 1958; Carson, 1962a). However, females do not respond to male courtship, and the one male examined proved to be sterile and most likely an X0 diploid (Carson, 1962a). The species is widespread both geographically and altitudinally (sea level to 4,000 feet), indicating a broad ecological range (Carson, 1962a). Moreover, this species can be successfully maintained in the laboratory (Murdy and Carson, 1959; Carson, 1962a).

Although obligate thelytoky is represented by only one species, tychoparthenogenesis is far more common (at least in the laboratory). Stalker (1951, 1952, 1954) in his pioneering work established that parthenogenesis existed in *Drosophila* and that it was a widespread phenomenon. Stalker first discovered parthenogenesis in crosses of a strain of *D. parthenogenetica* females (normally a bisexual species despite its name) with males of other species from the *D. cardini* species group. These "crosses" resulted in daughters identical with their mothers morphologically, chromosomally and physiologically. Subsequent tests with *D. parthenogenetica* virgin females revealed that parthenogenesis rather than hybridization was involved. This discovery motivated Stalker to examine a number of other species. The results are shown in Table I. As can be seen, some parthenogenetic development occurred in 21 of the 23 *Drosophila* species, and two species (*D. polymorpha* and *D. affinis*) gave rise to parthenogenetic adults.

Since Stalker's initial discovery and survey, other *Drosophila* have been found to be capable of tycoparthenogenesis, including *D. robusta* (Carson, 1961), *D. mercatorum* (Carson, 1962b, 1967a), *D. ananassae* (Futch, 1973), *D. pallidosa* (Futch, 1973), *D. hydei* (Templeton, 1979a), and *D. paulistorum* (L. Ehrman, personal communication).

Most of the above studies were performed on stocks that had been in the laboratory for various lengths of time and subject to an unknown degree of inbreeding. The possibility exists therefore that the capacity for parthenogenesis is a laboratory phenomenon that does not exist in natural populations. However, the data of Stalker (1954) shown in Table I on the F_1 and F_2 of wild-caught flies indicates this is not so. Direct evidence is provided by Carson (1961), Templeton *et al.* (1976a) and Templeton (1979a). Carson (1961) discovered tychoparthenogenetic development in seven out of twelve wild strains examined in *D. robusta*. Templeton *et al.* (1976a) examined the parthenogenetic capacity of a natural population of *D. mercatorum*, a species for which parthenogenetic adults had previously been obtained from laboratory stocks (Carson, 1967a). They discovered that for flies caught at the Kamuela, Hawaii garbage dump the probability of an unfertilized egg giving rise to a viable adult was 4×10^{-5}. Moreover, the probability of an unfertilized egg giving rise to an adult herself capable of parthenogenetic reproduction was 10^{-5}. This rate is sufficiently high as to allow the establishment of self-sustaining parthenogenetic lines directly from the natural population with no intervening artificial selection. The Kamuela *D. mercatorum* also inhabit patches of the cactus *Opuntia megacantha*. When the cactus patch populations were screened, the probability of an unfertilized egg giving rise to a viable adult was 10^{-5}, while the probability of it giving rise to an adult herself capable of parthenogenesis was 10^{-6} (Templeton, 1979a). These differences in parthenogenetic capacity between the Kamuela garbage dump and cactus patch populations of *D. mercatorum* are not altogether surprising in light of the fact that an isozyme survey revealed several significant allele frequency differences and a very different

TABLE I. Summary of rates of parthenogenetic development in various species of Drosophilidae. Rates for *D. parthenogenetica* are not included in this table. Members of the same species group are bracketed. From Stalker (1954).

Species	Total unfertilized eggs examined	Total dead embryos	Total dead larvae	Adults
Drosophila				
⎡ *cardini*[a]	52,850	3		
polymorpha[a,b]	37,629	109		2
neocardini[a]	11,439	1		
cardinoides[a]	30,777	3		
acutilabella[a]	16,463	11		
crocina[a]	23,682	14		
acutilabella X				
⎣ *cardinoides* F₁ hybrids[a]	6,231	1		
⎡ *macrospina*	2,655			
⎣ *funebris*	43,198	3		
⎡ *melanica*	5,199	1		
⎣ *nigromelanica*	5,243	1	1	
americana	18,165			
⎡ *tripunctata*	9,280	2		
⎣ *mediostriata*[a]	14,883	7		
robusta	10,706	2		
immigrans	8,153	1		
⎡ *transversa*	11,616	2		
⎣ *quinaria*	11,868	6		
putrida	8,431	4		
hydei	63,027	18	1	
Hirtodrosophila				
duncani	16,044	1		
Sophophora				
affinis	19,059	4		1
⎡ *melanogaster*	32,197	1		
melanogaster[c]	500,000		2	
⎣ *simulans*	13,872	1		
Scaptodrosophila				
victoria	1,292			
Scaptomyza				
adusta	2,340			
graminum	4,314	4		
Zaprionus				
vittiger[a]	10,167			

[a] Virgins from established laboratory stocks. In all other cases the virgins were F₁ or F₂ progeny of wild flies captured in the St. Louis, Missouri area. [b] These figures are for the unselected bisexual strains. [c] Approximate number of eggs, oviposition in half-pint bottles, unfertilized eggs not examined for presence of dead embryos.

population structure between the two. In addition to *D. mercatorum*, *D. hydei* inhabits the Kamuela cactus patches. This species was also screened for parthenogenesis even though there was no previous report of it producing parthenogenetic adults. However, the probability of an unfertilized *D. hydei* egg giving rise to a viable adult was 5×10^{-6}. Moreover, the parthenogenetic rates in *D. hydei* were not significantly different from those of the sympatric cactus-patch *D. mercatorum*. These results on natural populations when coupled with Stalker's original survey (Table I) strongly indicate that the capacity for parthenogenesis is widespread in the genus *Drosophila* and may be present in a majority of the species.

III. Mechanisms of Parthenogenesis and their Genetic Consequences

A. General Considerations

As mentioned in the Introduction, all parthenogenesis in *Drosophila* is automictic. Hence, the problem arises as to how the ploidy level is restored without fertilization. Studies on meiosis and egg structure in uninseminated eggs have revealed the essential mechanisms by which this problem is circumvented in the genus *Drosophila*.

Doane (1960) studied the completion of meiosis in uninseminated eggs of *Drosophila melanogaster*. A total of 93 eggs were cytologically examined in detail, and in all cases meiosis had gone to completion. Moreover, Doane gives evidence that this is the rule rather than the exception in the *Drosophila*. This means that the *Drosophila* are essentially preadapted to the first prerequisite of automixis—the production of unfertilized eggs that spontaneously undergo meiosis. Moreover, the mechanism for satisfying this prerequisite does not depend upon gynogenesis or some peculiar external trigger; rather, the eggs must simply remain uninseminated.

The second requirement for automixis is the restoration of the diploid (or polyploid) state. Doane (1960) noted that following maturation, the meiotic products in uninseminated *D. melanogaster* eggs tend to fuse with one another, resulting in diploid, triploid, and even tetraploid nuclei. Moreover, with the completion of meiosis, a cytoplasmic extension forms along which a pronucleus or fusion nucleus may sometimes pass to the center of the egg, where cleavage divisions are normally initiated. This pattern of post-meiotic fusions also appears to be widespread in the genus (Stalker, 1954; Murdy and Carson, 1959; Sprackling, 1960; Carson *et al.*, 1969; Counce and Ruddle, 1969; Carson, 1973; Futch, 1973, 1979). Hence the *Drosophila* seem to be generally preadapted to the second prerequisite of automixis—the restoration of diploidy or polyploidy—through the mechanism of post-meiotic nuclear fusion.

However, there is a great deal of heterogeneity in exactly what types of fusion events occur. For example, Counce and Ruddle (1969) studied strain differences in egg structure in *D. hydei* and some interspecific contrasts as well. The species and strains differed in a number of ways, including the behavior of the polar body nuclei. In one strain of *D. hydei*, the polar body nuclei quickly fused followed by the formation of a single triploid metaphase plate. In another strain, however, about 50% of the time only two nuclei fused to form a haploid/diploid double plate. In yet a third strain fusion of any of the polar nuclei was uncommon. Moreover, formation of cytoplasmic islands like those formed during cleavage were observed in unfertilized eggs from one *D. hydei* strain, but not in another. Counce and Ruddle (1969) concluded that these differences could affect the ability of the strains to evolve parthenogenesis.

The above studies indicate that *Drosophila* are in general preadapted to automictic parthenogenesis via post-meiotic nuclear fusions, but that considerable interspecific and interstrain variability exists in the precise pattern of post-meiotic fusion. The remainder of this section will examine the actual fusion and meiotic patterns that occur in parthenogenetic *Drosophila* and their genetic implications.

B. Central Fusion

Central fusion occurs when the pronuclei that separated at meiosis I fuse to restore diploidy (Fig. 1). Suppression of meiosis I is a genetic equivalent (Fig. 1), but is unknown in the *Drosophila*. On the other hand, central fusion occurs in many parthenogenetic *Drosophila*, including *D. mangabeirai*, the only naturally occurring parthenogenetic species in the genus.

Murdy and Carson (1959) have studied the cytological mechanism of central fusion in *D. mangabeirai*. In most *Drosophila* that have been examined, the meiotic divisions take place close to the dorsal surface of the egg, approximately one-third of the distance from the anterior pole. Moreover, the spindles at both meiotic divisions are oriented perpendicularly to the long axis of the egg, causing the four pronuclei to line up on this perpendicular (Fig. 2a). The innermost nucleus normally becomes the egg nucleus. However, in *D. mangabeirai*, there is a strong tendency for the first meiotic division to be parallel to the long axis of the egg, while the second division is perpendicular (Fig. 2b). This places the pronuclei that separated at meiosis I into the inner part of the egg. There, the two nuclei fuse, and cleavage divisions are initiated.

If no recombination occurs between a locus and its centromere, central

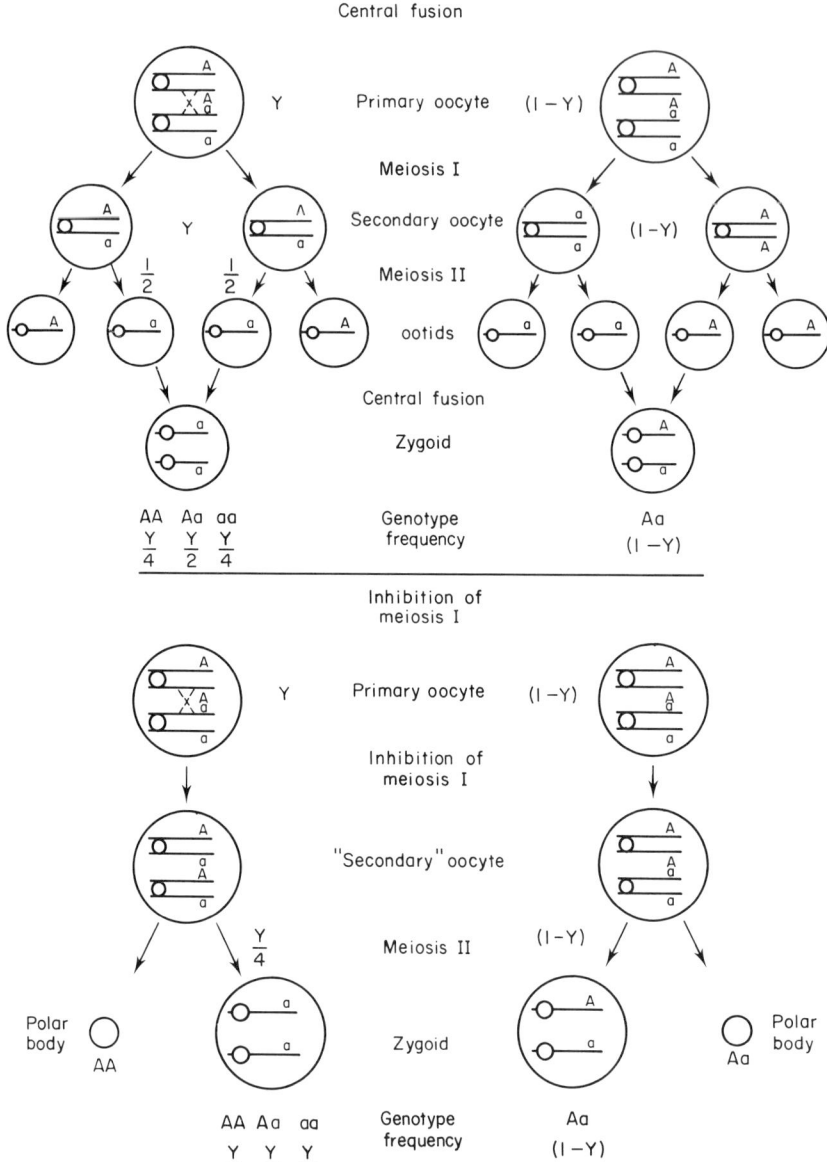

FIG. 1. Diagrammatic representation of the meiotic events of central fusion and inhibition of meiosis I. A single locus with two alleles (A and a) is considered where y is the probability of a recombination between the locus and its centromere (from Asher, 1970a).

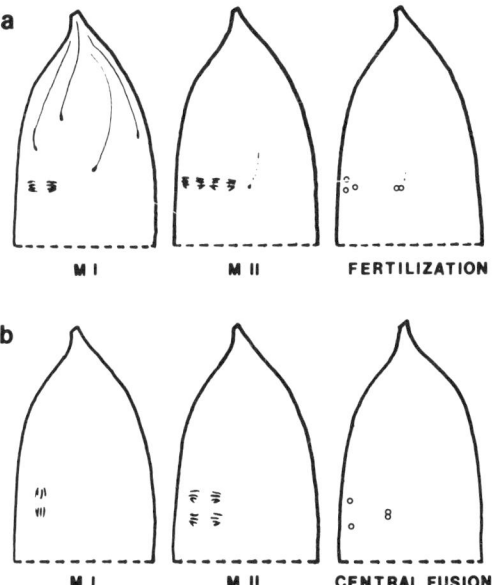

FIG. 2. Meiosis in *Drosophila melanogaster* under sexual reproduction (a) and in *Drosophila mangabeirai* under central fusion (b). MI depicts a longitudinal section of the meiosis I spindle orientation, MII a longitudinal section of the meiosis II spindle orientation, and finally a longitudinal section of the configurations of the haploid pronuclei either at the moment of fertilization (a) or central fusion (b). Modified from Murdy and Carson (1959).

fusion can maintain heterozygosity at a locus (Suomalainen, 1950, 1962). Moreover, heterozygosity can still be restored by central fusion even if recombination has taken place, but with a reduced probability (Fig. 1). Asher (1970a,b) has quantified these relationships and has shown that the probability of maintenance of heterozygosity increases with tighter linkage to the centromere according to the formula

$$\text{Prob. (heterozygosity : central fusion)} = 1 - y/2 \qquad (1)$$

where y is the probability of recombination between the locus and its centromere. (Note, the probability of recombination does *not* equal the recombination frequency, see Section VI, A.) If the linkage is absolute ($y=0$), central fusion can result in permanent heterozygosity (Carson, 1962a, 1967b). Indeed, *D. mangabeirai* has a complex of three heterozygous inversions that act as cross-over suppressors. As a result, a sizable portion of the *D. mangabeirai* karyotype is locked into a state of permanent heterozygosity, as has been confirmed by extensive studies on both wild-caught and lab-reared flies (Carson, 1962a, 1967b). *D. mangabeirai* clearly illustrates that automixis or agonoid thelytoky (Nur, 1979) does *not* necessarily promote homozygosity—a widespread erroneous notion. In fact, almost all cases of central fusion in obligate thelytokous populations are coupled with some mechanism of cross-over suppression, yielding permanent heterozygosity for all or most of the genome (Carson, 1967b). Hence, permanent heterozygosity seems to be the rule, not the exception, with this automictic mechanism.

C. Terminal Fusion

Under terminal fusion, the meiotic products that separated at meiosis II fuse to restore diploidy. This automictic mechanism is known to occur in *D. parthenogenetica* (Stalker, 1954) and *D. pallidosa* (Futch, 1979) and cannot be excluded as a possibility in *D. mercatorum* (Templeton and Rothman, 1973).

As Fig. 3 shows, and as Asher (1970a,b) has proven more quantitatively, the probability of maintenance of heterozygosity at a locus equals the probability of recombination between the locus and its centromere. Hence, tight linkage to the centromere insures rapid homozygosity. Moreover, since the probability of recombination normally asymptotes to 2/3 (Asher, 1970a,b), terminal fusion can never maintain permanent heterozygosity even for loci randomly recombining with their centromeres. In fact, central fusion at its worst maintains heterozygosity as well as terminal fusion at its best ($y=2/3$ in equation 1).

D. Gamete (Pronuclear) Duplication

Figure 4 is a diagrammatic representation of gamete or pronuclear duplication, the third and final automictic mechanism found in parthenogenetic *Drosophila*. Under this mechanism, meiosis produces a haploid egg nucleus which undergoes one or more cleavage divisions. Then, two of the haploid cleavage nuclei fuse to restore diploidy. The resulting diploid nucleus goes on to develop into the parthenogenetic individual. This is the primary mechanism found in *D. mercatorum* (Carson *et al.*, 1969; Templeton and Rothman, 1973; Carson, 1973), *D. ananassae* (Futch, 1973) and *D. pallidosa* (Futch, 1973, 1979).

As is obvious from Fig. 4, gamete duplication does more than promote homozygosity—it enforces it (Suomalainen, 1950)! Hence, all parthenogenetic progeny are totally homozygous regardless of the level of heterozygosity in their female parent. However, because meiosis is retained, the totally homozygous offspring of a heterozygous female will be genetically heterogeneous (Carson *et al.*, 1969; Templeton and Rothman, 1973). Once initiated, further gamete duplication results in clones of genetically identical flies, barring mutation.

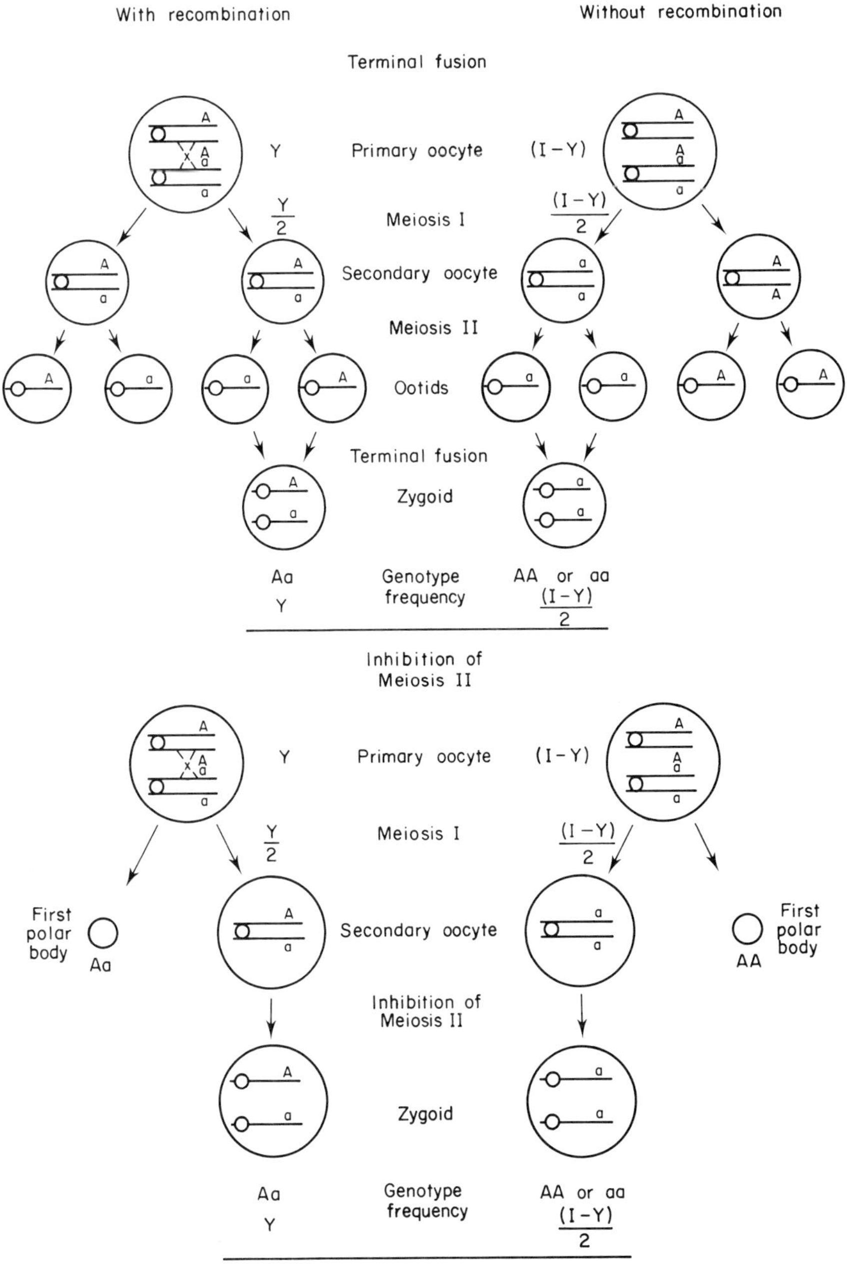

FIG. 3. Diagrammatic representation of the meiotic events of terminal fusion and inhibition of meiosis II. A single locus with two alleles (A and a) is considered where y is the probability of a recombination between the locus and its centromere (from Asher, 1970a).

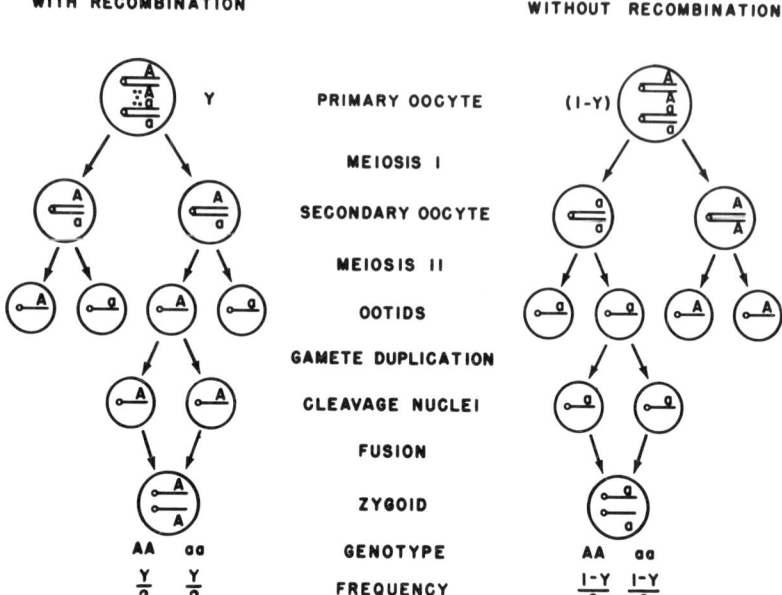

FIG. 4. Diagrammatic representation of the meiotic events of gamete (pronuclear) duplication. A single locus with two alleles (A and a) is considered where y is the probability of a recombination between the locus and its centromere.

E. Mixtures of Mechanisms

All examined cases of tychoparthenogenesis in *Drosophila* have revealed a mixture of automictic mechanisms occurring in the unfertilized eggs, even in the eggs derived from a single female. For example, in *D. parthenogenetica*, Stalker (1954) discovered that central and terminal fusion are about equally frequent in the unfertilized eggs of XXY diploid females. Moreover, if both central and terminal fusion occur in the same egg, a triploid is produced, which is also observed in *D. parthenogenetica*. These initial conclusions of Stalker have been strengthened and confirmed by Sprackling (1960).

Carson *et al.* (1969) discovered that although gamete duplication is the primary automictic mechanism in *D. mercatorum*, central and/or terminal fusion also occurs. Moreover, the amount of duplication versus fusion is partially under genetic control with the percent of fusion varying from 1% to 43% (Table II) depending upon the parthenogenetic strain (Carson, 1973; Annest, 1974). Templeton and Rothman (1973) estimated that the proportion of gamete duplication occurring in the parthenogenetic strain

TABLE II. The rate of production of X0 males and of pronuclear fusion in several parthenogenetic strains of *Drosophila mercatorum*. Rates are measured in terms of percent among the emerging adults. All adults not due to fusion arose through gamete duplication.

Strain[a]	% X0 Males[b]	% Fusion[c]
RSS-18-Im	0·27	11·9%
OB-2-Im	0·42	No Data
RSB-6-Im	0·31	No Data
RSB-7-Im	0·45	21·5%
RS-3-Im	0·72	0·3%
F_i of RSS-18-ImX RSB-7-Br	0·25[d]	43·2[d]

[a] Details of the derivations of the strains may be found in Carson (1967a). [b] Unless otherwise stated, all data on X0 males are from Spector (unpublished data). [c] Unless otherwise stated, all data on fusion rates are from Carson (1973). [d] Data from Annest (1974).

Iso-8-S-1-Im was between 94% (under the assumption that all remaining eggs underwent terminal fusion) and 96% (under the assumption that all remaining eggs underwent central fusion). Unfortunately, none of these studies directly investigated the type of fusion that occurs in *D. mercatorum*. However, unpublished data that were gathered during a selection experiment (Templeton et al., 1976b) can illuminate this problem. In this experiment, hybrids and partial hybrids between two parthenogenetic strains were bred (see Section VI, C and Figure 13). The hybrid and partially hybrid females were heterozygous for several isozyme loci, three of which were located on a metacentric autosome (Fig. 5). The heterozygous females were allowed to reproduce parthenogenetically, and the resulting parthenogenetic progeny were scored for their isozyme loci. As explained in previous sections, heterozygosity could only result from central or terminal fusion, but the probability of heterozygosity should decrease with increasing distance from the centromere under central fusion, but increase under terminal fusion. If a mixture of central and terminal fusion is occurring, the models of Asher (1970a), Templeton and Rothman (1973), and Templeton (1974) indicate that heterozygosity will decrease with increasing distance from the centromere only if the proportion of central fusion is twice as great as that of terminal fusion. The actual heterozygosities for these hybrids and

F_1 ♀ PARENT	HETEROZYGOSITY IN PARTHENOGENETIC F_2		
SO_1:	.000	.031	.007
SO_2:	.008	.027	.001
SO_3:	.022	.031	.009
SO_3':	.012	.025	.003
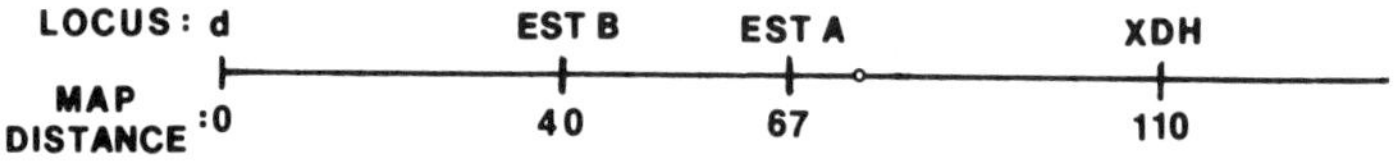 LOCUS: d	EST B	EST A	XDH
MAP DISTANCE: 0	40	67	110

FIG. 5. Heterozygosity levels found in the parthenogenetic progeny of hybrid and partial hybrid females between the unisexual stocks S-sl v pm vl-Im (derived from the stock S-sl v pm vl-Br12, a bridge stock to Iso-8-S-1-Im—see Fig. 10) and Iso-8-0-3-Im (See Fig. 13 for derivation of hybrids). The progeny were scored for three linked isozyme loci: esterase A (EST A), esterase B (EST B) and xanthine dehydrogenase (XDH). The map locations of these loci are indicated, using the visible marker droopy (d) as the origin. The centromere is located nearest to EST A, which consistently displays the highest levels of heterozygosity. This indicates a predominance of central fusion over terminal fusion by at least a 2:1 margin.

partial hybrids are also given in Fig. 5. The pattern in all cases clearly shows heterozygosity decreasing with increasing distance from the centromere, indicating that central fusion is at least twice as common as terminal fusion in these stocks. Unfortunately, the hybridization also invoked strong selection (Templeton et al., 1976b), so it would probably be best not to attempt point estimates of the fusion rates from these data.

D. ananassae and D. pallidosa also demonstrate the same pattern as D. mercatorum—a mixture of fusion and duplication (Futch, 1973). However, in D. pallidosa terminal fusion is more likely than central fusion (Futch, 1979). Hence, every case of tychoparthenogenesis in the genus that has been examined has consisted of a mixture of parthenogenetic mechanisms that are apparently at least under partial genetic control. This pattern implies that the exclusiveness of central fusion in D. mangabeirai is an evolved condition.

F. NON-DISJUNCTION AND PRODUCTION OF MALES

Stalker (1954) noted that 1·5% of the progeny of diploid virgin females of D. parthenogenetica are X0 sterile diploid males. He attributed this to non-disjunction of the X chromosome at the second meiotic division followed by central fusion (Fig. 6).

Carson (1962b) discovered a similar pattern in D. mercatorum, reporting

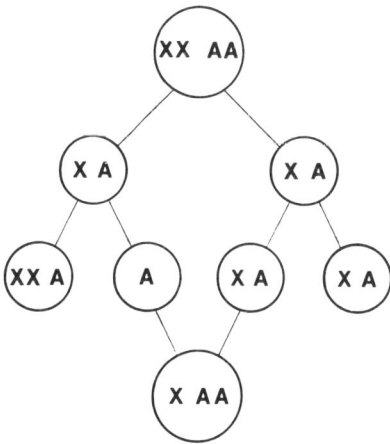

FIG. 6. Diagrammatic representation of the meiotic events leading to the production of an X0 sterile male under parthenogenesis. X refers to the X chromosome, A refers to the set of autosomes. As shown in this figure, non-disjunction of the X-chromosome at meiosis II followed by central fusion can result in an X0 sterile male.

from 1 to 2% X0 sterile diploid males depending on the strain. Similarly, Futch (1973, 1979) reported 0·5% to 0·2% of the flies in parthenogenetic *D. ananassae* and *D. pallidosa* to be sterile (presumably X0) males. The production of X0 males in *D. mercatorum* was examined in more detail by E. Spector (unpublished observations) who found that the rate of X0 male production varied from 0·27% to 0·72% (Table II). Annest (1974) found 0·25% X0 males among the parthenogenetic progeny of the hybrid *D. mercatorum* strain RSS-18-Im × RSB-7-7Br. Annest (1976) speculated that the production of X0 males in *D. mercatorum* could be like that of *D. parthenogenetica* since some central fusion is known to occur. However, the data presented in Table II does not support this hypothesis. If X0 male production is via Stalker's mechanism (Fig. 6), the rate of X0 males should increase with increasing fusion rates. However, Table II clearly indicates just the opposite is true—the ranks for X0 production are perfectly inverted from the ranks for fusion. Moreover, in the stock RS-3-Im, the X0 production rate is more than twice as large as the fusion rate. Hence, even if non-disjunction occurred in every meiosis, the majority of X0 males would still have to be produced from eggs undergoing gamete duplication. Unfortunately, X non-disjunction either before or during meiosis followed by gamete duplication results in either XXXX or 00 non-viable flies—not X0 males. Therefore, perhaps an X chromosome is lost during a post-meiotic cleavage division followed by fusion of the cleavage nuclei

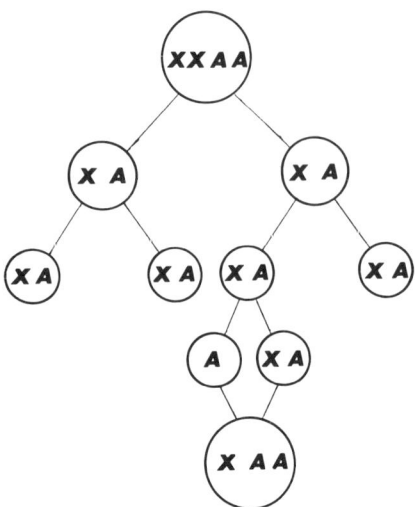

FIG. 7. Diagrammatic representation of the events leading to the production of an X0 sterile male under gamete duplication. X0 males can be produced if an X chromosome is lost or does not replicate during the post-meiotic cleavage divisions preceding fusion of the cleavage nuclei.

(Fig. 7). Annest (personal communication) has also speculated that the X0 males could result from a preferential non-duplication of one of the two X's. Unfortunately, no direct cytological evidence exists concerning these hypotheses.

Annest (1976) revealed another interesting aspect of parthenogenetic male production in *D. mercatorum*. He had set up some population cages with F_1 virgin females from the cross parthenogenetic RSB-7-Im (white) ♀♀ × S-sl v pm vl-Br13 ♂♂ (a sexual stock closely related to the parthenogenetic strain Iso-8-S-1-Im—see Templeton *et al.*, 1976b). By the seventh generation, fertile males comprised 10% of the population. Annest suspected contamination, so the cage was re-initiated by isolating some virgins from the old cage, but by the third generation in this second cage 40% of the flies were fertile males. A closer examination revealed that some virgin females produced fertile males while others did not. Moreover, the fertile males had Y chromosomes. Subsequently, Val Giddings (personal communication) isolated four virgin females which each gave rise to a fertile male. The males were then crossed to their parthenogenetic sisters to yield sexual stocks, while other sisters remained virgin to maintain a parthenogenetic strain. (The parthenogenetic strain has to be screened every generation since about 1 to 2% of the emerging flies are fertile males.) Annest (personal communication) then examined 60 metaphase smears of larval brains from one of Giddings' parthenogenetic lines. He discovered four XXYY and one XXYYYY metaphase smears. Based on these results, Annest speculated that one of the original hybrid females carried a Y chromosome followed by the events diagrammed in Fig. 8, leading to XXYY and XXYYYY females. A loss of an X in these flies will often result in fertile males, thus explaining the regular production of fertile males in these strains. In this sense, these particular strains of D. mercatorum are amphitokous.

Although this may seem rather unusual, such an amphitokous population could arise naturally. Wharton (1942, 1944) described a population of D. mercatorum in which the Y chromosome was translocated onto the dot-chromosome. Thus, males were apparently X0, but actually were XYY, and females were XXYY. If a parthenogenetic strain arose from this bisexual population, loss of an X would result in fertile XYY males.

These results demonstrate that parthenogenesis is not a dead-end that, once entered, can never be abandoned. Rather a population could evolve from sexual reproduction to a primarily thelytokous state and back again. Darevskii et al. (1977) have demonstrated that this type of phenomenon had indeed occurred in nature among parthenogenetic rock lizards. In this case, the production of rare (less than 0·1%) fertile males resulted in naturally occurring hybrids between two sympatric unisexual species. This type of residual bisexuality and occasional sexual reproduction may have very great evolutionary implications for certain parthenogenetic species and for constructing parthenogenetic phylogenies.

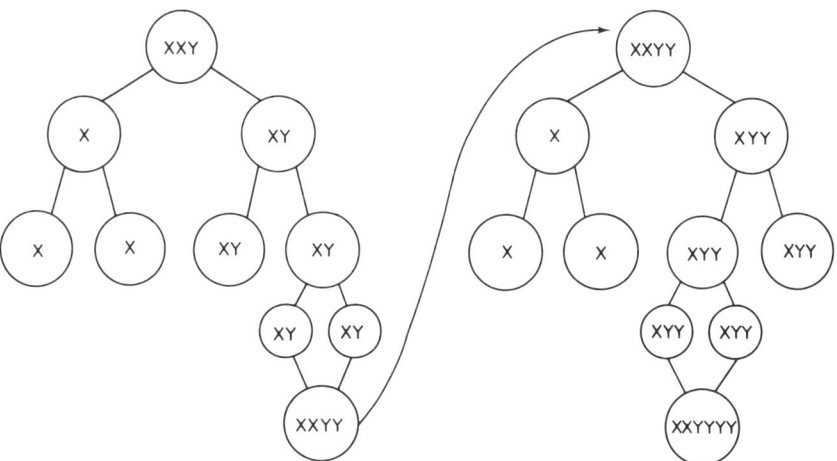

FIG. 8. Hypothetical route for the production of XXYY and XXYYYY females under gamete duplication beginning with an XXY female.

G. Sexual Reproduction and Parthenogenesis

As the above section implies, there are no absolute barriers between sexual reproduction and parthenogenesis. Indeed, since most parthenogenetic *Drosophila* are automictic diploids, they generally retain the capacity for sexual reproduction. In parthenogenetic *D. mercatorum* females, sexual reproduction can usually be obtained simply by placing males with females.

The option of sexual reproduction can be used to create sexual analogues of parthenogenetic strains. The system of Annest and Giddings discussed in Section III, E, results in sexual analogues in which all chromosomes are of parthenogenetic origin. Templeton *et al.* (1976b) utilized the absence of crossing-over in males coupled with visible autosomal markers to create flies in which all chromosomes but the Y of the males are of parthenogenetic origin (Fig. 9). A final scheme is the "bridge" technique of repeated backcrossing through female progeny (Carson, 1967a; Carson *et al.*, 1969; Carson and Snyder, 1972; Templeton *et al.*, 1976b) shown in Fig. 10. The bridge system allows new genetic markers to be introduced onto a parthenogenetic background. By increasing the number of backcrosses, the amount of the chromosome surrounding the marker that is of non-parthenogenetic origin can be minimized.

Once such sexual analogues have been bred, parthenogenetic strains may be hybridized or crossed with other bisexual stocks. Consequently, parthenogenetic *D. mercatorum* offers a rare advantage: clonal reproduction which allows the production of truly isogenic, totally homozygous stocks that may then be thoroughly characterized for their developmental, physiological, behavioral and fitness attributes; and sexual reproduction which allows the genetic basis of an attribute to be examined in a straight-forward manner.

IV. The Genetic Basis of Parthenogenesis

The capacity for tychoparthenogenesis is widespread throughout the genus *Drosophila*, but does it have any genetic basis within a species? The work of Counce and Ruddle (1969) on strain differences in egg structure and meiotic configurations offers indirect evidence for a genetic basis. More direct evidence relating to this question is now reviewed.

A. Evidence from Selection Experiments

One of the classic techniques for demonstrating a genetic basis for a trait is to see if the trait can respond to selection. Stalker (1954, 1956a,b) was the first to test if parthenogenetic capacity would respond to selection. Stalker

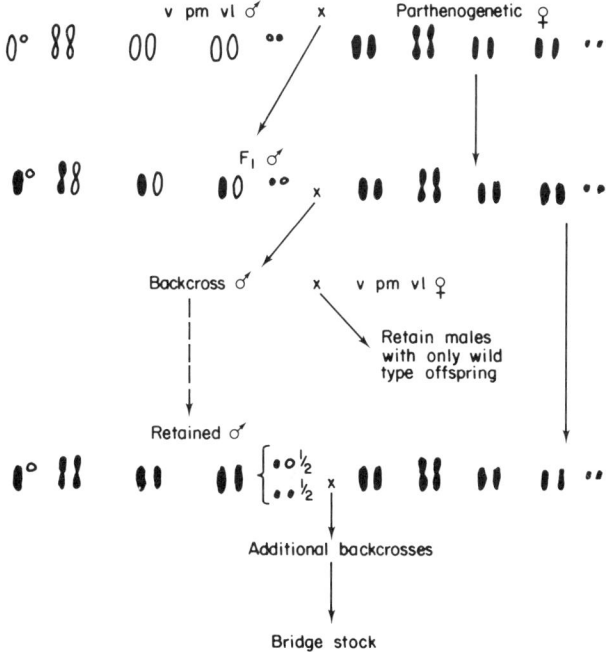

FIG. 9. Breeding scheme for generating bridge stocks that are sexual analogues to parthenogenetic stocks. Parthenogenetic females are first crossed to males from the sexual stock $v\ pm\ vl$. This stock has recessive visible markers on all three major autosomes (vermilion eyes on the metacentric (v), plum (pm) eyes on the acrocentric I, and veinless (vl) wings on the acrocentric II), leaving only the autosomal dot unmarked. The resulting F_1 males are then backcrossed to the parthenogenetic line. Since there is no crossing-over in males, the backcross males will be either homozygous for a parthenogenetic-type autosome or heterozygous for a marked autosome with equal probability for each autosome. A test cross reveals the genotype of the backcross males, and only those males are retained which are hemizygous for the parthenogenetic X chromosome and homozygous for all three parthenogenetic-type major autosomes. Besides the Y chromosome, only one of the dots is potentially of non-parthenogenetic origin in these retained males. At least eight additional backcrosses are made to the parthenogenetic line to insure, with high probability, homozygosity for the dot as well.

(1954) followed the fate of a unisexual strain of *D. parthenogenetica* and contrasted it with the unselected bisexual strain from which it was derived. Presumably, parthenogenesis itself would induce selective forces favoring further increases in the parthenogenetic capacity. Stalker found that the frequency of eggs initiating development had risen from 0·91% to 8·20% by the 17th unisexual generation and the survival to larval stage of those individuals that did initiate development increased from 8·96% to 18·88%.

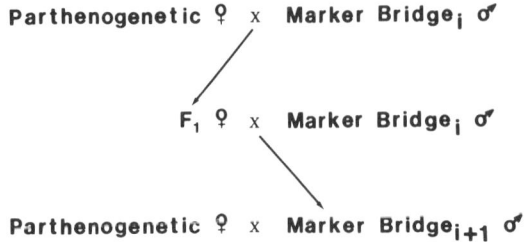

FIG. 10. Breeding scheme for generating bridge stocks that are coisogenic with a parthenogenetic stock except for an introduced marker and the chromosome segment immediately surrounding the marker. Note that each bridge cycle goes through an F_1 female, allowing recombination to occur. Hence, as the number of bridge cycles increases, the degree of coisogenicity with the parthenogenetic strain increases.

Combining these two factors, the overall rate of parthenogenesis (i.e., the number of viable larva/number of unfertilized eggs) increased about twenty-fold (from 0·08% to 1·55%).

The unisexual strains of *D. parthenogenetica* could respond to selection because they reproduce by a mixture of central and terminal fusion, mechanisms which generate genetic diversity as long as some heterozygosity persists (Figs 1 and 3). However, Stalker (1956a,b) speculated that the available heterozygosity had disappeared by the 17th generation since no further response to selection was observed. To restore genetic variability, Stalker made some backcrosses at unisexual generation 55 to males from the bisexual ancestral strain followed by additional unisexual and backcross generations. After this influx of new genetic variability, the parthenogenetic rate doubled (to 3·06% from 1·55%), principally due to the increase in the percentage of eggs initiating parthenogenetic development. These results not only clearly indicate a genetic basis for parthenogenesis in *D. parthenogenetica* but also demonstrate that a large (20-fold) and rapid response to selection is possible under strictly parthenogenetic reproduction during the early unisexual generations.

Stalker (1954) found a very different response to selection in unisexual strains of *D. polymorpha*. Unlike the gradual and continuous response with *D. parthenogenetica*, *D. polymorpha* appeared to have a rather abrupt genetic differentiation which caused a sudden rise in the parthenogenetic rate.

Carson (1961) discovered that parthenogenesis in *D. robusta* also had a genetic basis as inferred from a 2·5-fold increase in parthenogenetic rate in response to selection.

The largest response to selection has been obtained in *D. mercatorum*. Since *D. mercatorum* reproduces primarily by gamete duplication, genetic variability is rapidly lost. Hence, Carson (1967a) alternated unisexual and bisexual generations to obtain a 60-fold increase in parthenogenetic rate. However, Templeton et al. (1976a) observed a 1,000-fold increase over a single unisexual generation by using highly heterozygous females from natural populations as the source of genetic variability.

B. Evidence from Crosses Between Unisexual and Bisexual Strains

Stalker (1954) crossed unisexual females of *D. parthenogenetica* with unselected bisexual males. The resulting F_1 virgin females displayed a parthenogenetic rate intermediate to that of the two parents, indicating a genetic basis. Futch (1973) also used crosses between parthenogenetic *D. ananassae* females with bisexual mutant males to show that the parthenogenetic capacity is genetic. Carson (1967a) used such crosses to demonstrate a genetic basis for parthenogenesis in *D. mercatorum* and moreover showed that the capacity for parthenogenesis could be inherited through the male, thereby excluding maternal effects.

C. Evidence from Contrasts of Isogenic Lines

Carson (1967a) isolated several different isogenic lines of parthenogenetic *D. mercatorum*. Each line had its own characteristic parthenogenetic rate. Annest (1976) showed these parthenogenetic rates could be broken down into line specific rates for eggs initiating development, eggs hatching, and larval to adult survivorship. Moreover, Carson (1973) demonstrated that the proportion of gamete duplication versus fusion also was constant within a given isogenic line, but varied considerably between strains. These results indicate that not only is the overall parthenogenetic capacity under genetic control in *D. mercatorum*, but so is the specific type of parthenogenetic mechanism and developmental history.

D. Evidence from Screening Experiments

Carson (1961, 1967a) found that bisexual stocks of different geographic origin both in *D. robusta* and *D. mercatorum* differ in their chances of yielding parthenogenetic offspring. Similarly, Futch (1962) confirmed Stalker's original observations of parthenogenesis in a laboratory strain of *D. parthenogenetica* derived from Atlixco, Mexico, but found no parthenogenetic development whatsoever in a strain of *D. parthenogenetica* from Monteria, Colombia.

The above results indicate that there are genetic differences between geographic populations with respect to tychoparthenogenetic capacity, a situation similar to that reported in the moth *Solenobia triquetrella* by Seiler (1959). Templeton *et al.* (1976a) and Templeton (1979a) have shown that such heterogeneity exists even within a single geographic population. Templeton *et al.* (1976a) collected *D. mercatorum* females inhabiting the garbage dump at Kamuela, Hawaii and immediately screened their F_1 daughters (from matings that occurred in nature) for parthenogenesis. The parthenogenetic rates were not homogeneous among the 31 isofemale lines thus established; rather, the lines could clearly be divided into three "hot" lines with a rate of 10^{-4} (rate=number of viable adults/number of unfertilized eggs) and 28 remaining lines with a rate of 10^{-5}. In 1976 and 1978, collections of *D. mercatorum* were made in patches of *Opuntia megacantha* about 10 kilometers from the Kamuela garbage dump. The cactus patch populations differed from the garbage dump flies for allele frequencies at several isozyme loci. Moreover, the levels of polymorphism and individual heterozygosity were much smaller in the cactus patch flies (Templeton, 1979a). The 1976 and 1978 cactus patch results were homogeneous, but the overall cactus patch rate (10^{-6}) was significantly different from the overall garbage dump rate (10^{-5}), indicating microgeographic differentiation. In addition, as with the garbage dump population, the cactus patch lines were not homogeneous amongst themselves, but were clearly separable into "hot" lines with rates of 10^{-4} and remaining lines with a rate of 10^{-6}. The overall difference between the cactus patch population with the garbage dump population was due primarily to the lower rate in the "non-hot" lines, although a lower frequency of "hot" lines in the cactus patch flies also contributed. These data indicate that genetic variability exists within local populations for tychoparthenogenetic capacity in addition to microgeographic differences.

E. Components of the Genetic Basis of Parthenogenesis

The evidence reviewed above clearly establishes that parthenogenesis has a genetic basis, but this capacity can be subdivided into several components. First, in going from sexually reproducing flies to parthenogenetic flies, there is selection among the original sexual females favoring those producing eggs initiating parthenogenetic development and/or those laying the most unfertilized eggs. That genetic selection is indeed operating at this stage is indirectly indicated by the previously discussed existence of "hot" lines in *D. mercatorum*. Thus, out of a total of 53 parthenogenetic flies produced by screening 141 Kamuela isofemale lines, 34 of the parthenogenetic flies are derived from 8 lines—indicating a highly non-random distribution.

More direct evidence for selection at this stage is provided by an association between parthenogenesis and an X-linked recessive allele known as abnormal abdomen (*aa*). Abnormal abdomen is characterized by indistinct tergites and sternites, and an absence of bristles and pigmentation on the abdomen (Templeton and Rankin, 1978). The penetrance is influenced by several other loci, and its morphological effects are generally suppressed in its bearers, both bisexual and unisexual (Templeton and Rankin, 1978). The frequency of the *aa* allele in the cactus patch populations of *D. mercatorum* has been estimated to be 0·20 in 1976 (Templeton and Rankin, 1978) and 0·22 in 1978 (Templeton, unpublished data). Yet *aa* is fixed in 2/3 of the successful parthenogenetic strains derived from the cactus patch flies. Moreover, an examination of the parthenogenetic strains of *D. mercatorum* isolated previously by Carson (1967a) revealed that a majority of the strains of independent origin also were fixed for *aa*. Thus, in the very first transition from sexual flies to parthenogenetic adults, the frequency of *aa* increases dramatically, *despite* the fact that homozygosity for *aa* is often associated with deleterious viability effects (Templeton and Rankin, 1978).

A possible clue as to why this occurs is provided by data gathered during the 1976 collection. The egg laying capacities of each of the 20 isofemale lines from that collection were determined by direct count as part of the screen for parthenogenesis (Templeton, 1979a). Moreover, the frequency of *aa* for that year was estimated by taking a single female from each isofemale line and crossing her to males from an *aa* tester stock (Templeton and Rankin, 1978). These two pieces of data can be coupled to see if *aa* has any association with the egg laying capacity of the F_1 females derived from matings in nature. Any effects of *aa* will be underestimated with this design because some lines segregating for *aa* will be missed since only a single female is scored. This, of course, will obscure any differences between the isofemale lines that are scored as *aa* versus those scored as non-*aa*. The results of this contrast are shown in Fig. 11. The twelve non-*aa* lines have a very broad range of egg laying capacities, but are divided into low and high mode groups. The eight lines known to be segregating for *aa* are very homogeneous and have a significantly different distribution from the non-*aa* lines. Thus, scoring for *aa* is highly informative about egg production. Moreover, note that all but one of the *aa* lines fall in the upper mode of the non-*aa* category, and none in the lower mode. Consequently, *aa* is indeed associated with increased and stabilized egg production. In addition, each of the *aa* lines was scored for its penetrance of the morphological effects of *aa*. Table III presents the ranks of the eight *aa* lines with respect to egg production and with respect to penetrance of *aa*. No correlation exists, although the sample size is too small to detect anything but a very large effect. However, this implies that the suppressors and

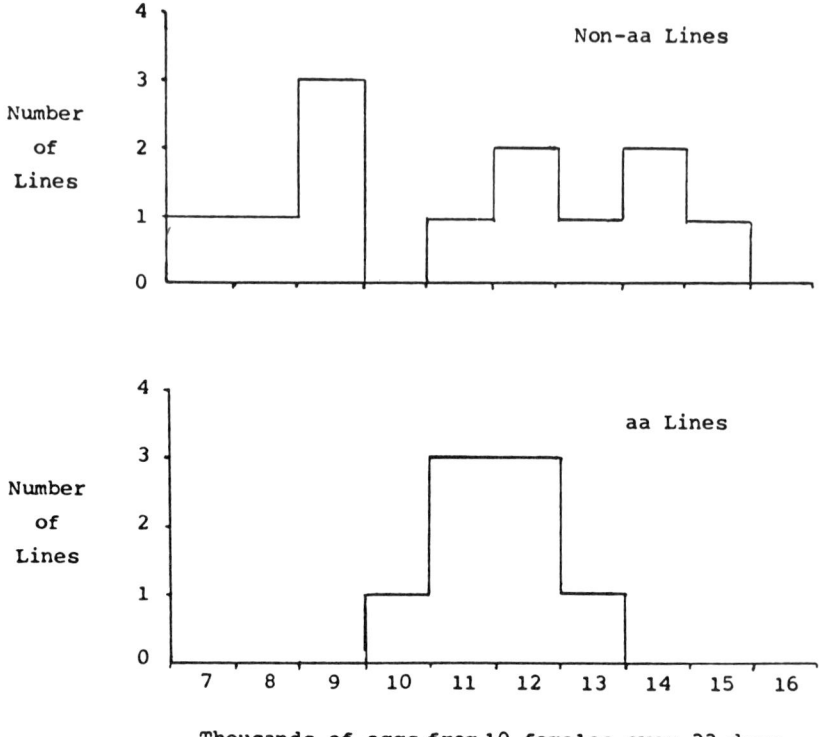

FIG. 11. The effect of abnormal abdomen (*aa*) upon the number of eggs produced by the isofemale lines collected from Kamuela in 1976.

TABLE III. Ranks of egg production and ranks of penetrance of abnormal abdomen (*aa*) in the eight *aa* isofemale lines from the 1976 Kamuela collection.

Line number	Egg rank	*aa* Penetrance rank
40	7	1
41	3	6
42	1	5
43	2	3
44	4	2
45	8	7
46	5	8
47	6	4

Spearman's Rank Correlation Coefficient = 0·05.

enhancers of the morphological attribute of *aa* do not strongly influence its association with increased egg production.

Because *aa* is associated with increased egg production, there would be selection favoring *aa* even if the parthenogenetic rates per egg for *aa* and non-*aa* bearing flies were the same. However, under this assumption and using the data in Fig. 11, the frequency of *aa* in the parthenogenetic flies should be increased from 0·20 to 0·28. Obviously, selection on egg laying capacity in the bisexual females is not a complete explanation for the increase in *aa*, although an increase in egg laying capacity in the parthenogenetic flies over their sexual ancestors is indeed observed (Templeton *et al.*, 1976a). Further studies need to be done to see if *aa* is related in any way to the probability of initiating parthenogenetic development.

Another gene or set of genes affecting the initiation of parthenogenetic development has been unmasked by C. Sing and R. Clark at the University of Michigan (unpublished data) with some breeding experiments performed on the parthenogenetic line K23-0-Im (Templeton *et al.*, 1976a). Sing and Clark introduced into K23-0-Im a small segment of the metacentric autosome of *D. mercatorum* surrounding the isozyme locus α-glycerophosphate dehydrogenase (α-GPDH) bearing an electrophoretic variant of α-GPDH derived from another parthenogenetic stock, S-α-Im. Both of these parthenogenetic stocks are non-*aa* lines, and moreover both are apparently adapted to parthenogenesis in a similar fashion (Templeton, 1979b). The probability of an egg from K23-0-Im successfully hatching and completing development to the adult state is 0·02 while that for S-α-Im is 0·06. However, among thirty parthenogenetic lines isolated from a female heterozygous at the α-GPDH locus and the surrounding chromosomal segment but otherwise identical to K23-0-Im, two lines displayed abnormally high probabilities of successful hatching of at least 0·10 and perhaps higher (Templeton, unpublished data). This is at least a fifty-fold increase beyond the original K23-0-Im rate and is the maximum recorded for *D. mercatorum*. There are two straightforward explanations for this phenomenon. First, a recombination occurred in the heterozygous chromosome region that resulted in a gene complex yielding a parthenogenetic rate far in excess of either parental type. In this regard, both high ranking stocks have the K23-type α-GPDH isozyme variant; hence, if they do have any S-derived alleles, it must have been through a recombinational event. Second, a new mutation occurred that increased the parthenogenetic rate of K23-0-Im by at least 15-fold. In this regard, it is interesting to note that both high ranking lines bear a new and previously unobserved eye-color. Studies are in progress to elucidate the genetic basis of this dramatic increase in the probability of successful initiation of parthenogenetic development.

Once parthenogenetic development has been initiated, selection will now occur favoring those genotypes that are viable under the genetic environment imposed by parthenogenetic reproduction (Annest and Templeton, 1978). Templeton et al. (1976a) and Templeton (1979a) present evidence that the initial increase in parthenogenetic capacity in *D. mercatorum* is primarily due to selection favoring a genotype viable under total homozygosity—the genetic environment imposed by gametic duplication. This selection in turn involves both the elimination of any recessive lethals that existed in the sexual ancestors as well as the creation of gene complexes resulting in viable developmental and physiological pathways under the new genetic environment. For example, fixation of the *aa* allele during the initial phase of selection in *D. mercatorum* would induce viability selection at this second stage favoring those genotypes that had fixed the suppressors of the morphological effects of *aa*. Indeed, this is just the pattern that is consistently observed in the Kamuela parthenogenetic flies. Direct experimental evidence on the selective forces accompanying the reproductive transition (Templeton, 1979b) indicates that selection for the creation of a viable gene complex is quantitatively as or more important than selection for elimination of recessive lethals.

This type of viability selection would presumably be less important in a fly produced by central fusion since the change in genetic environment would be less extreme. This supposition is supported by contrasting the work of Templeton et al. (1976a) and Templeton (1979a,b), in which a 1,000-fold increase in parthenogenetic rate can be primarily attributed to viability selection in a parthenogenetic system imposing total homozygosity, with the work of Stalker (1954), in which viability selection could explain only a little more than two-fold of an overall 20-fold increase in parthenogenetic rate in flies reproducing by a mixture of central and terminal fusion.

Once a viable parthenogenetic genotype attains adulthood, there is selection on the ability of that adult herself to be capable of parthenogenetic reproduction. This induces a second selective bottleneck in *D. mercatorum*, indicating that additional genes or sets of genes not directly related to viability are now the targets of selection (Annest and Templeton, 1978).

This second round of selection for parthenogenetic reproduction also involves different or additional genes from those involved in the initial selection for parthenogenetic reproduction in the original bisexual females. As mentioned earlier, a total of 34 viable parthenogenetic adults have been obtained from the "hot" bisexual lines from the Kamuela *D. mercatorum*, and 19 from the remaining "non-hot" lines. However, only five of the 34 (15%) parthenogenetic adults from the "hot" lines and two of the 19 (11%) from the "non-hot" lines could themselves reproduce parthenogenetically.

This difference is not significant, indicating parthenogenetic flies with "hot" bisexual ancestors are not more efficient at parthenogenetic reproduction than unisexual adults from "non-hot" lines. Moreover, the most vigorous parthenogenetic line from the 1974 and 1976 collections (the 1978 collections have not been fully analysed yet) came from a "non-hot" bisexual line. Thus, the genes conferring increased parthenogenetic capacity in the bisexual lines do not confer increased capacity or efficiency once the initial reproductive transition has been made.

Once selection has produced an array of genotypes that are both viable under the genetic conditions imposed by parthenogenesis and capable of parthenogenetic reproduction themselves, clonal selection will occur to enhance both viability and efficiency of parthenogenetic reproduction (Annest and Templeton, 1978). It is important to recall that automictic parthenogenesis, even gamete duplication, can create considerable genetic variability even among the descendants of the same ancestral bisexual female, thus allowing effective clonal selection. Recall also the work of Carson (1973) that different clones are characterized by different types of frequencies of parthenogenetic mechanisms. Consequently, the very mode of parthenogenesis is subject to clonal selection. This type of selection has been directly observed by Annest (1976). He initiated parthenogenetic population cages with flies characterized by very high fusion rates but observed that the high fusion rates were quickly lost in favor of gamete duplication. Selection for central fusion almost surely occurred in *D. mangabeirai* and would account for the unique meiotic configuration observed in that species that results in central fusion (Murdy and Carson, 1959).

As the above discussion indicates, the genetic basis for parthenogenesis in *Drosophila* is not simple—there is no gene for parthenogenesis. Rather, the experimental evidence indicates parthenogenesis is a complex polygenic trait which can be subdivided into many components affecting very diverse attributes of the fly.

V. Behavioral Genetics

A. Sexual Behavior and Isolation

Because parthenogenesis in *Drosophila* is almost always automictic, diploid and thelytokous, females are freed from the necessity of mating, yet are certainly physically capable of mating and reproducing sexually. This state leads to some unique opportunities for studying sexual behavior in *Drosophila*, and in addition raises an interesting evolutionary question: what happens to the sexual behavior of thelytokous females when they are freed from the constraint of mating to reproduce?

Wolfson (1958) was the first to experimentally address this question. Wolfson's work was inspired by unpublished data of Stalker. Stalker repeatedly backcrossed bisexual males to unisexual *D. parthenogenetica* females. These repeated backcrosses yielded a weak and semi-sterile bisexual strain, but it seemed that the fertility of the males, rather than the females, was being adversely affected. Stalker therefore hypothesized that during the unisexual generations selection favored gene complexes facilitating parthenogenetic reproduction which, when introduced into males, reduced fertility or otherwise interfered with sexual reproduction. Wolfson (1958) set out to test this hypothesis.

Wolfson first examined the genital structures of bisexual and unisexual females, but found no evidence for any regressive evolution in the unisexual strains. Similarly, a comparison of bisexual males with unisexual X0 males revealed no abnormalities beyond those expected from the X0 condition itself. However, she did find evidence for sterility in the repeated-backcross males. She then made behavioral comparisons in unisexual and bisexual strains and discovered that the unisexual strains had a wide range of sexual receptivity—both higher and lower than that of the bisexual strains. Thus, although some unisexual strains did show low levels of sexual receptivity, there was no evidence that selection favored a lowering of sexual receptivity in females that had reproduced parthenogenetically for over 30 generations.

Wei (1968) reached similar conclusions on studies of parthenogenetic *D. mercatorum*. She discovered considerable variability in the unisexual strains for sexual receptivity. Wei found no correlation between the time of isolation from the sexual population and the speed of mating, indicating this variability was due to chance fixations occurring during the initial transition from sexual reproduction to parthenogenesis rather than selection. Nevertheless, the results of Wei (1968) do clearly indicate that most parthenogenetic females of *D. mercatorum* have a lower propensity to mate than bisexual females. Carson *et al.* (1977) also observed that mating propensity was significantly lower in unisexual strains of *D. mercatorum* than in the bisexual controls; in some cases by a factor of 10. Moreover, three of four parthenogenetic lines derived from the same bisexual stock differed from each other as well as from their sexual ancestors. Also, two parthenogenetic strains isolated shortly before the mating experiments had extremely different mating propensities despite the fact they were derived from the same natural population at the same time. This pattern suggests chance fixation of genes affecting mating propensity during the initial transition to parthenogenesis. However, an explanation is still required for the systematic bias in favor of sexual isolation. The mating system in *Drosophila* is one of female choice, and this has often led to the evolution of very elaborate courtship rituals and mating actions that depend upon several behavioral and morphological attributes—the so-called "mate recognition system" (Paterson, 1978). As Paterson (1978) and Templeton (1979c) have argued, such a mate recognition system is normally subject to strong stabilizing selection and is generally polygenic. When the transition to parthenogenesis is made, the events described in Section IV, E, occur which obviously involve a great many genes influencing a great many attributes. Either through pleiotropic effects or linkage to genes involved in the mate recognition system, such an extensive and rapid polygenic alteration will almost always disrupt some aspect of the female's mate recognition system. Since the females do not mate to reproduce, this disruption is not counterbalanced by selection re-establishing a mate recognition system, thus leading to a systematic bias in favor of sexual isolation in parthenogenetic *Drosophila*.

This spontaneous occurrence of sexual isolation in parthenogenetic *Drosophila* has reached its most extreme expression in the stock K28-0-Im. This strain was isolated directly from the Kamuela garbage dump population of *D. mercatorum* (Templeton *et al.*, 1976a) and shows an extremely low propensity to mate when tested against males from the S bisexual strain—an old laboratory stock derived from El Salvador (Carson *et al.*, 1977). However, their isolation against other strains—including their bisexual ancestors—was even more extreme. No mating was observed in 30 minutes or even overnight with garbage dump bisexual males (Templeton, unpublished data). Consequently, an alternative design was used to measure mating propensity. Single K28-0-Im females were placed in a shell vial with two bisexual garbage dump males. After a week, the flies were transferred to a fresh vial for another week (only vials in which the female and at least one male remained alive for two weeks were used). With this design, only 10 out of 25 K28-0-Im females mated after two weeks versus 100% of the garbage dump females (Templeton, unpublished data). When K28-0-Im females were tested with the same design against males from the bridge stock S-sl v pm vl-Br15 the results were even more extreme—only 3% of the females mated (Templeton and Rankin, 1978; Templeton, unpublished data). This could not be attributed to geographical isolation (the S strains originated from El Salvador) because 97% of the bisexual garbage dump Kamuela females mated with S-sl v pm vl-Br15 males under the same experimental design. The sexual isolation between K28-0-Im and S-sl v pm vl-Br15 is so strong, that mating after one week of isolation with S-sl v pm vl-Br15 males can be scored as a yes/no trait. Chromosomal contrasts of K28-0-Im and S-sl v pm vl-Br15 were therefore bred to map the position of the genetic element(s) responsible for the sexual isolation. These contrasts revealed that a female will display this nearly absolute sexual isolation only when she is homozygous for the K28-type X chromosome *and* homozygous for the K28-type acrocentric II autosome (Templeton and Rankin, 1978; Templeton, unpublished data). Interestingly, males hemizygous for the K28-type X and homozygous for the K28-type acrocentric II autosome lacked motile sperm and were sterile, although they would court and copulate (Templeton and Rankin, 1978; Templeton, unpublished data). These results clearly show that very strong—essentially complete—pre-mating and post-mating isolation can arise spontaneously during tychoparthenogenetic reproduction of females from natural, bisexual populations.

In the above studies, no attempt was made to deliberately select for sexual isolation. However, parthenogenetic *Drosophila* can be used in studies of artificial selection for sexual isolation as well. Henslee (1966) discovered that the parthenogenetic strain RS-3-Im of *D. mercatorum* displayed significant sexual isolation from both R and S males (R and S being the sexual ancestors of RS-3-Im) whereas R and S females exhibited no sexual isolation from each other's males. Henslee then selected two females of RS-3-Im that did not mate in his experiments and allowed them to reproduce parthenogenetically. He then tested their parthenogenetic descendants for sexual isolation, and found that the degree of sexual isolation was substantially increased in one of the substrains. Thus, effective clonal selection was apparently still possible in this long established parthenogenetic strain which reproduces almost exclusively by gamete duplication (Table II).

Doerr (1967) further investigated whether artificial selection could increase the degree of sexual isolation in parthenogenetic *D. mercatorum*. Unlike Henslee, Doerr generated genetic variability by interspersing sexual outcrosses with the parthenogenetic generations. In this manner, she was able to greatly enhance sexual isolation.

The design of using both sexual and parthenogenetic reproduction was elaborated by Ikeda and Carson (1973). After just two cycles of selection, they obtained a parthenogenetic strain which was significantly less receptive to males from three different bisexual strains.

The experiments discussed in this section show that sexual isolation tends to appear spontaneously when the transition to parthenogenetic reproduction is made, and that the degree of sexual isolation can be enhanced rapidly by clonal selection occurring either shortly after or a long time after the last occurrence of sexual reproduction.

B. Other Behavioral Traits

Little work has been performed on other behavioral traits of parthenogenetic flies. Carson *et al.* (1977) note that the stock K28-0-Im displays a very low level of general activity while in food bottles or vials. In fact, one can open a bottle, even after vigorous shaking, and few if any flies will attempt to escape. For this reason, Dr. H. L. Carson nicknamed the K28-0-Im stock the "lazy girls." However, their laziness is conditional. When K28-0-Im females were placed into a wind tunnel with a gentle air current flowing through, they were quiet for a short time, then groomed vigorously, and then traveled upwind more than any other strain of *D. mercatorum* yet tested (J. S. Johnston, personal communication).

A. Speer (unpublished data) did a diallele analysis on the genetic components of wind response in three parthenogenetic/bridge stocks of *D. mercatorum* using the wind tunnel constructed by Dr. J. S. Johnston. Such stocks are ideal for a diallele analysis since the parental stocks are totally homozygous and isogenic. Wind response was measured by how far flies traveled upwind and downwind at various wind velocities. Speer concluded that wind response in these stocks is primarily genetic with only a small environmental variance. The genetic variance is primarily nonadditive at high wind velocities, but the additive component increases as wind velocity decreases.

VI. Population Genetics

A. Population Genetic Theory for Automictic Thelytoky

For years—and even today—many evolutionary biologists dismissed parthenogenesis, and particularly automixis (e.g., see Maynard Smith, 1978), as an evolutionary "dead-end." Although the phrase "dead-end" can be used with many different connotations, automixis is certainly not a dead-end in a microevolutionary sense (Suomalainen, 1962). However, this widespread prejudice perhaps explains why the first rigorous population genetic models for automictic thelytoky did not appear until the 1970's.

Asher (1970a,b) developed a one locus model for an automictic population reproducing by a fixed mixture of central and terminal fusion. Templeton (1972) and Templeton and Rothman (1973) extended this model to include gamete duplication, so the models are particularly appropriate for parthenogenetic *Drosophila*. As previously mentioned, almost all parthenogenetic *Drosophila* are automictic diploids that restore diploidy through terminal fusion, central fusion, gamete duplication, or a mixture of the three. Hence, the frequencies with which these three mechanisms occur in unfertilized eggs given an initiation of parthenogenetic development provides a general description of the reproductive behavior of most unisexual strains of *Drosophila*. Hence, let:

E_1 = proportion of eggs developing by terminal fusion;

E_2 = proportion of eggs developing by central fusion;

E_3 = proportion of eggs developing by gamete duplication;

$E_1+E_2+E_3 = 1$.

The genetic consequences of these parthenogenetic mechanisms for a single locus with two alleles (A and a) are as follows. First, if the female is homozygous (AA or aa), all of her parthenogenetic progeny will be an identical homozygote barring mutation regardless of the mechanism of restoring diploidy. A heterozygous female (Aa) can only produce heterozygous progeny under fusion (see Figs 1, 3 and 4), and from Figs 1 and 3, the probability of a heterozygous female producing a heterozygous parthenogenetic progeny is

$$K = E_1 y + E_2(1-y/2) \qquad (2)$$

where y = probability of recombination between the locus and its centromere:

$$\frac{2}{3} \cdot \frac{(e^{3kt}-1)(e^{2kt}-1)}{e^{(2+k)t}(e^{2kt}-1)}$$

k = coefficient of coincidence; t = uncorrected map distance between the locus and its centromere; and x = map distance = $kt(1-e^{-2t})/(1-e^{-2kt})$. Now let the genotypes have different fitnesses such that W_{ij} = the constant fitness of genotype ij, and let P_n = the frequency of AA at generation n; Q_n = the frequency of aa at generation n; and R_n = the frequency of Aa at generation n. Then, combining the fitness differences with equation (2) yields

$$P_{n+1} \propto (P_n + 0 \cdot 5(1-K)R_n)W_{AA}$$
$$Q_{n+1} \propto (Q_n + 0 \cdot 5(1-K)R_n)W_{aa}$$
$$R_{n+1} \propto KR_n W_{Aa}.$$

Heterozygotes will only be present at equilibrium if $W_{AA}, W_{aa} < KW_{Aa}$ with equilibrium frequency

$$R_{eq} = \left[\frac{(1-K)W_{AA}}{2(KW_{Aa}-W_{AA})} + \frac{(1-K)W_{aa}}{2(KW_{Aa}-W_{aa})} + 1\right]^{-1}. \qquad (3)$$

Equations (2) and (3) show that in a constant environment, heterozygotes can only be maintained by fitness heterosis unless $E_2 = 1$ (all central fusion) and $y = 0$ (absolute linkage to the centromere). The greater the amount of central fusion, the more likely it is for heterozygosity to be maintained. Equations (2) and (3) also illustrate the evolutionary interactions between linkage to the centromere and the various automictic mechanisms.

These basic one locus models have been extended in a variety of ways. Asher (1970a) performed simulations to study the effects of finite population size. He discovered that as population size decreased, the intensity of selection needed to maintain heterozygosity increased. Asher (unpublished data) extended his model to include coarse-grained heterogeneity in which fitnesses fluctuate either over generations or across spatial patches. With spatial heterogeneity, the effect of migration is also incorporated into the model. Asher demonstrated that heterozygosity could only be maintained if there is a type of average heterozygote fitness advantage, with the exact type of average being determined by the nature of the environmental heterogeneity and the migration pattern. Templeton (1974) investigated another type of environmental heterogeneity, density dependent selection, and once again showed that a type of heterozygote advantage must exist for heterozygosity to be maintained. Finally, Templeton and Rothman (1978) presented a fine-grained model in which individuals experience heterogeneity within their lifetimes. They showed that if the heterozygote had superior homeostatic capabilities, it may persist in the population even if it has inferior fitness responses under the most commonly occurring environments. Moreover, Templeton and Rothman (1978) considered the problem of competition between the parthenogenetic population and its bisexual ancestors. Once again, by stabilizing a genotype with superior homeostatic capabilities in extreme environments, the parthenogenetic population may persist and even sometimes drive its sexual ancestors to extinction.

Unlike models for sexual populations, the automictic one locus models are easily extended to two or more loci. Thus, multi-locus analogues of the one locus models discussed above may be found in Templeton (1972, 1974). An extremely important distinction emerges between automictic n-locus models and sexual models as one increases the number of loci—the fitness properties of an intact n-locus genotype play a dominant role in the evolutionary dynamics of automictic populations but usually do not do so in sexual populations. It is for this reason that population geneticists often distinguish between the additive (or sometimes multiplicative) and non-additive genetic components of fitness in polygenic models for sexual populations; only the additive components respond to selection. However, this distinction is both unnecessary and even misleading for parthenogenetic populations. Consequently, the genetic components being selected differ greatly under sexual and parthenogenetic reproduction even if the systems are initially identical in every way.

Recall from Section IV, E, that the very mixture of automictic mechanisms observed in a thelytokous population is under genetic control and hence subject to evolutionary change. In terms of equation (2), E_1, E_2 and E_3 can be modified by the action of selection. Templeton (1974) examined this problem and argues that, if an initial heterosis is present, selection favors an increase of E_2, the proportion of central fusion. Moreover, this process is self-reinforcing under the n-locus models: the more E_2 is increased, the more selection favors even further increases in E_2. Hence, if primed by an initial heterosis and if the necessary genetic variants exist, automictic populations tend to evolve central fusion as their primary or sole means of restoring diploidy. Moreover, within the clones having a high rate of central fusion, selection favors a reduction in the probability of recombination with the centromere. The fate of the genome under central fusion has been discussed in greater detail by Asher (1970a). He argues that selection will favor (1) lower efficiency of synapsis, (2) lower degree of interference, (3) production of smaller chromosomes, and (4) production of chromosomal rearrangements. Work with *Drosophila* indicates that genetic variants for (1) and (4) commonly occur, so perhaps routes (1) and (4) represent the most common evolutionary fates of genomes under central fusion. As explained in Templeton (1974), route (4) has occurred independently many different times, with a good example being *D. mangabeirai*. A good example of route (1) is the coccid studied by Nur (1979). Thus under this interpretation, the "gonoid" thelytoky of Nur is an evolved genomic state associated with central fusion. The models explain well the observed fact that among naturally occurring obligate automictic thelytokous organisms, the most common type of parthenogenesis is central fusion coupled with some form of partial or complete cross-over suppression.

Before ending this section, one final advantage of modeling automictic populations must be mentioned: statistical analogues to the theoretical models can usually be readily constructed (e.g., Templeton, 1972; Templeton and Rothman, 1973; Templeton *et al.*, 1976b; Annest and Templeton, 1978). Hence the parameters of the models are generally estimable and the predictions experimentally testable—rare attributes for most theoretical population genetic models.

B. Clonal Selection in Experimental Populations

Under gamete duplication, each parthenogenetic female, barring mutation, produces genetically identical offspring. Even under fusion, many heterozygous genotypes can be derived only from identical genotypes in the previous generation; once again, barring mutation. For example, in the two locus model given in Templeton (1974), the double heterozygote class is closed from input from all other genotypic classes although it can contribute offspring to all other genotypic classes. Consequently, in this sense, all double heterozygotes are clonally derived from one-another. Finally, when central fusion is coupled with absolute linkage to the centromere, as in the case of *D. mangabeirai*, the genotypes breed true and may be thought of as clones. In any of these cases, when fitness differences exist among clones, clonal selection will occur.

Reference to clonal selection has already been made in Sections IV, E, and V, A. The studies referred to there clearly establish the reality of clonal selection and also demonstrate that microevolutionary responses can occur even in old, long-established automictic lines, although the most dramatic and rapid effects generally occur within a few generations after sexual reproduction has been abandoned. I now examine some additional cases of clonal selection in populations of *D. mercatorum*.

Wei (1968) performed the first population cage experiments on parthenogenetic *D. mercatorum*. She made hybrids between two parthenogenetic stocks that differed at the X-linked locus spotless (*sl*) and, presumably, many other loci as well. She then set up both bisexual and unisexual population cages from the F_1 hybrids, with each cage having a starting frequency of *sl* of 0·5. This frequency did not change significantly over the course of a year in either cage, and there was no significant evidence for heterosis in the bisexual cage. According to the model of Templeton and Rothman (1973), heterosis in the unisexual population should at the very least slow down the rate of decrease of heterozygotes in the unisexual cage. The heterozygote class did indeed persist for nearly a year (see Table IV), and Wei (1968) felt this indicated heterosis at the spotless locus—but does it? Wei (1968) unfortunately did not estimate the percentage of fusion occurring in this population, but she did note that reproduction is primarily via gamete duplication. To be on the conservative side (for inferring heterosis), assume the percent of fusion in this stock is 40%, about the maximum ever recorded in *D. mercatorum* (Annest, 1974). As a further conservative assumption, assume all fusion is central fusion. Hence, $E_1=0$, $E_2=0.4$ and $E_3=0.6$ in terms of the model presented in Section VI, A. Finally, mapping experiments on the three X-linked visibles currently available in *D. mercatorum* revealed that *sl* had a central location on the X between white eyes and miniature wings, with miniature wings being slightly closer to *sl* with an uncorrected map distance of 28·4 centimorgans (Templeton, unpublished data). Since the X in *D. mercatorum* is a rod, the uncorrected map distance between the centromere and *sl* is at least 0·284 map units. Assuming $t=0.284$, once again a conservative assumption for inferring heterosis, $K=0.32$. Finally, the data given in Table IV is based on

TABLE IV. The decay of heterozygosity at the spotless (sl) locus in a unisexual cage of *Drosophila mercatorum*. The cages were initiated from F_1 virgin females obtained by hybridizing (through bridge stocks) the two parthenogenetic strains RSS-18-Im and OB-2-Im (Wei, 1968). The expected decay under neutrality with $K = 0.32$ is also given, as well as the decay under heterosis and $K = 0.32$, with the relative fitness of the heterozygote to the homozygotes being 2.2. See text for details.

Sample date	Estimated generation	Frequency sl/sl^+	Expected neutrality	Expected heterosis
Dec. 1965	0	1·000	1·000	1·000
Feb. 1966	3	0·091	0·033	0·189
Apr. 1966	6	0·082	0·000	0·057
Jun. 1966	9	0·018	0·000	0·019
Aug. 1966	12	0·007	0·000	0·007
Oct. 1966	15	0·038	0·000	0·002
Dec. 1966	18	0·000	0·000	0·001

bimonthly egg samples drawn from the cages (Wei, 1968). Within this period, three to five generations could have occurred. Once again, being conservative, assume three generations every two months. Then the decay of the heterozygous clone under neutrality is given by $(0.32)^n$ where n is the number of generations. This theoretical decay is given in Table IV. Obviously, the decay rate of heterozygotes is far less than predicted even under all of these conservative assumptions. Using a modification of the maximum likelihood statistical procedure given in Templeton and Rothman (1973), the heterozygote fitness would have to be 2·2 times that of the homozygotes to explain Wei's data under the assumption of $K = 0.32$ (Table IV)—and this is undoubtedly an underestimate! This may seem unrealistically high, particularly in light of the fact that sl displayed no significant heterosis in the bisexual line. However, recall that central fusion restores heterozygosity with high probability (at least 2/3) for *all* loci in the genome and that the original founders were undoubtedly heterozygous for a very large number of loci besides spotless. Moreover, any fly heterozygous for sl is derived from a line of descent reproducing exclusively by central fusion. Consequently, heterozygosity for sl in the unisexual cage is really marking an extensive genomic heterozygosity. Thus, the heterosis could involve a great many loci in the unisexual cage. This apparent heterosis would be accentuated even more by the fact that those homozygotes due to gamete duplication would have very low fitness on average, based on the experiments of Templeton *et al.* (1976b) and Templeton (1979b) that show large fitness declines in the totally homozygous recombinants between two different parthenogenetic strains. Thus, the absence of heterosis for *s1* in the bisexual cage in no way makes the large estimated genomic heterosis in the unisexual cage unlikely. The difference obtained between Wei's bisexual and unisexual cages illustrates therefore the difference in unit of selection encountered under sexual versus parthenogenetic inheritance.

Further evidence that the unisexual heterosis is not associated with the *s1* locus *per se* comes from the more extensive experiments on clonal selection of Annest (1976) and Annest and Templeton (1978). As in Wei's study, these experiments were initiated by hybridizing two parthenogenetic genomes and setting up both unisexual and bisexual population cages from the F_1. Annest (1976) initiated cages from seven distinct F_1 flies from seven genetically distinct hybridizations, none of which were the same two as Wei (1968) used but all of which resulted in heterozygosity for spotless and another X-linked locus, white eyes (*w*). Despite high initial fusion rates in some of these F_1 females, none of the cages displayed the pattern observed by Wei (1968), strengthening the hypothesis that spotless is primarily marking genomic phenomena and is not subject to single locus heterosis. Consequently, in Annest's cages, almost all the selection was due to competition between genetically diverse but homozygous clones.

Annest (1976) measured the life history consequences of this clonal selection in five cages. Of these five, two were controls in which females from a given parthenogenetic strain were crossed to males from a bridge stock for the same unisexual strain. Another cage was between two different parthenogenetic strains, RSS-18-Im and RSB-7-Im, which are closely related (for details of their derivation, see Carson, 1967a). Consequently, of the 5 cages examined, only two (cages 3 and 4) represented a hybridization between highly genetically distinct parental strains. Not surprisingly, these two cages gave the strongest evidence for a response to interclonal selection, and this is shown in Table V. As can be seen, both cages increased their fecundity significantly, with cage 4 displaying almost a 2-fold increase. Moreover, both cages significantly increased the number of eggs initiating development. Egg hatchability, given development has been initiated, was not increased in either cage, and larva-to-adult survivorship was only significantly increased in cage 3. However, both cages experienced a significant increase in total egg-to-adult survivorship. That these alterations in life history parameters were due to selection under parthenogenetic reproduction is shown by the same life history data gathered for virgin females extracted from the bisexual analogues of cases 3 and 4 (Table V). Unlike the unisexual females, the bisexual virgins have a decreased fecundity and percent eggs developing. This decrease was so drastic, it was not possible to obtain an adequate sample for measuring larval survivorship (the data given in Table V are based on 29 and 4 larvae respectively for

TABLE V. Some results of clonal selection in two unisexual cages (No. 3 and No. 4) of *Drosophila mercatorum* and their bisexual controls (from Annest, 1976). All cages were initiated from F_1 hybrids of genetically diverse parthenogenetic strains and their bridges. Life history measurements were made upon the initial F_1, upon the parthenogenetic flies emerging after a year from the unisexual cage (F_{i-n}), and upon virgin females emerging after a year from the bisexual cage (F_n).

Virgin females	Eggs/♀/day		% Eggs developing		% Eggs hatching		% Larva-to-adult survival		% Total survival	
	No. 3	No. 4	No. 3	No. 4	No. 3	No. 4	No. 3	No. 4	No. 3	No. 4
F_1	65·7	62·4	32·3	47·2	20·5	23·5	66·7	85·7	4·4	9·5
F_{i-n}	78·9[a]	120·0[a]	41·1[a]	70·2[a]	20·9	18·5	87·5[a]	90·5	7·5[a]	11·7[a]
F_n	53·7[a]	36·2[a]	18·5[a]	23·0[a]	18·2	23·5	100·0[a]	100·0	3·0	5·4

[a] Significantly different from F_1 at 0·05 probability level.

bisexual cages 3 and 4). Thus, the selective forces responsible for the increase in these parameters in the unisexual cages were absent in the bisexual cages.

Annest and Templeton (1978) analysed in more detail the unisexual cage 6 set up by Annest (1976). In this case the hybridization provided a total of 6 genetic markers with at least one marker on every major chromosome arm. Consequently, the outcome of the clonal selection could be monitored directly for a sizable portion of the genome. With this hybridization, there was no evidence for fusion, so all competition from the first generation on was between the genetically diverse homozygous clones created by meiosis in the original F_1 females. The very first generation yielded a highly non-random phenotypic distribution (see Fig. 12 for data on the five visible markers). This was attributed to an intense selective bottleneck favoring those genomes produced by the F_1 mothers that could initiate development and result in viable larvae and adults. Consequently, by the end of the first generation, all clones had proven themselves viable under total homozygosity. However, the population size decreased in going from generation 1 to 2 despite the fact that all lethal clones had already been eliminated. Thus, a second selective bottleneck was encountered, favoring those viable females capable of parthenogenesis themselves. Thus, by the end of the first two generations all clones had been eliminated that were incapable of surviving and reproducing under parthenogenesis and the genetic state of total homozygosity. From generation three on, there remained only those clones that had non-zero viabilities and parthenogenetic efficiencies. As a consequence, the population size increased dramatically in going from generation 2 to 3 and was more stable thereafter. As Fig. 12 illustrates,

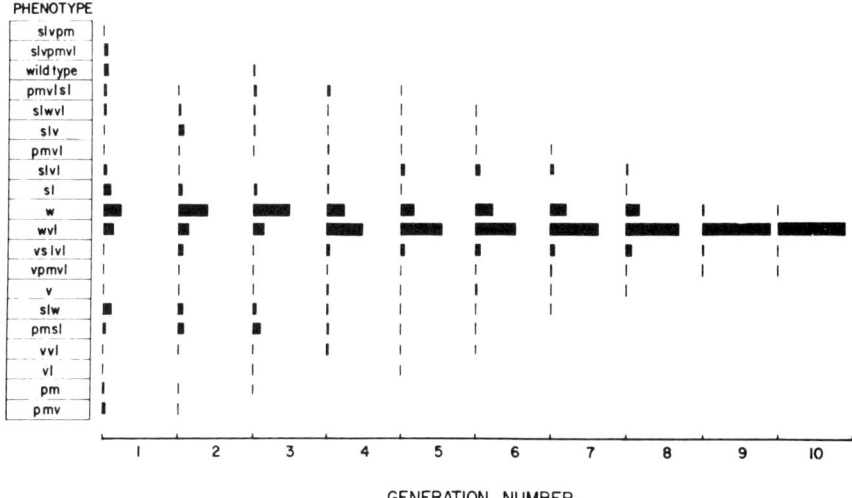

FIG. 12. Clonal selection among 20 phenotypes over 10 generations of parthenogenetic reproduction in a unisexual, discrete generation population cage. The first generation females were generated by automictic parthenogenesis in virgin mothers which were heterozygous for six gene markers. All phenotypes in subsequent generations represent one or more completely homozygous genotypes. The length of each bar is proportional to the frequency of each phenotype in each generation. From Annest and Templeton (1978).

clonal selection on the relative viabilities and parthenogenetic efficiencies was intense from generation 3 on and resulted in a rapid elimination of almost all but one genotypic class. A fitness analysis revealed that the selection occurring during this period primarily involved non-additive and non-multiplicative fitness components. Hence, selection was truly operating upon genomic properties and not upon the properties of individual loci *per se*.

C. The Unit of Selection in Experimental Populations

The unit of selection is that level of genetic organization to which fitness measures can be applied that combine additively or multiplicatively with the measures on other such units to yield an accurate prediction of the outcome of selection in a population. Other definitions exist, but this definition has the advantage of making the unit of selection experimentally estimable. Theoretically, this unit could vary from an individual locus to the entire, intact genotype of an individual. In most population genetic models, the unit of selection is assumed to be the locus. However, this assumption is based more on mathematical convenience than on solid evidence since measuring this unit is a very difficult task in most populations (Sing and

Templeton, 1975). However, the existence of *n*-locus theoretical models and statistical analogues for automictic populations greatly simplifies the task of measuring the unit of selection in parthenogenetic *Drosophila* (Sing and Templeton, 1975).

Templeton *et al.* (1976b) represents the first study measuring the unit of selection in *D. mercatorum*. Hybrids and partial hybrids between two parthenogenetic strains were bred (Fig. 13). Some of the resulting hybrid and partial hybrid females were mated to various bisexual stocks as controls, while other females remained virgin and reproduced parthenogenetically. As the degree of hybridity increased, the ability of the females to reproduce parthenogenetically decreased (Table VI). This indicated that the two parental genomes were incompatible with one another in a fitness sense under total homozygosity; therefore, recombination between the two genomes should engender strong selective forces. Note that this incompatibility cannot be explained by recessive lethals or any other single locus effect. All alleles entering these hybrids came from parental stocks that reproduced quite successfully under the constraint of total homozygosity. (Recall also the breakdown in parthenogenetic capacity in the bisexual cages presented in Table V despite the fact that both parental stocks were derived from highly successful unisexual strains.) Hence, the two genomes were said

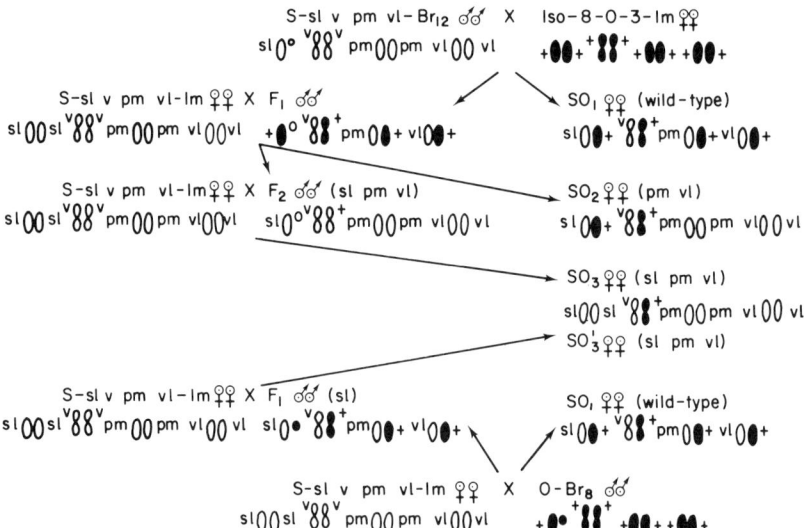

FIG. 13. Breeding scheme used to generate the experimental treatments in the unit of selection experiments of Templeton *et al.* (1976b). There are five major chromosome arms of approximately equal size with the metacentric "v"-marked autosome having two arms and the others only one. A small pair of dot autosomes is not shown in this diagram.

TABLE VI. Parthenogenetic capacities, as measured by the number of adult parthenogenetic offspring produced per female, in the parthenogenetic stock S-v pm vl-Im (derived from Iso-8-S-1-Im) and in hybrids and partial hybrids between S-v pm vl-Im and the parthenogenetic stock Iso-8-0-3-Im. The derivation of the hybrids and partial hybrids is shown in Fig. 13. Data are from Templeton et al. (1976b).

Type of female	Percent of genotype that is hybrid	Number of parthenogenetic offspring per female
S-v pm vl-Im	0	14·54
SO_3 and SO_3'	40	10·25
SO_2	60	5·36
SO_1 and SO_1'	100	1·63

to be coadapted to total homozygosity because they had gene complexes that confer high fitness when homozygous due to fitness interactions between loci. Insight into the nature of these interactions was gained by scoring the parthenogenetic progeny produced by the hybrid and partial hybrid females for several visible and isozyme loci.

These data could then be used to directly estimate the fitness of a multi-locus genotype. Next, the multi-locus genotype fitness was subdivided into various components to see if any unit of selection could be identified. The main conclusions of this study were as follows. (1) The parental genomes were indeed coadapted to total homozygosity since their high fitness attributes depended primarily upon non-additive and non-multiplicative fitness interactions between both linked and unlinked loci scattered throughout the genome. (2) The coadaptation was truly due to a selective response to the internal genetic environment of total homozygosity and not due to geographical coadaptation. (3) The unit of selection is a function of the intensity of selection: the greater the selection, the larger the unit. (4) Physical linkage relationships are also important determinants of the unit of selection, with selection preferentially operating upon linked blocks of genes. (5) Apparent selective neutrality may arise as an artifact in a coadapted genetic complex if the genetic markers used do not identify the unit of selection.

The above study raised an interesting question: namely, how did the coadapted genomes originate in the first place? Conceivably, they could have arisen at the inception of parthenogenetic reproduction or later, under exclusive parthenogenetic reproduction, due to the accumulation of mutants in unisexual clones. To answer this question, Templeton (1979b) performed a "before and after" unit of selection experiment upon a parthenogenetic stock isolated directly from the Kamuela natural population (the "after" contrast) and upon the bisexual females that gave rise to that parthenogenetic line (the "before" contrast). The fitness properties of the genomes extracted from the bisexual females showed no evidence whatsoever for coadaptation to total homozygosity, and almost all detected selection was additive at the level of individual marker loci. The fitness property of the one successful parthenogenetic genome these bisexual females did produce, however, was one of strong non-additive and non-multiplicative fitness interactions involving every major chromosome arm of the genome. Hence, the coadapted genomes evolve extremely rapidly as a result of the intense selective bottleneck encountered during the reproductive transition. Thus, the reproductive transition is also accompanied by a radical transition in the unit of selection. (Recall also the discussion of Wei's (1968) experiments in Section VI, B.)

The results obtained by Templeton (1979b) also indicated that the visible and isozyme markers used to monitor the reproductive transition were not the target of the observed selection. What loci, then, were being selected? During the course of the unit of selection experiments, Templeton (1979b) noted several developmental abnormalities that greatly lowered viability appearing in recombinants between the parthenogenetic lines. One of these, abnormal abdomen (see Section IV, E) has been examined for its genetic basis (Templeton and Rankin, 1978; Templeton, unpublished data). Currently, a minimum of six loci have been identified that underlie this complex, and certain combinations of alleles at these loci yield normal development whereas others do not. When what is known of the genetics of *aa* is overlaid upon the fitness pattern revealed in the unit of selection experiments (Templeton, 1979b), much (but not all) of the fitness pattern can be explained by the *aa* syndrome (Templeton and Rankin, 1978). In addition, studies have been initiated to elucidate the biochemical and physiological basis of abnormal abdomen and other developmental abnormalities that occur in the parthenogenetic recombinants (Templeton and Rankin, 1978). The option of parthenogenesis has already proved to be quite useful in pursuing this work. In this manner, it is hoped that the biochemical and physiological causes for coadaptation may be identified.

Sing *et al.* (1976) are taking an alternative approach to the unit of selection problem by starting with biochemical and physiological studies on parthenogenetic strains. Taking advantage of the fact that a large number of flies with the same genotype may be cloned, Sing *et al.* have been able to measure the pool sizes of 15 glycolytic intermediates and related chemicals in various parthenogenetic strains of *D. mercatorum*. From such data, they have discovered that isogenic strains have characteristic glycolytic profiles that are genetically determined. Moreover, they are accompanying these studies on intermediate pool sizes with electrophoretic and kinetic studies on the glycolytic and related enzymes themselves. In this manner, they can identify the specific loci that influence the glycolytic profile. For example, in 1973 Templeton (unpublished data) isolated two parthenogenetic strains that are coisogenic except for the enzyme locus α-glycerophosphate dehydrogenase. A glycolytic profile analysis of these two coisogenic strains revealed that they differed in pool sizes for several intermediates, including α-glycerol-phosphate and dihydroxyacetone, the intermediates directly affected by this enzyme (Sing, personal communication). Thus, this specific alteration of the glycolytic profile can be ascribed to the α-GPDH locus. As is evident, Sing *et al.* (1976) are starting with a system known to strongly interact biochemically and known to be important physiologically. Ultimately, it is hoped to be able to trace the effects of this system upward through the various biological hierarchies to the populational and evolutionary levels.

D. Total Homozygosity and Isozyme Variability

As mentioned earlier, a total of 53 parthenogenetic adults have been extracted directly from natural bisexual populations inhabiting the vicinity of Kamuela, Hawaii (Templeton *et al.*, 1976a; Templeton, 1979a). An electrophoretic survey was performed on both the bisexual population and 40 of the 53 parthenogenetic progeny. Considerable isozyme variability was discovered in the bisexual populations with 17 loci segregating out of a total of 29 (Templeton, 1979a; Templeton *et al.*, 1976a). Of the 40 parthenogenetic females studied, all but one of them proved to be totally homozygous for all isozyme loci. Hence, the primary automictic mechanism among the tychoparthenogenetic Kamuela females is gamete duplication. (The one fly not totally homozygous was heterozygous for esterase A, so she was most likely the product of central fusion, as argued in Section II, E, and Fig. 2). Consequently, almost the entire sample of parthenogenetic flies should have been subjected to the extremely intense selective bottleneck described by Templeton (1979b). However, the isozyme loci allele frequencies are not altered during the reproductive transition. The parthenogenetic genomes correspond quite well to a random sample of the sexual gene pool. This conclusion is based on single locus frequencies considered independently. The sample size is not large enough to consider multi-locus contrasts, so—recalling conclusion (5) from Section VI, C, that neutrality can be inferred if the unit of selection is not marked—selection still could be operating at a multi-locus level on the isozyme loci. However, there is evidence even against this interpretation. In addition to the "before and after" unit of selection experiments described in the previous section, similar experiments were performed on a second parthenogenetic line isolated at the same time and from the same population as the first unisexual line. Moreover, these two lines matched for all 17 segregating isozyme loci (Templeton, 1979b). Both were fixed for the most common allele at each locus, but still the probability of this homozygous genotype occurring at random was only 0·03. Nevertheless, when the coadaptation experiments were performed, it was discovered that these two parthenogenetic strains had adapted to total homozygosity in radically different fashions. Moreover, the second parthenogenetic line had apparently coadapted in a fashion completely compatible with a parthenogenetic strain whose bisexual ancestors came from El Salvador and from which it was quite distinct at many isozyme loci (Templeton, 1979b). Consequently, the isozyme loci were totally noninformative about the nature of the coadaptation and the radical fitness restructuring that occurred. This fact when coupled with the previously noted isozyme survey results on the 40 parthenogenetic flies indicates that isozyme loci are neutral with respect to the selective events on viability associated with the reproductive transition and total homozygosity. However, only 4 of these 40 parthenogenetic flies could themselves produce a self-sustaining parthenogenetic line, and these 4 flies had almost identical isozyme phenotypes (Clark *et al.*, 1981). This near identity was highly significant, so isozyme loci do seem to be under selection in this system, but the relevant fitness component is not viability.

VII. Speciation

A. The Origin of Parthenogenetic Species and the Evolution of Sex

Parthenogenetic *Drosophila* may serve as a useful experimental model for studying the origin of parthenogenetic species from their sexual ancestors. In this regard, Sections IV, D, IV, E, V, A and VI, A and B are particularly relevant. I now discuss in more detail how work on parthenogenetic *Drosophila* relates to the problem of the evolution of an obligate thelytokous population from a bisexual ancestor (note that cases involving interspecific hybridization are not considered here).

The necessary pre-requisite for the evolution of a thelytokous population is the existence of a tychoparthenogenetic capacity in the bisexual ancestors. The work with *Drosophila* shows that many sexual populations do indeed have a low level of parthenogenetic capacity and, most importantly, this capacity is inherited. However, several barriers may block the utilization of this capacity. The first is the opportunity and frequency with which unfertilized eggs are laid. This is a function of the ecology, population structure, abundance of males, and female behavior (e.g., some *Drosophila* females do not lay any eggs until after they have mated) of the sexual ancestors and of temporal fluctuations in these parameters. Given an unfertilized egg in which parthenogenetic development has been initiated, a viability selective bottleneck is encountered if the mode of parthenogenesis greatly alters the genetic environment. Such a bottleneck appears to be a major barrier in *D. mercatorum* (Templeton *et al.*, 1976a). Cuellar (1977) has hypothesized that inbreeding could preadapt a population for overcoming such a barrier, but his hypothesis is not supported by the *Drosophila* evidence. In *D. mercatorum*, the chances for producing a totally homozygous parthenogen increase as the level of heterozygosity increases in the bisexual females, both between (Templeton, 1979a) and within populations (Templeton *et al.*, 1976a). Moreover, it is easier to obtain totally homozygous parthenogens from outcrossed wild stocks than from inbred laboratory stocks (Templeton *et al.*, 1976a). An explanation for this apparent paradox is provided by the analysis of Templeton (1979b) which shows that selection for a gene complex yielding a viable developmental history under total homozygosity is quantitatively as or more important than selection against recessive lethals during the initial reproductive transition. Inbreeding will reduce the incidence of recessive lethals, but it does not necessarily mimic the selective environment associated with total homozygosity. Indeed, the *D. mercatorum* results strongly suggest that inbreeding cannot be equated with homozygous isogenicity, a frequent assumption in many *Drosophila* experiments. However, inbreeding does reduce the average amount of heterozygosity, which in turn reduces the amount of interclonal variability produced under gamete duplication. This in turn reduces the effectiveness of clonal selection for a coadapted gene complex, thereby lowering the chances for successfully passing the viability selective bottleneck (Templeton *et al.*, 1976a).

An alternative preadaptation hypothesis is suggested by the work of Nur (1971, 1972). Nur has shown that most natural obligate thelytokous populations with gamete duplication evolved from sexual ancestors with males that are cytologically or genetically haploid. Nur (1971) has argued that heterosis plays less of an evolutionary role in these populations. Hence, mutants are selected more for their effects under haploidy or effective homozygosity. Maynard Smith (1978) has also added that recessive lethals and deleterious genes are rarer in such haplo-diploid species, but in view of the *D. mercatorum* data, this effect is of secondary importance. However, the work with *D. mercatorum* does support the view that male haploidy effectively mimics the selective environment associated with total homozygosity. Templeton *et al.* (1976b) noted that the multilocus fitness properties revealed by the X-linked markers in the control sexual males from the unit of selection experiments (see Section VI, C) were homogeneous with the parthenogenetic female results but inhomogeneous with the results obtained with their sexually produced sisters. In others words, hemizygosity did mimic the selective environment associated with total homozygosity, which supports Nur's (1971) hypothesis.

After viable parthenogens are selected, a second genetically distinct selective bottleneck is encountered when the parthenogens reproduce parthenogenetically themselves (Annest and Templeton, 1978). This is followed by a phase of evolutionary change effected by clonal selection (Annest and Templeton, 1978). This clonal selection can be extremely effective under all automictic mechanisms for the first several generations after the abandonment of sex, and can continue to be very effective for several hundreds of generations in clones reproducing primarily by central fusion if the initial genome consisted of short (in a recombinational sense) chromosomes. After the initial genetic variability passed on by the sexual ancestors is exhausted, clonal selection can still occur as mutations continue to create a clonal diversity. More of this will be discussed later.

Because of the two selective bottlenecks, the change in the unit of selection and clonal selection, the early phases of the reproductive transition to automixis are extremely dynamic, and sets of genes rather rare in the sexual ancestors can readily go to fixation in the parthenogens (e.g., the genes underlying the abnormal abdomen syndrome in $D.$ $mercatorum$—Section VI, C). These alterations, even if they involve only a handful of loci, can have cascading effects through pleiotropy or linkage on many attributes of the parthenogens (Templeton, 1979b, 1980). The entire life history can be modified (Templeton, 1979b; Templeton and Rankin, 1978; Annest, 1976) and behavioral responses to environmental cues altered (Templeton and Rankin, 1978; Section V, B). Such changes could easily effect a rapid shift in niche that may be important in allowing the parthenogens to coexist with their sexual ancestors, particularly since the parthenogenetic efficiency in newly arisen thelytokous females is almost universally low. Moreover, sexual isolation tends to accompany the reproductive transition (Section V, A). Since crossing to bisexual males destroys the parthenogenetic capacity (Carson, 1967a), any unisexual clones that did tend to mate would be effectively eliminated from the thelytokous population, thereby enhancing reproductive isolation. This isolation would also increase the chances for coexistence of the sexual and thelytokous populations. Finally, parthenogenesis can stabilize a multi-locus genotype that confers superior homeostasis in extreme environments, a phenotype that is evolutionarily unimportant in an ancestral sexual population can come to rapidly dominate a thelytokous population if it depends upon heterozygosity at several unlinked or loosely linked loci or upon many non-additive effects. The stabilization of a homeostatic genotype can insure survival and even growth of the unisexual population despite competition from sexual ancestors

(Templeton and Rothman, 1978). Moreover, this would explain the common pattern observed in insects, that parthenogenetic populations generally inhabit broader and more ecologically diverse areas than their sexual ancestors (Suomalainen et al., 1976) and that many parthenogens, such as *D. mangabeirai* (Murdy and Carson, 1959; Carson, 1967b), seem to have broad ecological tolerances.

Finally, during these initial unisexual generations, the very mode of parthenogenesis itself can be selectively altered. Most tychoparthenogenetic populations reproduce by a mixture of automictic mechanisms, with the proportions of the mixture being genetically determined (Carson, 1973). The resulting automictic mode that is eventually stabilized is therefore determined in part by selection and in part by the type of genetic variability present in the initial unisexual population.

The stabilized automictic mode then alters selective forces operating upon genome structure, such as the favoring of chromosomal rearrangements under central fusion (Asher, 1970a; Templeton, 1974). One could argue that the rearrangements existed before parthenogenesis evolved, but even then their frequent incorporation into thelytokous populations is most likely influenced by selection. Moreover, recent evidence gathered by Ochman *et al.* (1980) implies chromosomal evolution can occur after parthenogenesis has evolved. Ochman *et al.* (1980) performed an isozyme survey on the parthenogenetic fly *Loncoptera dubia* and overlaid their results upon the cytological work of Stalker (1956b). Of the four chromosomal races of this species (Stalker, 1956b), Ochman *et al.* (1980) discovered that the most parsimonious explanation for the origin of at least two of them was to assume a single unisexual ancestor followed by some electrophoretic and chromosomal divergence. Since this fly reproduces by central fusion (Stalker, 1956b), selection would favor any structural heterozygosity that might arise (Templeton, 1974). Moreover, all parthenogenetic races of *L. dubia* have a diploid chromosome number of 4, whereas all their sexual relatives have 6. This state may also have evolved after the establishment of central fusion since selection would then favor a reduction in chromosome number (Asher, 1970a).

Central fusion would also cause selection to favor certain meiotic mutants, such as those causing failure of synapsis. In this manner, an automictic diploid population could evolve gonoid thelytoky (Nur, 1979; Goldstein and Triantaphyllou, 1978), a mechanism that is neither totally automictic or apomictic. This process could continue until true apomixis has evolved (Seiler, 1947; Stalker, 1956c; Lokki and Saura, 1980).

All of the above selective routes tend to create modes of parthenogenesis that faithfully reproduce the parental genotype, *barring mutation*. This last phrase is extremely important, since mutation is never fully barred and hence the potential for evolutionary change is never fully lost. Moreover, the theoretical work of Leigh (1970) shows there is generally selection against mutators that increase the mutability of the entire genome in sexual populations, but not in asexual or parthenogenetic populations. On the contrary, such general mutators will be favored in parthenogenetic populations until an optimal level is reached. This enhancement of the mutation rate would also increase the evolutionary potential of the parthenogenetic population. Eshel (1973) has also shown that clonal selection can very efficiently modify mutation rates, recombination rates, sexual behavior, etc. so as to increase the expected clone survival probability. That clonal selection can indeed operate in long established parthenogenetic strains has already been experimentally verified by Henslee (1966), as discussed in Section V, A. Although the considerations given above are supported by experimental and observational evidence from studies with *Drosophila* and other organisms, they have been virtually ignored in most accounts purportedly dealing with the evolution of obligate thelytoky from a single bisexual ancestor (thus excluding interspecific hybridization from consideration), for example in the books of Williams (1975) and Maynard Smith (1978). Indeed, Maynard Smith (1978) throughout his book dismisses automixis as "an unpromising starting point for the origin of a parthenogenetic variety" (page 44). With this and other dismissals, Maynard Smith (1978) ignores the fact that naturally occurring automictic populations have evolved independently many times in the insects (Suomalainen *et al*., 1976), and therefore the phenomenon does seem to deserve some sort of evolutionary explanation. More importantly, almost all cases of tychoparthenogenesis in insects have proven to be automictic, a fact recognized by Maynard Smith (page 43) but dismissed without further comment. This is unfortunate, because all obligate thelytokous populations with a non-hybrid origin would have had to arisen through thelytokous tychoparthenogenesis in a bisexual ancestor. Curiously, Maynard Smith states (page 63), "Cases in which an occasional egg develops without fertilization are not relevant, particularly if such development is automictic." However, Maynard Smith fails to explain how thelytoky can arise from a bisexual ancestor without tychoparthenogenesis, although he clearly applies his models to cases not involving interspecific hybridization.

If most instances of insect tychoparthenogenesis are automictic, why are many natural parthenogenetic populations of insects apomictic? One reason is that automictic populations tend to evolve, by mechanisms discussed in Asher (1970a) and Templeton (1974), towards modes of reproduction that are apomictic or genetically equivalent to apomixis. Thus, the evolutionary causes for *abandonment* of sex in many apomictic populations can only be understood in terms of automixis, with apomixis being a derived, secondary character not directly involved in the reproductive transition (but perhaps very important in maintaining the long term survival of the thelytokous population). For example, Maynard Smith (pages 45 and 46) notes that apomictic weevils are the only group of insects in which thelytoky is at all common; but he fails to note that Seiler (1947) long ago showed that these weevils most likely evolved from automictic ancestors. More recently, Lokki and Saura (1980) have discussed diploid automixis as the major route in the evolution of apomictic polyploidy in insects. Moreover, the work of Nur (1979) shows there is no clear distinction between automixis and apomixis as Maynard Smith assumes; rather the two grade into one another as expected under the evolutionary hypothesis presented here. Thus, the evolutionary transition from sex to parthenogenesis in insects is very commonly effected through automixis, making the *Drosophila* model of general importance in the insects. Indeed, the evolutionary flexibility that automixis provides during the initial transitional generations may be critical in the establishment of a parthenogenetic variety, a flexibility that would be absent if the population were initially apomictic. Hence, automixis represents a most promising start for a parthenogenetic variety.

The models of Maynard Smith (1978) and others, therefore, have little to do with the *evolution* of obligate thelytoky from a sexual ancestor; rather, they are more pertinent to the maintenance of an already evolved parthenogenetic system in the presence of competition from sexual species. However, there are serious flaws even in this regard. One of the basic assumptions of Maynard Smith's (1978) models is that there is a two-fold advantage to parthenogenesis since eggs are not "wasted" on males. Unfortunately, little observational evidence exists concerning this issue, and what little does exist does not support the idea (Lamb and Willey, 1979). For example, in *D. mangabeirai*, a parthenogenetic species that has obviously been selected for parthenogenetic development for quite some time, only 60% of the eggs hatch. Consequently, this long established parthenogenetic species has barely exceeded the break-even point in Maynard Smith's model. Moreover, it apparently drove its sexual ancestors to extinction. It is therefore hard to see the relevance of Maynard Smith's two-fold advantage in this case. In addition, parthenogenetic efficiencies are low in *every* study of newly arisen parthenogenetic populations known to me, making the hypothesized two-fold advantage irrelevant to the initial reproductive transition. Indeed, the experimental work described in this chapter suggests the real question should be how a newly evolved parthenogenetic population can survive in competition with its sexual ancestors despite a several fold *disadvantage* in the production of viable eggs? The irrelevancy of the two-fold advantage was accidentally demonstrated during a mite infection. Nine bridge stocks were infected. Recall that these stocks are genetically identical to totally homozygous parthenogenetic stocks but reproduce sexually. However, the largest parthenogenetic efficiency of any of these bridge stocks was 6% of the eggs developing to adulthood. The mites did not attack the *Drosophila* directly, but rather were 'media' mites, so all infected stocks that could not be discarded were kept at low densities and frequently changed. This eliminated the mites in most cases, but at the end of two months, four of the bridge stocks had reverted to exclusive parthenogenetic reproduction. The cause and effect of this reversion is unknown, but it probably does not involve the mites directly killing or sterilizing males as several exclusively sexual stocks went through the same procedure without extinction. Consequently, environmental conditions do exist that favor the abandonment of sex that *cannot* be explained by any two-fold advantage to parthenogenesis. In general, the experimental studies on parthenogenesis suggest that by the time a parthenogenetic system has been sufficiently perfected to the point where a majority of the eggs hatch, so much other evolutionary divergence has occurred between the unisexual population and the bisexual ancestors as to make the "two-fold advantage" of dubious ecological or genetic significance.

Another model concerning the maintenance of parthenogenetic populations is "Muller's ratchet" (Muller, 1964; Haigh, 1978; Maynard Smith, 1978). The idea is simple: under clonal reproduction, if k mutants occur in an individual, all progeny of that individual and their descendants will never have any less than k mutants. If the population is sufficiently small, it becomes probable that sometimes all individuals in the population will bear one or more newly arisen deleterious mutants and hence the entire next generation will have at least one more deleterious mutant than their parents. However, these models ignore the results of Eshel (1973) and Leigh (1970) that selection can modify mutation rates to optimize clone survivorship far more efficiently under unisexual reproduction than under sexual reproduction. Moreover, these models were developed for asexual haploids, and they do not always apply to parthenogenetic diploids. Recently, Heller and Maynard Smith (1979) have extended the ratchet model to selfing diploids. They show the ratchet can operate in the selfing population under their set of assumptions and state (without proof) that the ratchet may also be important in parthenogenetic populations. However, their model depends upon several assumptions that are known to be false for parthenogenetic populations. First, they assume each heterozygous locus decays into homozygosity independently of all others: an assumption completely invalid under automixis even when dealing with loci on different chromosomes (recall the discussion of Wei's work in Section VI, B). The effects of linkage and the inter-chromosomal correlation induced by various modes of automixis drastically reduces the effective number of independently segregating units. This fact alone makes the ratchet virtually inoperable. The second assumption is that individual loci behave as independent multiplicative fitness units. This assumption is also known to be false for parthenogenetic populations (Section VI, C). It is therefore questionable as to how much applicability Muller's ratchet has to automictic diploids. However, under apomixis or its genetic equivalent, Muller's ratchet could conceivably be effective since recombination is lost. Even then diploidy coupled with permanent heterozygosity buffers the organism against expressing deleterious mutations, making it more likely for the ratchet to be inoperable (Haigh, 1978; Maynard Smith, 1978). Moreover, once apomictic diploidy or central fusion has evolved, there is a tendency to evolve polyploidy (Stalker, 1956c; Lokki and Saura, 1980), further reducing the potential of the ratchet to operate.

B. PARTHENOGENESIS AS AN EXPERIMENTAL MODEL OF SPECIATION IN SEXUAL POPULATIONS

The founder effect/genetic revolution model has been applied extensively to *Drosophila* evolution, particularly the Hawaiian *Drosophila* (Carson, 1975). This model was originally formulated by Mayr (1954) who envisioned that the increased homozygosity associated with a founder effect could change the selective genetic environment so rapidly as to cause a "genetic revolution" throughout the genome. This revolution could be so radical and extensive that in a very short time the founder population would be essentially a different species from its ancestors. Templeton (1979b) argues that the reproductive transition in *D. mercatorum* mimics, albeit in an extreme form, all the essential events in Mayr's model. Thus, the natural bisexual population corresponds to the ancestors, a virgin female to a small founder group, and her totally homozygous parthenogenetic progeny to the highly homozygous founder population. Consequently, Templeton (1979b) analysed the data on the reproductive transition in this light. The analysis lead to the following conclusions. (1) Genetic revolutions are real phenomena. Drastic changes in the genetic environment can indeed lead to the evolution of a new set of balanced genes that is incompatible with the old balance. The revolution is observable in terms of morphology, development, life history, and behavior, and it can result in the erection of pre- and/or post-mating isolating mechanisms. (2) Isozyme loci are not the target of the "genetic revolution"; rather, genes having fundamental developmental or regulatory roles in the organisms are being selected. This supports Carson's (1975) view of an open and closed system with isozyme loci being part of the open system. Indeed, only a handful of loci need be involved. For this reason, the term genetic revolution is misleading since it connotes extensive changes throughout the genome (Mayr, 1954). A more neutral term such as "genetic transilience" would be more appropriate. (3) The *a priori* chances of a successful genetic transilience increase as the level of outcrossing and individual heterozygosity in the ancestors increases. These conclusions, and theoretical justifications for them, have been incorporated into a new model of speciation via the founder effect known as the founder effect–genetic transilience model (Templeton, 1980).

Because the founder effect–genetic transilience model requires the ancestors to be outcrossed (Conclusion 3 above), it has limited applicability to organisms that are subdivided into small, relatively inbred demes. However, a hybrid-genetic transilience model can be invoked in these circumstances. Suppose a founder population consists of hybrids between members of two different demes that have different coadapted gene complexes. Subsequent inbreeding in the founder population re-establishes the parental genetic environment, but recombination between the original coadapted gene complexes allows novel genetic responses to this genetic environment to be made. Occasionally, an entirely new coadapted gene complex will evolve that is adaptively incompatible with both parental types. Moreover, hybridization followed by inbreeding is often associated with increased mutation rates and chromosomal aberrations (e.g., Woodruff *et al.*, 1979). This effect when coupled with the inbreeding increases the chances for fixation of chromosomal rearrangements that could enhance isolating barriers, further protecting the new coadapted gene complex from additional recombination with parental complexes. Note, the chromosomal evolution occurring in this model is not the primary cause of the speciation as it is in the stasistipatric model (White, 1978). *D. mercatorum* can also serve as an excellent experimental model for this type of speciation. Different parthenogenetic strains now correspond to the different, highly inbred local demes of the model, and hybridization of two different strains followed by gamete duplication simulates the hybridization followed by inbreeding in the sexual cases. Indeed the experimental work of Annest (1976) already provides an example of this hybrid–genetic transilience model. As mentioned in Section VI, B, Annest had set up two cages from hybrid females of two genetically distinct *D. mercatorum* unisexual strains. After parthenogenesis was re-established, the strains quickly evolved with respect to the two X-linked markers he was following. In cage 3, the population quickly re-established one of the parental genotypic states for the X-linked loci. This result is not surprising in light of the results of Templeton *et al.* (1976b) and Templeton (1979b) that there is strong selection favoring the parental genotype over the recombinants. However, occasionally, a novel recombination will occur that is selectively advantageous. This apparently happened in cage 4 of Annest (1976). Unlike cage 3, clonal selection in this case quickly resulted in near fixation of a recombinant genotype for the X-linked markers. Moreover, cage 4 displayed the stronger divergence from the parental strains in life history characteristics with all but "percent eggs hatching" showing a superior value than that of either parental strain. In contrast, cage 3 was not significantly different or intermediate to the parental strains for all life history parameters. It would be very interesting to couple experiments like those of Annest (1976) with hybrid dysgenesis experiments like those of Woodruff *et al.* (1979). The combination should provide an excellent experimental system for studying the validity of the hybrid–genetic transilience model of speciation.

VIII. Prospects for Parthenogenetic *Drosophila*

Most of the experimental work performed with parthenogenetic *Drosophila* has had an evolutionary perspective simply because almost all the workers in this area have been evolutionary biologists. The results obtained so far and the questions they have raised indicate that parthenogenetic *Drosophila* will be an extremely useful tool in many future evolutionary studies. However, the potential for using parthenogenetic *Drosophila* in other disciplines of genetics and biology has rarely been utilized. For example, the gamete duplication system of *D. mercatorum* offers unique advantages for screening of mutants (Carson and Snyder, 1972), but these advantages have seldom been used. The work of Sing *et al.* (1976) illustrates well the advantages parthenogenetic *Drosophila* offer for certain types of biochemical work, and initial studies on juvenile hormone mentioned in Templeton and Rankin (1978) reveal the potential usefulness of parthenogenetic *Drosophila* as a tool in insect physiological research. Similarly, the developmental syndromes that appear in recombinant parthenogenetic lines (Templeton, 1979b) are exceedingly interesting since they are associated not with individual loci but with sets of loci, and not with mutant alleles but with alleles extracted from nature. Hence, studies on these syndromes could reveal much about how developmental pathways are controlled by sets of interacting genes. In conclusion, I can only agree wholeheartedly with the statement made by Petrunkevitch (1905) during the infancy of the science of genetics that "In this way, parthenogenesis, an important problem in itself, becomes at the same time a method for the study of other and greater problems."

References

ANNEST, J. L. (1974). Nuclear fusion in an outcrossed parthenogenetic strain of *Drosophila mercatorum*. Master's thesis, University of Hawaii, Honolulu.

ANNEST, J. L. (1976). Genetic response in unisexual and bisexual laboratory populations of *Drosophila mercatorum*. Ph.D. dissertation, University of Hawaii, Honolulu.

ANNEST, J. L. and TEMPLETON, A. R. (1978). Genetic recombination and clonal selection in *Drosophila mercatorum*. *Genetics* **89**, 193–210.

ASHER, J. H., JR. (1970a). Parthenogenesis and genetic variability. Ph.D. dissertation, University of Michigan, Ann Arbor.

ASHER, J. H., JR. (1970b). Parthenogenesis and genetic variability. II. One-locus models for various diploid populations. *Genetics* **66**, 369–391.

CARSON, H. L. (1961). Rare parthenogenesis in *Drosophila robusta*. *Amer. Nat.* **95**, 81–86.

CARSON, H. L. (1962a). Fixed heterozygosity in a parthenogenetic species of *Drosophila*. *Univ. Texas Publ.* **6205**, 55–62.

CARSON, H. L. (1962b). Selection for parthenogenesis in *Drosophila mercatorum*. *Genetics* **47**, 946.

CARSON, H. L. (1967a). Selection for parthenogenesis in *Drosophila mercatorum*. *Genetics* **55**, 157–171.

CARSON, H. L. (1967b). Permanent heterozygosity. *Evol. Biol.* **1**, 143–168.

CARSON, H. L. (1973). The genetic system in parthenogenetic strains of *Drosophila mercatorum*. *Proc. Natl. Acad. Sci. USA* **70**, 1772–1774.

CARSON, H. L. (1975). The genetics of speciation at the diploid level. *Amer. Nat.* **109**, 83–92.

CARSON, H. L. and SNYDER, S. H. (1972). Screening by parthenogenesis for induced mutations in *Drosophila mercatorum*. *Egypt. J. Genet. Cytol.* **1**, 256–261.

CARSON, H. L., WHEELER, M. R. and HEED, W. B. (1957). A parthenogenetic strain of *Drosophila mangabeirai* Malogolowkin. *Univ. Texas Publ.* **5721**, 115–122.

CARSON, H. L., WEI, I. Y. and NIEDERKORN, J. A., JR. (1969). Isogenicity in parthenogenetic strains of *Drosophila mercatorum*. *Genetics* **63**, 619–628.

CARSON, H. L., TERAMOTO, L. T. and TEMPLETON, A. R. (1977). Behavioral differences among isogenic strains of *Drosophila mercatorum*. *Behav. Genet.* **7**, 189–197.

CLARK, R. L., TEMPLETON, A. R. and SING, C. F. (1981). Studies on enzyme polymorphisms in the Kamuela population of *D. mercatorum*. I. Estimation of the level of polymorphism. *Genetics* **98**, 597–611.

COUNCE, S. J. and RUDDLE, N. H. (1969). Strain differences in egg structure in *Drosophila hydei*. *Genetica* **40**, 324–338.

CUELLAR, O. (1977). Animal parthenogenesis. *Science* **197**, 837–843.

DAREVSKII, I. S., KUPRIYANOVA, L. A. and BAKRADZE, M. A. (1977). Residual bisexuality in parthenogenetic species of rock lizards of the genus *Lacerta*. *Zh. Obschch. Biol.* **38**, 772–780.

DOANE, W. W. (1960). Completion of meiosis in uninseminated eggs of *Drosophila melanogaster*. *Science* **132**, 677–678.

DOERR, C. A. (1967). Artificial selection for sexual isolation within a species. Master's Thesis, Washington University, St. Louis, Mo.

ESHEL, I. (1973). Clone selection and the evolution of modifying features. *Theor. Pop. Biol.* **4**, 196–208.

FUTCH, D. G. (1962). Hybridization tests within the *cardini* species group of the genus *Drosophila*. *Univ. Texas Publ.* **6205**, 539–554.

FUTCH, D. G. (1973). Parthenogenesis in Samoan *Drosophila ananassae* and *Drosophila pallidosa*. *Genetics* **74**, s86–s87.

FUTCH, D. G. (1979). Intra-ovum nuclear events proposed for parthenogenetic strains of *Drosophila pallidosa*. *Genetics* **91**, s36–s37.

GOLDSTEIN, P. and TRIANTAPHYLLOU, A. C. (1978). Occurrence of synaptonemal complexes and recombination nodules in a meiotic race of *Meloidogyne hapla* and their absence in a mitotic race. *Chromosoma* **68**, 91–100.

HAIGH, J. (1978). The accumulation of deleterious genes in a population—Muller's Ratchet. *Theor. Pop. Biol.* **14**, 251–267.

HELLER, R. and MAYNARD SMITH, J. (1979). Does Muller's ratchet work with selfing? *Genet. Res.* **32**, 289–293.

HENSLEE, E. D. (1966). Sexual isolation in a parthenogenetic strain of *Drosophila mercatorum*. *Amer. Nat.* **100**, 191–197.

IKEDA, H. and CARSON, H. L. (1973). Selection for mating reluctance in females of a diploid parthenogenetic strain of *Drosophila mercatorum*. *Genetics* **75**, 541–555.

LAMB, R. Y. and WILLEY, R. B. (1979). Are parthenogenetic and related bisexual insects equal in fertility. *Evolution* **33**, 771–774.

LEIGH, E. G. JR. (1970). Natural selection and mutability. *Amer. Nat.* **104**, 301–306.

LOKKI, J. and SAURA, A. (1980). Polyploidy in insect evolution. *In*: "Polyploidy: Biological Relevance", (W. Lewis, ed.), Plenum Press, New York.

MALOGOLOWKIN, C. (1958). Sobre a genitalia dos drosfilideos. V. A genitalia masculin em "D. mangabeirai" (Diptera, Drosophiladae). *Rev. Brasil. Biol.* **18**, 443–445.

MAYNARD SMITH, J. (1978). *The Evolution of Sex.* Cambridge University Press, Cambridge.

MAYR, E. (1954). Change of genetic environment and evolution. In *Evolution as a Process*, J. Huxley, ed., Allen and Unwin, London.

MULLER, H. J. (1964). The relation of recombination to mutational advance. *Mutat. Res.* **1**, 2–9.

MURDY, W. H. and CARSON, H. L. (1959). Parthenogenesis in *Drosophila mangabeirai* Malog. *Amer. Nat.* **93**, 355–363.

NUR, U. (1971). Parthenogenesis in Coccids (Homoptera). *Am. Zool.* **11**, 301–308.

NUR, U. (1972). Diploid arrhenotoky and automictic thelytoky in soft scale insects (Lecaniidae: Coccoidea: Homoptera). *Chromosoma* **39**, 381–402.

NUR, U. (1979). Gonoid thelytoky in soft scale insects (Coccidae: Homoptera). *Chromosoma* **72**, 89–104.

OCHMAN, H., STILLE, B., NIKLASSON, M., SELANDER, R. K. and TEMPLETON, A. R. (1980). Evolution of clonal diversity in the parthenogenetic fly *Lonchoptera dubia*. *Evolution* **34**, 539–547.

PATERSON, H. E. H. (1978). More evidence against speciation by reinforcement. *S. Afr. J. Sci.* **74**, 369–371.

PETRUNKEVITCH, A. (1905). Natural and artificial parthenogenesis. *Amer. Nat.* **39**, 65–76.

SEILER, J. (1947). Die Zytologie eines parthenogenetischen Rüsselkäfers, *Otiorrhynchus sulcatus* F. *Chromosoma* **3**, 88–109.

SEILER, J. (1959). Untersuchungen über die Entstehung der Parthenogenese bei *Solenobia triquetrella* F.R. (Lepidoptera, Psychidae). I. Die Zytologie der Bisexuellen *S. triquetrella*, ihr Verhalten und ihr Sexual verhältris. *Chromosoma* **10**, 73–114.

SING, C. F. and TEMPLETON, A. R. (1975). A search for the genetic unit of selection. In *Isozymes in Genetics and Evolution*, C. L. Markart, ed. Academic Press, New York and London.

SING, C. F., BREWER, G. J. and WESTOVER, C. J. (1976). Glycolytic intermediate analysis in *Drosophila*. *Genetics* **83**, S71–S72.

SPRACKLING, L. S. (1960). The chromosome complement of the developing eggs produced by *Drosophila parthenogenetica* Stalker virgin females. *Genetics* **45**, 243–256.

STALKER, H. D. (1951). Diploid parthenogenesis in the *cardini* species group of *Drosophila*. *Genetics* **36**, 577.

STALKER, H. D. (1952). Diploid and triploid parthenogenesis in the *Drosophila cardini* species group. *Genetics* **37**, 628–629.

STALKER, H. D. (1954). Parthenogenesis in *Drosophila*. *Genetics* **39**, 4–34.
STALKER, H. D. (1956a). Selection within a unisexual strain of *Drosophila*. *Genetics* **41**, 662.
STALKER, H. D. (1956b). On the evolution of parthenogenesis in *Lonchoptera* (Diptera). *Evolution* **10**, 345–359.
STALKER, H. D. (1956c). A case of polyploidy in Diptera. *Proc. Natl. Acad. Sci. USA* **42**, 194–199.
SUOMALAINEN, E. (1950). Parthenogenesis in animals. *Adv. Genet.* **3**, 193–253.
SUOMALAINEN, E. (1962). Significance of parthenogenesis in the evolution of insects. *Ann. Rev. Entomol.* **7**, 349–366.
SUOMALAINEN, E., SAURA, A. and LOKKI, J. (1976). Evolution of parthenogenetic insects. *Evol. Biol.* **9**, 209–257.
TEMPLETON, A. R. (1972). Statistical models of parthenogenesis. Ph.D. dissertation, University of Michigan, Ann Arbor.
TEMPLETON, A. R. (1974). Density dependent selection in parthenogenetic and self-mating populations. *Theor. Pop. Biol.* **5**, 229–250.
TEMPLETON, A. R. (1979a). The parthenogenetic capacities and genetic structures of sympatric populations of *Drosophila mercatorum* and *Drosophila hydei*. *Genetics* **92**, 1283–1293.
TEMPLETON, A. R. (1979b). The unit of selection in *Drosophila mercatorum*. II. Genetic revolutions and the origin of coadapted genomes. *Genetics* **92**, 1265–1282.
TEMPLETON, A. R. (1979c). Once again, why 300 species of Hawaiian *Drosophila*? *Evolution* **33**, 513–517.
TEMPLETON, A. R. (1980). The theory of speciation via the founder principle. *Genetics* **94**, 1011–1038.
TEMPLETON, A. R. and RANKIN, M. A. (1978). Genetic revolutions and control of insect populations. In: *The Screwworm Problem*, R. A. Richardson, ed., University of Texas Press, Austin, pp. 83–111.
TEMPLETON, A. R. and ROTHMAN, E. D. (1973). The population genetics of parthenogenetic strains of *Drosophila mercatorum*. I. One locus model and statistics. *Theor. Appl. Genet.* **43**, 204–212.
TEMPLETON, A. R. and ROTHMAN, E. D. (1978). Evolution and fine-grained environmental runs. In *Foundations and Applications of Decision theory*, Vol. II, 131–183. Hooker, C. A., Leach, J. J. and McClennen, E. F. eds. Reidel, Dordrecht, Holland.
TEMPLETON, A. R., CARSON, H. L. and SING, C. F. (1976a). The population genetics of parthenogenetic strains of *Drosophila mercatorum*. II. The capacity for parthenogenesis in a natural, bisexual population. *Genetics* **82**, 527–542.
TEMPLETON, A. R., SING, C. F. and BROKAW, B. (1976b). The unit of selection in *Drosophila mercatorum*. I. The interaction of selection and meiosis in parthenogenetic strains. *Genetics* **82**, 349–376.
WEI, I. (1968). Mode of inheritance and sexual behavior in the parthenogenetic strains of *Drosophila mercatorum*. Master's thesis, Washington University, St. Louis, Mo.
WHARTON, L. T. (1942). Analysis of the repleta group of *Drosophila*. *Univ. Texas Publ.* **4228**, 23–53.
WHARTON, L. T. (1944). Interspecific hybridization in the repleta group. *Univ. Texas Publ.* **4445**, 175–193.
WHITE, M. J. D. (1978). *Modes of Speciation*. Freeman and Co., San Francisco.
WILLIAMS, G. C. (1975). *Sex and Evolution*. Princeton University Press, New Jersey.
WOLFSON, M. (1958). A study of unisexual and bisexual strains of *Drosophila parthenogenetica* with particular attention to a new microsporidian infection. Master's thesis, Washington University, St. Louis, Mo.
WOODRUFF, R. C., THOMPSON, J. N., JR. and LYMAN, R. F. (1979). Intraspecific hybridization and the release of mutator activity. *Nature, Lond.* **278**, 277–279.

26. *Drosophila melanogaster* Models for the Control of Insect Pests

MALCOLM FITZ-EARLE

Department of Biology
Capilano College
North Vancouver, B.C., Canada

and

DAVID G. HOLM

Department of Zoology
The University of British Columbia
Vancouver, B.C., Canada

I. Introduction	399
II. Theoretical Models for Population Suppression and Manipulation	402
III. Population Suppression and Replacement by Translocation Lines	403
IV. Population Replacement by Compound Autosome Strains	405
A. Meiotic Behaviour	405
B. Laboratory and Field Studies	408
V. Population Replacement by Compound; Free-Arm Stocks	410
VI. Controlling or Eradicating Factors	413
VII. Other Possible Genetic Mechanisms for Insect Control	414
VIII. Genetic Control Methods Applied to Pest Insects Other Than *Drosophila melanogaster*	416
A. Chromosome Rearrangements and Meiotic Drive	416
B. Desirable Factors	418
Acknowledgements	419
References	419

I. Introduction

The publication of *The Genetics and Biology of Drosophila* in three extensive volumes is testimony to the fact that *Drosophila melanogaster* has been and continues to be one of the most important tools available for genetic research. It should not be overlooked, however, that *D. melanogaster* is also a significant pest of a wide range of crops. It principally serves as a vector of undesirable microorganisms which affect such crops as tomatoes in processing (Mason et al., 1968), figs and oranges in production (T. Prout, personal communication) and grapes in wine manufacture (R. D. McMullen, personal communication). Perhaps paradoxically, therefore, D. melanogaster has become in recent years a favoured organism for the development of pest control strategies. Theoretical, experimental and field work with this insect are providing sophisticated improvements to chemical approaches for insect control. Such improvements have become necessary because of the widely accepted disadvantages of chemical control, namely the evolution of resistance to insecticides and the often unpredictable consequences of releasing biologically active chemicals into the environment.

Many of the alternatives to chemical control techniques fall under the umbrella of "biological control" in which predators, pathogens and parasites are released against a target pest. For the most part, development of biological control strategies against a particular pest requires the identification of a corresponding biological agent. The success of biological control is dependent upon the availability and effectiveness of a specific external agent.

This paper is primarily concerned with the use of D. melanogaster as a tool for developing alternative techniques that are genetically based and whereby an insect pest is controlled by members of its own species. While several methods fall into this category, it is in the more genetically complex methods that D. melanogaster has been used to maximum advantage.

Within the realm of genetic insect control a limited amount of work with D. melanogaster has been done on the genetically straightforward, but so far the most practical, method, that of sterile insect release. The technique involves sterilizing large numbers of insects of a given species and then releasing them in appropriate numbers into a region where that same species is a pest. The idea is that the sterilized insects will mate with their fertile counterparts who would leave reduced numbers of offspring (for a review of this technique see for example Whitten and Foster, 1975; Waterhouse et al., 1975).

Chemicals or radiation are used to sterilize the insects prior to release. However, there are potential environmental risks associated with the chemosterilants and the weakening effects of irradiation. Additionally, because the sterilized insects are only effective for a single generation, continued releases may be necessary. Nevertheless the sterile insect technique is the only genetic control method with a record of demonstrated successes in practical situations. The pioneering programme was that for the control of the screwworm, *Cochliomyia hominivorax*, a major pest of livestock in the southern United States and Mexico (Bushland,

1971). The programme has been effective for almost 20 years with savings of millions of dollars in hides and meat. Other prominent sterile insect programmes are for the suppression of the codling moth, *Laspeyresia pomonella* (Proverbs et al., 1969), the Mediterranean fruitfly, *Ceratitis capitata* (Serghiou, 1975) and the boll weevil, *Anthonomus grandis* (Cross, 1973).

No doubt in part because of the significance of *D. melanogaster* as a pest of tomatoes and in part because of its convenience as an experimental organism, several studies were conducted on the effects upon it of gamma radiation (Henneberry, 1963; Henneberry and McGovern, 1963) and the chemosterilants apholate (Mason and Smith, 1967; Henneberry et al., 1967) and tepa (Mason et al., 1973). These investigations led to successful field trials in which releases of apholate-sterilized (Mason et al., 1968) and tepa-sterilized flies (Mason et al., 1975) were able to suppress native flies in tomato plots.

Denell (1973) has suggested a method of producing sterile males whereby it may be possible to circumvent the deleterious effects of conventional methods of sterilization applied immediately prior to release. He suggested a scheme whereby X-linked sterilizing mutations may be generated in P_1 males and captured using appropriate balancer X chromosomes in females. The F_2 males not carrying the balancer are then screened for reduced fertility. In an experiment in which males were treated with ethyl methane sulphonate, 4·7% of lethal-free X chromosomes carried an induced male-sterilizing mutation. Some of these mutations were also temperature-sensitive, a fact that led to the proposal that large numbers of insects could be raised at the non-sterilizing temperature, prior to release at the sterilizing temperature. Also suggested in the paper is a genetic method for the generation of large numbers of sterile males to the exclusion of females, that is a female-kill system. To our knowledge, the idea of generating male-sterilizing mutations has not been pursued in sterile insect release programs for species other than *D. melanogaster*.

Other techniques generally considered to be genetic such as cytoplasmic incompatability, hybrid sterility and inherited semi-sterility, have been generally applied to specific non-*Drosophila* pests and will not be considered further here (for brief reviews see Fitz-Earle, 1976; Whitten and Foster, 1975).

The majority of the studies which have used, and continue to use, *D. melanogaster* fall broadly into two categories: the application of genetic load and population manipulation. This chapter will consider these approaches from the theoretical, genetical/cytological and experimental viewpoints, as well as in field applications.

II. Theoretical Models for Population Suppression and Manipulation

Insects bearing certain types of chromosomal rearrangements can have two possible effects when released into a population of the same species of insects carrying standard chromosomes. One of the effects stems from the fact that hybrids between certain classes of rearrangements and standards have reduced fitness, a situation leading to a reduction in overall fitness of the composite population. In addition it can be shown that, after a release, the impact of the reduced fitness is felt over subsequent generations since the surviving progeny of each generation are identical to those forming the initial composite population. The reduced fitness each generation is termed the genetic load (see Wallace, 1970) and load itself is a useful method of suppressing a pest population. Three examples of such a genetic insect control technique are to be found in hybrid sterility, inherited semi-sterility and the release of semi-sterile translocation heterozygotes, as in the sterile insect method (Waterhouse *et al.*, 1975).

In addition to imposing a load upon a pest population the release of a strain carrying rearranged chromosomes in homozygous condition can lead theoretically to the complete replacement of an entire population bearing standard chromosomes. The released rearrangements could provide a mechanism for the introduction of factors for control or eradication. The theoretical basis for insect control by population replacement was suggested initially by Serebrovsky (1940) in a paper published in Russian. His ideas were not recognized widely until C. F. Curtis translated the paper into English (Serebrovsky, 1969). Essentially the same ideas were recognized independently by Curtis (1968). Subsequently the theoretical basis for the Serebrovsky–Curtis concepts was expanded by Whitten (1971). In brief the procedure begins with the same situation as with the genetic load, namely that when insects homozygous for chromosomal rearrangements are mated to those carrying standard chromosomes, there may be a degree of reduction in viability of the hybrid offspring compared either to the rearrangements or to the standards, which is often referred to as negative heterosis. Consequently, when the overall fitness of the homozygote exceeds that of the heterozygote, the proportion of the two strains can be adjusted according to their relative viabilities, such that the combined population will be in a state of unstable genetic equilibrium (Li, 1955). If the proportion of the rearrangements exceeds the unstable equilibrium value, the population will be driven to fixation in favour of the rearranged strain. Conversely, if there is a deficit of rearrangements then the standards will be driven to fixation. Such a situation is often termed frequency-dependent selection. Thus, in a pest management programme, the above principle could be applied for the replacement of a target pest population carrying standard chromosomes by insects of the same species raised in the laboratory but carrying rearranged chromosomes. The number of released insects would be a function of the relative fitnesses of the standards and rearrangements, and in turn the rate of replacement would reflect the initial release ratio.

A detailed mathematical treatment of negative heterosis and its application to pest control has been developed by Prout (1977). Curtis and Hill (1971) have explored, through computer simulations, the particular case of population replacement for translocation strains, and a similar analysis for compound autosome and other strains has been developed (Barclay and Fitz-Earle, 1983).

III. Population Suppression and Replacement by Translocation Lines

Translocations, the chromosomal aberrations first to be examined in connection with insect regulation, when present as heterozygotes place a severe genetic load on the population and as homozygotes provide a system of chromosome replacement that serves as a vector for establishing within the population certain mutations more desirable to man. Focusing attention, in this section, on studies primarily restricted to *D. melanogaster*, it is important to review some features of meiotic and other properties of translocations that will influence the effectiveness of chromosome replacement and insect regulation. Since chromosome replacement involves frequency-dependent selection, this discussion is limited to reciprocal translocations of the autosomes and mainly chromosomes 2 and 3, the metacentric autosomes.

Translocation heterozygotes are semi-sterile as a function of producing two types of gametes: orthoploid gametes, arising from alternate disjunction, and aneuploid gametes, arising from adjacent disjunction (see Zimmering, 1976). In males, where there is no crossing over, it appears that both types are produced in approximately equal proportions (Robinson and Curtis, 1973; also see review by Roberts, 1976). In females, however, the ratio of orthoploid to aneuploid gametes is dependent on the position of the breakpoints (Zimmering, 1955, 1976; Roberts, 1976) and the effect translocations have upon the frequency of crossing over (Roberts, 1970). The average viability of translocation heterozygous females ranges from approximately 50% to 70% (Robinson and Curtis, 1973; Roberts, 1976). Assuming that, on average, translocation heterozygotes produce 50% aneuploid gametes, then a mixed population with translocation and standard chromosomes at equilibrium will have a reproductive capacity that is reduced by 42% (Serebrovsky, 1940; Curtis and Hill, 1971). Increased negative heterosis has also been demonstrated in a double translocation where fertility in the female is significantly lower than in the male (Robinson and Curtis, 1972). This reduction has been attributed to crossing over in the differential segments of the translocations thereby producing extra unbalanced gametes. It should also be noted that lethality, as a function of (segmental) aneuploidy, does not occur in the gametes (Muller and Settles, 1927; McCloskey, 1966; Lindsley and Grell, 1969) but at some postzygotic stage of development.

While frequency-dependent displacement of one chromosome type by another offers an attractive model, putting this model into practice is limited by the availability of suitable translocations that are homozygous viable. Unfortunately, the majority of reciprocal translocations between chromosomes 2 and 3 in *D. melanogaster* either have greatly reduced viability or are lethal as homozygotes. Lethality and reduced viability have been attributed to mutations induced at the site of interchange, position effects that influence the expression of genes in the vicinity of the breakpoint, or independent mutations induced on the translocated chromosomes at regions other than the points of breakage. While position effect has been offered as a more likely explanation (Patterson *et al.*, 1934; Ives and Fink, 1962), Ytterborn (1970) has found that the degree of homozygous lethality associated with radiation-induced translocations increases as a function of radiation dose.

Studies on homozygous viable 2;3 translocations in *D. melanogaster* have demonstrated that fitness measured under noncompetitive conditions does not necessarily indicate the success of frequency-dependent displacement. Robinson and Curtis (1973) used a homozygous viable, reciprocal 2;3 translocation with a recessive visible position effect. Under control (noncompetitive) conditions the translocation homozygote was found to be as fertile and as viable as the wild-type strain used in their study. Moreover, the overall fitness of the homozygote was far superior to that of the translocation heterozygote. Nevertheless, in a mixed population of translocation homozygotes and wild types at an initial ratio of 9:1, respectively, the population went to fixation within 9 generations in favour of the wild-type (standard) chromosomes. While the control crosses indicated that frequency-dependent selection would occur in favour of the translocations where it was in excess, the authors concluded that in a mixed population the viability of the homozygous translocation was reduced to a level below that of the semi-sterile heterozygote.

Competition studies between different translocation strains have given similar results. Two different translocation homozygotes were used to establish two separate populations, one at an initial ratio of 1:4, the other at 4:1 (Hossain *et al.*, 1974). However, in both populations a stable polymorphism was reached after ten generations. In a second experiment, two different homozygous 2;3 translocations were placed in a cage in equal proportions (Robinson, 1977). After 25 generations the frequency of the heterozygous double translocations was greater than 59%. Therefore, although translocation heterozygotes were theoretically at a disadvantage owing to reduced fertility, in a mixed population their fitness far exceeded that of either homozygote. Consequently, in all the above experiments, natural selection favoured the heterozygote and unstable equilibrium failed to be realized.

In a study by Reid and Wehrhahn (1976), 57 marker-free reciprocal 2;3 translocations were recovered in *D. melanogaster*. Twenty of these translocations were homozygous viable. By selecting translocations showing high fitness as homozygotes and high negative heterosis, they found two translocation strains that revealed potential replacement of wild-type when cages were established at initial ratios of 9:1 in favour of the translocation homozygotes. While their experiments were terminated after one month, and prior to completion, their findings at least offered some encouragement for further studies. As suggested by Robinson (1976), and as indicated by the last-mentioned study, translocations differ markedly and selection for those that express fitness sufficient to create negative heterosis is essential if replacement is to succeed.

Even in the absence of chromosome replacement the use of translocation heterozygotes is an effective means of depressing the population fertility. Moreover, by increasing the number of translocations, the genetic load in the population can reach a level where the population could be eliminated (Serebrovsky, 1940; Curtis, 1968; Curtis and Robinson, 1971). However, as pointed out by Curtis (1968) it is probably preferable to replace the standard population with a homozygous translocation bearing desirable genetic factors than to leave a habitat vacant for reinfestation by immigrants. It is also of interest to note that even if frequency-dependent displacement can be effected, the population will pass through a prolonged phase of depressed fertility, but will rise to normal as the population becomes fixed in favour of one or the other homozygote (Curtis and Hill, 1971).

IV. Population Replacement by Compound Autosome Strains

A. Meiotic Behaviour

Complete expression of unstable equilibrium through negative heterosis is best exemplified in mixed populations composed of compound autosome and standard strains. First generated in the laboratory of E. B. Lewis (Rasmussen, 1960), compound autosomes, like the commonly used reverse metacentric compound X in *D. melanogaster* (Lindsley and Grell, 1968), are chromosomes that carry two homologous arms attached to a common centromere. Compound autosomes (hereinafter referred to as "compounds") represent a special class of translocations that arise as interchanges between breaks either on opposite sides of homologous centromeres or on opposite sides of the centromere of sister chromatids (Bateman, 1968; Leigh and Sobels, 1970; Holm, 1976). Interchanges giving rise to compounds appear to be limited to breaks induced in the proximal heterochromatin (Gibson, 1977).

The repair of these breaks, through the joining of an acentric arm to an homologous centric free arm, results in a chromosome that is heterozygous for a proximal deficiency and which carries a duplication for the proximal region of the opposite arm. For many of the compounds examined, however, these duplications and deficiencies do not include essential genes (Hilliker and Holm, 1975; Holm, 1976).

A complement of compounds, i.e. *C(2L);C(2R)* or *C(3L);C(3R)*, represents a pair of metacentric heterologues that replace a pair of metacentric homologues (otherwise referred to as standards as compared in Fig. 1). While for most pairs of compounds there is little or no change in the full complement of genes, there is a distinct change in the events governing

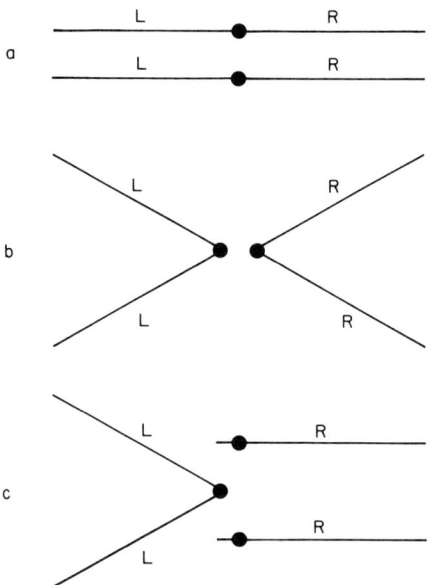

FIG. 1. (a) Unrearranged chromosome—"standard". (b) Compound left and compound right—"compound". (c) Compound left and free-arm right—"free arm".

the meiotic distribution of these chromosomes. Normally, their distribution in females is operationally analogous to homologues in that the majority of the gametes carry only one of the complementary pair. However, when the female genome carries (1) a free Y chromosome, or (2) a compound X, or (3) inversion heterozygosity in chromosomes other than the compounds, a large proportion (5 to 40%) of the meiotic products recovered are either disomic or nullosomic for the rearrangement (Baldwin and Chovnick, 1967; Holm et al., 1967; Grell, 1970; Holm and Chovnick, 1975; Holm, 1976).

In distinct contrast, most pairs of compounds assort randomly during spermatogenesis. Evidence for this comes from the findings that: (1) all four classes of sperm, compound left, compound right, disomic and nullosomic, are viable (McCloskey, 1966; Lindsley and Grell, 1969); (2) aneuploids for compounds are embryonic lethals (Scriba, 1967, 1969); and (3) the frequency of egg hatch for most compound strains approaches 25% (Scriba, 1967, 1969; Lutolf and Würgler, 1970; Lutolf, 1972; Holm and Chovnick, 1975; Holm, 1976). As illustrated by the Punnett Square in Fig. 2, since all aneuploid combinations of compounds are lethal, random assortment in males leads to no greater than 25% recovery of hatched eggs.

From crosses between standard and compound strains (Fig. 3), virtually all progeny produced are aneuploids in which lethality is expressed either in the embryo or in the first instar larval stage (Lutolf and Würgler, 1970; Lutolf, 1972). With the rare exception of progeny arising as products of nondisjunction in structurally normal standard strains (Gavin and Holm,

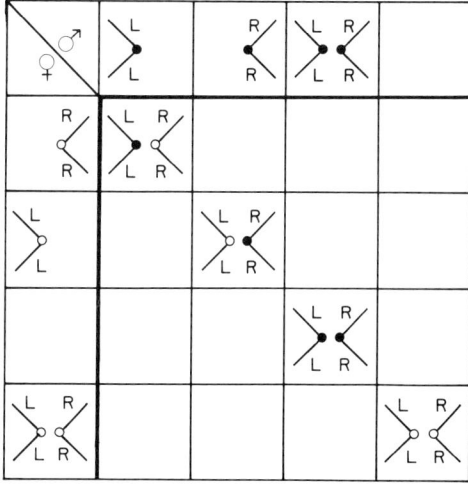

FIG. 2. The meiotic behaviour of a compound line. Only the surviving zygotes are depicted. The viability of a compound strain is approximately 25% that of standards.

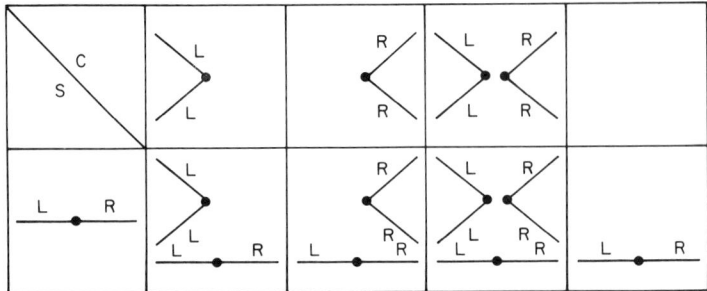

FIG. 3. The cross between a compound strain (C) and a standard line (S). All hybrids are lethal aneuploids.

1972; Gibson, 1977), spontaneous compound formation (Chadov, 1970; Gibson, 1977) or compound detachment (Hilliker and Holm, 1975; Chadov and Chadova, 1980), the optimum condition, complete hybrid lethality, has been realized for establishing an unstable equilibrium between two chromosomally different strains. Therefore, in a mixed population, standard and compound strains remain genetically isolated, not by any apparent mating barrier, but by a reproductive barrier. If the definition of fitness is restricted to the percentage of eggs that hatch, then the fitness of compound strains relative to standard (wild-type) strains is approximately 25% (Fitz-Earle et al., 1973; Holm and Chovnick, 1975). Consequently, in theory an unstable equilibrium will be established at a population ratio of four compounds to one standard. A shift to either side of this unstable equilibrium ratio will lead eventually to the fixation within the population of that chromosome complement which exceeds the equilibrium value.

B. Laboratory and Field Studies

The principle of population replacement of standards by compounds has been demonstrated for *D. melanogaster* under various conditions in the laboratory and in the field.

Compounds have been shown to displace standards in bottled cultures (Foster et al., 1972) and in population cages (Childress, 1972; McKenzie, 1976). However these experiments used discrete non-overlapping generations. To simulate more realistically actual field populations, generations should be continuous or overlapping. Indeed experiments which used this latter regime revealed effective displacement in favour of compounds (Fitz-Earle et al., 1973; Cantelo and Childress, 1975). Cage experiments with continuous generations showed for several different laboratory strains of compounds and standards that compounds could become fixed when the initial ratio of compounds to standards was at the theoretical unstable equilibrium value of 4:1 (Fitz-Earle, 1975). In no experiments initiated at lower ratios (less than 4:1) were the compounds ever fixed (Fitz-Earle and Holm, 1978).

In an attempt to duplicate more closely possible field situations, several compound strains were synthesized from native material. These native compounds were also successful in replacing laboratory standards in cages in the laboratory at initial ratios down to a minimum of 4:1 compounds to standards (Fitz-Earle *et al.*, 1975). However, when the same native compounds were established in cages with native standards, the minimum initial ratio for replacement was slightly higher at 5:1 (Fitz-Earle, 1975). By contrast, in a study by Cantelo and Childress (1975), laboratory compounds were unable to replace native standards in cages even when the release ratios were as high as 30:1. These experiments in the laboratory clearly suggest that to have the best chance of success in any control programme, compounds must be synthesized from material derived from recently caught members of the wild target population.

Field cage experiments have been conducted by several groups using a variety of environmental conditions. Native compounds were competed against laboratory standards in British Columbia, Canada, under two completely different sets of environmental conditions: temperate rain forest and semi-desert. For both environments displacement of standards was achieved in field cages established at ratios of 5:1 and higher (Fitz-Earle *et al.*, 1975). In another study, laboratory compounds and native standards were tested at an initial ratio of 100:1 in field cages in Maryland, U.S.A., but the compounds disappeared and only the standards survived (Cantelo and Childress, 1975). The failure of this trial was attributed, in part, to the inability of the released laboratory compounds to oviposit on the tomato diet provided in the cage. A third study involved the release of essentially laboratory compounds into a wine cellar (almost equivalent to a field cage) in Victoria, Australia in which there was already a population of native *D. melanogaster* (McKenzie, 1976). The compounds were neither able to establish themselves in the cellar, nor were they able to displace the standards. The demise of the released compounds was considered to be due to the competition by native immigrants to the cellar during the experiment (McKenzie, 1976). However, other possible causes for the disappearance of the compounds could have been competition from the native residents of the cellar or simply the inability of the released laboratory compounds to tolerate the environmental conditions in the wine cellar.

A few attempts have been made to test the principle of population replacement of standards by compounds under open field conditions. On-going cultures containing adults and all developing stages of laboratory-derived compounds of *D. melanogaster* were released into *Drosophila*-free piles of well-ripened fruit of the kinds known to act as hosts for fruitflies. The goal of the releases was to establish populations of compounds sufficiently numerous to resist invasions by standards, both native migrants and those introduced by man (Fitz-Earle *et al.*, 1975). However in none of the fruit piles was a colony of released laboratory compounds established successfully. The failure of these field trials was attributed to such factors as alteration of diet and dessication leading to field mortality, dispersion from the fruit piles and interference of the fruit piles by scavengers such as black bears, *Ursus americana* (Fitz-Earle *et al.*, 1975). In another study, laboratory compounds were released into open fields of tomatoes at two different sites such that the initial ratios of compounds to standards were estimated by trapping to be about 62:1 at one site and 91:1 at another site (Cantelo and Childress, 1975). The laboratory compounds were unable to displace the native standards under these field conditions, again most probably because of the inability of released flies to utilize tomatoes as an oviposition site.

For the most part the impact of ecological factors such as immigration, dispersion, competition and mortality from predators, lack of food utilization or other causes has not been quantified. A laboratory study of the effect of immigration on genetic control using compounds over discrete generations (McKenzie, 1977) has shown that immigration rates of less than 10% per generation by unmated migrants would have little impact. However, immigration rates of 0.5% could disrupt the control zone if the immigrants were inseminated females.

V. Population Replacement by Compound; Free-arm Stocks

The major disadvantage of compound autosome lines in *D. melanogaster* is their low egg hatch, which is a consequence of the random assortment of the rearrangements during male meiosis (see above). The hatch values are generally about 25% that of standards though there are some instances of hatch values as low as 19% and as high as 41% (Scriba, 1967, 1969; Evans, 1971; Fitz-Earle *et al.*, 1973; Clark and Sobels, 1973; Holm and Chovnick, 1975; Holm, 1976). The survival to adults is also further reduced to $18-22\%$ for compounds, as compared to $80-86\%$ for standards (Fitz-Earle *et al.*, 1973). However, it is to be noted that compound survivors appear to be as competitive in their mating as standards.

One type of chromosome rearrangement that to some extent circumvents the problem of reduced survival, yet retains all other properties necessary for population replacement, is the compound; free-arm combination (hereinafter termed "free arms"). Free-arm strains are characterized by having one pair of homologous arms attached as in a compound but the other pair of homologous arms unattached (Fig. 1 compares a compound left; free-arm right arrangement with a compound autosome and a standard). Free-arm lines were initially synthesized in *D. melanogaster* using the technique described in Grell (1970). Many free-arm lines of *D. melanogaster* have been generated subsequently, using laboratory and native material, with a view to testing their potential for population replacement (Fitz-Earle and Holm, 1976, 1978).

Genetic tests with a free-arm strain have indicated a 49% egg hatch (Fitz-Earle and Holm, 1976), as compared to 90–95% for standard lines (Fitz-Earle *et al.*, 1973; Holm and Chovnick, 1975; Holm, 1976). These findings are consistent with the expected meiotic behaviour of free-arm strains (as shown in Fig. 4), namely that the free arms disjoin regularly at metaphase I of meiosis whereas the compound assorts independently of them (Grell, 1970). Survival to adults from the same free-arm line was about 36% (Fitz-Earle and Holm, 1976) as contrasted to 80–86% for standard strains (Fitz-Earle *et al.*, 1973). The theoretical unstable equilibrium ratio between free arms and standards is in the range of 2·2–2·4:1 (based upon survival to adult values) as compared to the considerably higher ratios for compounds and standards of 3·8–5·0:1.

As expected, crosses of free-arm strains to standards (Fig. 5) fail to produce any viable hybrids, thus supporting the contention that the two chromosomal types are completely genetically isolated (Fitz-Earle and Holm, 1976). Unlike matings between compound and standard strains in which most hybrids fail to hatch, and those that do, die as first instar larvae, some of the hybrids from crosses between free-arm and standard strains

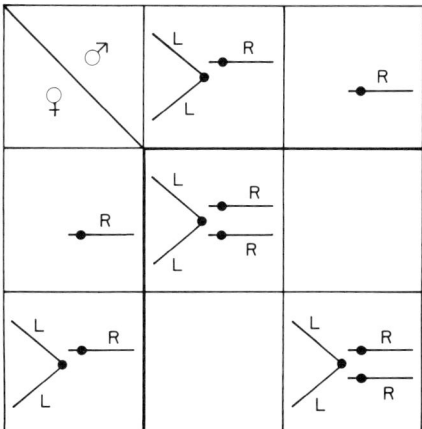

FIG. 4. The meiotic behaviour of a free-arm line. The viability of a free-arm strain is approximately 50% that of standards.

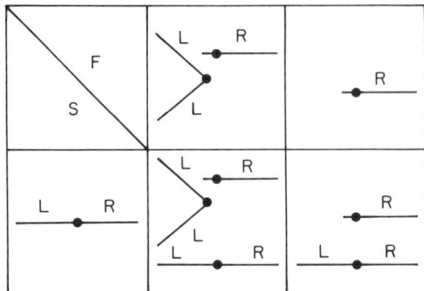

FIG. 5. The cross between a free-arm strain (F) and a standard line (S). All zygotes are aneuploid and die by the pupal stage.

survive until the late pupal stage before dying. These late-surviving progeny have been shown cytologically to be trisomic for a whole arm, that is their genomes contained a normal chromosome 2, a compound 2L and a free 2R, as well as their other complement of chromosomes. Such a situation had not been observed previously in *D. melanogaster* (Fitz-Earle and Holm, 1976). From the insect control perspective, the presence of trisomic hybrids unable to survive past the pupal stage suggested the potential for competition for food and space by a type of progeny that could not contribute to the adult population. That is, the hybrid larvae might serve to suppress the population by competing with free arm and standard larvae for available resources. However this system would offer no advantage in pests destructive at the larval stage.

Laboratory and native free-arm lines, when competed against laboratory or native standards at initial ratios to a minimum of 3:1, have gone to fixation in relatively short periods of time (Fitz-Earle and Holm, 1976 and 1978). These population cage experiments all suggested that free arms may provide a more efficient genetic test than compounds for population replacement of pest insects. However, one cage in which laboratory free arms and native standards were introduced at the ratio of 4:1 gave rather unexpected results that tended to cloud the optimism regarding free-arm strains. After 92 days the mixed population was still apparently fluctuating about the equilibrium frequency for free arms, a phenomenon that had not been observed in over 160 previous cage competitions. Subsequent analysis of flies from this cage revealed the existence of a strain in which the second chromosomes consisted of a free 2R and a standard second, on the right arm of which was attached a duplication for all of 2L (designated *Dp(2L;2)*) (Holm *et al.*, 1980). Though the origin of the *Dp(2L;2)* chromosome within the cage was unclear, genetic and cytological examinations revealed that it represented the reciprocal crossover product expected from an exchange such as that used to generate a $F(2R)$ (see Grell, 1970). Once generated, the $Dp(2L;2)/F(2R)$ individuals could mate with standards and, through recombination, produce heterozygous acrocentric (pericentric inversion) offspring *(In(2LR)A)* as well as additional standards. The $In(2LR)A$ heterozygous offspring could in turn mate with individuals from the free-arm strain to produce, through recombination, further $Dp(2L;2)/F(2R)$ progeny as well as free-arm progeny. Thus the duplication and inversion individuals were serving as a bridge between otherwise genetically isolated free-arm and standard strains; that is, producing a situation of unstable genetic isolation. This cage experiment suggested that if such inversion strains were present in the wild they would influence the success of the free-arm strategy for population replacement and hence insect control.

VI. Controlling or Eradicating Factors

Population replacement does not in itself constitute a means of population control, but it does provide a mechanism for introducing desirable factors into native populations. Since genetic factors that eradicate may create ecological imbalances, those that control would be preferable. For example, it might be advantageous to introduce a gene that renders a population innocuous to man, yet does not reduce its size. One method of control might be the introduction of a gene for refractoriness to a disease into a population that currently acts as a vector of that disease. If insects carried a compound autosome or other rearrangement, included in which was a gene for refractoriness, the degree of infectivity of the native population could be gradually reduced, while at the same time the numbers of insects would remain essentially constant.

Another possible pest management goal would be for the complete eradication of a pest insect of livestock, crops or man. In this circumstance the introduction of a lethal gene ("kill system") would reduce the population number and would be both ecologically and economically appropriate. An example would be the introduction of a conditional lethal gene, that permits survival under one set of conditions but causes death under another. A simple example of a conditional lethal might be a temperature-sensitive lethal (ts lethal) which allows for survival at one temperature (the permissive) but causes death at another (higher or lower) temperature (the restrictive). The approach would be to introduce insects, carrying chromosomal rearrangements coupled with a heat-sensitive ts lethal mutation, into the native population at the permissive temperature, e.g. in the springtime. Population replacement would proceed, also at the permissive temperature. Then, following complete replacement, but when the ambient temperatures had increased to the restrictive temperatures (e.g., in the summertime) for the heat-sensitive *ts* lethal mutation, death would begin to occur and subsequently the whole population would be eliminated. This approach to eradication, however, is limited to those species that have many generations per year. Temperature-sensitive lethal mutations have been studied extensively in *D. melanogaster* (for a recent review see Suzuki *et al.*, 1976). Strains of *D. melanogaster* have been generated such that an increase in temperature to the restrictive level can cause death at any stage of the life cycle. The principles of population replacement, followed by eradication with a *ts* lethal mutation have been demonstrated successfully with *D. melanogaster* under laboratory conditions in cages (Fitz-Earle and Suzuki, 1975).

Temperature-sensitivity is not the only form of conditional lethality that has been considered for population eradication. Where appropriate, non-diapausing mutants (Klassen *et al.*, 1970), non-feeding mutants, and behavioural traits affecting feeding (Smith, 1971) can all be considered as potentially useful eradicating factors. In any event the underlying principle remains the same: a desirable (controlling or eradicating) factor is combined with a chromosome rearrangement and a release is made under appropriate conditions and with appropriate numbers to effect replacement. It is important, however, that the desirable factor remains linked to the replacement system. For this reason, compound autosome strains, which are almost totally reproductively isolated from standard strains, provide a considerable advantage over translocation homozygotes which are not reproductively isolated from standards. Loss of the desirable factor from a translocation could be reduced significantly if the factor were tightly linked to the point of interchange. Only through the production of triploid female hybrids would a desirable mutation be lost from a compound autosome.

VII. Other Possible Genetic Mechanisms for Insect Control

Three other types of genetic procedures, which have been described in the *D. melanogaster* literature and which provide possible models for insect control application, are entire compound autosomes, meiotic drive and pericentric inversions.

Entire compound autosomes are rearrangements involving the translocation of, for example, free 2Rs to the distal tips of a compound 2L, that is *2R2L·2L2R*, and are designated as *C(2)EN*. The construction of such a strain has been described by Novitski (1976). Intuitively, it might be expected that the hatch within a strain carrying compound entire would be about 50% and the survival to adults somewhat less, as was found for free-arm stocks (Fitz-Earle and Holm, 1976, 1978). In fact for the entire compound strain marked with $c\ bw$ (i.e., $C(2)EN\ c\ bw$) (for a description of these and other markers see Lindsley and Grell, 1968) it has been shown that the hatch is 32·7% and the survival to adults 28·9% (Gethman, 1976), values that are only marginally superior to compound autosomes and significantly less than free arms. The efficacy of entire compound autosomes in population replacement and hence control or elimination has not been tested to this date (E. Novitski, personal communication).

Meiotic drive, in which a particular chromosome is transmitted preferentially (i.e., more than 50% of the time) to progeny, has been suggested as a possible transporting mechanism for desirable characteristics. Meiotic drive has been studied extensively in $D.\ melanogaster$ using segregation-distorter (SD) stocks (for a review see Hartl and Hiraizumi, 1976). The generally accepted basis for preferential recovery of the SD chromosome is that, unlike normal spermatogenesis in which there is equal recovery of the four meiotic products, in SD heterozygotes the SD causes degeneration of non-SD spermatids (Peacock et al., 1972). The suggestion has been made that because of high spermatid degeneracy in homozygotes (SD/SD), there would be sterility levels that could reduce the value of meiotic drive as a transport system (Foster and Whitten, 1974). Indeed Hartl (1973) has shown that various combinations of different SD chromosomes had reduced fertility (0–50% of controls). However, by coupling SD to a Y chromosome, through the generation of $Y;2$ translocations, the SD chromosome will be carried almost exclusively by heterozygous males. Upon analysing a series of newly generated $T(Y;2)SD$ chromosomes, Lyttle (1977) demonstrated that this pseudo-Y drive system could successfully lead to population extinction owing to the selective production of males. The stated advantage of this pseudo-Y meiotic drive over that of using compound autosomes for population replacement was the small initial release required to effect control. However the success of such a system depends to a large extent on target populations that are not already segregating for significant numbers of background suppressors of drive strength. Even in populations initially without detectable modifiers, after several generations polygenic suppressor variability was generated, sufficient to significantly reduce the strength of drive (Lyttle, 1979). Nevertheless, such an approach to population elimination appears to have some promise.

Ideally, of course, it would be desirable to have a drive directly linked to the Y chromosome, thereby eliminating the problem of reduced viability associated with translocations. Attempts to generate meiotic drive in $D.\ melanogaster$ led to the recovery of at least one meiotic drive second chromosome and one drive linked to the X chromosome (Novitski and Hanks, 1961). Both were moderate in effect, and both acted exclusively in males.

Roberts (1967) has shown in *D. melanogaster* that pericentric inversions (centromere included in inversion) when present in heterozygous condition in females can reduce egg hatch provided crossing over is allowed to occur. The studies showed that a relatively small inversion encompassing the markers *th* and *cu* on chromosome 3 caused little reduction in fertility compared to a paracentric inversion (centromere not included in inversion) control. By contrast, females heterozygous for another pericentric inversion covering the region of 3L proximal to the centromere through *ro* in distal 3R, in which there was considerable opportunity for crossing over, had fertility reduced to 50% that of the control. Such findings suggest the possibility for insect control by releasing homozygous pericentric inversion strains as a means of producing pericentric inversion heterozygotes which would be at a selective disadvantage. The system could be similar to other chromosomal rearrangements in that it could simply serve to place a load on the pest population or it could act as a transport system for controlling factors.

VIII. Genetic Control Methods Applied to Pest Insects Other Than *Drosophila melanogaster*

Several groups have been developing the kinds of genetic tools discussed in this chapter in pests other than *D. melanogaster*. The following is not intended as a review of the literature on genetic control of these pests but as a guide to the current applications of the models developed in *D. melanogaster* to selected pests.

A. Chromosome Rearrangements and Meiotic Drive

Several groups have been generating, in a variety of insects, reciprocal translocations with a view to applying them as either a load-inducing system or as a means for introducing desirable factors.

In part owing to its significance as a vector of African sleeping sickness, a form of trypanosomiasis, much attention has been given to generation of translocation lines in the tsetse fly *Glossina austeni* by Curtis and his co-workers (for a review see Curtis *et al.*, 1972; see also Robinson, 1976). Several other disease vectors have been studied extensively with a view to controlling them with translocations. They include the following mosquitos: *Culex tarsalis*, the vector of equine encephalitis (McDonald *et al.*, 1978); *Aedes aegypti*, the yellow-fever mosquito (for a review see Rai *et al.*, 1974; see also Uppal *et al.*, 1978); *Culex tritaeniorhynchus*, the main vector of Japanese encephalitis (Baker *et al.*, 1977); *C. pipiens fatigans*, the northern house mosquito, in some parts of the tropics the vector of the nematode

Wuchereria bancrofti, the causative agent of Bancroftian filiariasis (Curtis, 1976) and the malarial mosquitoes *Anopheles albimanus* (Rabbani and Kitzmiller, 1972) and *A. gambiae* (Akiyama, 1973). Translocations have also been generated in the housefly, *Musca domestica* (for a review see Wagoner *et al.*, 1974) and in the German cockroach, *Blattella germanica* (Ross and Cochran, 1973). Two examples of the field applications of translocations in disease vectors are the experiments of Laven *et al.* (1971) with *C. pipiens* in France and the trials of Grover *et al.* (1976) using *Aedes aegypti* in India.

Reciprocal translocations have also been synthesized in pests of agricultural and veterinary importance. Chromosome mutations genetically resembling translocations have been isolated by fertility reduction in the two-spotted spider mite, *Tetranychus urticae* (Feldmann, 1975). Likewise, translocations have been generated in the onion fly, *Hylemya antiqua*, using the semi-sterility method (Wijnands-Stäb and van Heemert, 1974; van Heemert and Wijnands-Stäb, 1975; Reid and McEwen, 1977). Using field cage populations of *H. antiqua*, Reid and McEwen (1979) have demonstrated that the presence of translocation individuals could lower the reproductive potential of the populations and that they continued their effect into the third generation. An attempt at generating compound autosomes in *H. antiqua* has been made using heterozygotes for a pericentric inversion (van Heemert, 1977). This procedure departs significantly from the reciprocal translocation technique used for *D. melanogaster* as described in Holm (1976), in that the pseudo-compound chromosomes that are generated share a considerable homology in their centromeric regions. Such pseudo-compounds have not yet been recovered in adults (van Heemert, 1977). However the author has suggested that the system would provide a source of material for a sterile male release program. Studies of all aspects of the biology of the Australian sheep blowfly, *Lucilia cuprina*, the most serious ectoparasite of sheep in that country, have been conducted over many years. The lack of adequate control of the pest by chemical or biological methods, however, stimulated research into genetic techniques (Foster and Whitten, 1974; Whitten *et al.*, 1975). One of the results of these investigations was the synthesis of a compound autosome line in this species (Foster *et al.*, 1976). The compound autosome strain resembled *D. melanogaster* compound strains in being genetically isolated from standards, but its hatch was significantly lower than that found in *D. melanogaster* (11% compared to an average of 25%) (Foster *et al.*, 1976). Other compound lines are being generated with a view to enhancing the fitness (G. G. Foster, personal communication). To date small-scale releases of the compound strain of *L. cuprina* as larvae have been made in a valley approximately 300 square kilometres in area. These preliminary trials suggested that the released insects were capable of surviving and mating in the wild and of competing well with native standards (Whitten et al., 1976; Foster et al., 1978).

Studies on meiotic drive and sex ratio distortion have provided some promise for the possible genetic control of the yellow fever mosquito, *Aedes aegypti*. The Y-linked sex ratio distorting gene M^D in *A. aegypti* causes X chromosome breakage during male meiosis and depleted sperm production (Wood and Newton, 1976). Some X chromosomes are more sensitive to the effects of M^D than others, a situation that is reflected in the geographic distribution of M^D throughout the world. The M^D system as a transporting mechanism for a linked marker has been demonstrated in a laboratory cage experiment (Wood et al., 1977). In addition, M^D in association with a double translocation has proved to be extremely effective for suppressing a population of *A. aegypti* in a large field cage (Curtis et al., 1976), in spite of the fact that in the native population used in the trial there was a moderate level (20%) of X chromosome resistance to M^D (Suguna et al., 1977). It was concluded that a low level of resistance to meiotic drive may not necessarily limit the success of this technique. In another study, attempts were made to detect meiotic drive in the Australian sheep blowfly, *Lucilia cuprina*. Irradiation treatment was delivered over 84 generations but no case of meiotic drive was detected in population cages (Foster and Whitten, 1974).

B. Desirable Factors

We would finally like to discuss a number of genetic factors which are desirable because they may be of value in control programs. Temperature-sensitive conditional lethal genes suitable for eradication of pests have been generated in a number of insects such as the wasp, *Habrobracon serinopae* (Smith, 1971), the housefly, *M. domestica* (McDonald and Overland, 1972) and the mosquito, *C. tritaeniorhynchus* (Sakai and Baker, 1974). In addition, a translocation homozygote in the housefly was also shown to be heat sensitive (McDonald and Overland, 1973). In *L. cuprina* it has been shown that at least two autosomal eye colour mutants (white-eye and topaz-eye) act as conditional lethals in that they survive well under laboratory conditions but have very low viability in the field. Such mutants have been combined with Y-autosome translocations associated with dieldrin-resistance as a means of producing a female kill system. Thus, native populations could be "swamped" with genetically sterile males as a control measure (Whitten et al., 1976). A potentially important example of nondiapause has been found in the boll weevil, *Anthonomus grandis* (McCoy et al., 1968). In the context of control (rather than eradication) genes for inability to act as a vector of a disease (refractoriness) are being actively sought in other insects. In recent years a strain of *Anopheles gambiae* refractory to *Plasmodium berghei berghei* has been isolated (Al Mashhadani, 1974). Such a strain is being combined with a Y-autosome translocation (itself carrying the gene for dieldrin resistance as a means of generating males only) in order to test in cages the concept of displacement of genes for susceptibility to *Plasmodium* by release of males from a refractory strain (Curtis, 1976).

Acknowledgements

The authors wish to thank W. W. Cantelo, C. F. Curtis, G. G. Foster, I. C. McDonald, J. A. McKenzie, R. D. McMullen, E. Novitski, T. Prout, J. A. K. Reid, and A. S. Robinson for supplying material, and Lynne Kroetlinger for typing the manuscript. We wish to acknowledge the valuable and constructive comments on the manuscripts by C. F. Curtis.

The preparation of this chapter was supported in part by N.S.E.R.C. of Canada grant A5853 to D.G.H.

References

Akiyama, J. (1973). Further isolations of translocations in *Anopheles gambiae* species A. *Trans. Roy. Soc. Trop. Med. Hyg.* **67**, 440–441.

Al Mashhadani, H. M. (1974). Selection of *Anopheles gambiae* species A for susceptibility and refractoriness to malaria parasites. *Trans. Roy. Soc. Trop. Med. Hyg.* **68**, 267–268.

Baker, R. H., Sakai, R. K., Saifuddin, U. T. and Ainsley, R. W. (1977). Translocations in the mosquito *Culex tritaeniorhynchus*. *J. Hered.* **68**, 157–166.

Baldwin, M. and Chovnick, A. (1967). Autosomal half-tetrad analysis in *Drosophila melanogaster*. *Genetics* **55**, 277–293.

Barclay, H. J. and Fitz-Earle, M. (1983). Genetic population replacement for insect control: A new method for estimating fitness and generation time of continuously-breeding competing strains. (In preparation)

Bateman, A. M. (1968). Nondisjunction and isochromosomes from irradiation of chromosome 2 in *Drosophila*. *In*: "Effects of Radiation on Meiotic Systems". pp. 63–70. International Atomic Energy Agency, Vienna.

Bushland, R. C. (1971). Sterility principle for insect control: Historical development and recent innovations. *In*: "Sterility Principle for Insect Control or Eradication". Pp. 3–14. International Atomic Energy Agency, Vienna.

Cantelo, W. W. and Childress, D. (1975). Laboratory and field studies with a compound chromosome strain of *Drosophila melanogaster*. *Theor. Appl. Genet.* **45**, 1–6.

Chadov, B. F. (1970). The spontaneous formation of the second isochromosomes in female *Drosophila melanogaster* with a normal and structurally altered genotype. *Genetika* **6**, (9), 170–172 (in Russian).

Chadov, B. F. and Chadova, E. V. (1980). Spontaneous interchanges in females of *Drosophila melanogaster*. Part 1: Formation of half-translocations in XX and XXY females. *Theor. Appl. Genet.* **56**, 161–173.

Childress, D. (1972). Changing population structure through the use of compound chromosomes. *Genetics* **72**, 183–186.

Clark, A. M. and Sobels, F. H. (1973). Studies on non-disjunction of the major autosomes in *Drosophila melanogaster*. I. Methodology and rate of induction by X rays for the compound second chromosome. *Mutation Res.* **18**, 47–61.

CROSS, W. H. (1973). Biology, control and eradication of the boll weevil. *Ann. Rev. Entomol.* **18**, 17–46.
CURTIS, C. F. (1968). Possible use of translocations to fix desirable genes in insect pest populations. *Nature, Lond.* **218**, 368–369.
CURTIS, C. F. (1976). Testing systems for the genetic control of mosquitoes. Proc. XV Int. Congr. Entomol. (Wash., D.C.). Pp. 106–116.
CURTIS, C. F. and HILL, W. G. (1971). Theoretical studies on the use of translocations for the control of tsetse flies and other disease vectors. *Theor. Pop. Biol.* **2**, 71–89.
CURTIS, C. F. and ROBINSON, A. S. (1971). Computer simulation of the use of double translocations for pest control. *Genetics* **69**, 97–113.
CURTIS, C. F., SOUTHERN, D. I., PELL, P. E. and CRAIG-CAMERON, T. A. (1972). Chromosome translocations in *Glossina austeni*. *Genet. Res.* **20**, 101–113.
CURTIS, C. F., GROVER, K. K., SUGUNA, S. G., UPPAL, D. K., DIETZ, K., AGARWAL, H. V. and KAZMI, S. J. (1976). Comparative field cage tests of the population suppressing efficiency of three genetic control systems for *Aedes aegypti*. *Heredity* **36**, 11–29.
DENELL, R. E. (1973). Use of male sterilization mutations for insect control programmes. *Nature, Lond.* **242**, 274–275.
EVANS, W. H. (1971). Preliminary studies on frequency of autosomal non-disjunction in females of *D. melanogaster*. *Dros Inf. Serv.* **46**, 123–124.
FELDMANN, A. M. (1975). The induction of structural chromosome mutations in males and females of *Tetranychus urticae* Koch (Acarina: Tetranychidae). *In*: "Sterility Principle for Insect Control 1974". Pp. 437–446. International Atomic Energy Agency, Vienna.
FITZ-EARLE, M. (1975). Minimum frequency of compound autosomes in *Drosophila melanogaster* to achieve chromosomal replacement in cages. *Genetica* **45**, 191–201.
FITZ-EARLE, M. (1976). Insect population control using genetic engineering. *Bull Entomol. Soc. Amer.* **22**, 11–14.
FITZ-EARLE, M. and HOLM, D. G. (1976). The application of compound autosomes to insect control including the first experimental successes with compound-fragment combinations. Proc. XV Intl. Congr. Entomol. (Wash., D.C.). Pp. 146–156.
FITZ-EARLE, M. and HOLM, D. G. (1978). Exploring the potential of compound; free-arm combinations of chromosome 2 in *Drosophila melanogaster* for insect control and the survival to pupae of whole-arm trisomies. *Genetics* **89**, 499–510.
FITZ-EARLE, M. and SUZUKI, D. T. (1975). Conditional mutations for the control of insect populations. *In*: "Sterility Principle for Insect Control 1974". Pp. 365–374. International Atomic Energy Agency, Vienna.
FITZ-EARLE, M., HOLM, D. G. and SUZUKI, D. T. (1973). Genetic control of insect populations. I. Cage studies of chromosome replacement by compound autosomes in *Drosophila melanogaster*. *Genetics* **74**, 461–475.
FITZ-EARLE, M., HOLM, D. G. and SUZUKI, D. T. (1975). Population control of caged native fruitflies in the field by compound autosomes and temperature-sensitive mutants. *Theor. Appl. Genet.* **46**, 25–32.
FOSTER, G. G. and WHITTEN, M. J. (1974). The development of genetic methods of controlling the Australian sheep blowfly, *Lucilia cuprina*. *In*: "The Use of Genetics in Insect Control". Ch. 2. (R. Pal and M. J. Whitten, eds.). Elsevier, Amsterdam.
FOSTER, G. G., WHITTEN, M. J., PROUT, T. and GILL, R. (1972). Chromosome rearrangements for the control of mosquitoes and other insect pests. *Science* **176**, 875–880.
FOSTER, G. G., WHITTEN, M. J. and KONOWALOW, C. (1976). The synthesis of compound autosomes in the Australian sheep blowfly *Lucilia cuprina*. *Can. J. Genet. Cytol.* **18**, 169–177.

FOSTER, G. G., WHITTEN, M. J., VOGT, W. G., WOODBURN, T. L. and ARNOLD, J. T. (1978). Larval release method for genetic control of the Australian sheep blowfly, *Lucilia cuprina* (Wiedemann) (Diptera: Calliphoridae). *Bull. Ent. Res.* **68**, 75–83.

GAVIN, J. A. and HOLM, D. G. (1972). Gamma ray induced nondisjunction and chromosome loss of chromosome 2 in females. *Dros. Inf. Serv.* **48**, 143–144.

GETHMAN, R. C. (1976). Meiosis in male *Drosophila melanogaster*. II. Nonrandom segregation of compound-second chromosome. *Genetics* **83**, 743–751.

GIBSON, W. G. (1977). Autosomal products of meiosis arising from radiation-induced interchange in female *Drosophila melanogaster*. Ph.D. Thesis, University of British Columbia.

GRELL, E. H. (1970). Distributive pairing: mechanism for segregation of compound autosomal chromosomes in oocytes of *Drosophila melanogaster*. *Genetics*, **65**, 65–74.

GROVER, K. K., SUGUNA, S. G., UPPAL, D. K., SINGH, K. R. P., ANSARI, M. A., CURTIS, C. F., SINGH, D., SHARMA, V. P. and PANIKER, K. N. (1976). Field experiments on the competitiveness of males carrying genetic control systems for *Aedes aegypti*. *Ent. Exp. and Appl.* **20**, 8–18.

HARTL, D. L. (1973). Complementation analysis of male fertility among the segregation distorter chromosomes of *Drosophila melanogaster*. *Genetics* **73**, 613–629.

HARTL, D. L. and HIRAIZUMI, Y. (1976). Segregation distortion. Ch. 15. In: "Genetics and Biology of Drosophila" Vol. 1b. Edited by M. Ashburner and E. Novitski. Academic Press, London.

VAN HEEMERT, C. (1977). Synthesis of compound chromosomes from a pericentric inversion in the onion fly *Hylemya antiqua*. *Nature, Lond.* **266**, 445–447.

VAN HEEMERT, C. and WIJNANDS-STÄB, K. J. A. (1975). Radiation induced semi-sterility for genetic control purposes in the onion fly *Hylemya antiqua* (Meigen). II. Induction, isolation and cytogenetic analysis of new chromosomal rearrangements. *Theol. Appl. Gen.* **45**, 349–354.

HENNEBERRY, T. J. (1963). Effects of gamma radiation on the fertility and longevity of *Drosophila melanogaster*. *J. Econ. Entomol.* **56**, 279–281.

HENNEBERRY, T. J. and MCGOVERN, W. L. (1963). Effects of gamma radiation on mating competitiveness and behavior of *Drosophila melanogaster* males. *J. Econ. Entomol.* **56**, 739–741.

HENNEBERRY, T. J., MASON, H. C. and MCGOVERN, W. L. (1967). Some effects of gamma radiation and apholate on the fertility of *Drosophila melanogaster*. *J. Econ. Entomol.* **60**, 853–857.

HILLIKER, A. J. and HOLM, D. G. (1975). Genetic analysis of the proximal region of chromosome 2 of *Drosophila melanogaster*. I. Detachment products of compound autosomes. *Genetics* **81**, 705–721.

HOLM, D. G. (1976). Compound Autosomes. In: "Genetics and Biology of Drosophila" Vol 1b. Edited by M. Ashburner and E. Novitski. Academic Press, London.

HOLM, D. G. and CHOVNICK, A. (1975). The compound autosomes of *Drosophila melanogaster*: The meiotic behavior of compound-3. *Genetics* **81**, 293–311.

HOLM, D. G., DELAND, M. and CHOVNICK, A. (1967). Meiotic segregation of C(3L) and C(3R) chromosomes in *Drosophila melanogaster*. *Genetics* **56**, 565–566 (Abstr.).

HOLM, D. G., FITZ-EARLE, M. and SHARP, C. B. (1980). Chromosome replacement in mixed populations of compound-2L; free-2R and standard strains of *Drosophila melanogaster*: An example of unstable genetic isolation. *Theor. Appl. Genet.* **57**, 247–255.

HOSSAIN, M. A., CURTIS, C. F. and JAFFE, W. P. (1974). Selection on the fertility of translocation heterozygotes in *Drosophila melanogaster*. I. The extent of the changes produced by selection. *J. Genet.* **61**, 205–217.

IVES, P. T. and FINK, G. R. (1962). Comparison of translocation and crossover chromosomes produced by gamma irradiation of *Drosophila* males. *Genetics* **47**, 963 (Abstr.).

KLASSEN, W., KNIPLING, E. F. and MCGUIRE, J. U. (1970). The potential for insect-population suppression by dominant conditional lethal traits. *Ann. Entomol. Soc. Amer.* **63**, 238–255.

LAVEN, H., COUSSERANS, J. and GUILLE, G. (1971). Inherited semisterility for control of harmful insects. III. A first field experiment. *Experientia* **27**, 1355–1357.

LEIGH, B. and SOBELS, F. H. (1970). Induction by X-rays of isochromosomes in the germ cells of *Drosophila melanogaster* males: Evidence for nuclear selection in embryogenesis. *Mutation Res.* **10**, 475–487.

LI, C. C. (1955). The stability of an equilibrium and the average fitness of a population. *Amer. Nat.* **89**, 281–296.

LINDSLEY, D. L. and GRELL, E. H. (1968). Genetic variations of *Drosophila melanogaster*. Carnegie Institute of Wash. Pub. No. 627.

LINDSLEY, D. L. and GRELL, E. H. (1969). Spermiogenesis without chromosomes in *Drosophila melanogaster*. *Genetics* **61**, (Suppl. 1). 69–78.

LUTOLF, H. U. (1972). Meiotic segregation of compound-3 chromosomes in *Drosophila*. *Genetica* **43**, 431–442.

LUTOLF, H. U. and WÜRGLER, F. E. (1970). Meiotische segregation von isochromosomen bei *Drosophila melanogaster*. *Arch. Julius Klaus-Stift. Vererb. Forsch.* **46**, 44–51.

LYTTLE, T. W. (1977). Experimental population genetics of meiotic drive systems I. Pseudo-Y chromosomal drive as a means of eliminating cage populations of *Drosophila melanogaster*. *Genetics* **86**, 413–445.

LYTTLE, T. W. (1979). Experimental population genetics of meiotic drive systems II. Accumulation of genetic modifiers of segregation distorter (*SD*) in laboratory populations. *Genetics* **91**, 339–357.

MCCLOSKEY, J. D. (1966). The problem of gene activity in the sperm of *Drosophila melanogaster*. *Amer. Nat.* **100**, 211–218.

MCCOY, J. R., LLOYD, E. P. and BARTLETT, A. C. (1968). Diapause in crosses of a laboratory and a wild strain of boll weevils. *J. Econ. Entomol.* **61**, 163–166.

MCDONALD, I. C. and OVERLAND, D. E. (1972). Temperature-sensitive mutations in the housefly: The characterization of heat-sensitive recessive lethal factors on autosome III. *J. Econ. Entomol.* **65**, 1364–1368.

MCDONALD, I. C. and OVERLAND, D. E. (1973). House fly genetics II. Isolation of a heat-sensitive translocation homozygote. *J. Hered.* **64**, 253–256.

MCDONALD, P. T., ASMAN, S. M. and TERWEDOW, H. A. (1978). An alternative method for isolating homozygotes of autosomal translocations in the mosquito *Culex tarsalis*. *Can. J. Genet. Cytol.* **20**, 581–588.

MCKENZIE, J. A. (1976). The release of a compound-chromosome stock in a vineyard cellar population of *Drosophila melanogaster*. *Genetics* **82**, 685–695.

MCKENZIE, J. A. (1977). The effect of immigration on genetic control. A laboratory study with wild and compound chromosome stocks of *Drosophila melanogaster*. *Theor. Appl. Genet.* **49**, 79–83.

MASON, H. C. and SMITH, F. F. (1967). Apholate as a chemosterilant for *Drosophila melanogaster*. *J. Econ. Entomol.* **60**, 1127–1130.

MASON, H. C., HENNEBERRY, T. J., SMITH, F. F. and MCGOVERN, W. L. (1968). Suppression of *Drosophila melanogaster* in tomato plots by the release of flies sterilized by apholate. *J. Econ. Entomol.* **61**, 166–170.

MASON, H. C., BALOCK, J. W., SMITH F. F. and GUEST, R. T. (1973). Mass sterilization of laboratory reared *Drosophila melanogaster* with residues of tepa. *J. Econ. Entomol.* **66**, 753–755.

MASON, H. C., BALOCK, J. W., GUEST, R. T., KWIETNIAK, R. T., SMITH, F. F., ANDERSON, H. V., DEBLOIS, L. D. and WIRTH, W. W. (1975). Suppression of *Drosophila melanogaster* in commercial tomato fields by area and direct field releases of tepa-sterilized adults. *J. Econ. Entomol.* **68**, 76–78.

MULLER, H. J. and SETTLES, F. (1927). The non-functioning of genes in spermatozoa. *Z. indukt. Abstamm. Vererb. Lehre* **43**, 285–301.

NOVITSKI, E. (1976). The construction of an entire compound two chromosome. Appendix to Ch. 13. *In*: "Genetics and Biology of Drosophila" Vol. 1b. Edited by M. Ashburner and E. Novitski. Academic Press, london.

NOVITSKI, E. and HANKS, G. D. (1961). Analysis of irradiated *Drosophila* populations for meiotic drive. *Nature, Lond.* **190**, 989–990.

PATTERSON, J. T., STONE, W., BEDICHEK, S. and SUCHE, M. (1934). The production of translocations in *Drosophila*. *Amer. Nat.* **68**, 359–369.

PEACOCK, W. J., TOKUYASU, K. T. and HARDY, R. W. (1972). Spermiogenesis and meiotic drive in *Drosophila*. Pp. 247–268. *In*: "Proc. Int. Symp. on the Genetics of the Spermatozoon". (R. A. Beatty and S. Glueckson-Waelsch, eds.).

PROUT, T. (1977). The formal theory of negative heterosis with applications to pest control and parapatric speciation. Syllabus, Part II. In: "Genetic Aspects of Pest Management". Entomology 207, University of California, Davis.

PROVERBS, M. D., NEWTON, J. R. and LOGAN, D. M. (1969). Codling moth control by release of radiation-sterilized moths in a commercial apple orchard. *J. Econ. Entomol.* **62**, 1331–1334.

RABBANI, M. G. and KITZMILLER, J. B. (1972). Chromosomal translocations in *Anopheles albimanus*. *Mosquito News* **32**, 421–432.

RAI, K. S., LORIMER, N. and HALLINAN, E. (1974). The current status of genetic methods for controlling *Aedes aegypti*. Pp. 119–132. In: "The Use of Genetics in Insect Control". Edited by R. Pal and M. J. Whitten, Elsevier, Amsterdam.

RASMUSSEN, I. E. (1960). Reports on new mutants. *Dros. Inf. Serv.* **34**, 53.

REID, J. A. K. and MCEWEN, F. L. (1977). Genetic insect control: chromosome rearrangements isolated using a simple system in the onion maggot, *Hylemya antiqua* (Diptera: Anthomyiidae). *Can. Entomol.* **109**, 1287–1291.

REID, J. A. K. and MCEWEN, F. L. (1979). Semi-sterile mutants in the onion maggot, *Hylemya antiqua* (Diptera: Anthomyiidae), population effects in field cages. *Can. Entomol.* **111**, 749–750.

REID, J. A. K. and WEHRHAHN, C. F. (1976). Genetic control of insect populations: Isolation and fitness determination of autosomal translocations. *Can. Entomol.* **108**, 1409–1415.

ROBERTS, P. A. (1967). A positive correlation between crossing over within heterozygous pericentric inversions and reduced egg hatch of *Drosophila* females. *Genetics* **56**, 179–187.

ROBERTS, P. A. (1970). Screening for X-ray-induced crossover suppressors in *Drosophila melanogaster*: Prevalence and effectiveness of translocations. *Genetics* **65**, 429–448.

ROBERTS, P. A. (1976). The genetics of chromosome aberrations. In: "Genetics and Biology of Drosophila" Vol. 1a. (Edited by M. Ashburner and E. Novitski). Academic Press, London and New York.

ROBINSON, A. S. (1976). Progress in the use of chromosomal translocations for the control of insect pests. *Biol. Rev.* **51**, 1–24.

ROBINSON, A. S. (1977). Translocations and a balanced polymorphism in a *Drosophila* population. *Genetica* **47**, 231–236.

ROBINSON, A. S. and CURTIS, C. F. (1972). Crossing over in a double translocation in Drosophila. *Can. J. Genet. Cytol.* **14**, 129–137.

ROBINSON, A. S. and CURTIS, C. F. (1973). Controlled crosses and cage experiments with a translocation in *Drosophila*. *Genetica* **44**, 591–601.

ROSS, M. H. and COCHRAN, D. (1973). German cockroach genetics and its possible use in control measures. *Patna J. Med.* **47**, 325–337.

SAKAI, R. C. and BAKER, R. H. (1974). Induction of heat-sensitive lethals in *Culex tritaeniorhynchus* by ethyl methane sulphonate. *Mosquito News* **34**, 420–424.

SCRIBA, M. E. L. (1967). Embryonale Entwicklungsstorungen bei Defizienz and Tetraploidie des 2. Chromosoms von *Drosophila melanogaster*. *Wilhelm Roux Archiv. Entwmech.* **159**, 314–345.

SCRIBA, M. E. L. (1969). Embryonale Entwicklungsstorungen bei Nullosomie und Tetrasomie des 3. Chromosoms von *Drosophila melanogaster*. *Devl. Biol.* **19**, 160–177.

SEREBROVSKY, A. S. (1940). On the possibility of a new method for the control of insect pests. *Zool. Zh.* **19**, 618–630 (In Russian.)

SEREBROVSKY, A. S. (1969). On the possibility of a new method for the control of insect pests. *In*: "Sterile-male Technique for Eradication or Control of Harmful Insects". Pp. 123–137. International Atomic Energy Agency (Vienna).

SERGHIOU, C. (1975). The sterile-male technique for the control of the Mediterranean fruitfly, *Ceratitis capitata* Wied. in the Mediterranean basin. *In*: "Sterility Principle for Insect Control 1974". Pp. 11–28. International Atomic Energy Agency (Vienna).

SMITH, R. H. (1971). Induced conditional lethal mutations for the control of insect populations. In: "Sterility Principle for Insect Control or Eradication". Pp. 453–465. International Atomic Energy Agency (Vienna).

SUGUNA, S. G., WOOD, R. J., CURTIS, C. F., WHITELAW, A. and KAZMI, S. J. (1977). Resistance to meiotic drive at the M^D locus in an Indian wild population of *Aedes aegypti*. *Genet. Res.* **29**, 123–132.

SUZUKI, D. T., KAUFMAN, T., FALK, D. and the U.B.C. Drosophila Research Group (1976). Conditionally expressed mutations in *Drosophila melanogaster*. In: "Genetics and Biology of Drosophila" Vol. 1a. Pp. 207–263. (M. Ashburner and E. Novitski, eds.). Academic Press, London and New York.

UPPAL, D. K., CURTIS, C. F. and SONI, V. K. (1978). Laboratory evaluation of a translocation double heterozygote for genetic control of *Aedes aegypti*. *Theor. Appl. Genet.* **51**, 153–157.

WAGONER, D. E., McDONALD, I. C. and CHILDRESS, D. (1974). The present status of genetic control mechanisms in the housefly *Musca domestica* L. In: "The Use of Genetics in Insect Control". Pp. 183–197. (R. Pal and M. J. Whitten, eds.). Elsevier, Amsterdam.

WALLACE, B. (1970). "Genetic Load: Its Biological and Conceptual Aspects". Prentice-Hall, Englewood Cliffs, N.J.

WATERHOUSE, D. F., LACHANCE, L. E. and WHITTEN, M. J. (1975). Use of autocidal methods, Ch. 26. In: "Theory and Practice of Biological Control". (C. Huffaker, P. DeBach and P. S. Messenger, eds.). Academic Press, New York and London.

WHITTEN, M. J. (1971). Insect control by genetic manipulation of natural populations. *Science* **171**, 682–684.

WHITTEN, M. J. and FOSTER, G. G. (1975). Genetical Methods of Pest Control. *Ann. Rev. Entomol.* **20**, 461–476.

WHITTEN, M. J., FOSTER, G. G., ARNOLD, J. T. and KONOWALOW, C. (1975). The Australian sheep blowfly, *Lucilia cuprina*. In: "Handbook of genetics" Vol. 3 (R. C. King, ed.). Plenum, New York.

WHITTEN, M. J., FOSTER, G. G., VOGT, W. G., KITCHING, R. L., WOODBURN, T. L.

and KONOWALOW, C. (1976). Current status of genetic control of the Australian sheep blowfly, *Lucilia cuprina* (Wiedemann) (Diptera: Calliphoridae). Proc. XV Int. Congr. Entomol. (Wash., D.C.).

WIJNANDS-STÄB, K. J. A. and VAN HEEMERT, C. (1974). Radiation-induced semi-sterility for genetic control purposes in the onion fly, *Hylemya antiqua* (Meigen). I. Isolation of semi-sterile stocks and their cytogenetical properties. *Theor. Appl. Genet.* **44**, 111–119.

WOOD, R. J. and NEWTON, M. E. (1976). Meiotic drive and sex ratio distortion in the mosquito *Aedes Aegypti*. Pp. 146–156. Proc. XV Int. Congr. Entomol. (Wash., D.C.).

WOOD, R. J., COOK, L. M., HAMILTON, A. and WHITELAW, A. (1977). Transporting the marker gene *re* (red eye) into a laboratory cage population of *Aedes aegypti* (Diptera: Culicidae), using meiotic drive at the M^D locus. *J. Med. Entomol.* **14**, 461–464.

YTTERBORN, K. H. (1970). Homozygous lethal effects of II–III translocations in *D. melanogaster*. *Dros. Inf. Serv.* **45**, 158.

ZIMMERING, S. (1955). A genetic study of segregation in a translocation heterozygote in *Drosophila*. *Genetics* **40**, 809–825.

ZIMMERING, S. (1976). Genetic and Cytogenetic Aspects of Altered Segregation Phenomena in Drosophila. In: "The Genetics and Biology of Drosophila". Vol. 1b, pp. 569–613. Edited by M. Ashburner and E. Novitski. Academic Press, London.

Addendum to Chapter 20
Factors Affecting Mutation Rates

R. C. WOODRUFF, B. E. SLATKO and J. N. THOMSON, JR.

The following is a list of the additional references given for this chapter.

BREGLIANO, J. C. and KIDWELL, M. G. (1982). Hybrid dysgenesis determinants. *In*: "Mobile Genetic Elements", (J. Shapiro, ed.), Academic Press, London, in press.
CARLOS, M. P. and MILLER, J. P. (1980). Transposable elements. *Cell* **20**, 579–595.
GOLDBERG, M. L., PARO, R. and GEHRING, W. J. (1982). Molecular cloning of the *white* locus region of *Drosophila melanogaster* using a large transposable element. *The EMBO Journal* **1**, 93–98.
LEUDERS, K., LEDER, A., LEDER, P. and KUFF, E. (1982). Association between a transposed α-globin pseudogene and retrovirus-like elements in the Balb/c mouse genome. *Nature* **295**, 426–428.
LEVIS, R. and RUBIN, G. M. (1982). The unstable w^{DZL} mutation of Drosophila is caused by a 13 kilobase insertion that is imprecisely excised in phenotypic revertants. *Cell* **30**, 543–550.
LEVIS, R., COLLINS, M. and RUBIN, G. M. (1982). FB elements are the common basis for the instability of the w^{DZL} and w^c Drosophila mutations. *Cell* **30**, 551–565.
LIM, J. K. (1979). Site-specific instability in *Drosophila melanogaster*: The origin of mutations and cytogenetic evidence for site specificity. *Genetics* **93**, 681–701.
LIM, J. K. (1980). Site-specific intrachromosomal rearrangements in *Drosophila melanogaster*: Cytogenic evidence for transposable elements. *Cold Spring Harbor Symp. Quant. Biol.* **45**, 553–560.
MAINX, F. (1964). The genetics of *Megaselia scalaris* Loew (Phoridae): A new type of sex determination in Diptera. *Amer. Nat.* **98**, 415–430.
MAINX, F. (1966). Die Geschlechtsbestimmung bei *Megaselia scalaris* Loew (Phoridae). *Z. indukt. Abstamm. VererbLehre* **98**, 49–60.
MAINX, F. and DOSCHEK, E. (1967). Der genetische austausch bei den beiden geschlechtern von *Megaselia scalaris*. *Molec. Gen. Genetics* **99**, 203–218.
MCCLEMENTS, W., DAHR, R., BLAIR, D., ENQUIST, L., OSKARSSON, M. and VANDE WOUDE, G. (1981). The long terminal repeat of Moloney Sarcoma virus. *Cold Spring Harbor Symp. Quant. Biol.* **45**, 699–705.
RAYMOND, J. D. and SIMMONS, M. J. (1981). An increase in the X-linked lethal mutation rate associated with an unstable locus in *Drosophila melanogaster*. *Genetics* **98**, 291–302.
RUBIN, G. M. and SPRADLING, A. C. (1982). Genetic transformation of *Drosophila* with transposable element vectors. *Science* **218**, 348–353.
RUBIN, G. M., BROREIN, W. J., JR., DUNSMUIR, P., FLAVELL, A. J., LEVIS, R., STROBEL, E., TOOLE, J. J. and YOUNG, E. (1981). *copia*-like transposable elements in the *Drosophila* genome. *Cold Spring Harbor Symp. Quant. Biology.* **45**, 619–628.

SCOBIE, N. and SCHAFFER, H. (1982a). A mutator factor in a strain of *Drosophila melanogaster*: Identification by use of mutation, reversion rates and male recombination. *Genetics* **101**, 417–429.

SCOBIE, N. and SCHAFFER, H. (1982b). The location of a mutator factor in a strain of *Drosophila melanogaster* by assaying male recombination. *Genetics* **101**, 405–416.

SHIMOTOHNO, K., MITZUTANI, S. and TEMIN, H. (1980). Sequence of retrovirus provirus resembles that of bacterial transposable elements. *Nature* **285**, 550–554.

SNYDER, M. P., KIMBRELL, D., HUNKAPILLAR, M., HILL, R., FRISTROM, J. and DAVIDSON, N. (1982). A transposable element that splits the promoter region inactivates a *Drosophila* cuticle protein gene. *Proc. Natl. Acad. Sci. U.S.A.* **79**, 7430–7434.

SPRADLING, A. C. and RUBIN, G. M. (1981). Drosophila genome organization: Conserved and dynamic aspects. *Ann. Rev. Genet.* **15**, 219–264.

SPRADLING, A. C. and RUBIN, G. M. (1982). Transposition of cloned P elements into *Drosophila* germ line chromosomes. *Science* **218**, 341–347.

STANFIELD, S. and LENGYEL, J. (1979). Small circular DNA of *Drosophila melanogaster*: Chromosomal homology and kinetic complexity. *Proc. Natl. Acad. Sci. U.S.A.* **76**, 6142–6146.

TCHURIKOV, N. A., ILYIN, Y. V., SKRYABIN, K. G., ANAIEV, E. V., BAYEV, A. A., KRAYEN, A. S., ZELENTSOVA, E. S., KULGVSKIN, V. V., LYUBOMIRSKAYA, N. V. and GEORGIEV, G. P. (1981). General properties of mobile dispersed genetic elements in *Drosophila melanogaster*. *Cold Spring Harbor Symp. Quant. Biol.* **45**, 655–665.

TRUETT, M. A., JONES, R. S. and POTTER, S. S. (1981). Unusual structure of the FB family of transposable elements in *Drosophila*. *Cell* **24**, 753–763.

WOODRUFF, R. C. and THOMPSON, J. N., JR. (1982). Genetic factors that affect rates of spontaneous mutation and chromosome aberrations in *Drosophila melanogaster*. *Cytogenet. Cell Genet.* **33**, 152–159.

YOUNG, M. W. and SCHWARTZ, H. E. (1981). Nomadic gene families in *Drosophila*. *Cold Spring Harbor Symp. Quant. Biol.* **45**, 629–640.

ZACHAR, Z. and BINGHAM, P. M. (1982). Regulation of *white* locus expression: The structure of mutant alleles at the *white* locus of *Drosophila melanogaster*. *Cell* **30**, 529–541.

Author Index

The numbers in italics refer to the reference pages at the end of the chapters where the full reference appears.

A

Abrahamson, S., 8, *31*, 40, 42, 43, 47, 50, 51, 52, *107*, *110*, *115*, *122*
Adams, W. T., 251, *271*
Agarwal, H. V., 418, *420*
Ahearn, J. N., 265, 268, *271*
Ainsley, R. W., 416, *419*
Akiyama, J., 417, *419*
Alexander, M. L., 44, 98, 106, *107*
Alexandrov, Y. N., 59, *112*
Ali, A. M. M., 319, *336*
Allen, P., 145, *151*
Al Mashhadani, H. M., 419, *419*
Almeida, J. C., 58, *118*
Altenburg, E., 42, 45, *107*
Altenburg, L. S., 42, *107*
Andersen, F. S., 311, *333*
Anderson, H. V., 401, *423*
Anderson, W. W., 26, *31*, 241, 246, 256, 263, *271*, 264, 324, *339*
Ando, Y., 53, 59, 73, *117*
Andrewartha, H. G., 287, *333*
Angus, D. S., 87, *107*, *110*
Annest, J. L., 354, 355, 357, 358, 363, 368, 369, 376, 377, 379, 380, 381, 388, 394, 395, *395*
Ansari, M. A., 417, *421*
Antonovics, J., 290, 294, 318, *333*, *340*
Antony, C., 242, 270, *276*
Anxolabehere, D., 251, *275*
Applewhite, P. B., 243, *280*
Arnold, J. T., 417, 418, *421*, *424*
Ashburner, M., 38, 88, *107*
Asher, J. H., Jr., 350, 352, 353, 355, 373, 375, 376, 389, 390, *396*
Asman, S. M., 416, *422*
Auerbach, C., 14, *31*, 39, 40, 47, 54, 62, *108*
Averhoff, W. W., 234, 263, 269, *271*
Ayala, F. J., 16, *34*, 38, 48, *110*, 127, 128, 129, *149*, 246, 251, 264, *271*, 286, 287, 288, 290, 292, 293, 294, 295, 296, 307, 308, 310, 311, 317, 320, 321, 322, 323, 324, 326, 328, 329, *333*, *334*, *337*, *339*, *340*

B

Baars, A. J., 58, *108*
Baillie, D., 88, *120*
Baker, B. S., 45, 50, 54, 62, 63, 108, *111*, *114*
Baker, R. H., 416, 418, *419*, *424*
Baker, R. N., 42, *107*
Baker, W. K., 47, *108*
Bakker, K., 286, 306, *334*
Bakradze, M. A., 359, *396*
Baldwin, M., 407, *419*
Ballantyne, G. H., 45, *108*
Balock, J. W., 401, *422*, *423*
Banerjee, J., 45, *108*
Barclay, H. J., 403, *419*
Barker, J. S. F., 55, *120*, 177, 183, 188, *217*, 262, 266, 268, 269, *271*, 286, 289, 291, 293, 304, 307, 308, 311, 312, 313, 314, 315, 318, 319, 320, 321, 322, 323, 325, 327, 328, 331, *334*, *335*, *339*
Barnes, B. W., 165, 166, 213, *215*, *217*
Barr, L. G., 129, *149*, 264, *271*
Bartell, R., 234, *281*
Bartlett, A. C., 418, *422*
Bastock, M., 225, 231, 236, *271*
Bateman, A. J., 248, 255, 257, 261, 263, 265, 266, *271*
Bateman, A. M., 406, *419*
Bauer, H., 50, *108*
Baumann, H., 241, *271*
Baumiller, R. C., 59, *108*
Beardmore, J. A., 323, *334*
Becker, J., 58, *118*
Bedichek, S., 404, *423*
Belgovsky, M. L., 106, *108*
Bell, A. E., 164, 178, *215*
Belloni, M., 60, *110*

Bennet-Clark, H. C., 230, 231, 232, 258, 267, 268, 271, 275
Bennett, J., 255, 281
Ben-Zeev, N., 21, 23, 32
Benzer, S., 242, 276
Berchtold, W., 47, 108, 123
Berg, R. L., 17, 31, 42, 53, 94, 98, 102, 106, 108, 109, 112, 135, 149
Bergendahl, J., 42, 107
Bingham, P. M., 39, 102, 103, 106, 109, 119, 139, 140, 143, 149, 153
Birch, L. C., 286, 287, 333, 334
Birley, A. J., 163, 219
Bishop, Y. M. M., 250, 271
Blaylock, B. G., 320, 334
Blijleven, W. G. H., 58, 108
Boam, T. B., 165, 183, 189, 204, 205, 220
Bodmer, W. F., 30, 34
Booker, R., 244, 280
Borowsky, R., 293, 334
Borstel, R. C. von, 106, 117
Bortolozzi, J., 64, 123
Bos, M., 193, 203, 215, 291, 307, 335
Bösiger, E., 263, 272
Bouletreau-Merle, J., 240, 241, 272
Bowman, J. T., 45, 109, 123
Bowman, K.O., 47, 114
Boyd, J. B., 54, 62, 63, 108, 109, 118
Bradley, B. P., 192, 193, 217
Brattsten, R., 55, 123
Braun, R., 62, 111
Breese, W. L., 169, 187, 189, 214, 215
Bregliano, J. C., 17, 31, 84, 98, 99, 100, 103, 109, 119, 130, 131, 132, 133, 134, 141, 143, 144, 149, 152
Breimer, D. D., 58, 108
Brewer, G. J., 384, 385, 395, 397
Bridges, C. B., 239, 272
Brittnacher, J. G., 16, 31
Brncic, D., 300, 306, 316, 335
Broadhead, R. S., 137, 149
Brokaw, B., 355, 356, 358, 360, 376, 378, 382, 383, 387, 394, 398
Brown, R. G. B., 225, 238, 258, 259, 272
Brown, W. L., 291, 335
Browning, L. S., 45, 107, 109
Brun, G., 58, 109
Bryant, P. 251
Bucci, M. J., 243, 280

Bucheton, A., 17, 31, 84, 98, 99, 100, 109, 119, 130, 131, 132, 133, 134, 141, 144, 145, 149, 152
Budnik, M., 306, 316, 335
Bühler, R., 272
Bulmer, M. G., 290, 335
Bundegaard, J., 16, 31, 241, 257, 280
Burdette, W., 59, 109
Burla, H., 261, 265, 272
Burnet, B., 227, 231, 232, 236, 237, 240, 241, 255, 262, 269, 272, 273, 291, 307, 335
Buruga, J. H., 304, 335
Bushland, R. C., 400, 419
Buzzati-Traverso, A. A., 292, 335
Byers, H. L., 55, 56, 57, 109

C

Caligari, P. D. S., 159, 166, 167, 189, 192, 196, 197, 214, 215, 215
Campbell, C., 246, 271
Cannon, G. B., 60, 109
Cantelo, W. W., 408, 409, 410, 419
Cardellino, R. A., 53, 86, 88, 89, 90, 110, 118, 124
Carlson, E. A., 44, 45, 110, 112
Carmody, G., 264, 272
Carpenter, A. T. C., 54, 62, 63, 108, 144, 152
Carson, G. L., 22, 34, 54, 113
Carson, H. L., 38, 70, 110, 148, 149, 164, 215, 251, 252, 256, 261, 265, 268, 269, 270, 271, 272, 281, 297, 300, 304, 330, 331, 335, 336, 337, 345, 346, 348, 349, 351, 352, 353, 354, 355, 356, 360, 362, 363, 364, 365, 367, 368, 369, 370, 371, 372, 379, 385, 387, 388, 389, 393, 395, 396, 397, 398
Castle, W. E., 156, 216
Catcheside, D. G., 171, 216
Chadov, B. F., 408, 419
Chadova, E. V., 408, 419
Chang, H., 230, 232, 272
Chejne, A. J., 269, 276
Chen, P. S., 241, 272
Chiang, H. C., 241, 273
Chigusa, S. I., 8, 9, 13, 14, 17, 21, 23, 24, 25, 26, 33, 89, 110, 117, 118
Childress, D., 408, 409, 410, 417, 419, 424
Chinnici, J. P., 268, 273

Chovnick, A., 45, 88, *108*, *120*, 407, 408, 410, 411, *419*, *421*
Christiansen, F. B., 16, *31*
Claringbold, P. J., 319, 320, 321, 322, *335*
Clark, A. M., 410, *419*
Clark, C., 60, *110*
Clark, D. J., 62, *110*
Clarke, B., 290, *335*
Clarke, C. A., 202, *216*
Clayton, G. A., 16, *31*, 171, 174, 175, 176, 177, 183, 188, *216*
Clayton, F. E., 330, 331, *337*
Cochran, D., 417, *424*
Cochran, W. G., 249, *281*
Cockerham, C. C., 9, 18, 26, *31*, *33*, 46, 88, 90, *117*
Cohet, Y., 264, *273*
Cole, L. C., 287, *335*
Coleman, T. P., 294, *335*
Colgan, D. J., 87, *110*
Collins, M., 39, 106, *110*
Comstock, R. F., 178, *216*
Connell, J. H., 290, 296, *335*
Connolly, K., 227, 228, 231, 232, 236, 237, 240, 241, 255, 262, 269, *272*, *273*
Cook, L. M., 418, *425*
Cook, R. M., 228, 232, 238, 240, 241, 242, *272*, *273*
Cooper, D. M., 304, 305, *335*, *336*
Cooper, K. W., 169, *216*
Cordeiro, A. R., 58, *110*, 261, 265, *272*
Corwin, H. O., 44, *121*
Coulter, F., 145, *151*
Counce, S. J., 348, 349, 360, *396*
Cousserans, J., 417, *422*
Cowling, D. E., 231, *273*
Coyne, J. A., 262, 269, *273*
Cozzi, R., 60, *110*
Craig-Cameron, T. A., 416, *420*
Craymer, L., 53, *110*
Cross, W. H., 401, *420*
Crossley, S. A., 237, 269, *273*
Crow, J. F., 4, 5, 6, 7, 8, 9, 13, 14, 16, 17, 21, 22, 23, 24, 26, 27, 28, 29, 30, *31*, *32*, *33*, *34*, *35*, 40, 43, 47, 61, 89, *110*, *118*
Cuellar, O., 387, *396*
Cullen, T. L., 58, *118*
Cummins, L. J., 262, *271*
Curtis, C. F., 402, 403, 404, 405, 416, 417, 418, 419, *420*, *421*, *424*

D

DaCunha, A. B., 64, *111*, 164, *220*, 252, 261, 265, *272*, *273*, 304, *335*, *336*
Daniel, G., 42, *107*
Dapples, C. C., 270, *282*
Darevskii, I. S., 359, *396*
Darlington, C. D., 190, 202, 212, 214, *216*
Darlington, P. J., Jr., 287, 288, *336*
Darwin, C., 255, 257, *273*, 286, *336*
Davenport, C. B., 224, *273*
David, J. R., 264, *273*
Davies, R. W., 166, 168, 187, 188, *216*
Davis, C. G., 319, *340*
Dawood, M. M., 325, *340*
Dawson, P. S., 6, *35*
DeBach, P., 287, 288, *336*
DeBenedictis, P. A., 308, 317, 318, *336*
Deblois, L. D., 401, *423*
DeJianne, D., 270, *280*
DeJongh, C., 8, *31*, 40, 43, *107*
Deland, M., 407, *421*
Delden, W. van, 251, *279*, 325, 327, *340*
DeMarco, A., 60, *110*
Demerec, M., 42, 50, 64, 65, 68, *108*, *110*
Denell, R. E., 75, *114*, 401, *420*
Dennis, L., 231, 232, *272*
d'Entrement, C. J., 264, *283*
Detleysen, J. A., 170, *216*
Dias Collazo, A., 264, *272*
Dietz, K., 418, *420*
Dijken, F. R. van, 262, 269, *273*
Doane, W. W., 348, *396*
Dobzhansky, Th., 2, 26, 28, 31, 38, 48, *110*, 125, 126, 127, 128, 129, 146, *149*, *150*, *151*, 161, 162, 164, 207, 211, *216*, 224, 237, 241, 244, 252, 256, 261, 264, 265, 266, 267, 268, 269, *271*, *272*, *273*, *274*, *277*, *278*, *283*, 303, 304, 320, 323, *334*, *335*, *336*
Doerr, C. A., 372, *396*
Dolan, R., 306, *336*
Domoto, T., 43, 44, *114*
Donald, C. M., 286, *336*
Doschek, E., 102, *116*
Dow, M. A., 239, 255, 265, 266, *274*, *281*
Drake, J. W., 39, *110*
Drew, R. T., 58, *118*
Dreyfus, A., 265, *273*
Dubinin, N. P., 42, 52, 98, *110*

Dudgeon, E., 305, *336*
Duncan, G. T., 251, *271*
Durrant, A., 172, *216*
Dusenbery, R., 62, *121*
Dyer, K., 42, 54, 61, *119*

E

East, P. D., 304, *334*
Eastwood, L., 255, *272*, *273*
Edwardes, P. M. J., 209, *218*
Eeken, J. C. J., 93, *110*, *121*
Eggleston, P., 100, *110*, 142, *150*
Ehrenfeld, J. G., 129, *149*, 264, *271*, 294, 295, 308, *334*
Ehrlich, P. R., 286, *336*
Ehrman, L., 129, 130, 148, *150*, 234, 241, 243, 244, 246, 249, 250, 251, 255, 260, 261, 262, 263, 264, 265, 266, 267, 268, 269, *271*, *272*, *274*, *279*, *280*
Eibl-Eibesfeldt, I., 224, *274*
Eisenbud, M., 58, *118*
El-Dabbagh, H., 325, *340*
Elens, A. A., 245, *274*
El-Helw, M. R., 319, *336*
Ellis, L. B., 243, *274*
El-Tabey Shehata, A. M., 304, *335*
El-Wakil, H. M., 325, *340*
Emmerich, M., 58, *118*
Engels, W. R., 16, 17, *31*, *34*, 47, 53, 63, 79, 94, 95, 102, 103, *109*, *110*, *111*, 128, 130, 135, 136, 137, 138, 139, 140, 142, 144, *149*, *150*
Eoff, M., 244, 267, 268, *274*, *275*
Erickson, J., 53, *111*
Erk, F. C., 308, *336*
Eshel, I., 390, 392, *396*
Esposito, M., 54, 62, 63, *108*
Esposito, R., 54, 62, 63, *108*
Evans, A. T., 68, *113*
Evans, W. H., 410, *420*
Ewing, A. W., 230, 231, 232, 234, 257, 258, 260, 267, 268, *271*, *275*

F

Fahey, T. M., 94, 102, *120*, 135, *153*
Fahmy, M. J., 42, 43, 45, *111*
Fahmy, O. G., 42, 43, 45, *111*

Falconer, D. S., 175, 192, *216*, 260, *275*
Fales, H. M., 241, *272*
Falk, D., 414, *424*
Falk, R., 21, 22, 23, *31*, *32*, 45, *116*
Faro, S. H., 270, *280*
Farrow, R., 291, 307, *335*
Fausto-Sterling, A., 136, 137, *153*
Feeley, S. M., 80, *115*, 135, 142, 144, *151*
Feldmann, A. M., 417, *420*
Fellows, D. P., 301, *336*
Felton, A. A., 127, *153*
Fielding, C. J., 137, *150*
Fienberg, S. E., 250, *271*
Fink, G. R., 404, *422*
Fisher, R. A., 16, *31*, 170, 214, *216*, *217*, 257, *275*
Fitz-Earle, M., 401, 403, 408, 409, 410, 411, 412, 414, *419*, *420*, *421*
Fleuriet, A., 58, 59, *111*, *114*, 131, 145, *152*
Ford, E. B., *217*
Ford, H., 290, 294, *333*
Foster, G. G., 400, 401, 408, 415, 417, 418, *420*, *421*, *424*
Futch, D. G., 346, 348, 352, 353, 356, 357, 363, *396*
Foureman, P. A., 53, *111*
Fowler, G. L., 238, 240, 257, *275*
Franca, Z. M., 64, *111*
Frankham, R., 177, 183, 188, *217*
Friedman, L. D., 14, 23, *32*
Fuerst, P. A. 241, *275*
Fujikawa, K., 44, *111*
Fujioka, N., 71, 72, 73, 74, *117*
Fukatami, A., 268, *275*
Futuyma, D. J., 38, *111*, 327, 328, *337*

G

Galkovskaya, K. F., 105, *118*
Game, J., 62, *111*
Garrido, M. C., 64, *111*
Gatti, M., 50, 62, 63, *108*, *111*
Gause, G. F., 285, 288, 293, 332, *337*
Gavin, J. A., 407, *421*
Geer, B. W., 60, *111*, 236, 263, *275*
Gehring, W. J., 39, *112*
Geigy, R., 137, *150*
Gerasimova, T. I., 43, 91, 92, *112*

Gershenson, S. M., 59, *112*
Gerstenberg, M. V., 74, 79, *116*, 145, *151*
Gethman, R. C., 415, *421*
Gibo, D. L., 327, *337*
Gibson, J. B., 170, 192, 193, 204, 208, *217*, 220, 262, 269, *283*
Gibson, W. G., 406, 408, *421*
Gill, D. E., 296, *340*
Gill, R., 408, *420*
Gilpin, M. E., 294, 295, 308, 329, *334, 337, 340*
Glass, B., 43, 44, 45, 50, 62, *112*
Goldschmidt, R., 56, 65, 66, 67, *112*
Goldstein, L., 292, *337*
Goldstein, P., 389, *396*
Goldstein, R. B., 230, 267, *278*
Golino, M., 62, *109*
Golubovsky, M. D., 59, 91, 98, 99, 102, 106, *112*, 136, 140, *150*
Gomatam, J., 294, *335*
Goodman, D., 328, *337*
Goodman, P., 45, *112*
Gould, S. J., 106, *112*
Goux, J. M., 251, *275*
Grace, D., 45, *112*
Graf, U., 40, 47, 53, 54, 62, 63, *112, 123*
Grant, B., 195, *217*, 262, 269, *273, 275*
Green, M. M., 39, 43, 45, 53, 54, 62, 63, 64, 75, 77, 87, 91, 92, 98, 102, 106, *108, 109, 112, 113, 115, 118, 119, 120*, 140, *150*, 236, 263, *275*
Greenberg, R., 6, *32*
Greenspan, R. C., 270, *276*
Greenspan, R. J., 234, 241, *276*
Grell, E. H., 47, 50, *116*, 404, 406, 407, 411, 413, 415, *421, 422*
Gromko, M. H., 241, 256, 257, 270, *275, 280*
Grossfield, J., 229, 236, 237, 268, *275*
Grove, J., 301, *338*
Grover, K. K., 417, 418, *420, 421*
Guest, R. T., 401, *422, 423*
Guille, G., 417, *422*

H

Haake, J., 40, *118*
Hadler, N. M., 164, *217*
Hager, H., 255, *281*
Haigh, J., 392, 393, *396*
Hallinan, E., 415, *423*
Hainsberger, L., 58, *118*
Haldane, J. B. S., 2, 30, *32*, 140, *150*
Hall, J. C., 234, 241, 242, 243, 244, 270, *275, 276, 277, 281*
Halle, F. S. von, 40, 47, 51, *115, 122*
Hallstrom, I., 58, *116*
Hamilton, A., 418, *425*
Hanks, G. D., 415, *423*
Hanson, S., 270, *278*
Hardin, G., 287, *337*
Hardy, D. E., 256, 261, 270, *272*
Hardy, R. W., 415, *423*
Harris, P., 62, *109*
Harrison, B. J., 165, 170, 181, 183, 184, 189, *218*, 262, *278*
Hartl, D. L., 73, 90, *113*, 415, *421*
Harvey, P. H., 178, *216*
Hastings, P., 106, *117*
Hazra, S. K., 45, *108*
Heck, K. L., 299, *337*
Hedrick, P. W., 328, *337*
Heed, W. B., 256, *276*, 300, 301, 304, 330, 331, *336, 337, 338*, 345, *396*
Heemert, C. van, 417, *421, 425*
Hellack, J. J., 87, *113*
Heller, R., 392, *397*
Henderson, S. A., 50, 53, 81, 84, 85, 86, *113*, 137, 141, *150, 153*
Henneberry, T. J., 400, 401, *421, 422*
Henslee, E. D., 372, 390, *397*
Herskowitz, I., 60, *113*
Hertwech, H., 235, *276*
Hildreth, P. E., 54, 57, *113*, 189, *217*
Hill, W. G., 403, 405, *420*
Hilliker, A. J., 406, 408, *421*
Himoe, E., 42, *107*
Hinton, C. W., 50, 64, 100, 101, *113*
Hinton, T., 68, *113*
Hiraizumi, Y., 26, *32*, 49, 64, 73, 74, 75, 77, 78, 79, 85, 86, 90, 102, *113, 116, 120*, 135, 140, 150, *151*, 415, *421*
Hisey, B. N., 87, *113*
Hoar, D., 88, *120*
Hodosh, R. J., 233, 255, 259, 261, *276, 281*
Hodson, A. C., 241, *273*
Hoenigsberg, H. F., 263, 269, *276*, 300, *337*
Holland, P. W., 250, *271*
Holm, D. G., 51, *114*, 406, 408, 409, 410, 411, 412, 414, 417, *420, 421*

Holm, R. W., 286, *336*
Holthausen, C. F., 65, *119*
Hoogmoed, M. S., 193, 194, 196, 197, 201, 202, *219*
Hortogagji-German, E., 269, *276*
Hosgood, S. M. W., 309, *337*
Hossain, M. A., 404, *421*
Hotta, Y., 238, 241, 242, 243, *276*
Hsu, T. C., 237, *282*
Huang, S. L., 45, 50, *114*
Hubby, J. L., 163, *217*
Hull, F. H., 178, *217*
Hunt, W. G., 106, *114*
Huston, M., 290, *337*
Hutchinson, G. E., 255, *276*, 287, 288, 290, *337*
Huxley, J. S., 224, *276*

I

Ikebuchi, M., 71, 72, 73, 74, *117*
Ikeda, H., 372, *397*
Inagaki, E. T., 43, 44, *111*, *114*
Inoue, Y., 73, *117*
Isackson, D. R., 75, *114*
Ising, G., 102, *114*
Ito, K., 41, 53, 59, 71, 72, 73, 74, 90, *114*, *115*, *117*, 146, *151*
Ivanov, Yu. N., 91, 98, 102, *112*, 140, *150*
Ives, P. T., 42, 43, 44, 54, 55, 57, 60, 61, 64, 67, 68, 80, 81, 84, 105, *113*, *114*, *115*, *119*, 135, *151*, 404, *422*
Iwatsubo, T., 242, 270, *276*

J

Jacobs, M. E., 236, 239, 247, 255, 265, *276*
Jaenike, J., 298, *337*
Jaffe, W. P., 404, *421*
Jallon, J.-M., 238, 241, 242, 243, 270, *276*
Jenkins, J. B., 44, *114*
Jinks, J. L., 168, 177, 184, *218*
Johannsen, W., 156, 168, *217*
Johnson, F. M., 26, *34*
Johnson, L. L., 259, *283*
Johnson, N. A., 94, 102, *120*, 135, *153*
Johnson, T. K., 75, *114*
Johnston, D., 45, *114*
Johnston, J. S., 256, *276*
Jollos, V., 43, 57, *114*

Jones, L. P., 177, 183, 188, *217*
Jong, G. de, 287, 307, *336*
Jonsson, U. B., 238, 239, 241, 257, *277*
Judd, B. H., 8, *32*, 39, 102, 103, *109*
Jupin, N., 59, *114*
Justice, K. E., 294, *337*

K

Käfer, E., 14, *32*
Kaneshiro, K. Y., 248, 265, *271*, *276*, 330, 331, *337*
Karess, R. E., 39, 106, *114*
Karlsson, L. J. E., 262, 269, *271*
Kastenbaum, M. A., 47, *114*
Kastritsis, P. A., 265, 268, *274*
Kaufman, T. C., 8, *32*, 414, *424*
Kaufmann, B. P., 50, 54, *108*, *114*
Kawanishi, M., 265, 268, *276*, *283*, 300, *337*, *341*
Kazini, S.J., 418, *420*, *424*
Kearney, M., 240, 241, 255, *272*, *273*
Kearsey, M. J., 100, *110*, 142, 145, *150*, *151*, 165, 166, 194, 213, 214, *215*, *217*
Kegel, G., 58, *118*
Kellogg, F. F., 233, *276*
Kernaghan, R. P., 130, *150*
Kerr, S., 239, 255, 256, *278*
Kessler, S., 207, *217*, 243, 268, *274*, *276*
Kharmac, I. S., 105, *118*
Kidwell, J. F., 17, 18, *32*, 53, 63, 68, 89, 80, 81, 84, 95, 105, *114*, *115*, 130, 134, 135, 136, 137, 138, *149*, *151*, 241, *275*
Kidwell, M. G., 17, 18, *32*, 39, 53, 63, 68, 79, 80, 81, 84, 86, 88, 95, 100, 103, 105, 106, *109*, *114*, *115*, *119*, *122*, 130, 131, 132, 133, 134, 135, 136, 137, 138, 139, 140, 141, 142, 143, 144, *149*, *151*, *152*, *153*, 170, *217*
Kikkawa, H., 64, *115*
Kilbey, B., 14, *31*
Kimura, M. J., 4, 7, 29, 30, *31*, *32*, 304, *338*
King, J. L., 7, 30, *32*
Kinst, M., 22, *34*
Kircher, H. W., 301, *337*, *338*
Kitagawa, O., 26, *32*, 264, *279*
Kitching, R. L., 418, *424*
Kitzmiller, J. B., 417, *423*
Klassen, W., 414, *422*
Klomp, H., 311, *338*

Knapp, E. P., 304, *335*, *336*
Knight, G. R., 16, *32*, 183, 188, 207, *216*, *217*, 269, *276*
Knipling, E. F., 414, *422*
Kodani, M., 52, *121*
Koepfer, H. R., 270, *283*
Kojima, K., 46, *122*, 194, 214, *217*
Konowalow, C., 417, 418, *420*, *424*
Koopman, K. F., 164, 207, *217*, 268, *277*
Koref-Santibañez, S., 258, 261, 262, 263, 265, 267, 269, *276*, *277*
Kratz, F. L., 58, *115*
Kreber, R. A., 17, *31*, 53, 94, 102, *109*, 135, *149*
Kröner, B., 255, *281*
Kruckeberg, J. F., 256, *282*
Kruijt, J. P., 251, *279*
Kupriyanova, L. A., 359, *396*
Kvelland, I., 45, *121*, 238, *277*, *283*
Kwietniak, R. T., 401, *423*
Kyriacou, C. P., 270, *276*, *277*

L

Labeyrie, V., 253, *277*
Lachaise, D., 304, 331, *338*
LaChance, L. E., 400, 402, *424*
Lamb, R. Y., 391, *397*
Lamy, R., 60, *115*
Lande, R., 256, *277*
Langley, C. H., 26, *34*, 41, 42, 74, 75, 85, 86, 90, *113*, *115*, *119*
Latter, D. B. H., 198, *219*
Laven, H., 417, *422*
Lavige, J. M., 17, *31*, 84, 98, 99, 100, *109*, *119*, 130, 131, 132, 134, 141, 144, 145, *149*, *151*, *152*
Lawlor, L. R., 325, 328, *338*
Lecher, P., 132, *151*
Lee, W. H., 42, *123*
Lee, W. R., 40, 47, 51, 55, *115*, *122*
Lefevre, G. Jr., 45, 50, *115*, 238, 239, 241, 257, *277*
Leigh, B., 406, *422*
Leigh, E. G. Jr., 389, 392, *397*
Lemeunier, F., 38, 88, *107*
LeMoli, F., 244, 267, *277*
Leng, E. R., 156, *217*
Leuders, K., 104, *115*
Levenbook, L., 241, *272*

Levene, H., 246, 247, 248, 249, 264, 267, *277*
Levine, R. P., 42, *115*
Levins, R., 286, 298, *338*, *340*
Levitan, M., 69, 70, *115*
Lewis, E. B., 45, 47, 50, 51, 52, 53, *107*, *115*, *116*
Lewis, K., 59, *108*
Lewontin, R., C., 2, *32*, 38, *116*, 127, *153*, 163, *217*, 246, *277*, 320, 323, 325, *336*, *338*
L'Heritier, Ph., 17, *31*, 58, 59, 98, 99, *109*, *116*, 130, 131, 144, *149*, *152*, 161, *217*, 291, 292, *338*
Li, C. C., 402, *422*
Li, W., 28, *32*
Lifschytz, E., 45, *116*
Lim, J. K., 8, 17, *32*, *34*, 84, 94, 102, *120*, 140, *153*
Lindgren, D., 55, 56, *116*
Lindsay, S. L., 305, *338*
Lindsley, D. L., 47, 50, 62, *116*, *119*, 404, 406, 407, 415, *422*
Linney, R., 213, *217*
Liu, C. P., 8, *32*
Lloyd, E. P., 418, *422*
Logan, D. M., 401, *423*
Lokki, J., 389, 390, 391, 393, *397*, *398*
Long, C. E., 238, *277*
Lorimer, N., 416, *423*
Lotka, A. J., 288, *338*
Lutolf, H. U., 407, *422*
Lutz, F. E., 228, *277*
Lyman, R. F., 41, 43, 53, 67, 84, 87, *123*, 394, 395, *398*
Lyttle, T. W., 415, *422*

M

McAndrew, F. T., 233, 255, *276*
MacArthur, R. H., 288, 296, *338*
McBride, G., 177, *218*
McCloskey, J. D., 404, 407, *422*
McCoy, J. R., 418, *422*
McDonald, I. C., 417, 418, *422*, *424*
McDonald, L. W., 265, *279*
McDonald, P. T., 416, *422*
MacDowell, E. C., 156, *218*
McEwen, F. L., 417, *423*
McGovern, W. L., 400, 401, *421*, *422*

McGuire, J., 414, *422*
Machin, D., 318, *340*
McKenzie, J. A., 408, 409, 410, *422*
McKnight, R. H., 52, *121*
McPhee, C. P., 182, *218*
Madalena, F. E., 177, 188, *218*
Magnusson, J., 58, *116*
Maier, P., 54, *123*
Mainardi, M., 244, 267, *277*
Mainland, G. B., 146, *151*
Mainx, F., 102, *116*
Maliuta, S., 59, *112*
Malogolowkin-Cohen, Ch., 246, 249, 261, 264, 265, *272*, *277*, 345, *397*
Mampell, K., 50, 68, 69, 84, *116*
Mangan, R. L., 296, 301, 302, 303, 331, *338*
Manning, A., 224, 225, 230, 231, 234, 237, 239, 240, 257, 258, 259, 260, 265, 267, 270, *271*, *275*, *277*, *278*, *281*
Markow, T. A., 238, 239, 255, 256, 270, *277*, *278*
Markowitz, E. H., 40, 43, *116*
Marques, E. K., 58, *110*
Martin, D. W., 77, 78, 102, *116*, 140, *151*
Maruyama, T., 7, 21, 23, *32*, *33*
Mason, H. C., 400, 401, *421*, *422*, *423*
Mason, J. M., 62, *116*
Masri, A. M., 299, *340*
Mather, K., 159, 160, 163, 164, 165, 166, 167, 168, 169, 170, 171, 172, 173, 174, 177, 178, 179, 180, 181, 183, 184, 186, 187, 189, 190, 192, 196, 197, 200, 201, 202, 203, 206, 209, 212, 214, 215, *215*, *216*, *218*, 262, *278*
Matthews, K. A., 74, 77, 78, 79, 86, 102, *116*, 140, 145, *151*
May, R. M., 290, *338*
Maynard Smith, J., 209, *218*, 255, *278*, 325, 328, *338*, 373, 387, 390, 391, 392, 393, *397*
Mayr, E., 210, *218*, 224, 244, 261, 265, 266, 267, 268, *278*, 393, *397*
Mendelson, D., 53, *116*
Merle, J., 241, *278*
Merrell, D. J., 245, 246, 248, 263, 265, 266, *278*, 290, 292, 308, 318, *338*
Mertz, D. B., 305, *338*
Mettler, L. E., 8, 9, 13, 14, 17, 26, *33*, 89, *110*, *117*, *118*, 195, *217*, 262, *275*
Meyer, H. U., 6, 8, *31*, *35*, 40, 42, 43, *107*

Michaelis, A., 39, *119*
Migamoto, T., 43, 44, *111*, *114*
Milani, R., 236, 265, *278*
Milkman, R. D., 7, 30, *33*, 241, *278*
Miller, R. S., 287, 288, 289, 297, 307, 312, 313, 315, *338*, *340*
Millicent, E., 204, *218*
Miller, D. D., 230, 232, 261, 265, 267, *272*, *278*
Milne, A., 286, *339*
Minamori, S., 53, 59, 70, 71, 72, 73, 74, *116*, *117*, 146, *151*
Minamori, Y., 71, 72, 73, 74, *117*
Minawa, A., 163, *219*
Misro, B., 189, *219*
Mitchell, H. K., 60, *123*
Mitchell, J. A., 16, 22, 23, *33*
Miyamoto, T., 44, *117*
Moawad, H., 319, *336*
Mohn, G. R., 58, *108*
Montgomery, S. L., 304, *339*
Moore, C. H., 164, 178, *215*
Moore, J. A., 308, 309, 312, 318, 326, *339*
Morgan, K., 106, *117*
Moriwaki, D., 64, *117*, 268, *275*
Morris, J. A., 175, 176, 177, 183, 188, *216*
Morton, L., 263, 269, *280*
Morton, N. E., 5, 28, *33*
Moth, J. J., 307, 314, 319, 322, *339*
Mourad, A. M., 299, 325, *340*
Mourão, C. A., 294, 311, 322, 323, 324, *339*
Mukai, T., 8, 9, 10, 13, 14, 17, 18, 21, 23, 24, 25, 26, 27, *31*, *33*, *35*, 46, 48, 53, 86, 88, 89, 90, *110*, *117*, *118*, *122*, *124*, 172, 174, *219*
Muller, H. J., 4, 5, 28, *33*, 40, 42, 43, 47, 49, 55, 57, 60, 62, *109*, *118*, 126, *151*, 269, *279*, 392, 397, 404, *423*
Mulley, J. C., 331, *334*, *339*
Murata, M., 22, 26, *33*, *34*
Murdy, W. H., 345, 348, 349, 351, 369, 389, *397*
Muretov, G. D., 105, *118*
Murray, D. C., 88, *122*, 141, *153*
Myer, H. V., 43, *118*

N

Nakamura, A., 53, 59, 73, *117*
Nakao, Y., 44, *117*

Narda, R. D., 234, *279*
Narise, T., 318, *339*
Natarajan, A. T., 58, *108*
Neel, J. V., 64, 65, 66, *118*
Nei, M., 28, *32*, 126, *151*
Neill, W. E., 296, *339*
Nellett, S. M., 94, 102, *120*, 135, *153*
Nawberne, P. M., 60, *118*
Newton, J. R., 401, *423*
Newton, M. E., 418, *425*
Newman, J., 305, *339*
Nguyen, T. D., 54, 62, 63, *108*, *109*, *118*
Nicholson, A. J., 287, *339*
Nicoletti, B., 62, *119*
Niederkorn, J. A. Jr., 348, 353, 354, 360, *396*
Niklasson, I., 389, *397*
Nill, A., 74, 75, 85, 86, *113*
Nishimori, T., 44, *111*
Nissani, M., 242, *279*
Novitski, E., 22, *34*, 147, *153*, 414, 415, *423*
Novy, J. B., 80, *115*, 135, 136, 138, 142, 144, *151*
Nur, U., 345, 352, 376, 387, 389, 391, *397*

O

Ochman, H., 389, *397*
O'Donald, P., 251, 257, *279*
O'Hara, E., 244, 267, *279*
Ohnishi, O., 8, 9, 13, 14, 22, 24, 25, *34*
Ohta, A. T., 261, 264, 265, 268, *279*
Olembo, R. J., 304, *335*
Olenov, J. M., 42, 105, *118*
Olivieri, G., 60, *110*
Oliveira, W. de, 304, *335*
Osgood, C. J., 62, *109*
Oshima, C., 26, *35*
Osman, H. E. S., 177, *219*
Oster, I. I., 40, 44, 47, 49, 87, 88, *118*, *121*, *123*
Ostertag, W., 40, *118*
Overland, D. E., 418, *422*

P

Paine, R. T., 290, *339*
Palomino, H., 307, *336*
Pandey, J., 21, *34*
Paniker, K. N., 417, *421*
Papworth, D. G., 42, 54, 61, *119*
Park, T., 286, 305, *339*
Parker, D. R., 40, *118*
Parker, S. L., 313, *339*
Paro, R., 39, *112*
Parry, D. M., 62, *118*
Parsons, P. A., 244, 250, 260, 262, 264, 266, 268, *274*, *279*, 309, 310, *337*, *339*
Paterson, H. E. H., 371, *397*
Patterson, J. T., 125, 129, 146, 147, *151*, 229, 240, 252, 265, *279*, 404, *423*
Patty, R. A., 230, 267, *278*
Pavan, C., 252, 261, 265, *272*, *273*, 303, *336*
Pavlovetz, M. T., 105, *124*
Pavlovsky, O. A., 164, 207, *216*, 241, 256, *274*, 323, *334*, 336
Paxman, G. J., 16, *34*
Payne, F., 156, 164, *219*
Peacock, W. J., 415, *423*
Pearl, R., 313, *339*
Pelecanos, M., 84, 85, 95, 97, *121*, *124*
Palisson, A., 13, *31*, 84, 98, 99, 100, *109*, *119*, 130, 131, 133, 134, 141, 142, *149*, *152*
Pell, P. E., 416, *420*
Pendleburg, W. W., 241, *275*
PennaFranca, E., 58, *118*
Periquet, G., 137, 145, 146, *152*
Peters, R. H., 287, 289, *339*
Petit, C., 244, 246, 249, 250, 251, 255, 264, 266, *274*, *279*
Petrow, H., 58, *118*
Petrunkevitch, A., 395, *397*
Phaff, H. J., 304, *335*, *336*
Philip, U., 64, *118*, 237, *279*
Phillips, J. C., 156, *216*
Picard, G., 17, *31*, 84, 98, 99, 100, *109*, *118*, *119*, 130, 131, 132, 133, 134, 140, 141, 142, 144, 145, *149*, *152*
Pimentel, D., 206, *220*, 291, *339*
Pimpinelli, S., 62, 63, *108*, *111*
Pipkin, S. B., 252, *279*, 304, *340*
Plough, H. H., 55, 57, 58, 65, *119*
Plunka, S. M., 244, 267, *279*
Plus, N., 58, 59, *109*, *112*, *114*
Podger, R. N., 286, 312, 313, 315, 318, 322, *334*
Pot, W., 251, *279*
Potter, J. H., 244, 267, *279*
Powell, J. R., 129, *150*, 263, 269, *280*
Prakash, S., 127, *152*, 255, 264, *280*
Pratt, N. P., 240, *281*

Preston, C. R., 16, 13, *31*, *34*, 79, 94, 95, 102, *111*, 135, 136, 137, 138, 142, 144, *150*
Probber, J., 234, 246, *274*
Promtov, 56
Propping, P., 39, *119*
Prout, T., 16, 26, *34*, 192, *219*, 241, 257, *280*, 403, 408, *420*, *423*
Proverbs, M. D., 401, *423*
Pruzan, A., 243, 244, 250, 251, 267, *279*, *280*
Pruzan-Hotchkiss, A., 270, *280*
Purdom, C. E., 42, 54, 57, 61, *119*
Putwain, P. D., 318, *340*
Pyle, D. W., 241, 256, 257, 270, *275*, *280*

Q

Quaid, M., 239, 255, 256, *278*
Quinn, W. G., 244, *280*

R

Rabbani, M. G., 417, *423*
Racine, R. R., 42, 90, *119*
Radding, C., 62, *119*
Rahat, A., 21, 23, *32*
Rai, K. S., 416, *423*
Raisbeck, J. A., 87, *107*
Ramel, C., 58, *114*, *116*
Rankin, M. A., 365, 371, 372, 384, 388, 395, *398*
Rao, M. S., 87, *119*
Rasmussen, I. E., 405, *423*
Rasmuson, M., 165, 177, 183, *219*
Ratnayake, W. E., 43, *119*
Raymond, J. D., 135, *153*
Raymond, J. O., 94, 102, *120*
Reddi, O. S., 87, *119*
Reddy, G. M., 87, *119*
Reed, E. W., 292, *340*
Reed, S. C., 236, *280*, 292, *340*
Reed, T. E., 30, *34*
Reed, W. E., 236, *280*
Reeve, E. C. R., 164, 183, 185, 188, 193, *219*
Reid, J. A. K., 405, 417, *423*
Rendel, J. M., 56, 197, 198, *219*, 226, 228, 236, 263, *280*
Reno, D., 60, *111*
Resnick, M., 62, *111*
Richardson, R. H., 234, 263, 269, *271*, 298, 299, *340*
Richmond, R. C., 241, *280*, 329, *340*
Rieger, R., 39, *119*
Rinehart, R. R., 54, *119*
Ringo, J. M., 233, 237, 239, 240, 255, 261, 264, 265, 269, 270, *276*, *280*, *281*, *283*
Ripoll, P., 54, 62, *108*
Ritterhof, R., 43, 44, 45, 62, *112*
Rizki, M. T. M., 319, *340*
Roberts, E., 170, *216*
Roberts, P. A., 50, 51, 87, *119*, 403, 416, *423*
Robertson, A., 16, *31*, *32*, 171, 174, 176, 177, 182, 183, 188, 207, *216*, *217*, *218*, *219*, 269, *276*, 306, *336*
Robertson, F. W., 164, 183, 185, 188, 193, 213, *219*
Robinson, A. S., 403, 404, 405, 416, *420*, *423*, *424*
Robinson, H. F., 178, *216*
Rockwell, R. F., 229, *275*
Rohlf, F. J., 248, 249, 250, *281*
Rosenfeld, A., 144, *152*
Roser, F. X., 58, *118*
Ross, M. H., 417, *424*
Roth, P., 62, *111*
Rothman, E. D., 352, 353, 354, 355, 373, 375, 376, 377, 378, 388, *398*
Rubin, G. M., 39, 103, 106, *109*, *110*, *114*, *119*, 139, 140, 143, *149*, *153*
Ruddle, N. H., 348, 349, 360, *396*
Russell, J. S., 301, *337*, *338*

S

Sage, R. D., 106, *119*
Saifuddin, U. T., 416, *419*
Sakai, R. C., 416, 418, *419*, *424*
Salas, S. P., 329, *340*
Sameoto, D. D., 307, 315, *340*
Sammeta, K. P. V., 286, *340*
Sandler, L., 54, 62, 63, *108*, *119*, 144, *152*
Sang, J. H., 308, *336*
Sankaranarayanan, K., 40, *119*
Sano, K., 23, *33*
Santos, E. P., dos, 164, *220*
Saraswathy, T., 88, *119*
Saura, A., 389, 390, 391, 393, *397*, *398*
Schaefer, G. B., 81, 85, *122*
Schaefer, R. E., 88, *122*, 136, 137, 141, *153*
Schaffer, H. E., 46, 88, 90, 120, *122*, 250, *281*

Schalet, A., 9, *34*, 40, *119*
Scharloo, W., 93, 194, 196, 197, 201, 202, 203, *215*, *219*, 262, 269, *273*
Schilcher, F. von, 230, 232, 239, 242, 255, 257, 265, 266, 270, *274*, *276*, *281*
Schiller, R., 70, *115*
Schneiderman, H. A., 240, *281*
Schuitema, K. A., 196, *219*
Schwarz, H. E., 103, *124*
Schwer, W. A., 253, 255, *282*
Sciandra, R. J., 255, *281*
Scobie, N., 88, *120*
Scott, E. L., 305, *339*
Scowcroft, W. R., 198, *219*
Scriba, M. E. L., 407, 410, *424*
Sears, J. W., 261, 265, 267, *281*
Seiler, J., 364, 389, 391, *397*
Selander, R. K., 106, *114*, *119*, 389, *397*
Sen, B. K., 177, *219*
Sen, S. K., 45, *108*
Serebrovsky, A. S., 402, 403, 405, *424*
Serghiou, C., 401, *424*
Setlow, R. B., 62, *109*
Settles, F., 404, *423*
Sewell, D., 236, 237, *273*
Shakarnis, V. F., 40, *120*
Sharma, V. P., 417, *421*
Sharp, C. B., 412, *421*
Shaw, K. E. S., 62, *109*
Shear, C., 62, *121*
Sheldon, B. L., 55, 57, *120*, 198, *219*
Sheldon, E. W., 16, 22, *34*
Shen, M. W., 8, *32*
Shepherd, S. H. Y., 75, 91, 92, 98, *113*
Sheppard, P. M., 170, 202, *216*, *220*
Shimizu, A., 73, *117*
Shimotohno, K., 104, *120*
Shiomi, H., 53, 59, 73, *117*
Shorey, H. H., 230, 234, *281*
Sidorov, B. N., 52, *110*
Siegel, R. W., 244, 270, *276*, *281*
Silagi, I. S., 264, *272*
Simberloff, D., 296, *340*
Simmons, A. S., 246, 249, 264, *277*
Simmons, J. R., 45, *123*
Simmons, M. J., 16, 17, 21, 22, 23, 24, *33*, *34*, 84, 94, 102, *120*, 135, 140, *153*
Simpson, G. G., 160, *220*
Sinclair, D., 91, 93, *120*
Sing, C. F., 346, 355, 356, 358, 360, 363, 364, 367, 368, 371, 376, 378, 381, 382, 383, 384, 385, 387, 394, 395, *397*, *398*
Singer, K. M., 88, *120*
Singh, D., 417, *421*
Singh, K. R. P., 417, *421*
Singh, R. S., 127, *153*
Sismanidis, A., 157, 180, 181, 182, 188, 189, *220*
Skibinski, D. O. F., 203, *220*
Skirzipek, K. H. von, 255, *281*
Slatko, B. E., 68, 74, 75, 76, 77, 78, 79, 85, 86, 87, 90, 93, 102, *113*, *116*, *120*, *123*, 140, *151*, *153*
Slizynski, B. M., 39, *120*
Smith, D. B., 26, *34*
Smith, F. F., 400, 401, *422*, *423*
Smith, K., 62, *109*
Smith, L. H., 156, *220*
Smith, P. D., 44, 54, 62, 63, *108*, *120*, *121*
Smith, R. H., 414, 418, *424*
Smith-Gill, S. J., 296, *340*
Smouse, P. E., 298, 299, *340*
Snedecor, G. W., 249, *281*
Snyder, L. A., 8, *32*, 44, *121*
Snyder, R., 62, *121*
Snyder, S. H., 360, 395, *396*
Sobels, F. H., 47, 51, 55, 58, 93, *110*, *121*, *122*, *123*, 406, 410, *419*, *422*
Sochacka, J. H. M., 87, *121*, 141, *153*
Sokal, R. R., 248, 249, 250, *281*
Sokoloff, A., 297, 304, 312, *335*, *340*
Sokoloski, E. A., 241, *272*
Solar, E., del, 258, 261, 262, 265, 269, *277*, *281*, 307, 336
Soliman, M. H., 309, 310, *340*
Soni, V. K., 416, *424*
Southern, D. I., 416, *420*
Southin, J. L., 44, *110*, *121*
Souza, H. L. de, 164, *220*
Spassky, B., 26, *31*
Spencer, W. P., 38, 41, 47, 48, 50, 98, 106, *121*
Spickett, S. G., 190, *220*
Spiess, E. B., 38, *121*, 238, 244, 246, 251, 253, 255, 256, 260, 266, 271, *274*, *281*, *282*
Spiess, L. D., 251, *282*
Spieth, H. T., 225, 226, 228, 229, 233, 234, 235, 236, 237, 238, 239, 244, 245, 247, 254, 255, 256, 257, 258, 259, 260, 261, 265, 266, 267, 268, 269, 270, *272*, *282*

Spofford, J. B., 52, *121*
Sprackling, L. S., 348, 354, *397*
Spradling, A., 103, *119*
Stalker, H. D., 88, *121*, 164, *220*, 247, 261, 265, *282*, 345, 346, 347, 348, 352, 354, 356, 360, 362, 363, 368, 389, 393, *397*, *398*
Stamatis, N., 85, 95, 96, *121*, *124*
Stebbins, G. L., 38, 48, *110*
Stern, C., 22, *34*, 41, 52, *121*
Stevens, W. L., 47, *121*
Stille, B., 389, *397*
Stone, W. S., 147, *151*, 240, 256, 261, 265, 270, *272*, *279*, 404, *423*
Streams, F. A., 206, *220*
Streisinger, G., 245, 265, *282*
Strickberger, M. W., 323, *340*
Strömnaes, O., 45, *121*, 238, *283*
Sturtevant, A. H., 105, 106, *121*, 147, *153*, 156, *220*, 224, 225, 226, 228, 229, 233, 260, 268, *283*
Suche, M., 404, *423*
Sugimoto, K., 72, *117*
Suguna, S. G., 417, 418, *420*, *421*, *424*
Suomalainen, E., 352, 353, 373, 389, 390, *398*
Sustare, D., 259, *283*
Suzuki, D. T., 88, *120*, 408, 409, 410, 411, 414, *420*, *424*
Sved, J. A., 16, 17, 18, 30, *32*, *34*, 53, 63, 80, 81, 84, 88, 95, 105, *115*, *121*, *122*, 130, 135, 136, 137, 138, 141, *151*, *153*
Sziber, P. P., 244, *280*

T

Tajima, F., 17, *35*
Takamura, T., 264, *279*
Tan, C. C., 263, 265, *283*
Tantawy, A. O., 195, *220*, 299, 309, 310, 325, *340*
Tatsukawa, K., 59, *117*
Tayel, A. A., 195, *220*
Tchurikov, N. A., 103, 122
Teissier, G., 161, *217*, 291, 292, 338
Temin, R. G., 6, 8, 17, 21, 22, 24, 26, 28, *31*, *35*, 40, 43, 61, *110*
Templeton, A. R., 126, *153*, 346, 352, 353, 354, 355, 356, 358, 360, 363, 365, 367, 368, 369, 370, 371, 372, 373, 375, 376, 377, 378, 379, 380, 381, 382, 383, 384, 395, 386, 387, 388, 389, 390, 393, 394, 395, *396*, *397*, *398*
Teramoto, L. T., 370, 371, 372, *396*
Ter Kuile, A., 193, 194, 196, 197, 201, 202, *219*
Terwedow, H. A., 416, *422*
Thoday, J. M., 163, 165, 170, 180, 183, 189, 190, 192, 195, 197, 201, 202, 203, 204, 205, 208, 210, 211, *217*, *218*, *220*, 262, 269, *283*
Thompson, J. N., Jr., 41, 43, 50, 53, 63, 67, 79, 81, 84, 85, 86, 87, 90, 102, 103, 105, 106, *113*, *122*, *123*, 137, 141, *150*, *153*, 163, *220*, 394, 395, *398*
Tidwell, T., 264, *272*
Timofeeff-Ressovsky, N. W., 56, 57, 106, *122*
Tiniakov, G. G., 101, *122*
Tobari, I., 22, 24, *34*, 35
Tobari, Y. N., 46, 64, *117*, *122*
Tokuyasu, K. T., 415, *423*
Toll, G. L., 304, *334*
Tompkins, L., 270, *276*
Townsend, J. I., 129, *153*
Tracey, M. L., 128, 129, *149*, 264, *271*
Triantaphyllou, A. C., 389, *396*
Trippa, G., 62, *119*
Trivers, R. L., 256, *283*
Truett, M. A., 103, *122*
Tsacas, L., 304, *338*

U

Ullrich, R., 264, *272*
Uphoff, D., 22, *34*
Uppal, D. K., 416, 417, 418, *420*, *421*, *424*

V

Valadares, M., 101, *122*
Valencia J. I., 42, *118*
Valencia, R. M., 40, 42, 47, 51, *115*, *118*, *122*, *123*
Valentine, J. W., 38, 48, *110*
Veiga-Neto, A. J., 58, *110*
Vetukhiv, M., 211, *221*
Voelker, R. A., 42, 46, 86, 87, 88, 90, 97, *119*, *122*, 135, *153*
Vogel, E., 47, 51, 55, 58, *108*, 121, *122*, *123*
Vogt, W. G., 418, *421*, *424*
Volterra, V., 288, 332, *341*

W

Waddington, C. H., 196, 198, 207, *217, 220,* 262, 263, 269, *276, 277*
Waddle, F. R., 87, 88, *123*
Wagner, R. P., 60, *123,* 304, *341*
Wagoner, D. E., 417, *424*
Waldron, I., 230, *283*
Wallace, B., 4, 8, 20, 23, 24, *35,* 38, 40, 41, 42, 43, 48, 61, 62, 89, *123,* 150, 164, 207, 211, *220, 221,* 237, 261, 265, 269, *283,* 315, 316, 317, 330, 332, *341,* 402, *424*
Warren, D. C., 164, 178, *215*
Warters, M., 146, *153*
Wasserman, M., 270, *283,* 331, *335*
Watanabe, T. K., 26, *35,* 42, 89, *118, 123,* 265, 268, *276, 283,* 300, *337, 341*
Waterhouse, D. F., 400, 402, *424*
Wattiaux, J. M., 245, *274*
Wehrhahn, C. F., 405, *423*
Wei, I. Y., 348, 353, 354, 360, 370, 373, 378, 379, 384, *396, 398*
Weisbrot, D. R., 306, *341*
Westover, C. J., 384, 385, 395, *397*
Wharton, L. T., 147, *153,* 359, *398*
Wheeler, M. R., 147, *151,* 224, 229, *283,* 345, *396*
White, M. J. D., 394, *398*
Whitelaw, A., 418, *424, 425*
Whitten, M. J., 400, 401, 402, 408, 415, 417, 418, *420, 421, 424*
Widders, P. R., 304, *334*
Wigan, L. G., 164, 171, 173, 174, 186, *218, 221*
Wijnands-Stäb, K. J. A., 417, *421, 424*
Wijnstra, J. G., 196, *219*
Wilkinson, C., 55, *123*
Willey, R. B., 391, *397*
Williams, G. C., 390, *398*
Williams, W. R., 145, *151*
Williamson, D. L., 58, 70, *115, 123*
Wills, C., 7, 30, *35*
Wilson, E. O., 255, *283,* 291, 335
Winchester, A. M., 45, *114*
Winter, F. L., 156, *221*
Wirth, W. W., 401, *423*
Wit, C. T. de, 290, 315, *336*
Witt, A. A., 285, 293, *337*
Wolfson, M., 370, *398*
Wolstenholme, D. R., 204, *221*
Wood, D. F., 259, 264, 265, 270, *281, 283*
Wood, R. J., 418, *424, 425*
Woodburn, T. L., 418, *421, 424*
Woodruff, D. S., 106, *112*
Woodruff, R. C., 40, 41, 43, 45, 47, 50, 51, 53, 63, 64, 67, 79, 81, 84, 85, 86, 87, 90, 102, 103, 105, 106, *113, 121, 122, 123,* 137, 141, *150, 153,* 394, 395, *398*
Woods, R. A., 291, 307, *335*
Woolf, C. M., 248, *283*
Wright, S., 28, *31,* 156, 157, 170, *216, 221*
Würgler, F. E., 8, *31,* 40, 43, 47, 51, 53, 54, 62, 63, *107, 112, 115, 122, 123,* 407, *422*

Y

Yaeger, P., 238, *277*
Yamazaki, T. 24, 25, *33*
Yamaguchi, O., 17, 26, *33, 35,* 48, 86, 87, 88, 89, 90, *117, 123, 124*
Yannopoulos, G., 50, 84, 85, 86, 95, 96, 97, 102, *121, 124,* 136, 140, 145, *153*
Yeh, H. J. C., 241, *272*
Yoon, J., 59, *109*
Yoshikawa, I., 21, 23, 24, 25, 26, *33, 35*
Ytterborn, K. H., 404, *425*
Yule, G. U., 250, *283*

Z

Zacharopoulou, A., 84, 85, 96, 97, *124*
Zakharov, I. K., 98, *112*
Zamb, T., 62, *111*
Zeiger, E., 60, *118*
Zeither, R. R., 241, *278*
Zeleny, C., 156, *221*
Zijlstra, J. A., 58, *108*
Zimmering, S., 40, 47, 51, 53, 54, 63, *115, 118, 122, 123, 124,* 291, 292, 324, 341, 403, *425*
Zouros, E., 264, 270, *283*
Zuitin, A. I., 105, *124*
Zweep, A., 196, *219*

SUBJECT INDEX

Compiled by R. C. Woodruff and B. Woodruff.

Page numbers in bold type indicate pages on which figures and tables appear.

A

Activity of parthenogenic strains, 372–373
Adaptiveness, of mating behavior, 251–257
Adult, density effects on competition, 311–317
Age
  and fecundity of males, 238–239
  and *GD* sterility, 138
  and mating propensity, 238–239
  of medium and competition, 318–319
  and mutation rates, 60–61, **61**
  and sexual receptivity, 239–242
Aneuploidy
  in crosses with compound autosomes, 407–408, **408**
  in crosses with compound; free arm strains, 411–412, **412**
Antennae
  and auditory stimuli, 231
  and olfactory stimuli, 233
  and sexual isolation, 267
Amphitoky, 344
Aristae, and auditory stimuli, 231
Arrhenotoky, 344
Assortative mating, selection for, 207–209
Automictic thelytoky
  population genetic models for, 373–376
  population genetic theory for, 373–376

B

Biological control (*see also* Pest control)
  and competition, 330–333
Body length, and induced mutations, **15**
Bristles (*see* Chaetae)

C

Canalization, 196–199

Central fusion, 349–352, **350–351**
  and permanent heterozygosity, 352
  and recombination, 349–352, **350–351**
Chaetae
  abdominal, selection for number 164–165, **164**, 173–179, **174**, **176**, **179**, 183–184, **184**, 192
  number
    and fertility, 186–187
    and induced mutations, **15**
    and sexual isolation, 262
  scutellar, selection for number 156–157, **164**, 164–165, 180–182, **181**, **182**, 197–198
  sternopleural
    number and fitness, 213–214, **213**
    selection for number **164**, 165–168, **166**, 192–193, **193**, 195, 201–202, 204–205, **205**, 213, **213**
Chemoreceptors, and mating behavior, 235
Chromosomal contamination, 99, 133–134, 142
Chromosome
  breakage
    in *P-M* hybrids, 135
    measured by crossover suppression, 51
    methods to measure spontaneous rates of, 50–52
    precautions in measuring, 52–54
    rates by cytogenetic screens, 50
    screens by position effect, 52
    screens using compound autosomes, 51–52
    spontaneous rates, **47**
  rearrangements (*see also* Deficiencies, Inversions, Translocations and Transpositions and Mutant Index)
    and control of pest insects other than *D. melanogaster* 416–418
    and hybridization 394
    induced by mutators, 65–66, 68–70, 75,

Chromosome—*continued*
  rearrangements—*continued*
    82–84, 85–88, **86**, 93–97, **95**, **97**, 98–104
    and pest control, 402–403, 416–418
    replacement and frequency-dependent selection, 402–403
Coefficient
  of isolation, 248
  of mating success, 249–250
Coexistence, 290–291
  in laboratory populations, 291–296, **293**, **295**
Community matrix, 298–299, **299**
Competition, 285–333
  among Hawaiian species, 298–299, **299**, 303
  among Sonoran Desert species, 300–303, **301–302**
  categories of, 286–287, **288**
  coexistence, 290–291
  competitive ability, 320–328, **322**, **324**, **326**
    changes in two-species populations, 325–328, **326**
    evolution of, 326–328
  components of 286–287, **288**
  contest and scramble, 287, **288**
  defined, 286
  detection and importance of, 296–297
  direct evidence for, 300–303, **301–302**
  ecology of 305–320, **310**, **314**, **316**
  elimination of one species, 289–290
  environmental factors affecting 309–320, **310**, **314**
  for food, 305–307
  for what? 305–307
  frequency dependent, 317–318
  implications in evolution, ecology and biological control, 330–333
  indirect evidence for, 298–300, **299**
  in laboratory populations, 291–296, **293**, **295**
    first studies, 291
    methods to study, 291–292
  interspecific (*see* Interspecific competition)
  involving more than two species, 328–330, **329**
  larval biotic residues, 306
  larval density effects, 311–317
  larval dispersal, 308

Competition—*continued*
  light, 319–320
  medium age, 318–319
  microhabitat divergence, 303–305
  in natural populations, 296–305
  nature of, 286–287, **288**
  niche heterogeneity, 308
  niche separation, 303–305, 330–331
  niche specification and heterogeneity of resources, 308
  oviposition site preference 304–305, 308
  population density, 307, 311–317, **314**
  population fitness, 320–328, **323**, **324**, **326**
  possible outcome of, 289–291
  radiation, 320
  temperature, 309–311, **310**
  yeast preference, 304, 319
Competitive exclusion, 287–290
  defined, 287–289
  aneuploidy in crosses with, 407–408, **408**
  assortment in females and males, 407, **407**
  and chromosome breakage screens, 51–52
  described, 405–406
  dominant lethality, 407–408, 410
  how formed, 405–406, **406**
  meiotic behavior of, 405–408, **406–408**
  and population replacement, 405–410, **406–408**
  in laboratory and field studies, 408–410
  survival of stocks, 410
Compound; free arm
  aneuploidy in crosses with, 411–412, **412**
  described **406**, 410–411
  dominant lethality, 411
  meiotic behavior of, 411, **411**
  and population replacement, 410–413, **411–412**
Contest, 287, **288**
Copulation, 229
  duration of, 229
  effect, 240
  sexual receptivity, 240
  termination of, 229
CO_2 sensitivity, by sigma virus, 58–59
Courtship behavior (*see* Mating behavior)
Courtship displays (*see* Mating behavior)
Courtship songs (*see also* Stimuli, auditory)
  pulse length and interpulse interval, 232
  of sibling species, **231**
  sine and pulse, 230, 232, **231**

SUBJECT INDEX lxvii

Craymer effect, 53
Crossover suppression, and chromosome breakage, 51
Cyclohexamide, treatment and mating behavior, 243
Cy-killer, 59–60
 and distortion of segregation, 59–60
 and mutations, 59–60
Cytogenetic screens, for chromosome breakage, 50
Cytoplasm Y-chromosome interaction and hybrid sterility, 130
Cytotype, 134–146
 defined, 139
 switching, 139

D

Darkness, and mating behavior, 236–237
Deficiencies (*see also* Chromosome rearrangements and Mutant Index)
 mutator induced, 68, 70, 88–91, 93–100, **97**
 spontaneous, 39
Density, modification of sexual selection, 266–267
Development, time and induced mutations, **15**
Digestive tract, microsomal enzymes, 55
Discrimination index, 250
Dispersed load, 5–6
DNA elements (*see* Transposable DNA elements)
DNA repair, 62–63
Dominance, 19–29, 214
 and selection, 214

E

Ecological adaptation, and sexual isolation, 269
Ecology
 and competition, 330–333
 of competition, 305–320, **310**, **314**, **316**
Eggs
 development and SF sterility, 132
 hatchability
  and clonal selection, 379–381, **380**

Eggs—*continued*
 hatchability—*continued*
  in compound autosome stocks 407–408, 410
  in compound; free arm stocks, 411
  and induced mutations, **15**
  and population density, 314–315
  and SF sterility, 132–133
 laying capacities
  and *abnormal abdomen* 364–367, **366**
  and parthenogenesis, 364–367, **366**
 productivity and induced mutations, **15**
 uninseminated and meiosis, 348
Entire compound autosomes
 described, 414
 and pest control, 414–415
Epistasis
 and fitness, 29–30
 as a load-reducing mechanism, 29–30
 and mutation load, 6–8, 29–30
Ethylmethanesulphonate (EMS)
 induced mutations, 14–16, **15**, 21, 23–25
 heterozygous expression, 21–25
Evolution
 chromosomal and parthenogenesis, 389
 and competition, 330–333
 of competitive ability, 326–328
 of sex, 386–393
 of sexual isolation, 268–270
 of a thelytokous population, 386–387
Exclusion, in laboratory populations, 291–296, **293**, **295**
Experience
 and mating preferences, 243–244
 and sexual selection, 267
Exploitation, component or category of competition, 286–287, **288**

F

Fat bodies, microsomal enzymes, 55
Fate mapping, 242–243, **243**
 methods of, 242
 of mating behavior foci, 242–243, **243**
Fecundity
 and age of males, 238–239
 and clonal selection, 379–381, **380**
 and mating experience, 239
 and population density, 311–317, **314**

Feeding sites, 251–252
Fertility, and chaetae number, 186–187
Fitness
　Darwinian defined, 320
　and epistasis, 29–30
　heterozygous expression of dominant mutations, 19–29
　heterozygous expression of spontaneous mutations, 23–25
　and induced mutations, 14–16, **15**, 23–29
　and mutation rate, 1–19
　mutations
　　nature of, 18
　　in noncoding portion of genes 18
　and partial dominance mutants, 27–28
　population
　　defined, 320
　　measured by competitive ability, 320–328, **322, 324, 326**
　　methods to measure, 320–321
　and population density, 315–316
　and spontaneous mutations, 23–25
　and sternopleural chaeta number, 213–214, **213**
　of translocation bearing stocks, 403–405
　and truncation selection, 30
Founder effect, 393–395
Free arms (see Compound; free arms)
Frequency-dependent mating, 246, 251
　measure of, 251

G

Gamete duplication, 353, **354**
　and homozygosity, 353, **354**
　and recombination, 353, **354**
Gause's Principle (see Competitive exclusion)
Genetic drift, and sexual isolation, 269
Genetic load, 402
Genetic pest control (see Pest control)
Genetic revolution, 393–395
Genetic transilience, 393–395
Genetic variability, for parthenogenesis, 363–364
GD hybrids, genetic changes in, 135
GD sterility (see also Hybrid sterility and Sterility, intraspecific hybrid)

GD sterility, 134–146, **141, 143**
　and age, 138
　comparison with SF sterility, 140–146, **141, 143**
　defined, 135
　F_2 sterility, 137–138
　paternal contributor (P), maternal contributor (M), and neutral (Q) strains 134–146, **141, 143**
　physiology and morphology of, 136–137
　strain variability, 134–135
　and temperature, 137–138, **141**
　temperature sensitivity period, 138, **141**
Glycolytic profile, 384–385

H

Hardy-Weinberg Law, 4
Hawaiian species
　competition among, 298–299, **299**, 303
　mating behavior, 233, 239, 254, 260
Heterozygosity, levels in parthenogenetic progeny, 355–356, **356**
Hybrid dysgenesis, 16–18, 27, 63–100, **75–76, 80, 82–84,** 102–104, 394–395
　defined, 17
　effect of viability, 17–18, 27
　measure of mutation rates for viability, 16–18, 27
　P-M system, 79–81, 93–95, 102–104
　reciprocal cross effect, 80–81, 81–88, **82–84,** 86–88
　suppressor of, 79
　and transmission ratio distortion, 18, 27
Hybrid release, 87–88
Hybrid sterility (see also GD sterility, SF sterility, and Sterility, intraspecific hybrid)
Hybrid sterility, 125–148
Hybrid zone, and mutation rates, 106
Hybridization, and increased mutation rates and chromosome aberrations, 394
Host plant, of Sonoran Desert species, 300–303, **301**

I

I factors
　chromosome contamination, 133–134

SUBJECT INDEX   lxix

I factors—*continued*
   hereditary transmission, 133
   mobile genetic elements, 133
I strains, 131–146, **141**, **143**
I–R system
   comparison with *P–M* system, 140–146, **141**, **143**
   and *SF* sterility, 131–146, **141**, **143**
Immigration, and population replacement by compound autosomes, 410
Inbreeding
   effect on selection, 168–170, **169**, 192
   effect of variability, 168–170, **169**
   and lethal mutation elimination, 28–29
   and sexual isolation, 263
Insect pest control (*see* Pest control)
Insemination reaction, 240–241
Interchromosomal effect and male recombination, 79
Interference, component or category of competition, 286–287, **288**
Interspecific competition, 285–303
Intersexual selection, 255–257
Inversions (*see also* Chromosome rearrangements and Mutant Index)
   mutator induced, 70, 75, 88–91, 93–97, **95**, **97**
   pericentric, and pest control, 416
Isolation estimate, 248
Isolation index, 247–249
Isozymes
   mutation rates, 8
   variability in parthenogenic strains, 385–386.

J

Joint isolation index, 248
Juvenile hormone (JH), and sexual receptivity, 240

L

Larvae, density effects on competition, 311–317
Learning, and mating behavior, 244
Lek behavior, 233, 239
Lethal equivalent, 5

Lethals, 8–9, 20–30
   dominant, 8–9
   and inbreeding, 28–29
   number of loci that can mutate to, 8
Life history
   consequences of clonal selection, 379–381, **380–381**
   and mating behavior, 251–254
Light
   and competition, 319–320
   and mating behavior, 236–237
   and sexual isolation, 268
Load, mutations (*see*, Mutation load)
Longevity, and induced mutations, **15**
Lotka–Volterra model, 288–289

M

M component (*see also M* strain)
   hereditary transmission of, 139–140
M cytotype, 94–95, 102–104
M strains (*see also M* components)
M strains, 134–146, **141**, **143**
   distribution of, 135–136, 143–144
Macroevolution, of mating behavior, 257–260, **259**
Maize, selection in, 155–156
Male mating advantage, 234
Male recombination, 52–53, 73, 75, **75–76**, 77–91, **80**, **82–84**, **86**, 91–100, 102–104, 135–141
   clusters of, 75, 80
   and interchromosomal effect, 79
   mutator induced, 73, 75, **75–76**, 77–91, **80**, **82–84**, **86**, 91–100, 102–104
   in *P–M* hybrids, 135
   selection for, 80–81, **80**
   transmitted by injection, 141
Malpighian tubules, microsomal enzymes, 55
Mating behavior 169–170, 223–270, 369–373
   acceptance response, 225, **227–228**, 239–242
   adaptiveness of, 251–257
   ancestral pattern, 253–254
   auditory stimuli, 230–232, **231**
   basic nature of, 225–229, **226–228**
   chemoreceptors, 235

Mating behavior—*continued*
  courtship tracking, 238
  decapitated females, 236
  displays, speed of, 258–260
  experience and preferences, 243–244
  fate mapping foci, 242–243, **243**
  female, 239–242
  first appearance of, 237–242
  first report, 224
  historical development, 224–225
  housing conditions of adults, 243–244
  lack of wings and aristae, 231–232
  life history, 251–254
  light and dark, 236–237
  male displays and mating propensity, 238–239
  mating propensity, 238–239
  microevolutionary and macroevolutionary changes, 257–260, **259**
  models of, 251
  olfactory stimuli, 233–234
  in parthenogenic strains, 369–373
  pheromone release, 233–234
  phylogeny, 257–261
  ontogeny, 237–244, **243**
  rejection responses, 225, **227–228**, 239–242
  and selection, 169–170
  sequences of displays, 258–260, **259**
  sexual activity, measurement of, 244–251
  sexual receptivity, 239–242
    and age, 239–242
    after copulation, 240
    and juvenile hormone (JH), 239–242
    and ovarian development, 240
    and paragonial gland, 240–241
    presence of sperm in sperm receptacles, 240–241
  signaling movements, 225–229, **226–228**
  stimuli involved in, 229–237, **231**
  tactile stimuli, 234–236
  types of, 239
  visual stimuli, 236–237
Mating speed, and induced mutations, **15**
Meiosis
  behavior of compound autosomes, 405–408, **406–408**
  behavior of compound; free arm line, 411, **411**
  in intraspecific hybrids, 130
  in uninseminated eggs, 348

Meiotic drive
  and control of pest insects other than *D. melanogaster*, 418
  and pest control, 415
  pseudo-Y, 415
Methods
  autosomal lethal mutation rates, 48–49
  chromosome breakage screens
    by crossover suppression, 51
    by cytogenetic screens, 50
    by position effect, 52
    using compound autosomes, 52
  clonal selection, 377–378
  competition in laboratory populations, 291–292
  derive expectations for response to selection, 175
  dominant visible mutation rates, 49–50
  fate mapping, 242
  free and homozygotic potential variability, 165–170, **165**
  frequency-dependent mating propensity, 251
  generate bridge stocks, 360, **361–362**
  interspecific competitive ability, 320–324, **324**
  minor viability mutation rates 9–10, **9**
  mutation rates, 8–19, **9**, 39, **40–47**, 47–54
  nonrandom mating, 247–251
  pest control by sterile insect release, 400–401
  population fitness, 320–324, **324**
  producing sterile males, 401
  recessive visible mutation rates, 49
  selection of abdominal chaeta number, 175–176
  selection experiments, 161–162
  sex-linked lethal mutation rates, 47–48
  sexual activity, 244–251
  sexual isolation, 247–251
  spontaneous rates of chromosome breakage, 50–52
  translocation rates, 50–51
  wind response, 373
  unit of selection, 382–385, **382**
Methyl esters, and minority male mating advantage, 234
Microevolution, of mating behavior, 257–260, **259**

SUBJECT INDEX

Microsomal enzymes, 55, 58
   location of 55
Minority male mating advantage, 234, 246, 256
   and methyl esters of straight-chain carboxylic acids, 234
Mobile DNA elements (*see* Transposable DNA elements)
Modifiers, of mutator activity, 105–106
Models, of mating behavior, 251
Morphology, of *GD* sterility, 136–137
Muller's ratchet, 392–393
Mutants, mutagen sensitive, 62–63
Mutation
   clusters and mutator activity, 66, 68, 75, 91–93, 98–100
   in coding and noncoding portion of genes, 18
   conditional lethal gene and pest control, 413–414
   effects on selection, 171–175, **174**
   electrophoretic
      rate of 90–91
      and subspecies of *D. equinoxialis*, 129
      and subspecies of *D. pseudoobscura*, 127–128
      and subspecies of *D. willistoni*, 128–129
   fitness, nature of, 18
   heterozygous expression, 19–29
   induced, 14–16, **15**, 23–25
      abdominal bristle number, **15**
      body length, **15**
      development time, **15**
      egg hatchability and productivity, **15**
      and fitness, 14–16, **15**, 23–29
      heterozygous expression, 20–30
      longevity, **15**
      mating speed, **15**
      phototaxis, **15**
      viability, 14–16, **15**, 23–29
   insertion of *P* elements, 139–140
   isozymes, rates for, 8
   lethals (*see*, Lethals)
   male sterilizing and pest control, 401
   methods
      to measure autosomal lethal rates, 48–49
      to measure dominant visible rates, 49–50

Mutation—*continued*
   methods—*continued*
      to measure recessive visible rates, 49
      to measure sex-linked lethal rates, 47–48
   minor viability, 9–14, **10**, 20–30
      methods to measure rates, 9–10, **9**
   from natural populations, 25–27
   and natural radiation, 58
   partial dominance and fitness, 27–28
   in *P–M* hybrids, 135
   polygenic and radiation, 174–175
   potential, 54
      extrinsic factors affecting, 54–60, **56–58**
      intrinsic factors affecting, 60–63, **61**
   rate
      and age, 60–61, **61**
      defined, 39
      and fitness, 1–19
      in hybrid zones, 106
      and hybridization, 394
      measurement of, 8–19, **9**, 39, **40–47**, 47–54
      and mutagen sensitive mutants, 62–63
      and nutrition, 60
      and parthenogenesis, 389–390
      per cell division, 55
      precautions in measuring, 52–54
      and population size, 105
      and sex, **40–47**, 61–62
      spontaneous electrophoretic, 46
      spontaneous forward, **43–45**
      spontaneous lethals, **40–43**
      spontaneous lethals in females, **40–43**
      spontaneous lethals in males, **40–43**
      spontaneous reverse, **45–46**
      and temperature, 55, **56–58**
   in *SF* sterile hybrids, 131
   spontaneous and dominance, 23–25
   temperature-sensitive conditional lethal and pest control in wasp, housefly and mosquito, 418
   temperature-sensitive male sterilizing, 401
   temperature sensitive and pest control, 401, 413–414
   by transposable DNA elements, 102–104

SUBJECT INDEX

Mutation—*continued*
  unstable
    induced by mutators, 65, 91, 93–95, 98–100, **99**, 102
    and *P* element insertions, 140
    and variability, 171–175, **174**
  viability
    nature of, 18
    temperature effect, 24
  and viruses, 58–59
  visibles
    induced by mutators, 65–69, 74–75, 91, **92**, 94–95, **95**, 98–104, **99**
    rates for 9
Mutation load, 1–30
  defined, 4–5
  dominance, 19–29
  and epistasis, 6–8, 29–30
  heterozygous expression of mutations, 19–29
  theory, 2–8
Multiple insemination (*see* Remating)
Mutator, 63–101, **64**, **73**, **75–76**, **80**, **82–84**, **86**, **92**, **95**, **97**, **99**
  AW, *JH*, *OYW* and *Raleigh* lines, 88–91
  chromosome breakage factor of Levitan, 69–70
  Cranston, Harwich and South Amherst lines, 79–81, **80**, **82–84**
  delta genetic system, 70–74, **73**
  and distorted sex ratio, 71
  distorted transmission frequency, 74, 77, 93–95
  in *D. ananassae*, 100–101
  in *D. persimilis*, 68–69
  first report of, 64–65
  Florida line of Demerec, 64–65
  Florida line of Voelker, 97–98
  function and Y chromosome, 69
  in germ cells only, 65, 85, 89
  in germ and somatic cells, 66, 68, 69
  Haifa^{12} line, 91–93, **92**
  hi of Ives, 67–68
  inbred Oregon line of Neel, 65–66
  induced
    chromosome rearrangements, 65–66, 68–70, 75, **82–84**, 85–91, **86**, 93–97, **95**, **97**, 98–104
    clusters of mutations, 66, 68, 75, 91–93, 98–100

Mutator—*continued*
  induced—*continued*
    deficiencies, 68, 70, 88–91, 93–100, **97**
    inversions, 70, 75, 88–91, 93–97, **95**, **97**
    male recombination, 73, 75, **75–76**, 77–91, **80**, **82–84**, **86**, 91–97, 102–103
    sterility, 65–68, 79–81, 93–100
    translocations, 70, **82–84**, 88–91, 93–97, **97**
    transpositions, 70, 88–91, 93–97, **97**
    visible mutations, 65–69, 74–75, 91, **92**, 94–95, **95**, 98–104, **99**
  as infective agent, 69
  I–R system, 98–100
  Luminy and B.2 lines, **82–84**, 98–100
  male specific, 70
  modifiers of, 105–106
  31.1 MRF line, **82–84**, 95–97, **97**
  MR *T-007* of Hiraizumi 74–79, **75–76**, **82–84**
  and nondisjunction, 79, 98–100
  N.S.W. line, **82–84**
  OK1 line, 81–88, **82–84**, 86
  other names for, 63
  π_2 line **82–84**, 93–95, **95**
  and population size, 105
  potential, 90
  px bl line of Goldschmidt, 66–67
  sex specificity, 65–67
  in sexual, asexual and parthenogenetic populations, 389–390
  temperature sensitivity, 74, 93–100
  transposable, 77, 97–100, 102
  and unstable mutations, 65, 91, 93–95, 98–100, **99**, 102
  U.S.S.R. lines, 98, **99**
  W8D line, **82–84**
Mycoplasma-like factor, and hybrid sterility, 130

N

N strains, 131–146, **141**, **143**
Natural populations
  competition in, 296–305
  Cy killer, 59–60
  distribution of *I*, *R* and *SN* strains 131–132, 143–144

SUBJECT INDEX lxxiii

Natural populations—*continued*
 mutations and overdominance, 25–27
 mutators 63–101, **64**, **73**, **75–76**, **80**, **82–84**, **86**, **92**, **95**, **97**, **99**
 parthenogenesis in 346, **347**, 348
 spontaneous mutation rates, **41–42**
 spread of *I* and *P* elements, 144
 strains of *P–M* system, 135–136, 143–144
 and virus induced mutations, 59
Negative heterosis, 402–404
Niche
 heterogeneity, 308
 separation, 303–305, 330–331
Nondisjunction, 79, 98–100, 131, 135, 356–360, **357–359**
 in mutator line, 79, 98–100
 in *P–M* hybrids, 135
 in *SF* sterile hybrids, 131
Nonrandom mating, measures of, 247–251
Nutrition, and mutation rates, 60

O

Ontogeny, and mating behavior, 237–244, **243**
Outbreeding
 effect on selection, 168–170, **169**, 192
 effect on variability, 168–170, **169**
Ovaries
 development and sexual receptivity, 240
 in *GD* sterile hybrids, 136–137
Overdominance, 19–29
 of natural population mutations, 25–27
Oviposition sites, 251–252, 304–305, 308

P

P component (*see P* strain, *P* element or *P* factor)
P cytotype, 94–95, 102–104
P element (*see also P* factor and *P* strain)
P element, 94–95, 102–104, 139–140
 insertion mutations, 139–140
 number of copies, 139
 not in *M* strains, 139
 size of, 139

P factor (*see also P* element and *P* strain)
 hereditary transmission of, 138–139
 insertions and unstable mutations, 140
 transposable elements, 140
P strain (*see also P* factor and *P* element)
P strains, 134–146, 141, 143
 distribution of, 135–136, 143–144
P–M system, 134–146, 141, 143
 comparison with *I–R* system, 140–146, *141, 143*
Paragonial gland, and sexual receptivity, 240–241
Parthenogenesis, 343–395
 and *abnormal abdomen*, 364–365, 366
 agonoid, 345
 ameiotic, 344–345
 automictic thelytoky, 373–376
 automixis, 344
 and behavioral genetics, 369–373
 capacities of, 382–383, **383**
 and central fusion, 349–352, **350–351**
 and chromosomal evolution, 389
 classification of, 344–345
 and clonal selection, 376–381, **378**, **380–381**
 defined, 343–344
 and egg laying capacities, 364–367, **366**
 facultative, 344
 first discovered, 345–346
 and gamete duplication, 353, **354**
 general activity of, 372–373
 and generation of bridge stocks, 360, **361–362**
 generative, 344
 genetic basis of, 360–369, **366**
  from contrasts of isogenic lines, 363
  from crosses between unisexual and bisexual strains, 363
  from screening experiments, 363–364
  by selection, 360–365
 genetic consequences, 348–360, **350–351**, **353–359**
 genetic variability for, 363–364
 gonoid, 345
 incidence of, 345–348, **347**
 and isozyme variability, 385–386
 mechanisms of 348–360, **350–351**, **353–359**
 meiotic, 344–345

Parthenogenesis—continued
  by mixtures of mechanisms, 354–356, **355–356**
  as a model of speciation, 393–395
  and mutation rates, 389–390
  in natural populations, 346, **347**, 348
  and nondisjunction, 356–360, **357–359**
  obligate, 344–345
  origin of, 386–393
  and population genetics, 373–386, **378**, **380–382**
  prospects for, 395
  rate of production of XO males, 354, **355**, 356–360, **357–359**
  in rock lizards, 359
  selection for, 360–365
  and selection for sexual isolation, 372
  sexual analogues, 360, **361–362**
  and sexual isolation, 369–373
  and sexual reproduction, 360
  and speciation, 386–395
  and terminal fusion, 352, **353**
  and the unit of selection, 381–385, **382–383**
  and viability selection, 368
  and wind response, 373
Persistence, of a gene in nature, 27
Pest control (*see also* Biological control)
Pest control, 399–419
  of boll weevil, 401
  and chromosome rearrangements, 402–403
  of codling moth, 401
  by conditional lethal gene, 413–414
  desirable genetic factors, 418–419
  by a gene for refractoriness to a disease, 413
  and male sterilizing mutation, 401
  of Mediterranean fruitfly, 401
  by meiotic drive, 415
  and pericentric inversions, 416
  by population replacement, 402–403
  of screwworms, 400–401
  by segregation-distorter, (*SD*), 415
  sterile insect release, 400–401
  by use of entire compound autosomes, 414–415
Pheromone
  and lek behavior, 233
  and mating behavior, 233–234
  where released, 233

Phototaxis, and induced mutation, 15
Phylogeny
  and male courtship behavior, 260–261
  and mating behavior, 257–261
  and sexual isolation, 260–261
  of *GD* sterility, 136–137
  of *SF* sterility, 132
Population
  replacement
    by compound autosome strains, 405–410, **406–408**
    by compound; free arm stocks 410–413, **411–412**
    controlling or eradicating factors, 413–414
    and immigration, 410
    by laboratory and field studies, 408–410, 412–413
  suppression and manipulation
    theoretical models for, 402–403
    by translocation lines, 402–403
  density
    and adult viability, 314–315
    and competition, 307, 311–317, **314**
    and egg hatchability, 314–15
    and fecundity, 311–317, **314**
    and fitness, 315–316
  genetics
    models for automictic thelytoky, 373–376
    and parthenogenesis, 373–386, **378**, **380–382**
    theory for automictic thelytoky, 373–376
Position-effect
  and reduced viability of translocations, 404
  screens for chromosome breakage, 52
Positive assortative mating (*see* Sexual isolation)
Pronuclear duplication (*see* Gamete duplication)
Pseudogenes, 104

Q

Q strains, 134–146, **141**, **143**
  defined, 135
  distribution of 135–136, 143–144

SUBJECT INDEX   lxxv

R

R-strains, 131–146, **141, 143**
  reactivity of 133–134
Radiation
  and competition, 320
  induced mutations, 14–16, **15**, 23–25
  and heterozygous expression, 20–25
  and polygenic mutations, 174–175
Rare male mating advantage (*see* Minority male mating advantage)
Recombination
  and central fusion, 349–352, **350–351**
  and gamete duplication, 353, **354**
  genetic control of, 170–171
  and selection, 170–171, 179–182, 188–190
  and terminal fusion, 352, **353**
  and variability, 170–171, 179–182, 188–190
Remating, 241, 256–257
Reproductive isolation, (*see* Sexual isolation)
Resource level, 288
Retrovirus, 104

S

Scramble, 287, **288**
Segregation distortion (*see also* Transmission ratio distortion)
Segregation distortion, 26–27, 415
Segregation-distorter (*SD*) and pest control, 415
Selection, 30, 80–81, **80**, 155–215, **166, 193–194, 196, 213, 262, 268,** 360–363, 368, 376–381, **378, 380–381,** 402–403
  abdominal chaeta number 164–165, **164,** 173–179, **174, 176, 179,** 183–184, **184,** 192
  assortative mating, 207–209
  clonal, 376–381, **378, 380–381**
    and egg hatchability, 379–381, **380**
    and fecundity, 379–381, **380**
    life history consequences of, 379–381, **380–381**
    and survival, 379–381, **380**
  conditional, 159
  correlated responses and recombination, 188–190

Selection—*continued*
  directional (R) 158–162, **160,** 175–190, **176,** 179–**182, 184**
    early generations, 175–190, **176,** 179–**182, 184**
    later generations, 182–186, **184**
    methods to derive expectations of, 175
  disruptive (D) 158–162, **160,** 199–210, **205–206**
    assortative (D^+) 200, 203–206, **205**
    disassortative (D^-) **196,** 200–203
    and isolation, 206–209
    random (D^R) 200, 202–203
    sexual isolation, 262
    types of 199–201
  and dominance, 214
  early experiments, 155–156
  effects of inbreeding and outbreeding, 168–170, **169**
  effects of mutation, 171–175, **174**
  eye facet number, 156, 196–197
  frequency dependent and chromosome replacement, 402–403
  hard, 159
  in maize, 155–156
  for male recombination, 80–81, **80**
  mating behavior, 169–170
  methods, 161–162
  number of parents, 177–178
  for parthenogenesis, 360–363, 368
  recombination, 170–171
  scutellar chaeta number, 156–157, **164,** 164–165, 180–182, **181, 182,** 197–198
  sexual isolation, 262, 268
  soft, 159
  stabilizing (S) 158–162, **160,** 191–199, **193–194, 196**
    additive genetic variation, 191–195, **193–194**
  step-wise, 157
  sternopleural chaeta number **164,** 165–168, **166,** 192–193, **193,** 195, 201–202, 294–205, **205,** 213, **213**
  and temperature, 196–197
  thorax length, 185–186, 193–194
  towards fixed optima, 203
  truncation and selection, 30
  types, 158–162, **160**

Selection—*continued*
 types of character, 212–215, **213**
 unconditional, 159
 viability and parthenogenesis, 368
 wing vein length, 193–194, **194**, **196**, 202–203
Sex
 evolution of, 386–393
 and mutation rates, **40–47**, 61–62
 and mutator activity, 65–67, 70
Sex ratio distortion
 and control of pest insects other than *D. melanogaster*, 418
 and mutators, 71, 93–95
 in *P–M* hybrids, 135
Sexual attractiveness, 241–242
Sexual behavior (*see* Mating behavior)
Sexual isolation, 125–148, 223–270, 369–373
 and antennae, 267
 and chemoreceptivity, 267
 in crosses between geographic populations, 263–264
 degree at various stages of population differentiation, 261–265
 and ecological adaptation, 269
 evolutionary origin of, 268–270
 and genetic drift, 269
 in inbred lines, 263
 in interspecific hybrids, 270
 and light, 268
 modified by temperature, 266
 in mutant stocks, 263
 in parthenogenic strains, 369–373
 and phylogeny, 260–261
 role of males and females, 265–266
 selection for, 268
 and selection for chaeta number, 262
 in selection lines, 262
 in subspecies, 264–265
 and visual displays, 267–268
 and wing vibration, 267
Sexual reproduction, and parthenogenesis, 360, **361–362**
Sexual selection, 254–257, 266–268, 372
 environmental factors influencing, 266–268
 and experience, 267
 modified by density, 266–267
 and parthenogenesis, 372
 subdivision of, 255

SF sterility (*see also* Hybrid sterility and Sterility, intraspecific hybrid)
SF sterility, 131–146, **141**, **143**
 comparison with *GD* sterility, 140–146, **141**, **143**
 defined, 131
 distribution of *I*, *R*, and *N* strains 131–132
 and egg development, 132
 and egg hatchability, 132–133
 and failure to transmit by injection, 141
 hereditary transmission of reactivity, 133–134
 inducer (*I*), reactive (*R*), and neutral (*N*) strains, 131–146, **141**, **143**
 I–R system, 131–146, **141**, **143**
 physiology of, 132
 strain variability, 131
 and temperature, 132–133, **141**
Sibling species, courtship songs of, **231**
Sonoran Desert species, competition among, 300–303, **301–302**
 host plants, 300–303, **301**
Speciation
 founder effect/genetic revolution model, 393–395
 founder effect-genetic transilience model, 393–395
 and parthenogenesis, 386–395
 parthenogenesis as a model of, 393–395
Sperm, in receptacles and sexual receptivity, 240–241
Spermatids, in intraspecific hybrids, 129
Spermatogenesis, in intraspecific hybrids, 128
Sterility
 intraspecific hybrid (*see also GD* sterility, Hybrid sterility, and *SF* sterility)
 intraspecific hybrid, 125–148
  atrophie gonadique, 145–146
  cytoplasm-Y chromosome interaction, 130
  in *D. equinoxialis*, 129
  in *D. melanogaster*, 130–146
  in *D. micromelanica*, 147
  in *D. paulistorum*, 129–130
  in *D. peninsularis*, 147
  in *D. pseudoobscura*, 127–128
  in *D. tropicalis*, 129
  in *D. willistoni*, 128–129

Sterility—*continued*
 intraspecific hybrid—*continued*
  in *D. willistoni* group 128–130
  and meiosis, 130
  and mycoplasma-like factor, 130
  in *D. pallidipennis* group, 146
  spermatogenesis, 128
  in subspecies of *D. macrospina* 146
  in subspecies of *D. pseudoobscura*, 127–128
Sterility
 mutator induced, 65–68, 79–81, 93–100
 of translocation heterozygotes, 403–404
Stimuli
 auditory (*see also* Courtship songs)
 auditory, 230–232, **231**
  mutants that affect, 231
  via antennae, 231
 olfactory, 233–234
  and male mating advantage, 234
  receptors, 233
 tactile, 234–236
 visual, 236–237
  in mutants, 236–237
Stochastic loss hypothesis, 144
Subspecies
 of *D. equinoxialis*, 129
 of *D. macrospina*, 146
 of *D. paulistorum*, 129–130
 of *D. pseudoobscura*, 127–128
 of *D. tropicalis*, 129
 of *D. willistoni*, 128–129
 sexual isolation, 264–265
Suppressor, of hybrid dysgenesis, 79
Survival
 and clonal selection, 379–381, **380**
 of compound autosome stocks, 410

T

Temperature
 and competition, 309–311, **310**
 effect on mutation rates, 55, **56–58**
 effect on viability mutations, 24
 and *GD* sterility, 137–138, **141**
 modification of sexual selection, 266
 and selection, 196–197
 sensitive male sterilizing mutations, 401

Temperature—*continued*
 sensitivity of mutators, 74, 93–100
 and *SF* sterility, 132–133, **141**
Terminal fusion, and recombination, 352, 353
Testes
 in *GD* sterile hybrids, 137
 microsomal enzymes, 55
Thelytoky, 344–345
Thorax, selection for length, 185–186, 193–194
Translocations (*see also* Chromosome rearrangements and Mutant Index)
 average viability of heterozygotes, 403–404
 homozygous and fitness, 404–405
 methods to measure frequency of, 50–51
 mutator induced, 70, **82–84**, 88–91, 93–97, **97**
 and population suppression and manipulation, 402–403
 semi-sterility of, 403–404
 spontaneous rate, **47**
 viability and position effect, 404
Transmission ratio distortion (*see also* Segregation distortion)
Transmission ratio distortion, 18, 26–27, 135
 in *P–M* hybrids, 135
Transposable DNA elements, 102–104, 140
 and hybrid dysgenesis, 102–104
 insertion mutations, 102–104
 P factors, 140
 size and number of, 103
Transpositions (*see also* Chromosome rearrangements and Mutant Index)
 of mutator, 77, 97–100, 102
 mutator induced, 70, 88–91, 93–97, **97**
Transvection effect, 53
Tychoparthenogenesis, 344–346, 390

U

Unit of selection, 381–385, **382–383**
 defined, 381
 and glycolytic profiles, 384–385
 method to measure, 382–385, **383**
 and parthenogenesis, 381–385, **382–383**

V

Variability (see also Variation)
  free, 165–170, **169**
    methods to measure, 165–168
  homozygotic potential, 165–170, **169**
    methods to measure, 165–170, **169**
  inbreeding and outbreeding effects, 168–170, **169**
  and mutation, 171–175, **174**
  and recombination, 170–171, 179–182, 188–190
  release of, 168–170, **169**
Variation (see also Variability)
  hidden, 163–168, **164, 166**
  kinds, 162–163
  quantitative, 163
Viability
  and EMS induced mutations, 14–16, **15**, 23–29
  heterozygous expression, 19–29
  and hybrid dysgenesis, 17–18, 27
  mutations, 9–29, **9**
    nature of, 18
    in noncoding portion of genes, 18
    and temperature, 24
  and natural population mutations, 25–27
  and population density, 314–315
  and radiation induced mutations, 14–16, **15**, 20–25
  and spontaneous mutations, 23–25
  of translocation heterozygotes, 403–404

Viruses (see also retrovirus)
  C picornavirus, and mutation, 59
  and mutations, 58–59
  poliomyelitis, and mutations, 59
  Rous sarcoma, and mutations, 59
  sigma
    CO_2 sensitivity, 58–59
    and mutation rate, 59
    transmission of, 58–59
  Tipula iridescent, and mutations, 59

W

Wind response
  method to measure, 373
  in parthenogenic strains, 373
Wing vein, selection for length, 193–194, **194, 196**, 202–203
Wings, vibration and sexual isolation, 267

X

XO males, rate of production in parthenogenetic strains, 354, **355**, 356–360, **357–359**
XXYY females, 358–359, **359**
XXYYYY females, 358–359, **359**

Y

Y chromosome, cytoplasm interaction and hybrid sterility, 130
Yeast, preferred by Drosophila, 304, 319

INDEX OF GENETIC VARIATIONS

In the *Index of Genetic Variations* mutations are indexed alphabetically by their full names; their standard abbreviations follow in brackets. Except in a few cases alleles are not separately indexed.

Following the index of mutations chromosome abberrations are indexed in the order of Lindsley and Grell's "Genetic Variations of *Drosophilia melanogaster*" (1968, Carnegie Institution, Washington), i.e.: Deficiencies (*Df*), Duplications (*Dp*), Inversions (*In*), Rings (*R*), Translocations (*T*) and Transpositions (*Tp*). There then follow, in order, Balancers and special chromosomes, Compound chromosomes (*C*), Free chromosomes (*F*), X–Y combinations and Y derivatives.

We have adhered to Lindsley and Grell's practice as closely as possible for abbreviations and for the orders within sections. We have avoided extensive cross indexing and all but a few questions of synonymy. Unless noted all mutations are of *D. melanogaster*.

Mutations

A

Abnormal abdomen (*A*), 98
abnormal abdomen (*aa*) (*mercatorum*), 365–368, 384, 388
abnormal oocyte (*abo*), 145
achaete (*ac*), 43, 49
Acid phosphatase (*Acph*), 242
Alcohol dehydrogenase (*Adh*), 46, 91, 104
Aldehyde oxidase (*Aldox*), 46
Alkaline phosphatase (*Aph*), 46
amnesiac, 244
Amylase (*Amy*), 91
arc (*a*), 49, 66
aristaless (*al*), 49, 51, 75, 76, 77, 84, 91, 93, 231, 232
atrophie gonadique (*ag*), 137, 145, 146

B

Bar (*B*), 41, 47, 53, 66, 85, 156
Beadex (*Bx*), 45
bithorax (*bx*), 66, 103
black (*b*), 49, 51, 66, 75, 76, 77, 81, 84, 96, 101
blistered (*bs*), 66, 67
blistery (*by*), 44

bobbed (*bb*), 40, 41, 66, 67
bow (*bow*), 66
bran *see* arc
bright (*bri*) (*ananassae*), 100
Bristle (*Bl*), 81
brown (*bw*), 44, 47, 49, 50, 51, 71, 74, 81, 84, 85, 96, 237, 415

C

cacophony, 243
cardinal (*cd*), 66
carmine (*cm*), 43, 49, 91, 92
carnation (*car*), 43, 45, 49, 92
celibate (*cel*), 243
cinnabar (*cn*), 44, 49, 51, 64, 74, 75, 76, 77, 81, 84, 91, 92, 93, 96, 145
claret (*ca*), 49, 51, 61, 75, 76
claret (*ca*), (*ananassae*) 100, 101
clot (*cl*), 84
coitus interruptus (*coi*), 243
comma, *see dumpy*
crossover-suppressor of Gowen (*c(3)G*), 93
cubitus interruptus-dominant (ci^D), 193, 194, 202
curled (*cu*), 44, 49, 61, 66, 75, 76, 416
Curly (*Cy*), 9, 10, 11, 12, 59, 60, 71, 74, 81, 84, 91, 96, 145

curved (c), 49, 64, 66, 75, 76, 77, 84, 415
cut (ct), 43, 45, 49, 91, 92
cyclic AMP phosphodiesterase (cAMP), 104

D

dachs (d), 66
delta, 53, 64, 70, 71, 72, 73, 74, 90
Delta-a (Da), 72, 74
Dichaete (D), 66, 156
Dipeptidase-A (Dip-A), 91
divergent (dv), 67
droopy (d) (mercatorum), 356
dumpy (dp), 44, 49, 67, 75, 76, 77, 81, 84, 88, 96
dusky (dy), 43
dwarf (dw), 66

E

ebony (e), 49, 61, 66, 75, 76, 84, 207, 237, 239
echinus (ec), 43, 45
Esterase-6 (Est-6), 46
Esterase-A (Est-A) (mercatorum), 356
Esterase-B (Est-B) (mercatorum), 356
Esterase-C (Est-C), 46
eyeless (ey), 45

F

forked (f), 41, 43, 45, 46, 49, 51, 74, 84, 85, 92, 101, 103, 170
fringed (fr), 67
fruitless, 243

G

gap (gp), 67
garnet (g), 43, 46, 49, 92
Glued (Gl), 170
Glutamine oxaloacetic transaminase-2 (Got-2), 91
α-glycerophasphate dehydrogenase (αGpdh), 46, 91
α-glycerophosphate dehydrogenase (mercatorum), 367, 385
grandchildless (gs) (subobscura), 137

H

Hairless (H), 67
hairy (h), 44, 49, 61, 67, 75, 76
heldup (hdp), 95
Hexokinase-C (Hex-C), 91
high (hi), 64, 67, 68, 105

I

Isocitrate dehydrogenase (Idh), 46

K

Killer of prune (K-pn), 46

L

lethals, (see individual l( )'s)
l(1)J1, 41
l(2)gl, 91
light (lt), 76
Lobe (L), 66, 81, 84, 96
lozenge (lz), 43, 101, 103

M

Malate dehydrogenase-1 (Mdh-1), 46
Malate dehydrogenase-2 (Mdh-2), 91
meiotic-S282 (mei-S282), 62, 64
miniature (m), 43, 46, 92, 101
miniature (m) (mercatorum), 377
Minutes (M), 45, 49, 64, 66
Minutes (M) (ananassae), 100, 101
Minutes (M) (persimilis), 69, 82
morula (mr), 49
Mutator-recombination-23.5 (Mr-23.5), 96
Mutator-recombination-31.1 (Mr-31.1), 64, 83, 84, 95, 96, 97, 145
Mutator-recombination-102 (Mr-102), 64
Mutator-recombination-Haifa12 (Mr-h12), 64, 91, 92
Mutator-recombination-n1 (Mr-n1), 64, 92
Mutator-recombination-OK1 (Mr-OK1), 53, 64, 81, 82, 83, 84, 85, 86, 87, 88, 92
Mutator-recombination-s1 (Mr-s1), 64, 92
Mutator-recombination-Texas 007 (Mr-T007), 64, 74, 75, 76, 77, 78, 79, 83, 84, 85, 86, 92
Mutator-recombination-W8D (Mr-W8D), 83, 84, 86

INDEX OF GENETIC VARIATIONS

mutator of forked (*mu-f*), 64, 65, 68
multiple wing hairs (*mwh*), 85

N

Notch (*N*), 43, 66

O

Octanol dehydrogenase (*Odh*), 46
orange (*or*) (*pseudoobscura*), 243
outstretched (*os*), 43

P

P factors, 53, 63, 64, 82, 83, 84, 93, 94, 95
pink (*p*), 44, 50, 51
plexus (*px*), 49, 66, 67, 75, 76, 77, 84
plexus (*px*) (*ananassae*), 100, 101
plum (*pm*) (*mercatorum*), 356, 361, 371, 381, 382, 383
Plum (*Pm*), 9, 59, 60, 81, 84, 96
prune (*pn*), 43, 49, 92
purple (*pr*), 44, 49, 64, 74, 75, 76, 77, 84, 91, 92, 93, 145

R

raspberry (*ras*), 40, 43, 49, 74, 91, 92
Recovery Disrupter (*RD*), 53, 64
rosy (*ry*), 46, 103
rough (*ro*), 416
roughoid (*ru*), 45, 49, 61, 75, 76
ruby (*rb*), 44, 49, 91, 92
rudimentary (*r*), 4

S

sable (*s*), 44
scarlet (*st*), 45, 49, 50, 51, 61, 75, 76
scarlet (*st*) (*simulans*), 314, 315, 319
scute (*sc*), 46, 49, 51, 197, 198
Segregation Distorter (*SD*), 64, 73, 74, 415
sepia (*se*), 66, 207
silver (*svr*), 66, 67
singed (*sn*), 44, 64, 74, 91, 92, 94, 95, 98, 103, 136, 140
Ski (*Si*), 66
speck (*sp*), 49, 51, 75, 76, 77, 84
spotless (*sl*) (*mercatorum*), 371, 377–379, 381, 382

Star (*S*), 44
Sternopleural (*Sp*), 84
straw (*stw*), 76, 207
straw (*stw*) (*ananassae*), 100, 101
stripe (*sr*), 45, 49, 61
Stubble (*Sb*), 170
stuck (*sk*), 243

T

thread (*th*), 45, 49, 61, 75, 76, 231, 416
Tufted (*Tft*), 64, 93

V

veinless (*vl*) (*mercatorum*), 356, 361, 371, 381, 383
veinlet (*ve*), 51, 84, 96
vermilion (*v*), 40, 41, 44, 46, 49, 91, 92, 103, 237, 263
vermilion (*v*) (*mercatorum*), 356, 361, 371, 381, 382, 383
vermilion (*v*) (*simulans*), 314
vestigial (*vg*), 49, 88, 207, 231, 232

W

white (*w*), 40, 41, 44, 46, 49, 82, 84, 85, 91, 92, 103, 140, 170, 180, 237, 263, 292, 314, 315, 318, 319
white (*w*) (*mercatorum*), 356, 377, 379, 381
white (*w*) (*micromelanica*), 147

X

Xanthine dehydrogenase (*Xdh*) (*mercatorum*), 356

Y

yellow (*y*), 40, 41, 44, 46, 49, 53, 65, 66, 74, 84, 85, 91, 92, 98, 101, 103, 170, 180, 242, 263, 314, 315, 318, 319
yellow (*y*) (*pseudoobscura*), 263
yellow (*y*) (*subobscura*), 263

Z

zeste (*z*), 82, 140

Duplications (*Dp*)
Dp(2L;2), 412, 413

Inversions (*In*)
In(1)asc, 40
In(1)dl-49, 40, 41
In(1)EN, 40
In(1)FM6, 48
In(1)FM7, 48
In(1)sc^{4L}sc^{8R}+S, 48
In(1)sc^8, 40, 41
In(1)w^{m4}, 40
In(2L)NS, 67
In(2L)A, 413
In(2LR)bw^V, 89
In(2R)bw^{VDe2}, 84
In(2LR)O, 48
In(2LR)SM1, 48

Rings (*R*)
R(1)2, 40
R(1)X, 53

Translocation (*T*)
T(Y;2), 51
T(Y;2)SD, 415
T(Y;2;3), 51
T(Y;3), 53

Balancer and special chromosomes
Basc, 40, 48, 61, 80, 81, 82, 84
CyO, 48
FM6, 48
FM7, 48
Inscy, 48
SM1, 48
TM1, 48
TM3, 48

Compounds (*C*)
C(1), 92, 93, 406
C(1)DX, 49, 74, 84, 85
C(1)M3, 49
C(1)X, 65
C(2)EN, 414, 415
C(2L)RM, 51, 52, 406
C(2R)RM, 51, 52, 406
C(3L)RM, 406
C(3R)RM, 406

Free chromosomes (*F*)
F(2R), 413

X–Y combinations
X.Y, 51
X.Y^S, 41

Marked Y chromosomes
B^{S}Yy$^+$, 57, 53, 83, 84, 85
bw$^+$Yy$^+$, 47
y$^+$.Y, 40
Y^{bb}, 40

INDEX OF NON-DROSOPHILID TAXA

A

Aedes aegypti, 416, 417, 418
Anopheles albimanus, 417
Anopheles gambiae, 417, 418
Anthonomus grandis, 401, 418

B

Blattella germanica, 417

C

cactus, 300, 346, 348, 364, 365
Candida pulcherima, 319
Carnegiea gigantea, 301
Ceratitis capitata, 401
Cestrum parqui, 297
Cheirodendron, 303
coccid, 345, 376
Cochliomyia hominivorax, 400–401
Culex pipiens, 416, 417
Culex tarsalis, 416
Culux tritaeniorhynchus, 416, 418

F

Ficus, 331

G

Gecarcinus ruricola, 297
Glossina austeni, 416

H

Habrobracon serinopae, 418
hamster, 104
Hylemya antiqua, 417

L

Laspeyresia pomonella, 401
lizards, 359
Loncoptera dubia, 389
Lophocereus schotti, 297, 301
Lucilia cuprina, 417, 418

M

Machaerocereus gummosus, 301
man, 18
mice, 104
microsporidia, 58
mites, 391–392
Musca domestica, 417, 418
mycoplasma, 58, 130
Myoporum sandwicense, 331

N

Nadsonia elongata, 319
Nicotiana rustica, 169

O

Opuntia, 331, 332
Opuntia lindheimeri, 305
Opuntia megacantha, 346, 364

P

Pachycereus pringlei, 301
Panaxia dominula, 170
Papillio dardanus, 202
Pediculopsis, 169
Phaseolus vulgaris, 156
Picornavirus, 59
Pisonia, 298
Plasmodium berghei, 418, 419
polio virus, 59
Primula, 202

R

rat, 156
rickettsiae, 58
Rous sarcoma virus, 59

S

Saccharomyces cerivisiae, 307, 319
Sapindus, 298, 299
Schizosaccharomyces pompe, 319
sigma virus, 58
Solenobia triquetrella, 364
spirochaetes, 58
Stenocereus thurberi, 301

T

Tetranychus urticae, 417
Tipula iridescent virus, 59

tomato, 410
Tribolium, 305

U

Ursus americana, 410

W

weevils, 391
Wuchereria bancrofti, 417

Z

Zea, 156

INDEX OF DROSOPHILID TAXA

As explained on p. vi the editors have attempted to make the nomenclature of drosophilid taxa uniform with that of Chapter 1 by Wheeler (volume 3a). Valid names are in roman type. Valid genera and subgenera are in bold face. The following abbreviations are used: g.: genus; s.g.: subgenus.

C

Chymomyza g., 224

D

Drosophila s.g., 347
Drosophila acanthoptera, 229
D. acutilabella, 347
D. adiostola, 233
D. affinis, 261, 267, 298, 346, 347
D. albomicans, 226
D. aldrichi, 297, 305, 331, 332
D. algonquin, 261, 265, 267
D. ambigua, 232
D. americana, 266, 347
D. ananassae, 50, 64, 100, 101, 232, 244, 267, 316, 346, 352, 356, 357, 363
D. athabasca, 232, 298
D. auraria, 237
D. buzzatii, 297, 305, 331, 332
D. carcinophila, 297
D. cardini, 346, 347
D. cardinoides, 347
D. crocina, 347
D. crucigera, 229, 233, 246
D. duncani, 347
D. enigma, 229
D. equinoxialis, 129, 264
D. equinoxialis caribbensis, 129
D. erecta, 231
D. flavopilosa, 297
D. fulvimacula, 147
D. fulvimacula flavorepleta, 147
D. funebris, 50, 146, 241, 258, 291, 292, 308, 316, 317, 318, 325, 347
D. gaucha, 261, 265
D. grimshawi, 233, 237, 239, 240, 246, 261, 264
D. hamifera, 235
D. heedi, 330, 331
D. heteroneura, 148
D. hydei, 234, 346, 347, 348
D. imaii, 265
D. immigrans, 229, 258, 264, 347
D. imparisetae, 298, 299
D. kambysellisi, 298, 299
D. kepulauana, 226
D. macrospina, 146, 347
D. macrospina limpiensis, 146
D. macrospina ohioensis, 146
D. mangabeirai, 345, 349, 351, 352, 356, 369, 376, 377, 389, 391
D. mauritiana, 231
D. mediostriata, 347
D. melanica, 147, 347
D. melanogaster, 8–19, 20–27, 29–30, 39–54, 56–68, 67, 70–101, 102–106, 126, 130–146, 155–157, 159, 161–168, 170–190, 192–199, 201–209, 213, 214–215, 227, 230–232, 234, 235, 236, 237, 238, 239, 240, 241, 242, 243, 244, 252, 255, 256, 258, 259, 260, 262, 263, 264, 265, 266, 267, 269, 270, 291, 292, 297, 299, 300, 306, 307, 308, 309, 310, 311, 312, 313, 314, 315, 316, 318, 319, 320, 322, 323, 324, 325, 326, 327, 328, 329, 330, 331, 347, 348, 351, 399–419
D. mercatorum, 147, 164, 346, 348, 349, 352, 354, 355, 356, 357, 358, 359, 360, 361, 362, 364, 365–368, 370–372, 373, 377, 378–388, 393, 394
D. mercatorum pararepleta, 147
D. mesophragmatica, 258

Drosophila (*cont.*)
- D. mettleri, 301, 302
- D. micromelanica, 147
- D. mimica, 298, 299
- D. miranda, 312
- D. mojavensis, 264, 301, 302, 305
- D. mulleri, 305, 331
- D. nasuta, 228, 235, 256, 258, 261
- D. nebulosa, 58, 311, 316, 319, 324, 326, 327, 329, 330
- D. neocardini, 347
- D. nigromelanica, 347
- D. nigrospiracula, 252, 256, 301, 302, 303
- D. obscura, 235, 258, 259, 267, 268
- D. orphnopeza, 246
- D. pachea, 297, 300, 301
- D. paenehamifera, 235
- D. pallidifrons, 226
- D. pallidipennis, 129, 130, 146
- D. pallidipennis centralis, 146
- D. pallidosa, 244, 267, 346, 352, 356, 357
- D. parthenogenetica, 164, 346, 352, 354, 356, 357, 361, 362, 363, 370
- D. paulistorum, 129, 147, 164, 207, 236, 258, 261, 264, 265, 267, 268, 346
- D. pavani, 261, 265
- D. peninsularis, 147
- D. persimilis, 50, 64, 68, 69, 82, 106, 164, 207, 232, 236, 238, 244, 261, 265, 266, 267, 268, 297, 305, 312
- D. polymorpha, 346, 347, 262
- D. pseudoobscura, 106, 126, 127, 128, 164, 195, 207, 211, 212, 232, 236, 238, 243, 244, 246, 256, 259, 261, 262, 263, 264, 265, 266, 267, 268, 269, 291, 292, 293, 294, 296, 297, 300, 305, 306, 307, 310, 311, 312, 316, 317, 319, 323, 324, 326, 329, 330
- D. pseudoobscura bogotana, 127, 128
- D. pullipes, 264
- D. putrida, 347
- D. quinaria, 261, 265, 267, 347
- D. repleta, 147
- D. robusta, 69, 70, 164, 255, 346, 347, 362, 363
- D. serrata, 264, 292, 293, 310, 311, 323, 324, 326, 327, 329
- D. silvarentis, 330, 331
- D. silvestris, 148
- D. simulans, 64, 103, 106, 231, 232, 234, 237, 244, 246, 252, 258, 259, 260, 264, 265, 266, 267, 270, 292, 297, 299, 300, 307, 308, 309, 310, 311, 312, 313, 314, 315, 316, 318, 319, 320, 322, 323, 325, 326, 327, 331, 347
- D. subobscura, 64, 137, 228, 237, 240, 246, 255, 263, 265
- D. sulfurigaster, 235, 256
- D. teissieri, 231
- D. transversa, 347
- D. triauraria, 246
- D. tripunctata, 347
- D. tropicalis, 129, 264
- D. varipennis, 235
- D. victoria, 347
- D. virilis, 64, 235, 265, 266, 267, 305, 316, 317, 325
- D. willistoni, 58, 64, 127, 128, 164, 195, 258, 261, 264, 265, 292, 293, 294, 300, 311, 316, 317, 319, 323, 329, 330, 345
- D. willistoni quechua, 128
- D. yakuba, 231

H

Hirtodrosophila s.g., 347

L

Lissocephala g., 331

S

Scaptodrosophila s.g., 347
Scaptomyza g., 224, 253, 347
- Scaptomyza adusta, 347
- Scaptomyza graminum, 347

Sophophora s.g., 345, 347

Z

Zaprionus g., 331, 347
- Zaprionus vittiger, 347